Electromagnetics

Theory, Techniques, and
Engineering Paradigms

The photograph on the cover represents the Cassini–Huygens spacecraft, a joint NASA–ESA–ASI deep-space mission to image the surface and study the atmosphere of Saturn and its moons, in particular Titan.

The spacecraft will reach Saturn in the year 2004 and is equipped with a large number of electromagnetic sensors, operating in the microwave, infrared, and visible range. On top of the spacecraft is a 4-meter parabolic dish with a dichroic subreflector operating in the S, X, Ku, and Ka bands.

As James Clark Maxwell wrote,

> *We, that is, all the work we have done*
> *As waves in ether shall forever run*
> *In ever widening spheres through heaven*
> *Beyond the sun*

Electromagnetics
Theory, Techniques, and Engineering Paradigms

Giorgio Franceschetti
University of Naples
Naples, Italy, and
University of California at Los Angeles
Los Angeles, California

Plenum Press • New York and London

Library of Congress Cataloging-in-Publication Data

Franceschetti, Giorgio.
  Electromagnetics : theory, techniques, and engineering paradigms /
Giorgio Franceschetti.
     p.   cm.
  Includes bibliographical references and index.
  ISBN 0-306-45527-7
  1. Electromagnetics. 2. Electromagnetics--Industrial
applications. 3. Electromagnetic theory. I. Title.
  QC760.F73 1997
  621.3--dc21   97-13659
                                                              CIP

ISBN 0-306-45527-7

© 1997 Plenum Press, New York
A Division of Plenum Publishing Corporation
233 Spring Street, New York, N. Y. 10013

http://www.plenum.com

All rights reserved

10 9 8 7 6 5 4 3 2 1

No part of this book may be reproduced, stored in a retrieval system, or transmitted in any form or by any means, electronic, mechanical, photocopying, microfilming, recording, or otherwise, without written permission from the Publisher

Printed in the United States of America

To Professors
Gaetano Latmiral and Charles H. Papas
who profoundly changed
my way of thinking

Preface

During the last twenty years the lifestyle of a large portion of the inhabitants of our planet has changed dramatically. This would never have been possible without the massive use of electronic and photonic technology, telecommunications, and computers. These disciplines are designed to code, transmit, detect, decode, and process signals and related information, and can be broadly addressed as information science and technology. In the sophisticated society in which we live and operate, this science is diffused transversely and plays a major role in almost every human activity.

Information science and technology is the basis of a powerful industry that does not suffer the shortcomings of more traditional human enterprises. Information is a renewable source and its control and processing rely on software codes, which are a creation of the mind, and on related hardware, incredibly sophisticated but made out of simple, abundant materials.

The rate of change and transformation of this industry is the highest mankind has ever experienced, and it requires not only the replacement of technologies but also a continuous updating of expertise to keep up with the rapid transformation. There is no doubt that this calls for a change in university training, to avoid students graduating at an already obsolete level.

There appears to be an increasing consensus that university courses in applied science should concentrate on basic issues, so that the student can gain a very broad scientific background, a sort of hard core that will remain valid throughout her/his professional career. Among others, one of the basic disciplines is electromagnetics, which plays an important role in telecommunications, solid-state physics, remote sensing, electromagnetic compatibility, health hazards, and safety standards, to list some of the most popular areas. This book is an attempt to filter out of this broad range of material those topics most often encountered, and it can be used to provide the student with a basic background in applied electromagnetics during a two-semester graduate course.

The volume contains nine chapters. The first two are devoted to the essentials of electromagnetic theory. The next four, Chapters 3–6, introduce tools, methods, and procedures for studying electromagnetic fields. The last three present again all the material from the perspective of the user, the applied scientist, who must rely on simple, sound models for the design of electromagnetic components and systems. Knowledge of theory and related techniques is believed to be a most desirable prerequisite for the subsequent overview of the whole subject from an engineering standpoint.

A few details about the contents of each chapter now follow. Chapter 1 presents electromagnetic theory: Maxwell's equations, energy theorems, constitutive relationships, strictly in the time-domain. Many existing books on applied electromagnetics prefer the steady-state sinusoidal approach, but it is my understanding that this somehow abstract presentation does not favor the physics. The time-domain is a much more natural environment; also, recent broadband applications of the electromagnetic field strongly suggest its study.

Chapter 2 is ancillary to the first chapter. It presents simple solutions of Maxwell equations for propagation and radiation, always in the time domain. Space and time discontinuities are considered, as well as moving sources. The case studies examined in this chapter, together with the basic material presented in the first chapter, should provide the reader with a bit more than just a flavor of electromagnetic theory.

Chapter 3 is the first devoted to techniques used in solving the electromagnetic equations. It exploits Maxwell equations in the transformed spaces: the frequency, wavenumber, and frequency–wavenumber domains. Constitutive relations in frequency and wavenumber space, pulse propagation in dispersive media, and transient radiation are discussed. In addition the field representation is introduced, including the important issues of plane-wave expansion and its asymptotic evaluation.

Chapter 4 addresses the problem of narrowband signals, largely used in telecommunications. Phasor fields are treated, exploiting polarization and coherence of the field, and providing solutions for guided propagation in complex media and radiation from elementary sources, apertures, and arrays.

Chapter 5 deals with electromagnetic equations in the high-frequency regime. The transition from the wave-field to the ray-field description is used to derive optical and quasi-optical solutions for the propagation and scattering problems. Both canonical (half-plane) and application-oriented (guided propagation, reflectors) cases are discussed.

Chapter 6 introduces some of the numerical techniques that form the basis of the electromagnetic solver packages: the method of moments, as well as the finite element and finite difference methods. No attempt is made to generate numerical codes; a book cannot compete with specialized, commercially available packages. However, the basic information provided by the chapter is in favor of numerical-technique understanding and usage, allowing a less blind use of electromagnetic solvers.

Chapters 7–9 are devoted to propagation, radiation, and scattering from the application point of view. All the material previously treated is revisited and organized from a different perspective: simple models are employed, leading to equivalent circuits and (possibly scalar) performance parameters, to be used in the design of electromagnetic components and systems.

Propagation (Chapter 7) is rephrased in terms of transmission line models and equivalent circuits of microwave junctions, and covers waveguides, striplines, fibers, and related circuits.

Radiation (Chapter 8) is described in terms of simple transmitting and receiving parameters such as the gain and effective area, and covers wire antennas, apertures, reflectors, and arrays. Wave techniques as well as ray techniques (where appropriate) are exploited. Sampling techniques for efficient computation of the radiated field are presented as a first step in the synthesis problem.

Cavities and scattering (Chapter 9) are studied by means of equivalent circuits, modal expansion, and statistical tools in the case of rough surfaces.

All the chapters are organized in a number of sections and subsections, the latter developing details and case studies. This arrangement compares favorably with the conventional setup in which the exercises at the end of each chapter are often either simply numerical or so difficult as to scare the reader rather than interest him. A summary and very few references are added to each chapter, the aim being to stimulate curiosity and introduce the reader to a wider scientific arena. In the modern world of simply accessible large databases, exhaustive references can be obtained at the touch of one's fingers on any particular issue, and no conventional book can match this. Accordingly, the mission of the book and its inherent philosophy have been focused on providing the reader with the ability to select appropriate keywords in order to explore the available huge databases, and with the basic grammar to read, understand, and profit from the retrieved material. This has been the object in writing this book.

Giorgio Franceschetti

Contents

CHAPTER 1. Fundamentals

1.1. Maxwell Equations 1
    1.1.1. The Current Density Equation 4
    1.1.2. The Independence of Maxwell Equations 4
    1.1.3. The Lentz–Neumann Law 4
    1.1.4. Polarization and Magnetization 6
    1.1.5. Field Sources 8
    1.1.6. Source Power 9
1.2. Constitutive Relationships 10
    1.2.1. Nonlinear Media 14
    1.2.2. Linear Anisotropic Media 14
    1.2.3. Anisotropic Media Classification 16
    1.2.4. Linear Dispersive Media. I 17
    1.2.5. Linear Dispersive Media. II 19
    1.2.6. Chiral Media 20
    1.2.7. Fields at Space and Time Boundaries 20
1.3. Energy and Momentum 22
    1.3.1. Electromagnetic Energy for Nonlinear Media 26
    1.3.2. Poynting's Theorem for Anisotropic Media 26
    1.3.3. Poynting's Theorem for Dispersive Media 27
    1.3.4. Lossless and Lossy Media 28
    1.3.5. The Radiation Pressure 28
1.4. Initial and Boundary Conditions 30
    1.4.1. Electric and Magnetic Perfect Conductors 32
    1.4.2. The Radiation Condition 33
    1.4.3. The Edge Condition 35
1.5. Symmetry Properties 37
    1.5.1. Image Theory 41

1.5.2. The Duality Theorem for Inhomogeneous Media 42
1.6. Summary and Selected References 42
References 43

CHAPTER 2. Elementary Solutions

2.1. Plane Waves 45
    2.1.1. Lossy Media 48
    2.1.2. Anisotropic Media 50
2.2. Plane Waves at Discontinuity Boundaries 54
    2.2.1. Conductive Media 61
    2.2.2. Bounded Waves 62
2.3. Radiation from Prescribed Sources 65
    2.3.1. Elementary Sources 68
    2.3.2. Magnetic Sources 72
    2.3.3. Free-Space Green's Functions 74
    2.3.4. Vector and Scalar Potentials 76
    2.3.5. Radiation from Moving Sources 78
2.4. Summary and Selected References 81
References 82

CHAPTER 3. Spectral Domains

3.1. Preliminary Considerations 83
    3.1.1. Distributions and Dirac Functions 84
3.2. The Frequency Domain 86
    3.2.1. Properties of Transformed Fields and
           Related Quantities 89
    3.2.2. Conductive Media 92
    3.2.3. Polar Dielectrics 93
    3.2.4. Magnetized Plasma 93
    3.2.5. Plane-Wave Propagation in Dispersive Media 96
    3.2.6. Plane-Wave Propagation in a Plasma Medium 101
    3.2.7. Plane-Wave Propagation in a Conductive Medium . . 102
    3.2.8. Plane Wave at a (Space) Discontinuity Boundary . . . 104
    3.2.9. Radiation in Dispersive Media 107
    3.2.10. Scalar Green's Function Evaluation 109
3.3. The Wavenumber Domain 110

    3.3.1. Radiation from Prescribed Sources 112
3.4. The Wavenumber–Frequency Domain 115
    3.4.1. Space Dispersive Media. Compressible Plasma 116
    3.4.2. Radiation in Isotropic Homogeneous Media 118
    3.4.3. The Resonant Wave Solution 119
    3.4.4. Guided-Wave Representation 123
3.5. The Field Representation 126
    3.5.1. Fields and the Plane-Wave Spectrum Relationship . . . 128
    3.5.2. Asymptotic Evaluation of the Far Field 130
    3.5.3. Radiation from Apertures 133
    3.5.4. Gaussian Beams 135
3.6. Summary and Selected References 138
    References 140

CHAPTER 4. **Narrowband Signals and Phasor Fields**

4.1. Narrowband Signals 141
    4.1.1. Phasor Evaluation 145
    4.1.2. Linear Operations on Phasors 145
    4.1.3. Response of a Linear Time-Invariant Circuit
           to a Narrowband Signal 147
    4.1.4. Quadratic Averages 148
    4.1.5. Power and Phasors 149
    4.1.6. Bandlimited Signals 151
    4.1.7. Almost Bandlimited Signals 154
4.2. Complex Vectors 155
    4.2.1. Scalar Product for Complex Vectors and Some
           of Their Properties 160
    4.2.2. Polarization States 162
    4.2.3. Stokes Parameters and the Poincaré Sphere 163
    4.2.4. Field Coherence 168
4.3. Maxwell Equations in Phasor Form 170
    4.3.1. Poynting's Theorem 171
    4.3.2. The Energy Theorem 173
    4.3.3. Uniqueness 175
    4.3.4. Image Theory 176
    4.3.5. Duality Transformation 176
    4.3.6. Reciprocity 176
    4.3.7. The Equivalence Theorem 178
4.4. Plane-Wave Propagation 179

4.4.1. Propagation of a Gaussian Wavepacket
in a Plasma Medium 184
4.4.2. Information Scrambling through a Dispersive Channel 186
4.4.3. Plane-Wave Propagation in a Magnetized Plasma . . . 188
4.5. Guided Propagation 193
4.5.1. Oblique Incidence on a Dielectric Half-Space 195
4.5.2. Guided Propagation along a Dielectric Slab 201
4.5.3. Propagation inside a Parallel-Plate Guide 205
4.5.4. Guided Propagation along Cylindrical Structures 206
4.6. Radiation from Prescribed Sources 210
4.6.1. The Elementary Electric Dipole 212
4.6.2. The Elementary Magnetic Dipole 215
4.6.3. The Elementary Huygens Source 217
4.6.4. Radiation from Planar Sources 220
4.6.5. Radiation from Linear Arrays 222
4.7. Summary and Selected References 225
References 226

CHAPTER 5. High-Frequency Fields

5.1. Asymptotic Form of Maxwell Equations 227
5.1.1. The Transport Equation 232
5.1.2. Rays in a Homogeneous Medium 233
5.1.3. Ray Propagation in a Layered Medium 235
5.1.4. Polarization Change along a Ray 238
5.2. Ray Properties 239
5.2.1. Reflector Antennas 245
5.2.2. Lens Antennas 246
5.2.3. Guided Propagation 248
5.3. The Ray Coordinate System 251
5.3.1. High-Frequency Propagation in a Homogeneous
Environment 255
5.3.2. Ray Amplitude at Reflection Boundaries:
The Two-Dimensional Case 256
5.3.3. Ray Amplitude at Reflection Boundaries:
The Three-Dimensional Case 258
5.4. Asymptotic Form of Field Representations 260
5.4.1. Scattering by a Conducting Half-Plane 266
5.4.2. The Edge Ray 272
5.4.3. Transition Functions 275
5.4.4. The Slope Diffraction Coefficient 277

- 5.4.5. The Lateral Ray 279
- 5.4.6. The Creeping Ray 284
- 5.5. Summary and Selected References 292
- References 293

CHAPTER 6. The Numerical Domain

- 6.1. General Considerations 295
  - 6.1.1. Matrix Equations 298
  - 6.1.2. Matrix Inversion 298
  - 6.1.3. Eigenvalue Computation 299
  - 6.1.4. Matrix Condition 301
- 6.2. The Method of Moments 303
  - 6.2.1. The Electromagnetic Field Integral Equations 305
  - 6.2.2. Scattering by a Metal Strip 308
- 6.3. The Finite Element Method 310
  - 6.3.1. Elements and Element Bases 314
  - 6.3.2. Guided-Wave Propagation 320
  - 6.3.3. Absorbing Boundary Conditions 321
- 6.4. The Finite Difference Method 324
  - 6.4.1. Stability and Numerical Dispersion 326
  - 6.4.2. FDM in the Frequency Domain 328
- 6.5. Summary and Selected References 329
- References 330

CHAPTER 7. Engineering Topics: Propagation

- 7.1. General Considerations 331
- 7.2. Transmission Lines 332
  - 7.2.1. The Telegraphists' Equations 336
  - 7.2.2. Reflection Coefficient and Impedance 340
  - 7.2.3. Matching 345
  - 7.2.4. Multisection Transmission Lines 347
  - 7.2.5. Nonuniform Transmission Lines 350
  - 7.2.6. Multiconductor Transmission Lines 353
  - 7.2.7. Transmission Line Generators 355
- 7.3. Equivalent Transmission Lines: Two-Dimensional Structures 356
  - 7.3.1. Multilayer Propagation 359
  - 7.3.2. Transverse Resonance 361

7.3.3. Propagation along a Grounded Slab 362
7.4. Equivalent Transmission Lines: Three-Dimensional
Structures 364
   7.4.1. The Rectangular Waveguide 370
   7.4.2. The Circular Waveguide 373
   7.4.3. The Coaxial Cable 376
   7.4.4. Mode Orthogonality and Power Flux 378
   7.4.5. Waveguide Excitation 381
   7.4.6. Waveguide Losses 383
   7.4.7. The Inhomogeneous Rectangular Waveguide 386
   7.4.8. The Fiber 388
7.5. Planar Guiding Configurations 398
   7.5.1. The Effective Dielectric Constant 399
7.6. Equivalent Circuits 400
   7.6.1. Computation of Matrix Entries 403
   7.6.2. Junction Matrix Properties 405
   7.6.3. Shift of Port Position 408
   7.6.4. The Three-Port Junction 409
   7.6.5. The Four-Port Junction 410
   7.6.6. The Directional Coupler 413
   7.6.7. Periodic Structures 418
   7.6.8. Obstacles in Waveguides 421
7.7. Summary and Selected References 425
References 426

CHAPTER 8. Engineering Topics: Radiation

8.1. Transmitting and Receiving Antennas 427
   8.1.1. Reciprocity Theory and Antennas 429
8.2. Parameters of the Transmitting Antenna 432
   8.2.1. The Radiation Parameters of the Elementary
         Loop Antenna 435
   8.2.2. The Radiation Parameters of the Elementary
         Huygens Source 436
   8.2.3. Input Resistance of Elementary Antennas 436
   8.2.4. Antenna Beamwidth 437
   8.2.5. Mechanical Forces on Antennas 439
8.3. Parameters of the Receiving Antenna 440
   8.3.1. Power and Polarization Matching 443
   8.3.2. The Radio Link Equation 444
   8.3.3. Effective Area of Elementary Antennas 444

	8.3.4.	Noise Temperature of the Antenna	445
8.4.	Wire Antennas	447
	8.4.1.	Short Antennas	452
	8.4.2.	The Half-Wave Dipole Antenna	453
	8.4.3.	The Traveling-Wave Antenna	457
	8.4.4.	Mutual Impedance	459
8.5.	Aperture Antennas	463
	8.5.1.	The Rectangular Aperture	465
	8.5.2.	The Circular Aperture	470
	8.5.3.	The Patch Antenna	473
8.6.	Reflector Antennas	476
	8.6.1.	The Parabolic Dish	479
	8.6.2.	Computation of the Reflector Radiation Diagram via the Current Integration Method	483
	8.6.3.	Computation of the Reflector Radiation Diagram via Optical Techniques	487
8.7.	Arrays	491
	8.7.1.	Array with Uniform Excitation	494
	8.7.2.	Array with Tapered Excitation	496
	8.7.3.	The Binomial Array	497
	8.7.4.	Sum and Difference Patterns	498
8.8.	Summary and Selected References	499
	References	500

CHAPTER 9. Engineering Topics: Scattering

9.1.	Interior Resonance	501
	9.1.1.	The Parallelepiped Cavity	506
	9.1.2.	The Loaded Coaxial Cavity	508
	9.1.3.	Multimode Cavities	511
	9.1.4.	Open Cavities	513
	9.1.5.	Energy Decay in the Cavity	515
	9.1.6.	Equivalent Circuit of the Cavity	516
9.2.	Exterior Resonance	519
	9.2.1.	Cylindrical Coordinates	523
	9.2.2.	Spherical Coordinates	527
9.3.	Exterior Scattering via Asymptotic Techniques	532
	9.3.1.	Rough Surfaces	533
	9.3.2.	Scattering by a Planar Rough Surface	536
9.4.	Summary and Selected References	544
	References	545

APPENDIXES

A. Vector Analysis 547
B. Dyadic Analysis 553
C. Useful Integrals and Series 555
D. Special Functions and Asymptotic Evaluations 557

INDEX 565

Electromagnetics

Theory, Techniques, and
Engineering Paradigms

1
Fundamentals

1.1. MAXWELL EQUATIONS

Macroscopic electromagnetic phenomena are described by a set of equations named after James Clerk Maxwell.[1] In any time-invariant reference system of coordinates the differential form of Maxwell equations reads as follows:

$$\nabla \times \mathbf{e} = -\frac{\partial \mathbf{b}}{\partial t}, \qquad (1.1a)$$

$$\nabla \times \mathbf{h} = \frac{\partial \mathbf{d}}{\partial t} + \mathbf{j}, \qquad (1.1b)$$

$$\nabla \cdot \mathbf{d} = \rho, \qquad (1.1c)$$

$$\nabla \cdot \mathbf{b} = 0. \qquad (1.1d)$$

Equations (1.1) are in vector form and apply to any reference frame of coordinates.

The single scalar and five vectors that appear in the equations are *observables*, i.e., measurable quantities, each a function of position vector $\mathbf{r}$ and time t, e.g., $\mathbf{e} = \mathbf{e}(\mathbf{r}, t)$; see Table 1.1. Measurement units are clearly needed; these are also presented in Table 1.1 with reference to the SI international system. Other observables introduced later in this section are also included in Table 1.1.

Equations (1.1) involve pure electromagnetic quantities and do not describe mutual interaction forces. These are provided by the Lorentz[2] force density equation:

$$\mathbf{f} = \rho \mathbf{e} + \mathbf{j} \times \mathbf{b}. \qquad (1.2)$$

---

1. James Clerk Maxwell: Edinburgh (UK), 1831–Cambridge, 1879.
2. Hendrik Antoon Lorentz: Arnhem (Holland), 1853–Haarlem, 1928.

TABLE 1.1. Electromagnetic Observables and Their Corresponding Measurement Units

Observable	Name	Unit
e	Electric field	volt/m
h	Magnetic field	ampere/m
d	Electric induction	coulomb/m^2
d	Magnetic induction	weber/m^2 or tesla
j	Electric current density	ampere/m^2
ρ	Electric charge density	coulomb/m^3
f	Density force	newton/m^3
ϕ	Electric induction flux	coulomb
ψ	Magnetic induction flux	weber
i	Current	ampere
χ	Magnetic charge density	weber/m^3
k	Magnetic current density	volt/m^2
ε	Permittivity	farad/m
μ	Permeability	henry/m
σ	Conductivity	siemens/m
p	Polarization	coulomb/m^2
m	Magnetization	weber/m^2
ρ_S	Electric charge surface density	coulomb/m^2
χ_S	Magnetic charge surface density	weber/m^2
j$_S$	Electric current linear density	ampere/m
k$_S$	Magnetic current linear density	volt/m
s	Poynting vector	watt/m^2
c	Wave speed	m/s
ζ	Intrinsic resistance	ohm (Ω)

Maxwell equations in their differential form (1.1) express the space and time rate of change of electromagnetic quantities. An alternative form of the equations is obtained by considering field values over appropriate space and/or time intervals; this is also convenient for conducting measurements, which necessarily implies space–time averages. With reference to Fig. 1.1, this is readily accomplished by volume integration of Eqs. (1.1c–d):

$$\oiint_S dS\,\mathbf{d}\cdot\hat{\mathbf{n}} = \iiint_V dV\rho = q, \tag{1.3}$$

$$\oiint_S dS\,\mathbf{b}\cdot\hat{\mathbf{n}} = 0 \tag{1.4}$$

[see Eq. (A.24) of Appendix A (Gauss[3] theorem)] and surface integration of

---
3. Karl Friedrick Gauss: Brunswick (Germany), 1775–Göttingen, 1836.

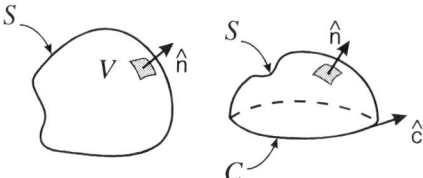

FIGURE 1.1. Volume and surface integration of Maxwell equations. The unit vector $\hat{n}$, on the left, points outside the volume V and the unit vector $\hat{c}$, on the right, is oriented counterclockwise with respect to $\hat{n}$: a corkscrew rotating as $\hat{c}$ proceeds along $\hat{n}$.

Eqs. (1.1a-b):

$$\oint_C d c \mathbf{e} \cdot \hat{\mathbf{c}} = - \iint_S dS \frac{\partial \mathbf{b}}{\partial t} \cdot \hat{\mathbf{n}} = - \frac{d\psi}{dt}, \qquad (1.5)$$

$$\oint_C d c \mathbf{h} \cdot \hat{\mathbf{c}} = \iint_S dS \left(\frac{\partial \mathbf{d}}{\partial t} + \mathbf{j} \right) \cdot \hat{\mathbf{n}} = \frac{d\phi}{dt} + i \qquad (1.6)$$

[see Eq. (A.23) (Stokes[4] theorem)], where surfaces and lines are sufficiently regular that normal $\hat{n}$ and tangent $\hat{c}$ unit vectors can be defined unambiguously.

Equations (1.3) and (1.4) express the Coulomb[5] law: the flux of induction throughout a closed surface S bounding a volume V equals the total charge inside the volume; and there are no separated magnetic charges (they can only appear in pairs, positive and negative).

Equation (1.5) expresses the Lentz–Neumann[6] law: the line integral of the electric field along a closed line C bounding an (open) surface S equals (but for the sign) the time rate of change of the magnetic induction flux ψ through the surface. Note the exchange of integral and $\partial/\partial t$ operators in Eq. (1.5); this is discussed in Section 1.1.3. For stationary fields Eq. (1.5) coincides with Kirchhoff's[7] second law.

Equation (1.6) expresses the Ampère–Faraday[8] law: the line integral of the magnetic field along a closed line C bounding an (open) surface S equals the sum of the current i and the time rate of change of the electric induction flux ϕ through the surface. The current and rate of change of electric induction flux are on the same footing; see also Section 1.1.1. For the time derivative operator, see also Section 1.1.3.

4. George Gabriel Stokes: Skreen, County Sligo (Ireland), 1819–Cambridge, 1903.
5. Charles Augustin de Coulomb: Angulême (France), 1736–Paris, 1806.
6. Emilij Christianovic Lentz: Dorpart, now Tartu (Russia), 1804–Rome, 1865.
   Franz Ernest Neumann: Joachimsthal (Germany), 1798–Köenigsberg, 1895.
7. Gustav Robert Kirchhoff: Köenigsberg (Germany), 1824–Berlin, 1887.
8. André Marie Ampère: Lyon (France), 1775–Marsiglie, 1836.
   Michael Faraday: Newington (UK), 1791–Hampton Court, 1867.

1.1.1. The Current Density Equation

By taking the divergence of Eq. (1.1b) we obtain

$$\nabla \cdot \left[\mathbf{j} + \frac{\partial \mathbf{d}}{\partial t} \right] = 0,$$

which states that the vector $\mathbf{j} + \partial \mathbf{d}/\partial t$ is solenoidal. Moreover, the use of Eq. (1.1c) yields

$$\nabla \cdot \mathbf{j} + \frac{\partial \rho}{\partial t} = 0, \qquad (1.7)$$

which is referred to as the current density continuity equation.

The integral form of Eq. (1.7) is readily obtained by integration over a volume V as depicted in Figure 1.1:

$$\oiint_S dS \mathbf{j} \cdot \hat{\mathbf{n}} + \frac{dq}{dt} = 0. \qquad (1.8)$$

In Eq. (1.8), the integral is the total current flux i out of the surface S and q is the total charge inside the volume.

Equation (1.8) states the conservation of charge: an increase (decrease) of charge inside the volume V calls for an inward (outward) current flux through the surface S. For stationary fields Eq. (1.8) coincides with Kirchhoff's first law.

1.1.2. The Independence of Maxwell Equations

Maxwell equations may not all be independent. By taking the divergence of Eq. (1.1a) we obtain

$$0 = \nabla \cdot \frac{\partial \mathbf{b}}{\partial t} = \frac{\partial}{\partial t} \nabla \cdot \mathbf{b},$$

which implies that $\nabla \cdot \mathbf{b}$ is independent of time. If the fields are equal to zero before a given time, then $\nabla \cdot \mathbf{b} = 0$ for all times, thus recovering Eq. (1.1d).

1.1.3. The Lentz–Neumann Law

In the derivation of Eq. (1.5) the integral and time derivative operators have been exchanged, which is certainly allowed if the surface S and the (closed) contour C (see Fig. 1.1) are stationary. If this condition is relaxed, the

FUNDAMENTALS

total time derivative of the induction flux ψ is given by

$$\frac{d}{dt}\iint_S dS\mathbf{b}\cdot\hat{\mathbf{n}} = \frac{d\psi}{dt}$$

$$= \iint_S dS\frac{\partial \mathbf{b}}{\partial t}\cdot\hat{\mathbf{n}} + \lim_{\Delta t \to 0}\frac{\iint_{S_2} dS\mathbf{b}\cdot\hat{\mathbf{n}} - \iint_{S_1} dS\mathbf{b}\cdot\hat{\mathbf{n}}}{\Delta t},$$

where $S_2 = S(t + \Delta t/2)$, $S_1 = S(t - \Delta t/2)$, and $\mathbf{b} = \mathbf{b}(t)$. Computation of the incremental ratio that appears in the equation is now in order.

With reference to Fig. 1.2, let each point of the surface S move with velocity $\mathbf{v}$, so that each point of the surface S_1 undergoes a displacement $\mathbf{v}\Delta t$ after time Δt. Let S_3 be the surface joining the two curves C_1 and C_2, so that the total surface $S_1 + S_2 + S_3$ encloses the volume $\Delta V = S_1\mathbf{v}\cdot\hat{\mathbf{n}}\Delta t$. For any vector $\mathbf{a}$ we have

$$\iiint_{\Delta V} dV\nabla\cdot\mathbf{a} = -\iint_{S_1} dS\mathbf{a}\cdot\hat{\mathbf{n}} + \iint_{S_2} dS\mathbf{a}\cdot\hat{\mathbf{n}} + \Delta t\oint_C dc(\hat{\mathbf{c}}\times\mathbf{v})\cdot\mathbf{a}$$

$C \to C_1$, C_2 as $\Delta t \to 0$, and the line integral takes care of the flux of $\mathbf{a}$ outside the surface S_3. By letting $dV = dS\mathbf{v}\Delta t\cdot\hat{\mathbf{n}}$, we have

$$\frac{1}{\Delta t}\left[\iint_{S_2} dS\hat{\mathbf{n}}\cdot\mathbf{a} - \iint_{S_1} dS\hat{\mathbf{n}}\cdot\mathbf{a}\right] = \iint_S dS(\mathbf{v}\nabla\cdot\mathbf{a})\cdot\hat{\mathbf{n}} + \oint_C dc(\mathbf{a}\times\mathbf{v})\cdot\hat{\mathbf{c}}.$$

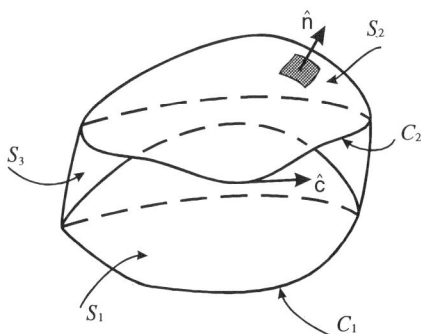

FIGURE 1.2. The Lentz–Neumann law in the presence of a moving integration surface.

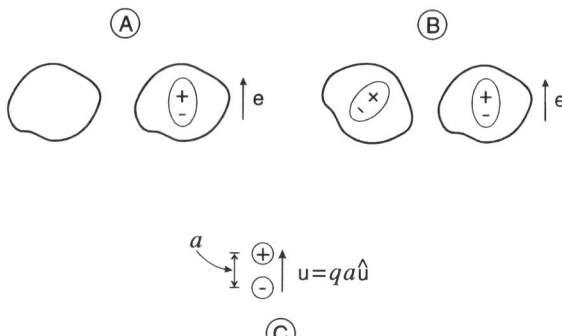

FIGURE 1.3. (A) A neutral molecule becomes polarized because the applied electric field displaces positive and negative charges. (B) A polarized molecule is oriented by the applied electromagnetic field. (C) The dipole moment definition.

If **a** is identified with **b**, the magnetic induction, then Eq. (1.5) reduces to

$$\oint_C dc[\mathbf{e} + \mathbf{v} \times \mathbf{b}] \cdot \hat{\mathbf{c}} = -\frac{d\psi}{dt},$$

because $\nabla \cdot \mathbf{b} = 0$.
The Ampère–Faraday law, Eq. (1.6) becomes

$$\oint_C dc[\mathbf{h} - \mathbf{v} \times \mathbf{d}] \cdot \hat{\mathbf{c}} = \frac{d\phi}{dt} + \iint_S dS(\mathbf{j} - \rho\mathbf{v}) \cdot \hat{\mathbf{n}}.$$

1.1.4. Polarization and Magnetization

Free electrons (or, more generally, ionized atoms or molecules) are the physical counterpart of the electric charge density. However, in addition to these *free charges* there are in nature *bonded charges*; these usually appear in couples and can be either excited or oriented by an applied electric field (see Fig. 1.3).[9] For each charge couple we define an electric dipole moment $\mathbf{u} = q a \hat{\mathbf{u}}$, where the orientation is toward the positive charge. An ensemble of dipoles is described by an electric dipole moment density, or *polarization* **p**. Also, magnetic materials can be modeled in terms of a magnetic dipole moment density, or *magnetization* **m**. It is important to assess how these vectors appear in Maxwell equations.

9. Free charges move over distances large with respect to atomic or molecular spacings, while bonded charge movement is limited to these spacings.

FUNDAMENTALS

With reference to the electric case and Fig. 1.4, consider a charge q within a volume ΔV, so that the (average) charge density is $\rho = q/\Delta V$. We let q remain constant and $\Delta V \to 0$, so that

$$\rho = \lim_{\Delta V \to 0} \frac{q}{\Delta V} = q\delta(\mathbf{r}),$$

where $\delta(\mathbf{r})$ is the (three-dimensional) Dirac function (see also Section 3.1.1):

$$\iiint \delta(\mathbf{r})dV = 1.$$

Similarly, two charges of opposite sign and closely spaced along the z-axis correspond to the charge density:

$$\rho_P = q\delta(x)\delta(y)\left[\delta\left(z - \frac{a}{2}\right) - \delta\left(z + \frac{a}{2}\right)\right]$$
$$\approx -qa\delta(x)\delta(y)\dot{\delta}(z) = -\nabla \cdot [\mathbf{u}\delta(\mathbf{r})] = -\nabla \cdot \mathbf{p}, \qquad (1.9)$$

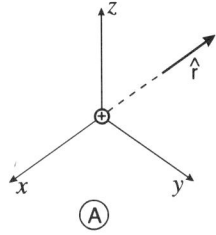

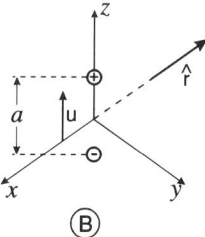

FIGURE 1.4. Electric (A) charges and (B) dipoles, and their equivalence as far as the associated fields and inductions are concerned.

where $\mathbf{u} = qa\hat{\mathbf{z}}$ is the dipole moment, $\dot{\delta}(z) = d\delta/dz$ and $\mathbf{p} = \mathbf{u}\delta(\mathbf{r})$ is the dipole moment density.

Equation (1.9) shows that a dipole moment density $\mathbf{p}$ is equivalent to a charge density $\rho_P = -\nabla \cdot \mathbf{p}$. In addition, the current density equation (1.7) shows that this dipole moment density corresponds also to a (polarization) current density $\mathbf{j}_P$ given by

$$\nabla \cdot \left(\mathbf{j}_P - \frac{\partial \mathbf{p}}{\partial t} \right) = 0,$$

i.e.,

$$\mathbf{j}_P = \frac{\partial \mathbf{p}}{\partial t}. \tag{1.10}$$

Accordingly, Maxwell equations (1.1b–c) become

$$\nabla \times \mathbf{h} = \frac{\partial (\mathbf{d} + \mathbf{p})}{\partial t} + \mathbf{j}, \quad \nabla \cdot (\mathbf{d} + \mathbf{p}) = \rho,$$

which show that the vectors $\mathbf{d}$ and $\mathbf{p}$ are exactly on the same footing and cannot be distinguished one from the other at macroscopic level. It may be convenient to take $\mathbf{d} + \mathbf{p} \to \mathbf{d}$ for the electric and similarly $\mathbf{b} + \mathbf{m} \to \mathbf{b}$ for the magnetic[10] induction, so that Maxwell equations can be written in the form of Eqs. (1.1) without introducing the new vectors $\mathbf{p}$ and $\mathbf{m}$, the charge density ρ in Maxwell equation (1.1c) corresponding to the free charge only.

1.1.5. Field Sources

Fields, inductions, charge and current densities appearing in Eqs. (1.1) are *unknown* functions of space and time that should satisfy Maxwell equations together with initial and boundary conditions (see Section 1.4). However, some of these functions may be prescribed in given space regions and time intervals. These *known* functions are referred to as *field* sources; usually, these sources are charges and currents.

Although not explicitly stated, we refer to a physical model where space–time evolution of the sources (amplitude, position, and movement of charges) is specified and sustained by external agents (mechanical, chemical, electrical, etc.) which are not influenced by the electromagnetic field generated by the sources themselves. These are *ideal* sources, at variance with more realistic *real* sources, whose behavior is more or less dependent on the field

---

[10]. For traditional reasons it is often written $\mathbf{b} \to \mathbf{b} + \mu_0 \mathbf{m}$, which implies a different measurement unit for the magnetization vector.

FUNDAMENTALS

they generate. However, in spite of this idealization, the use of ideal sources, or simply sources, has proved to be very fruitful. Accordingly, we decompose the charge and current terms appearing in Maxwell's equations (1.1) as follows:

$$\mathbf{j} \to \mathbf{j} + \mathbf{j}_0, \qquad \rho \to \rho + \rho_0, \tag{1.11}$$

where $\mathbf{j}_0$ and ρ_0 are the prescribed part, i.e., the sources, while $\mathbf{j}$ and ρ are dependent variables describing the observables, electric charge and current densities induced by the electromagnetic field. Should the sources be prescribed in terms of dipoles (see Section 1.1.4), we obtain additional forcing terms $\partial \mathbf{p}_0/\partial t$ and $-\nabla \cdot \mathbf{p}_0$.

The considered sources rely on the existence of electric charges in nature, the electrons, but some sources can be efficiently modeled in terms of (equivalent) magnetic charges (see Section 2.3.2). We can introduce magnetic charge χ and current $\mathbf{k}$ densities related by a conservation equation similar to Eq. (1.7),

$$\nabla \cdot \mathbf{k} + \frac{\partial \chi}{\partial t} = 0, \tag{1.12}$$

whose prescribed parts, χ_0 and $\mathbf{k}_0$, play the role of magnetic sources. Maxwell's equations take the more symmetric form

$$\begin{cases} \nabla \times \mathbf{e} = -\dfrac{\partial \mathbf{b}}{\partial t} - \mathbf{k}, & (1.13\text{a}) \\[6pt] \nabla \times \mathbf{h} = \dfrac{\partial \mathbf{d}}{\partial t} + \mathbf{j}, & (1.13\text{b}) \\[6pt] \nabla \cdot \mathbf{d} = \rho, & (1.13\text{c}) \\[6pt] \nabla \cdot \mathbf{b} = \chi & (1.13\text{d}) \end{cases}$$

while the force density equation becomes

$$\mathbf{f} = \rho \mathbf{e} + \chi \mathbf{h} + \mathbf{j} \times \mathbf{b} - \mathbf{k} \times \mathbf{d}. \tag{1.14}$$

1.1.6. Source Power

Consider a prescribed charge density ρ_0 moving with a prescribed velocity $\mathbf{v}_0$, so that the current density source $\mathbf{j}_0 = \rho_0 \mathbf{v}_0$ is given. These sources generate an electromagnetic field $(\mathbf{e}, \mathbf{h})$ which exerts the force density [see Eq. (1.2)]

$$\mathbf{f} = \rho_0 \mathbf{e} + \mathbf{j}_0 \times \mathbf{b} = \rho_0 (\mathbf{e} + \mathbf{v}_0 \times \mathbf{b})$$

upon the sources. The displacement $\mathbf{v}_0 \Delta t$ of the elementary charge $\rho_0 \Delta V$ in volume ΔV and within time Δt is related to the work:

$$\mathbf{f} \cdot \mathbf{v}_0 \Delta V \Delta t = \rho_0 \mathbf{v}_0 \cdot \mathbf{e} \Delta V \Delta t = \mathbf{j}_0 \cdot \mathbf{e} \Delta V \Delta t,$$

because $\mathbf{v}_0 \times \mathbf{b} \cdot \mathbf{v}_0 = 0$. This equation provides the work that the field exerts upon the sources within the space–time interval $\Delta V \Delta t$. Accordingly, the quantity

$$p_0 = -\mathbf{j}_0 \cdot \mathbf{e} \qquad (1.15)$$

is the power density that the sources deliver to the field. For magnetic sources we would similarly obtain

$$p_0 = -\mathbf{h} \cdot \mathbf{k}_0. \qquad (1.16)$$

1.2. CONSTITUTIVE RELATIONSHIPS

Maxwell equations (1.1) involve a number of unknowns larger than the number of equations; further information is needed to determine their solution. This information is provided by additional equations, the *constitutive relations*, that describe interaction of fields and matter from a macroscopic viewpoint and couple other (mechanical, thermal, etc.) scalar and vector fields to the electromagnetic ones.

From this viewpoint a most general field should be introduced, with the number of unknowns equal to the number of independent equations; this field would involve electromagnetic as well as nonelectromagnetic observables. However, a different approach can be adopted: the constitutive relations are used to express some of the electromagnetic observables in terms of the remaining ones, so that only Maxwell equations with the appropriate number of unknowns are finally used.

It is customary to represent inductions and currents in terms of fields:

$$\mathbf{d} = \mathscr{D}[\mathbf{e}, \mathbf{h}], \quad \mathbf{b} = \mathscr{B}[\mathbf{e}, \mathbf{h}], \quad \mathbf{j} = \mathscr{J}[\mathbf{e}, \mathbf{h}], \qquad (1.17)$$

where $\mathscr{D}[\cdot]$, $\mathscr{B}[\cdot]$, and $\mathscr{J}[\cdot]$ are functionals which depend upon the medium in which the electromagnetic field is considered and upon the fields themselves, in general.

The media can be categorized according to the functional appearing in Eq. (1.17). To exploit this point we define $\mathbf{d}, \mathbf{b}, \mathbf{j}$ as *output* variables while $(\mathbf{e}, \mathbf{h})$ are the *input* ones, and we refer to the first of relations (1.17).

FUNDAMENTALS   11

The medium is said to be *linear* if a linear combination of inputs produces the same linear combination of corresponding outputs:

$$\mathbf{d}_1 = \mathscr{D}[\mathbf{e}_1, \mathbf{h}_1] \quad \text{and} \quad \mathbf{d}_2 = \mathscr{D}[\mathbf{e}_2, \mathbf{h}_2]$$

imply that

$$\mathbf{d}_1 + \mathbf{d}_2 = \mathscr{D}[\mathbf{e}_1 + \mathbf{e}_2, \mathbf{h}_1 + \mathbf{h}_2].$$

The functional $\mathscr{D}[\cdot]$ is correspondingly linear (i.e., its structure is independent of the input variables) and homogeneous. In addition, we require it to be also continuous, under the physical constraint that the output depends continuously on the input. Linear media are considered in the following.

In its simplest form, the functional $\mathscr{D}[\cdot]$ reduces to proportionality:

$$\mathbf{d}(\mathbf{r}, t) = \varepsilon(\mathbf{r}, t) \cdot \mathbf{e}(\mathbf{r}, t) + \chi(\mathbf{r}, t) \cdot \mathbf{h}(\mathbf{r}, t). \tag{1.18}$$

Note that ε and χ are 3×3 matrices, while inductions and fields are 3×1 column vectors; each (space) component of the output depends on all the three (space) components of the input. Each of the matrices appearing in Eq. (1.18) describes some of the electromagnetic local properties of the medium; it is desirable that this description depends only on the medium properties, and not on the adopted system of coordinates.[11] This is possible only if the coordinate system is space–time invariant, i.e., a fixed cartesian system whose unit vectors are constant. This is assumed in the following.

Equation (1.18) states that the output at space–time $(\mathbf{r}, t)$ depends *only* on the value of the input at the *same* point $\mathbf{r}$ and *same* time t. These media are referred to as *local* or *nondispersive*.[12]

Examination of Eq. (1.18) shows that the induction $\mathbf{d}$ depends on both the electric $\mathbf{e}$ and magnetic $\mathbf{h}$ fields. These media are referred to as *bianisotropic*. Simpler media are the *anisotropic* ones (see also Section 1.2.3),

$$\mathbf{d}(\mathbf{r}, t) = \varepsilon(\mathbf{r}, t) \cdot \mathbf{e}(\mathbf{r}, t), \tag{1.19a}$$

$$\mathbf{b}(\mathbf{r}, t) = \mu(\mathbf{r}, t) \cdot \mathbf{h}(\mathbf{r}, t), \tag{1.19b}$$

$$\mathbf{j}(\mathbf{r}, t) = \sigma(\mathbf{r}, t) \cdot \mathbf{e}(\mathbf{r}, t), \tag{1.19c}$$

where ε, μ, and σ are the *permittivity*, *permeability*, and *conductivity*, respectively. These media are considered below.

A medium is *isotropic* if a rotation of the input implies the *same* rotation of the output. Let R be the rotation matrix, with $R \cdot \tilde{R} = I$, where I is the

---

11. The coordinate system is defined by the unit vectors, which in general change their orientation as we move in the space.
12. They are said to possess no heredity when reference is made to time only.

diagonal unit matrix. With reference to the permittivity of an isotropic medium we have

$$\mathbf{d} = \varepsilon \cdot \mathbf{e}, \qquad \mathbf{R} \cdot \mathbf{d} = \mathbf{R} \cdot \varepsilon \cdot \mathbf{e} = \mathbf{R} \cdot \varepsilon \cdot \widetilde{\mathbf{R}} \cdot \mathbf{R} \cdot \mathbf{e},$$

and isotropy requires that

$$\mathbf{R} \cdot \varepsilon \cdot \widetilde{\mathbf{R}} = \varepsilon, \qquad \text{i.e.,} \qquad \mathbf{R} \cdot \varepsilon = \varepsilon \cdot \mathbf{R},$$

which can be verified only if $\varepsilon = I\varepsilon$ because $\mathbf{R}$ is not symmetric ($\mathbf{R} \neq \widetilde{\mathbf{R}}$). The constitutive relation attains the simplest form:

$$\mathbf{d}(\mathbf{r}, t) = \varepsilon(\mathbf{r}, t)I \cdot \mathbf{e}(\mathbf{r}, t) = \varepsilon(\mathbf{r}, t)\mathbf{e}(\mathbf{r}, t), \tag{1.20a}$$

where input and output are collinear. Similarly

$$\mathbf{b}(\mathbf{r}, t) = \mu(\mathbf{r}, t)\mathbf{h}(\mathbf{r}, t), \tag{1.20b}$$

$$\mathbf{j}(\mathbf{r}, t) = \sigma(\mathbf{r}, t)\mathbf{e}(\mathbf{r}, t). \tag{1.20c}$$

The medium parameters ε, μ, and σ are scalars; input and output are collinear. In free space, $\varepsilon \to \varepsilon_0 = 8.854 \times 10^{-12}$ [farad/m] and $\mu \to \mu_0 = 1.256 \times 10^{-6}$ [henry/m].

A medium is *homogeneous*[13] if a time–space translation of the input implies the *same* time–space translation of the output:

$$\mathbf{d}(\mathbf{r}, t) = \varepsilon(\mathbf{r}, t) \cdot \mathbf{e}(\mathbf{r}, t),$$

$$\mathbf{d}(\mathbf{r} - \mathbf{r}_0, t - t_0) = \varepsilon(\mathbf{r}, t) \cdot \mathbf{e}(\mathbf{r} - \mathbf{r}_0, t - t_0),$$

i.e.,

$$\varepsilon(\mathbf{r}, t) = \varepsilon(\mathbf{r} - \mathbf{r}_0, t - t_0).$$

Accordingly, the electromagnetic parameters do not depend on $\mathbf{r}$ and t.

The integral[14] is the next, more involved, form for the functional $\mathscr{D}[\cdot]$. For an anisotropic medium

$$\mathbf{d}(\mathbf{r}, t) = \int d\mathbf{r}' \int dt' \mathbf{g}(\mathbf{r}, \mathbf{r}'; t, t') \cdot \mathbf{e}(\mathbf{r}', t'), \tag{1.21}$$

where $\mathbf{g}(\mathbf{r}, \mathbf{r}_0; t, t_0) \cdot \hat{\mathbf{e}}$ is the *unit response* of the medium, i.e., the output due to

---

13. These media are also said to be time-invariant, or stationary, and space-invariant.
14. A different type of constitutive relation is considered in Section 1.2.6.

the unitary input $\hat{e}\delta(\mathbf{r}' - \mathbf{r}_0)\delta(t' - t_0)$ localized at $\mathbf{r}' = \mathbf{r}_0, t' = t_0$. Note that the dimensions of $g(\cdot)$ are [farad/(m$^4 \cdot$s)].

Equation (1.21) states that the output at the space–time $(\mathbf{r}, t)$ depends on the values of the input throughout the space–time interval. Media described by Eq. (1.21) are defined as *dispersive* and can be anisotropic as well as bianisotropic. Transition to the nondispersive case is readily accomplished by letting

$$\mathbf{g}(\mathbf{r}, \mathbf{r}'; t, t') \to \varepsilon(\mathbf{r}', t')\delta(\mathbf{r} - \mathbf{r}')\delta(t - t').$$

Isotropy requires the unit response to be scalar if the medium is spatially nondispersve[15]: $\mathbf{g}(\mathbf{r}; t, t') \to g(\mathbf{r}; t, t')\mathbf{I}$, and a rotation of the input produces the *same* rotation of the output. Note, however, that input and output are not necessarily collinear, as in the case of a local medium.

Causality and *finite velocity* of propagation for the input set a constraint on $g(\cdot)$:

$$\mathbf{g}\left(\mathbf{r}, \mathbf{r}'; t < t' + \frac{|\mathbf{r} - \mathbf{r}'|}{c}, t'\right) = 0, \tag{1.22}$$

where c is the velocity of light in vacuo.[16] Homogeneity requires that

$$\mathbf{g}(\mathbf{r}, \mathbf{r}'; t, t') \to g(\mathbf{r} - \mathbf{r}', t - t'), \tag{1.23}$$

so that a space–time translation of the input $\mathbf{e}(\mathbf{r} - \mathbf{r}_0, t - t_0)$ produces the *same* space–time translation of the output:

$$\int d\mathbf{r}' \int dt' \mathbf{g}(\mathbf{r} - \mathbf{r}', t - t') \cdot \mathbf{e}(\mathbf{r}' - \mathbf{r}_0, t' - t_0)$$

$$= \int d\mathbf{r}' \int dt' \mathbf{g}([\mathbf{r} - \mathbf{r}_0] - [\mathbf{r}' - \mathbf{r}_0], [t - t_0] - [t' - t_0]) \cdot \mathbf{e}(\mathbf{r}' - \mathbf{r}_0, t' - t_0)$$

$$= \mathbf{d}(\mathbf{r} - \mathbf{r}_0, t - t_0).$$

This is the necessary and sufficient condition for the space–time translation invariance of the medium, and the output–input relationship reduces to a convolution. Note further that space-dispersive media are necessarily time-dispersive, too, due to the finite velocity of propagation; the response of neighboring points arrives at a later time, thus producing time dispersion.

15. Only for spatially nondispersive media can the rotation matrix be taken inside the integral of Eq. (1.21) and the same reasonings leading to Eqs. (1.20) be used.
16. Even in matter a portion of the field (*precursors*) propagates with the speed of light in vacuo; see Sections 3.2.5 and 3.2.6.

In summary, constitutive relations are functionals that express the inductions **d** and **b** and the density current **j** in terms of the fields **e** and **h**; see Table 1.2. Maxwell's equations (1.1a–b) now form a system of two (vector) equations in the two unknowns **e** and **h**. As far as the remaining equations are concerned, Eq. (1.1d) is a consequence of Eq. (1.1a) (see Section 1.1.2) and Eq. (1.1c) specifies the charge density (see Section 1.1.1).

1.2.1. Nonlinear Media

An example of constitutive relationships governing nonlinear media is the following:

$$\mathbf{d} = \mathbf{d}_0 + \varepsilon \mathbf{e} + a \mathbf{e} \cdot \mathbf{e}\mathbf{e}.$$

The medium is nondispersive, homogeneous if $\mathbf{d}_0$, ε, and a are constant with respect to **r** and t, and anisotropic (unless $\mathbf{d}_0 = 0$) because a rotation of **e** (e.g., around itself) does not induce an identical rotation of **d**. Linearity is recovered if both $\mathbf{d}_0$ and a are zero.

1.2.2. Linear Anisotropic Media

As a simple example of anisotropy we consider a medium of free electrons of density n [electrons/m^3], biased by a dc magnetic field $\mathbf{h}_0$ (see Fig. 1.5). The electrons can move under the action of an applied field; stationary positive charges at rest assure that the medium is macroscopically neutral.

The constitutive relations are

$$\mathbf{j} = -nq\mathbf{v}, \qquad (1.24)$$

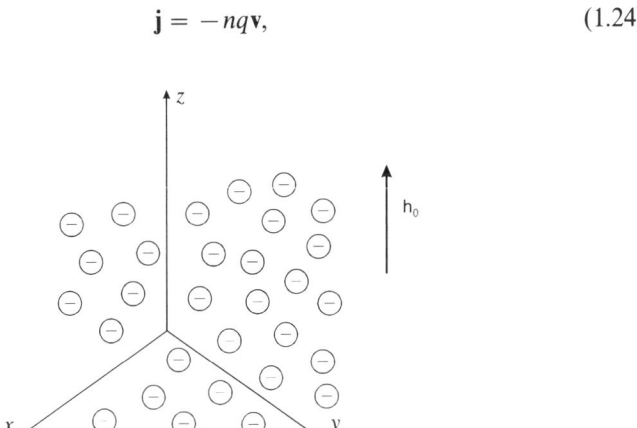

FIGURE 1.5. A medium of free electrons biased by a uniform dc magnetic field pointing along the z-axis.

FUNDAMENTALS

TABLE 1.2. Linear Media Classification (reference is made to the electric response)

Media	Local $\mathbf{d} = \varepsilon \cdot \mathbf{e}$		Time-Dispersive $\mathbf{d} = \int dt'\, \mathbf{g}(t,t') \cdot \mathbf{e}(t')$		Space-Dispersive $\mathbf{d} = \int d\mathbf{r}'\, \mathbf{g}(\mathbf{r},\mathbf{r}') \cdot \mathbf{e}(\mathbf{r}')$	
	Nonhomogeneous	Homogeneous	Time Variant	Time Invariant	Nonhomogeneous	Homogeneous
Anisotropic	$\boldsymbol{\varepsilon}(\mathbf{r},t)$	$\boldsymbol{\varepsilon}$	$\mathbf{g}(t,t')$	$\mathbf{g}(t-t')$	$\mathbf{g}(\mathbf{r},\mathbf{r}')$	$\mathbf{g}(\mathbf{r}-\mathbf{r}')$
Isotropic	$\varepsilon(\mathbf{r},t)$	ε	$g(t,t')$	$g(t-t')$	$g(\mathbf{r},\mathbf{r}')$	$g(\mathbf{r}-\mathbf{r}')$

and the Newton[17] force equation

$$m\frac{d\mathbf{v}}{dt} + \xi m\mathbf{v} = -q\mathbf{e} - q\mu_0 \mathbf{v} \times (\mathbf{h} + \mathbf{h}_0). \tag{1.25}$$

In Eqs. (1.24) and (1.25), $-q$ and m are the charge and mass of the electron, $\mathbf{v}$ its velocity, and the force term is provided by Eq. (1.2) augmented by the phenomenological friction term $-\xi m\mathbf{v}$, which accounts for the electron interaction; ξ can be interpreted as the number of electron collisions per unit time.

The case of a very large electron density n is of interest. This situation implies that the collision frequency is very large, too, and the term $md\mathbf{v}/dt$ can be neglected. Assuming further than the fields are vanishingly small, we neglect the term $\mathbf{v} \times \mathbf{h}$ (*linearization*), solve Eq. (1.25) for $\mathbf{v}$, and substitute in Eq. (1.24):

$$\mathbf{j} = -qn\mathbf{v} = \boldsymbol{\sigma} \cdot \mathbf{e},$$

$$\boldsymbol{\sigma} = \sigma \begin{vmatrix} a & -b & 0 \\ b & a & 0 \\ 0 & 0 & 1 \end{vmatrix},$$

where $\sigma = nq^2/m\xi$ is the usual conductivity of the material,

$$a = \frac{1}{1 + (\omega_c/\xi)^2}, \qquad b = \frac{\omega_c/\xi}{1 + (\omega_c/\xi)^2},$$

and

$$\omega_c = \frac{q\mu_0 h_0}{m} \tag{1.26}$$

is the *cyclotron* angular frequency of the electrons in the magnetic dc field $\mathbf{h}_0$.

This medium is nondispersive and exhibits the anisotropic conductivity $\boldsymbol{\sigma}$. Note that the latter is referred to the coordinate system of Fig. 1.5. A rotation of coordinates via the matrix $\boldsymbol{R}$ would change the representation of $\boldsymbol{\sigma}$ in the new one, $\boldsymbol{R} \cdot \boldsymbol{\sigma} \cdot \tilde{\boldsymbol{R}}$, $\tilde{\boldsymbol{R}}$ being the transpose of $\boldsymbol{R}$.

1.2.3. Anisotropic Media Classification

Let us consider a nondispersive anisotropic dielectric of permittivity ε. The *eigenvectors* $\mathbf{u}_i$ of the matrix ε represent the solution of the linear

---

17. Isaac Newton: Woolsthorpe, Lincolnshire (UK), 1642–Kensington, Middlesex, 1727.

FUNDAMENTALS

homogeneous system

$$(\boldsymbol{\varepsilon} - \varepsilon_i \boldsymbol{I}) \cdot \mathbf{u}_i = 0, \qquad (1.27)$$

where the eigenvalues ε_i are obtained by setting the determinant of Eq. (1.27) equal to zero:

$$\det(\boldsymbol{\varepsilon} - \varepsilon_i \boldsymbol{I}) = 0.$$

For a symmetric matrix, $\boldsymbol{\varepsilon} = \tilde{\boldsymbol{\varepsilon}}$, the eigenvalues ε_i are real, while the eigenvectors are orthogonal[18] and can be normalized:

$$\tilde{\mathbf{u}}_i \cdot \mathbf{u}_j = \delta_{ij}.$$

Then, we can form a matrix $\boldsymbol{R}$ by using the eigenvectors as rows. This matrix is unitary, $\boldsymbol{R} \cdot \tilde{\boldsymbol{R}} = \boldsymbol{I}$, and represents the rotation of coordinate axes which renders the matrix $\boldsymbol{\varepsilon}$ diagonal:

$$\boldsymbol{R} \cdot \boldsymbol{\varepsilon} \cdot \tilde{\boldsymbol{R}} = \begin{vmatrix} \varepsilon_1 & 0 & 0 \\ 0 & \varepsilon_2 & 0 \\ 0 & 0 & \varepsilon_3 \end{vmatrix}. \qquad (1.28)$$

The dielectric is said to be *biaxial* if the three eigenvalues are all different, and *uniaxial* if two of the eigenvalues are equal. Examples of a biaxial medium are some crystals; uniaxial media are provided by some magnetized plasmas (see Section 3.2.4).

An anisotropic medium whose properties are described by a symmetric matrix, e.g., $\boldsymbol{\varepsilon} = \tilde{\boldsymbol{\varepsilon}}$, is said to be *reciprocal*.

1.2.4. Linear Dispersive Media. I

As an example of a constitutive relation for a linear (time) dispersive medium, let us consider a dielectric material composed of randomly oriented electric dipoles. The applied field tends to line up the dipoles, at variance with the thermal agitation. The constitutive relations are given by (see Section 1.1.4)

$$\mathbf{d} = \varepsilon_0 \mathbf{e} + \mathbf{p},$$

and by the balance equation

$$\frac{d\mathbf{p}}{dt} = -\frac{1}{\tau}\mathbf{p} + \frac{\varepsilon_p}{\tau}\mathbf{e},$$

---

[18]. Should two eigenvalues be equal, a linear combination of the corresponding eigenvectors can be rendered orthogonal.

where **p** is the aligned dipole moment density; in unit volume and unit time the electric field lines up $\varepsilon_p \mathbf{e}/\tau$ dipoles, while $\mathbf{p}/\tau$ dipoles are brought back to a random orientation by the thermal agitation. The constants ε_p [farad/m] and τ [s] characterize the material, and the latter is clearly temperature-dependent.

Polarization and electric field are aligned. The unit response is a scalar given by

$$\frac{dg_p}{dt} + \frac{1}{\tau} g_p = \frac{\varepsilon_p}{\tau} \delta(t), \qquad g(t < 0) = 0. \tag{1.29}$$

For $t > 0$ we have

$$g_p(t) = A \exp\left(-\frac{t}{\tau}\right),$$

and the constant A is obtained by integrating Eq. (1.29) along a time interval $\Delta t \to 0$ across $t = 0$:

$$g_p(t) = \frac{\varepsilon_p}{\tau} \exp\left(-\frac{t}{\tau}\right) U(t).$$

The response $g_p(t)$ is depicted in Fig. 1.6.

We have

$$\mathbf{d}(t) = \varepsilon_0 \mathbf{e}(t) + \int dt g_p(t - t') \mathbf{e}(t'),$$

and the full electric unit response for the medium is given by

$$g(t) = \varepsilon_0 \delta(t) + \frac{\varepsilon_p}{\tau} \exp\left(-\frac{t}{\tau}\right) U(t). \tag{1.30}$$

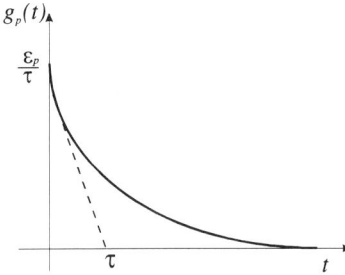

FIGURE 1.6. Polarization unit response of a dipolar dielectric.

FUNDAMENTALS

For instance, if $\mathbf{e}(t) = \mathbf{e}_0 U(t)$, we get (see Fig. 1.7)

$$\mathbf{d}(t) = \{\varepsilon_0 + \varepsilon_p[1 - \exp(-t/\tau)]\}\mathbf{e}_0 U(t).$$

1.2.5. Linear Dispersive Media. II

A *plasma* is a medium composed of unbonded positive and negative charges whose spatial average is zero. In its simplest model the mass of positive charges (ions) is much larger than the mass of negative ones (electrons), so that the former can be considered at rest.

Let n be the number of electrons per unit volume. The constitutive relations are

$$\mathbf{j} = -nq\mathbf{v},$$

together with the Newton force equation (1.25):

$$m\frac{d\mathbf{v}}{dt} + \xi m\mathbf{v} = -q\mathbf{e} - q\mu_0 \mathbf{v} \times \mathbf{h}. \quad (1.31)$$

In the linearized case (see Section 1.2.2) we neglect the term $\mathbf{v} \times \mathbf{h}$ and we further have $d/dt = \partial/\partial t + (\mathbf{v} \cdot \nabla) \to \partial/\partial t$. The vectors $\mathbf{v}$ and $\mathbf{e}$ are parallel, the

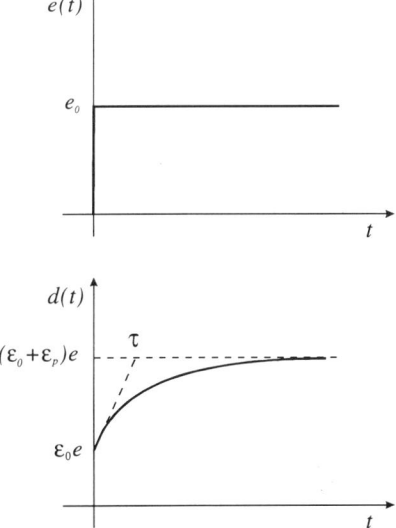

FIGURE 1.7. Electric response of a dipolar dielectric to a step excitation.

medium is isotropic, and the current unit response [see Section 1.2.4 for the solution of Eq. (1.31)] is

$$\mathbf{j}(t) = \int dt' g(t-t') \mathbf{e}(t'),$$

with

$$g(t) = \frac{nq^2}{m} \exp[-\xi t] U(t). \tag{1.32}$$

1.2.6. Chiral Media

Interesting media whose constitutive relationship cannot be cast in the form of an integral are the chiral media:

$$\begin{cases} \mathbf{d} = \varepsilon \mathbf{e} - \eta \dfrac{\partial \mathbf{b}}{\partial t}, \\ \mathbf{b} = \mu \mathbf{h} + \eta \dfrac{\partial \mathbf{e}}{\partial t}. \end{cases}$$

These constitutive relations apply to many classes of sugar solutions, amino acids, DNA, and other natural substances.

1.2.7. Fields at Space and Time Boundaries

Let us consider nonhomogeneous media, so that the constitutive relations may change as a function of space or time. In idealized conditions this change can be abrupt across space or time boundaries. In the former case we have two media of different characteristics separated by a smooth surface; in the latter, the properties of the medium change suddenly at a given time (e.g., due to very fast ionization). The solution of Maxwell equations can be obtained in either of the two regions, and matching conditions are needed on the boundaries.

For the spatial case, we refer to Fig. 1.8. Two media, 1 and 2, are separated by a smooth surface S. Fields and inductions in the two media are $(\mathbf{e}_1, \mathbf{h}_1; \mathbf{d}_1, \mathbf{b}_1)$ and $(\mathbf{e}_2, \mathbf{h}_2; \mathbf{d}_2, \mathbf{b}_2)$, respectively. We consider a small volume ΔV across the boundary, make use of Eq. (1.3), and let $\Delta l \to 0$. This yields

$$\int_{\Delta S} dS(\mathbf{d}_2 - \mathbf{d}_1) \cdot \hat{\mathbf{n}} = 0$$

for any ΔS, which implies continuity of the normal component $\hat{\mathbf{n}} \cdot \mathbf{d}$ of the electric induction at the media boundary. We have assumed the charge density

FUNDAMENTALS 21

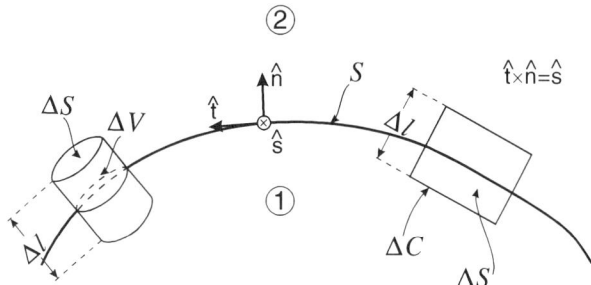

FIGURE 1.8. Field matching conditions at a spatial boundary.

ρ to be always finite inside the volume, so that $q \to 0$ when $\Delta V \to 0$. However, this is no longer true if there is a localized charge distribution over the boundary surface S, so that ρ is infinite there. If ρ_S denotes the surface charge density, i.e., the charge over the surface per unit area, we obtain

$$(\mathbf{d}_2 - \mathbf{d}_1) \cdot \hat{\mathbf{n}} = \rho_S. \qquad (1.33)$$

Equation (1.33) provides the matching conditions for the electric induction: its normal component is continuous across the boundary, unless electric charges are present on it.

From Eqs. (1.8) and (1.4) supplemented with magnetic charges [see Eq. (1.13d)], we similarly obtain

$$(\mathbf{j}_2 - \mathbf{j}_1) \cdot \hat{\mathbf{n}} = -\frac{\partial \rho_S}{\delta t}, \qquad (1.34)$$

$$(\mathbf{b}_2 - \mathbf{b}_1) \cdot \hat{\mathbf{n}} = \chi_S. \qquad (1.35)$$

We now consider a small surface ΔS across the boundary, make use of Eq. (1.6), and let $\Delta l \to 0$. Hence

$$\int_{\Delta C} dc(\mathbf{h}_1 - \mathbf{h}_2) \cdot \hat{\mathbf{t}} = \int_{\Delta C} dc(\mathbf{h}_1 - \mathbf{h}_2) \times \hat{\mathbf{n}} \cdot \hat{\mathbf{s}} = 0$$

for any ΔC, which implies continuity of the tangential component $\hat{\mathbf{n}} \times \mathbf{h} \times \hat{\mathbf{n}}$ of the magnetic field. Again, we have assumed both $\mathbf{j}$ and $\mathbf{d}$ to be finite over the surface ΔS. When a *localized* current is present over the boundary surface S, so that $\mathbf{j}$ is infinite there, we have

$$\hat{\mathbf{n}} \times (\mathbf{h}_2 - \mathbf{h}_1) = \mathbf{j}_S, \qquad (1.36)$$

where j_S is the electric current *linear* density, i.e., $j_S \cdot \hat{s}$ is the current flux per unit length over the surface along $\hat{s}$. Equation (1.36) provides the matching condition for the magnetic field: its tangential component is continuous across the boundary, unless an electric current is present on it.

From Eq. (1.5) supplemented by the magnetic current density k, we similarly obtain

$$\hat{n} \times (e_2 - e_1) = -k_S, \tag{1.37}$$

where k_S is the magnetic current *linear* density.

Note that surface charges and currents are not independent, being related by Eqs. (1.7) and (1.12).

In the case of time nonhomogeneous media, we integrate Eq. (1.1b) across a time interval $\Delta t = t_2 - t_1$ and let $\Delta t \to 0$, to obtain

$$d(t_2) - d(t_1) = 0, \tag{1.38}$$

which implies continuity of the electric induction across the time boundary. Similarly, from Eq. (1.1a), we derive

$$b(t_2) - b(t_1) = 0, \tag{1.39}$$

which implies continuity of the magnetic induction.

1.3. ENERGY AND MOMENTUM

Let us define the *Poynting*[19] (also referred to as the Umov[20]–Poynting) vector

$$s = e \times h. \tag{1.40}$$

We compute $\nabla \cdot s$ by means of Eq. (A.11) and make use of Eqs. (1.1a–b) to yield

$$\nabla \cdot s + e \cdot \frac{\partial d}{\partial t} + h \cdot \frac{\partial b}{\partial t} + j \cdot e = -j_0 \cdot e, \tag{1.41}$$

where the current density has been split into its induced j and source j_0 parts.

---

19. John Henry Poynting: Moutron (UK), 1852–Birmingham, 1914.
20. Nikolaj Aleksevic Umov: Simbirsk, now Ul'Janovik (Russia), 1846–Moscow, 1915.

The term

$$p_0 = -\mathbf{j}_0 \cdot \mathbf{e} \tag{1.42}$$

is the power density delivered by the (electric) sources to the field; see Eq. (1.15).

Let $\mathbf{j} = \sigma\mathbf{e}$, $\sigma \geq 0$ (see Section 1.3.4), so that

$$p_J = \sigma\mathbf{e} \cdot \mathbf{e} = \sigma\mathbf{e}^2. \tag{1.43}$$

The time integral

$$\int_{t_1}^{t_2} dt\, p_J(t) > 0, \qquad \sigma > 0,$$

from one $[\mathbf{e}(t_1), \mathbf{h}(t_1)]$ to another $[\mathbf{e}(t_2), \mathbf{h}(t_2)]$ state of the electromagnetic field depends on all the intermediate values of the field. The integral remains positive even if the two states coincide. This transformation is irreversible and p_j is readily recognized as the power density dissipated in the conducting medium (Joule[21] heat).

For a nondispersive, isotropic, time-invariant medium, $\mathbf{d} = \varepsilon\mathbf{e}$ and $\mathbf{b} = \mu\mathbf{h}$ with ε and μ independent of the time coordinate, and we obtain

$$\mathbf{e} \cdot \frac{\partial \mathbf{d}}{\partial t} + \mathbf{h} \cdot \frac{\partial \mathbf{b}}{\partial t} = \frac{\partial}{\partial t}\left[\frac{1}{2}\varepsilon\mathbf{e}^2 + \frac{1}{2}\mu\mathbf{h}^2\right] = \frac{\partial w}{\partial t}. \tag{1.44}$$

The function $w = w_e + w_m$ is a *state function*, depending only on the state variables $(\mathbf{e}, \mathbf{h})$, and is readily recognized, in the static case, as the sum of the electric $w_e = \varepsilon\mathbf{e}^2/2$ and magnetic $w_m = \mu\mathbf{h}^2/2$ energy densities. The time integral

$$\int_{t_1}^{t_2} dt\, \frac{\partial w}{\partial t} = w(t_2) - w(t_1)$$

from one to another state of the electromagnetic field depends only on their values at the endpoints of the time interval. The integral is zero if the two states coincide and this transformation is reversible. It is reasonable to identify w as the energy density of the electromagnetic field.

---

21. James Prescott Joule: Salford (UK), 1818–Sale, Cheshire, 1889.

We write Eq. (1.41) as

$$\nabla \cdot \mathbf{s} + \frac{\partial w}{\partial t} + p_j = p_0, \tag{1.45}$$

and integrate it over a fixed volume V bounded by a surface S (see Fig. 1.1):

$$\oiint dS \mathbf{s} \cdot \hat{\mathbf{n}} + \frac{dW}{dt} + P_j = P_0, \tag{1.46}$$

where

$$W = \iiint dVw$$

is the total electromagnetic energy inside V; similarly P_j and P_0 are the total dissipated and delivered powers in the same volume V.

Equation (1.46) states the conservation of energy if we read $\mathbf{s} \cdot \hat{n}$ as the electromagnetic energy flux per unit area orthogonal to $\hat{\mathbf{n}}$. According to this model, the energy associated with the electromagnetic field is distributed within, and propagated through, the medium. In any given space and time interval, the energy delivered by the sources equals the sum of the energy dissipated, transferred outside, and stored inside the volume. Note that the assumption of a time-invariant nondispersive medium has been crucial throughout the derivation.

We now consider the vector $\mathbf{d} \times \mathbf{b}$, compute its time derivative, and employ Eqs. (1.1a–b) to obtain

$$\frac{\partial}{\partial t}(\mathbf{d} \times \mathbf{b}) = -\mathbf{d} \times \nabla \times \mathbf{e} - \mathbf{b} \times \nabla \times \mathbf{h} - \mathbf{j}_0 \times \mathbf{b}, \tag{1.47}$$

where the conductivity of the medium is assumed to be zero. We further add and subtract $\mathbf{e}\nabla \cdot \mathbf{d}$, and add $\mathbf{h}\nabla \cdot \mathbf{b}$ (which is zero) to the right-hand side of Eq. (1.47), and finally make use of Eq. (1.1c) to derive

$$\frac{\partial}{\partial t}(\mathbf{d} \times \mathbf{b}) + (\mathbf{d} \times \nabla \times \mathbf{e} - \mathbf{e}\nabla \cdot \mathbf{d}) + (\mathbf{b} \times \nabla \times \mathbf{h} - \mathbf{h}\nabla \cdot \mathbf{b}) = -(\rho_0 \mathbf{e} + \mathbf{j}_0 \times \mathbf{b}). \tag{1.48}$$

For a nondispersive homogeneous medium $\mathbf{d} = \varepsilon \mathbf{e}$, and $\mathbf{b} = \mu \mathbf{h}$, with ε and μ being independent of space coordinates. Equations (A.9) and (B.35) enable

FUNDAMENTALS

the second term appearing in Eq. (1.48) to be written as

$$\varepsilon(\mathbf{e} \times \nabla \times \mathbf{e} - \mathbf{e}\nabla \cdot \mathbf{e}) = \frac{1}{2}\varepsilon\nabla(\mathbf{e}\cdot\mathbf{e}) - \varepsilon\nabla\cdot(\mathbf{ee}) = \nabla\cdot\left(\frac{1}{2}\varepsilon\mathbf{e}^2\mathbf{I} - \varepsilon\mathbf{ee}\right) = \nabla\cdot\mathbf{T}_e,$$

where **ee** is the matrix [see Eq. (B.21)]

$$\mathbf{ee} \to \begin{vmatrix} e_x e_x & e_x e_y & e_x e_z \\ e_y e_x & e_y e_y & e_y e_z \\ e_z e_x & e_z e_y & e_z e_z \end{vmatrix}.$$

Similarly,

$$\mu[\mathbf{h} \times \nabla \times \mathbf{h} - \mathbf{h}\nabla\cdot\mathbf{h}] = \nabla\cdot\left[\frac{1}{2}\mu\mathbf{h}^2\mathbf{I} - \mu\mathbf{hh}\right] = \nabla\cdot\mathbf{T}_m.$$

The matrix

$$\mathbf{T} = \mathbf{T}_e + \mathbf{T}_m \tag{1.49}$$

transforms as a tensor under rotation of coordinates and is referred to as the *Maxwell stress tensor*.

By using the definition (1.49) and (1.40), Eq. (1.48) becomes

$$\frac{\partial}{\partial t}\frac{\mathbf{s}}{c^2} + \nabla\cdot\mathbf{T} = \mathbf{f}_0, \tag{1.50}$$

where

$$\mathbf{f}_0 = -(\rho_0\mathbf{e} + \mathbf{j}_0 \times \mathbf{b}) \tag{1.51}$$

is the force density exerted by the sources and

$$c^2 = \frac{1}{\varepsilon\mu}, \tag{1.52}$$

c being recognized in Chapter 2 as the propagation speed of the electromagnetic field.

We integrate Eq. (1.50) over a fixed volume V bounded by a surface S (see Fig. 1.1):

$$\mathbf{F}_0 = \frac{d}{dt} \iiint dV \frac{\mathbf{s}}{c^2} + \oiint dS \hat{\mathbf{n}} \cdot \mathbf{T}, \qquad (1.53)$$

where $\mathbf{F}_0$ is the volume integral of $\mathbf{f}_0$. Equation (1.53) states conservation of the impulse if we read $\mathbf{s}/c^2$ as the momentum density of the electromagnetic field and $\hat{\mathbf{n}} \cdot \mathbf{T}$ as a pressure along $\hat{\mathbf{n}}$. According to this model a momentum density $\mathbf{s}/c^2$ associated with the electromagnetic field is diffused in, and propagated through, the medium In any given space and time interval, the impulse of the forces exerted by the sources of the field equals the sum of the electromagnetic momentum inside the volume and of the impulse transferred by the field outside the volume. Note that the assumption of a space-invariant nondispersive medium has been crucial throughout the derivation.

1.3.1. Electromagnetic Energy for Nonlinear Media

The definition of the electromagnetic energy density w is based upon the constitutive relations $\mathbf{d} = \varepsilon \mathbf{e}$ and $\mathbf{b} = \mu \mathbf{h}$, so that the medium is linear. This assumption may be relaxed. For instance, if we consider the nondispersive nonlinear medium of Section 1.2.1 with $\mathbf{d}_0 = 0$, we have

$$\mathbf{e} \cdot \frac{\partial \mathbf{d}}{\partial t} = \frac{\partial}{\partial t} \left[\frac{1}{2} \varepsilon \mathbf{e} \cdot \mathbf{e} + \frac{1}{4} a (\mathbf{e} \cdot \mathbf{e})^2 \right].$$

1.3.2. Poynting's Theorem for Anisotropic Media

Let us consider a nondispersive time-invariant anisotropic medium such that $\mathbf{d} = \boldsymbol{\varepsilon} \cdot \mathbf{e}$. We have

$$\frac{\partial}{\partial t}(\tilde{\mathbf{e}} \cdot \boldsymbol{\varepsilon} \cdot \mathbf{e}) = \frac{\partial \tilde{\mathbf{e}}}{\partial t} \cdot \boldsymbol{\varepsilon} \cdot \mathbf{e} + \tilde{\mathbf{e}} \cdot \frac{\partial}{\partial t}(\boldsymbol{\varepsilon} \cdot \mathbf{e})$$

$$= \tilde{\mathbf{e}} \cdot \tilde{\boldsymbol{\varepsilon}} \cdot \frac{\partial \mathbf{e}}{\partial t} + \tilde{\mathbf{e}} \cdot \frac{\partial}{\partial t} \boldsymbol{\varepsilon} \cdot \mathbf{e} = \tilde{\mathbf{e}} \cdot \frac{\partial}{\partial t} [(\tilde{\boldsymbol{\varepsilon}} + \boldsymbol{\varepsilon}) \cdot \mathbf{e}];$$

see also Eq. (B.8). For a reciprocal medium $\tilde{\boldsymbol{\varepsilon}} = \boldsymbol{\varepsilon}$,

$$\frac{1}{2} \frac{\partial}{\partial t} (\tilde{\mathbf{e}} \cdot \boldsymbol{\varepsilon} \cdot \mathbf{e}) = \tilde{\mathbf{e}} \cdot \frac{\partial}{\partial t} (\boldsymbol{\varepsilon} \cdot \mathbf{e}) = \tilde{\mathbf{e}} \cdot \frac{\partial \mathbf{d}}{\partial t},$$

FUNDAMENTALS

$w_e = \tilde{\mathbf{e}} \cdot \boldsymbol{\varepsilon} \cdot \mathbf{e}/2$, and the formulation (1.45) is still valid.

1.3.3. Poynting's Theorem for Dispersive Media

For a dispersive medium it is not usually possible to define an electromagnetic energy density w, so that the formulation (1.45) is not attainable. This is the price paid for modeling the problem in terms of Maxwell equations only. As an alternative, we may derive a generalized Poynting's theorem if Maxwell equations *and* the constitutive equations are considered together, thus introducing a field with a number of components larger than the electromagnetic ones.

As an example, we consider the medium of Section 1.2.5 and the equations

$$\begin{cases} \nabla \times \mathbf{e} = -\mu_0 \frac{\partial \mathbf{h}}{\partial t}, \\ \nabla \times \mathbf{h} = \varepsilon_0 \frac{\partial \mathbf{e}}{\partial t} - nq\mathbf{v}, \\ 0 = m\frac{\partial \mathbf{v}}{\partial t} + m\xi\mathbf{v} + q\mathbf{e}. \end{cases}$$

From the first two equations we have

$$\nabla \cdot \mathbf{s} + \frac{\partial}{\partial t} w_{em} - nq\mathbf{e} \cdot \mathbf{v} = 0,$$

with $w_{em} = \varepsilon_0 \mathbf{e}^2/2 + \mu_0 \mathbf{h}^2/2$. Now substituting $q\mathbf{e}$ from the third equation yields

$$\nabla \cdot \mathbf{s} + \frac{\partial}{\partial t}(w_{em} + w_k) + p_k = 0, \tag{1.54}$$

where

$$w_k = \frac{1}{2} nm\mathbf{v}^2 \tag{1.55}$$

is the kinetic energy density of the particles, and

$$p_k = n\xi\mathbf{v}^2 \tag{1.56}$$

is the dissipated power density due to electron collisions.

A similar procedure leads to the definition of the energy density

$$w = w_{em} + \frac{1}{2}\frac{1}{\varepsilon_p}\mathbf{p}^2$$

and dissipated power density

$$p_p = \frac{\tau}{\varepsilon_p}\left(\frac{\partial \mathbf{p}}{\partial t}\right)^2$$

for the dispersive medium of Section 1.2.4.

1.3.4. Lossless and Lossy Media

A nondispersive medium is *passive* if

$$\varepsilon > 0, \quad \mu > 0, \quad \sigma \geq 0, \quad (1.57)$$

which implies $w > 0$ and $p_J \geq 0$. The medium is *lossless* if $\sigma = 0$ ($p_J = 0$) and *lossy* if $\sigma > 0$ ($p_J > 0$).

Let us consider a shielded volume with no sources inside: in Eq. (1.46) $P_0 = 0$, as well as the flux of the Poynting vector. For a lossless medium, $P_J = 0$, $dW/dt = 0$, so that $W \geq 0$ remains constant in time. If the medium inside the cavity is lossy, $P_J > 0$, $dW/dt < 0$, so that $W \geq 0$ decreases with time to zero, as well as the fields. All this is consistent with the identification of w as the energy density of the field and p_J as the dissipated power density in the medium.

The situation $\varepsilon > 0$, $\mu > 0$, and $\sigma \leq 0$ is now in order. In the shielded cavity, $dW/dt > 0$ and W increases in time. This is a possible model of an *active* medium, which provides energy to the field.

The assumption $\varepsilon < 0$ and $\mu < 0$ leads to nonphysical results, and should be discarded. For instance, in the lossy shielded cavity, $|W|$ and the fields would increase with no bound in time.

1.3.5. The Radiation Pressure

Let us consider an electromagnetic field with electric field **e** oriented along the x-axis and magnetic field **h** along the y-axis, as depicted in Fig. 1.9. The Poynting vector **s** is oriented along the z-axis. Let us further assume that the fields are functions of coordinates z and t only, namely, $\mathbf{e} = \mathbf{e}(z, t)$; this is a plane wave (see Section 2.1).

FUNDAMENTALS

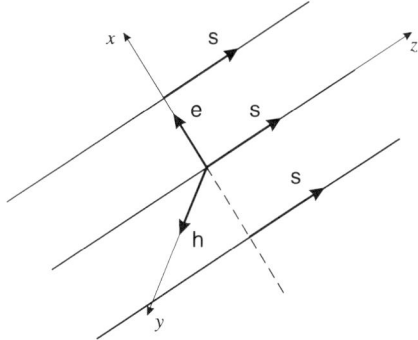

FIGURE 1.9. Introduction of radiation pressure.

Assume the medium to be homogeneous, isotropic, nonconductive, nondispersive, and the field sources to be situated outside the considered region. The projection of Eq. (1.50) along $\hat{z}$, together with Eq. (1.45), gives

$$\begin{cases} \dfrac{1}{c^2}\dfrac{\partial s}{\partial t} + \dfrac{\partial w}{\partial z} = 0, & (1.58a) \\ \dfrac{\partial s}{\partial z} + \dfrac{\partial w}{\partial t} = 0, & (1.58b) \end{cases}$$

because

$$\hat{z}\cdot(\nabla \cdot \boldsymbol{T}) = \hat{z}\cdot(\nabla \cdot \boldsymbol{T}) = \frac{\partial t_{zz}}{\partial z},$$

see Eqs. (B.28) and (B.4), and $t_{zz} = w$.

On differentiating Eq. (1.58b) with respect to z and substituting Eq. (1.58a), we immediately obtain

$$\frac{\partial^2 s}{\partial z^2} - \frac{1}{c^2}\frac{\partial^2 s}{\partial t^2} = 0,$$

which is the wave equation (see Section 2.1); the Poynting vector (as well as the energy density) propagates along z with constant velocity c. According to this model, the field carries energy in the direction of the Poynting vector with speed c. If the field hits a totally absorbing screen, so that the energy flux is not perturbed and all the electromagnetic energy disappears into the screen, there is a constant rate of change of electromagnetic momentum:

$$\frac{\mathbf{s}}{c^2}c = \frac{\mathbf{s}}{c}, \tag{1.59}$$

which is identified as the *radiation pressure*.

1.4. INITIAL AND BOUNDARY CONDITIONS

We consider a linear non-dispersive time-invariant medium and the electromagnetic field generated by *prescribed sources* $\mathbf{j}_0(\mathbf{r}, t)$. This field should be a solution of Maxwell equations together with the constitutive relationships. We assume this solution to exist and we wish to determine the conditions that assure uniqueness of the solution. These conditions are classified under two categories: *initial conditions* are related to the values of the field everywhere in the volume of interest at the initial time, while *boundary conditions* are related to the values of the field, at any time after the initial one, upon the surface bounding the considered volume.

We first consider the *interior problem*, where the space of interest is a finite volume V bounded by a surface S (see Fig. 1.10A). If the solution is not unique, let us consider two of these solutions and their difference $(\mathbf{e}, \mathbf{h})$. For linearity, this difference is the solution of Maxwell's equations driven by the source difference, which is zero; both the initial and boundary conditions are zero as well.

The formulation (1.46) of Poynting's theorem, relative to the volume V, becomes

$$\oiint_S dS \mathbf{s} \cdot \hat{\mathbf{n}} + \frac{dW}{dt} + P_J = 0,$$

where

$$\mathbf{s} \cdot \hat{\mathbf{n}} = (\mathbf{e} \times \mathbf{h}) \cdot \hat{\mathbf{n}} = \mathbf{e} \cdot (\mathbf{h} \times \hat{\mathbf{n}}) = (\hat{\mathbf{n}} \times \mathbf{e}) \cdot \mathbf{h}.$$

If boundary conditions are stated in terms of the tangential component of the electric $\hat{\mathbf{n}} \times \mathbf{e}$ or magnetic $\hat{\mathbf{n}} \times \mathbf{h}$ field on the surface S, then the normal component of the Poynting vector $\mathbf{s} \cdot \hat{\mathbf{n}}$ is zero there, and we are lead to the conclusion that

$$\frac{dW}{dt} \leqslant 0.$$

FUNDAMENTALS

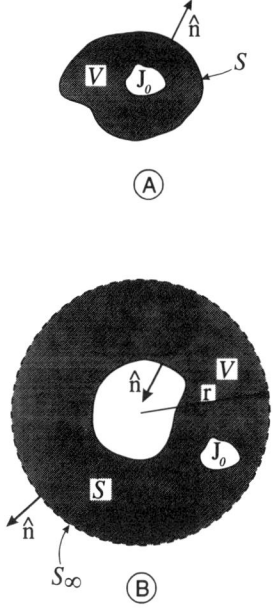

FIGURE 1.10. Definition of the (A) interior and (B) exterior problems.

The energy W is a nonnegative quantity (see Section 1.3.4) and is zero at the initial time. Accordingly, W remains equal to zero, as well as the field. The two postulated solutions coincide and uniqueness is assured.

The *exterior problem* is depicted in Fig. 1.10B where the space of interest is the unbounded volume outside the surface S (which may shrink to zero). We first limit the volume V by the surface $S + S_\infty$, where S_∞ is a large sphere[22] whose radius $R \to \infty$. We proceed as in the previous case and consider the field (**e**, **h**), the difference between two solutions of Maxwell equations with identical sources and the same initial and boundary conditions. The latter are given on S in terms of the tangential components of the electric or magnetic field. The Poynting vector $\mathbf{s} = \mathbf{e} \times \mathbf{h}$ of the difference field again has a zero component normal to this surface and its flux is zero. On the other hand, the fields on S_∞ are identically zero, if the sources have been switched on at a finite, say zero, time and $R > ct$, t being the observation time (causalty and finite velocity of propagation of the field); the flux of the Poynting vector across S_∞ is zero for the difference field (as well as for the original two fields). From Eq. (1.46) we

---

22. For different source geometries, the choice of the surface S_∞ may be different. For two- or one-dimensional problems, cylindrical or planar surfaces S_∞ are appropriate.

again obtain

$$\frac{dW}{dt} \leq 0,$$

and uniqueness is fulfilled. A different approach is examined in Section 1.4.2.

1.4.1. Electric and Magnetic Perfect Conductors

A perfect electric conductor is specified by the constitutive parameter $\sigma \to \infty$. The electric field inside the conductor is everywhere zero, otherwise an infinite current density would be induced by any finite field. Maxwell's equations state that the dynamic magnetic induction $\mu\mathbf{h}$ is also zero [see Eq. (1.1a)].

The tangential electric field is continuous at the perfect conductor boundary; it is zero inside, so that the boundary condition is

$$\hat{\mathbf{n}} \times \mathbf{e} = 0. \tag{1.60}$$

For a regular boundary surface

$$\nabla \cdot (\hat{\mathbf{n}} \times \mathbf{e}) = -\hat{\mathbf{n}} \cdot \nabla \times \mathbf{e} = \hat{\mathbf{n}} \cdot \frac{\partial \mathbf{b}}{\partial t},$$

because $\mathbf{e} \cdot \nabla \times \hat{\mathbf{n}} = 0$, and integration over a small volume across the surface (see Fig. 1.8 and Section 1.2.7) leads to the conclusion that

$$\hat{\mathbf{n}} \cdot \frac{\partial \mathbf{b}}{\partial t} = 0$$

over the surface. The normal component of the *dynamic* magnetic induction is zero over, and continuous across, the boundary.

Boundary condition (1.60) is appropriate for perfect (electric) conductors, and the solution of Maxwell's equations provides fields $\hat{\mathbf{n}} \times \mathbf{h}$ and inductions $\hat{\mathbf{n}} \cdot \mathbf{d}$ which are nonzero on the boundary and zero inside the conductor. The electric surface current (1.36) and charge (1.33) densities are induced by the field on the boundary itself.

Most metals are a good approximation of perfect electric conductors over the entire radio-frequency spectrum (see Section 3.2.2 and Table 1.3). All the field components remain continuous at the boundary, but $\hat{\mathbf{n}} \times \mathbf{h}$ and $\hat{\mathbf{n}} \cdot \mathbf{d}$ exhibit a fast space-varying behavior, which approaches a discontinuous one as $\sigma \to \infty$.

FUNDAMENTALS 33

TABLE 1.3. Conductivity of Several Metals

Metal	Conductivity σ [siemens/m]
Silver, 99.98% pure	6.14×10^7
Copper, annealed	5.80×10^7
Copper, hard drawn	5.65×10^7
Gold, pure drawn	4.10×10^7
Aluminum, commercial hard drawn	3.54×10^7
Magnesium	2.17×10^7
Tungsten	1.81×10^7
Zinc	1.74×10^7
Nickel	1.28×10^7
Iron, 99.98% pure	1.00×10^7
Steel	$1.00\text{--}0.5 \times 10^7$
Lead	0.48×10^7
Mercury	0.10×10^7

The dual of the electric perfect conductor is the magnetic perfect conductor, characterized by an infinite magnetic conductivity. Its boundary condition is

$$\hat{\mathbf{n}} \times \mathbf{h} = 0; \tag{1.61}$$

on the boundary, $\hat{\mathbf{n}} \cdot \mathbf{e} = 0$ as well. The field $\hat{\mathbf{n}} \times \mathbf{e}$ and the induction $\hat{\mathbf{n}} \cdot \mathbf{b}$ are nonzero on the boundary, and zero inside the (perfect) magnetic material; the magnetic surface current (1.37) and charge (1.35) densities are induced by the field on the boundary.

1.4.2. The Radiation Condition

It may be convenient to transform an exterior problem to an equivalent interior one, which leaves the solution unchanged.

The unbounded exterior volume is limited by a surface S far from the sources, which are all located at finite distance from the origin (see Fig. 1.11). The medium in this far region is assumed homogeneous, isotropic, nondispersive, and lossless. We take as boundary condition on S the *radiation condition*[23] (1.62):

$$\mathbf{e} \times \mathbf{h} = eh\hat{\mathbf{n}}, \tag{1.62a}$$

i.e., the fields are mutually orthogonal and orthogonal to $\hat{\mathbf{n}}$ and $\mathbf{e}$ and $\mathbf{h}$ exhibit

---

23. Physical implications of the radiation conditions are further examined in Section 2.1.

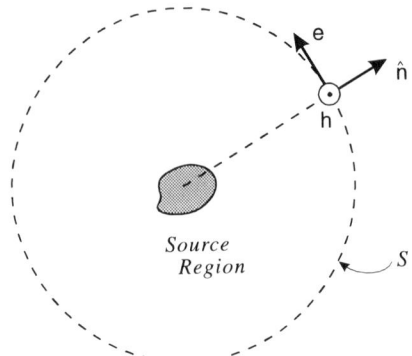

FIGURE 1.11. The radiation condition.

identical asymptotic behaviors, i.e., their amplitudes are proportional:

$$\mathbf{e} = \zeta \mathbf{h} \times \hat{\mathbf{n}}, \qquad \zeta = \sqrt{\mu/\varepsilon}, \tag{1.62b}$$

ζ being the *intrinsic resistance*[24] of space. For free space, $\zeta = \sqrt{\mu_0/\varepsilon_0} = \zeta_0 = 377\,[\Omega] \approx 120\pi\,[\Omega]$.

When conditions (1.62) are verified,

$$\mathbf{s} = s\hat{\mathbf{n}} = \frac{\mathbf{e}^2}{\zeta}\hat{\mathbf{n}} = \zeta \mathbf{h}^2 \hat{\mathbf{n}}, \qquad \mathbf{e}^2 = \mathbf{e}\cdot\mathbf{e}, \qquad \mathbf{h}^2 = \mathbf{h}\cdot\mathbf{h},$$

and the flux of the Poynting vector across S is a nonnegative quantity. If $\mathbf{e}, \mathbf{h}$ is the difference field (see Section 1.4) and the sources and initial conditions are identically zero, then uniqueness is verified for the exterior problem.

It is interesting to examine the consequences of radiation conditions (1.62) on the field (not the difference field) behavior at large distances from the sources. In this far region

$$w = \frac{1}{2}\varepsilon \mathbf{e}^2 + \frac{1}{2}\mu \mathbf{h}^2 = \frac{1}{c}s, \qquad c = 1/\sqrt{\varepsilon\mu},$$

because $e = \zeta h$, and the Poynting equation (1.45) becomes

$$\nabla \cdot \mathbf{s} + \frac{1}{c}\frac{\partial s}{\partial t} = 0.$$

---

24. The quantity ζ is also referred to as the intrinsic impedance of space.

FUNDAMENTALS

For a spherical surface,[25] $\hat{\mathbf{n}} = \hat{\mathbf{r}}$ and $\mathbf{s} = s\hat{\mathbf{r}}$. Expanding the divergence operator in spherical coordinates [see Eq. (A.47)], we obtain

$$\frac{\partial(r^2 s)}{\partial r} + \frac{1}{c}\frac{\partial}{\partial t}(r^2 s) = 0,$$

whose (r, t)-dependent part of the solution is readily seen by inspection (see also Section 2.1) to be of the type

$$r^2 s(r, t) = f(r - ct).$$

Note that the possible dependence of s on the other spherical coordinates (θ, ϕ) has been omitted.

The function $f(r - ct)$ depends on the difference $(r - ct)$ and describes propagation along the $\hat{\mathbf{r}}$-axis with velocity c (see Section 2.1). The omitted (θ, ϕ) dependence is related to the sources. The factor $1/r^2$ accounts for the geometrical spread of power: at large distances from the sources, the fields $\mathbf{e}$ and $\mathbf{h}$ both contain the geometrical decaying factor $1/r$. Accordingly, the radiation condition implies that

$$\mathbf{e} \sim O(1/r), \quad \mathbf{h} \sim O(1/r), \quad \mathbf{e} - \zeta \mathbf{h} \times \hat{\mathbf{n}} \sim o(1/r), \quad \text{as } r \to \infty. \quad (1.63)$$

1.4.3. The Edge Condition

Let us consider a perfectly conducting metal wedge of edge angle Θ, as depicted in Fig. 1.12. The appropriate boundary condition on the wedge surface is given by Eq. (1.60), i.e., $\hat{\mathbf{n}} \times \mathbf{e} = 0$ on S. However, this condition must be supplemented by an additional one enforcing the behavior of the field in the vicinity of the edge $r = 0$. As a matter of fact, a linear charge density can be accumulated on, and a current can flow along, the (ideal, sharp) edge, thereby rendering infinite the charge surface and the current linear densities there. We expect a divergence of the components of the electric and magnetic fields, in the plane orthogonal to the z-axis, as these components approach the edge, because they are related to charge surface and current linear densities by conditions (1.33) and (1.36).

In a region very close to the induced charges and/or currents on the edge, the field exhibits an essentially static behavior,[26] which can be deduced by

---

25. See the footnote on p. 31.
26. This region should be small compared to the characteristic temporal change of the field times the speed of light; see Section 2.3.1.

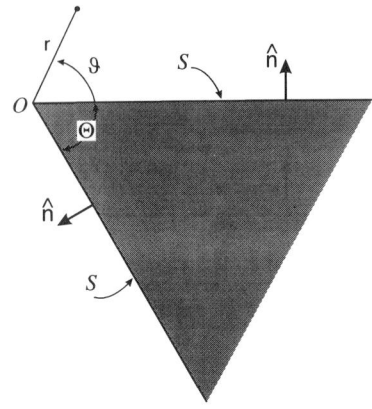

FIGURE 1.12. A perfectly conducting metal wedge. The z-axis is orthogonal to the drawing.

considering static scalar and vector potentials. For the electric field we have

$$\mathbf{e} \approx -\nabla \phi, \qquad \nabla^2 \phi = 0,$$

and the equation of the static potential is for $\partial/\partial z = 0$

$$\frac{1}{r}\frac{\partial}{\partial r} r \frac{\partial \phi}{\partial r} + \frac{1}{r^2}\frac{\partial^2 \phi}{\partial \theta^2} = 0,$$

with reference to the cylindrical coordinates of Fig. 1.12. It is readily seen by the method of separation of variables that the appropriate (r, θ) dependence for the potential is of the type

$$\phi \sim r^\nu \sin \nu\theta,$$

where the value of ν is determined by the condition that the potential is zero not only for $\theta = 0$, but also for $\theta = 2\pi - \Theta$:

$$\nu(2\pi - \Theta) = n\pi, \qquad n = 1, 2, \ldots.$$

The value $n = 1$ provides the divergent part of the field transverse to the edge, $\hat{\mathbf{z}} \times \mathbf{e}$, as $r \to 0$:

$$|\hat{\mathbf{z}} \times \mathbf{e}| \sim r^{\nu - 1} = \frac{1}{r^{(\pi - \Theta)/(2\pi - \Theta)}}. \qquad (1.64)$$

FUNDAMENTALS

Equation (1.64) provides the *edge condition*, and is also appropriate for the transverse part $\hat{z} \times \mathbf{h}$ of the magnetic field. For $\Theta = 0$ (metal half-plane) the transverse field behavior is of type $r^{-1/2}$. For $\Theta = \pi$ the field divergence disappears, as expected, because there is no edge at all; but also for $\Theta > \pi$ the field is regular (as a matter of fact, zero) as $r \to 0$. In this case the field is present inside, and not outside, the wedge.

It is noteworthy that Eq. (1.64), also referred to as the Meixner[27] condition, assures a finite stored energy within any small volume around the edge.

1.5. SYMMETRY PROPERTIES

In this section we consider a number of symmetry properties of the electromagnetic field: time and space reflection, electric and magnetic duality. The medium is nondispersive. We start from a field $(\mathbf{e}, \mathbf{h})$, a solution of Maxwell's equations with prescribed sources, as well as initial and boundary conditions. We apply a suitable transformation to this field, thus getting a new field $(\mathbf{e}', \mathbf{h}')$. Then, we determine sufficient conditions such that $(\mathbf{e}', \mathbf{h}')$ is again a solution of Maxwell's equations.

Let us consider a time-invariant lossless medium and the *time-reversal* transformation:

$$t' = -t. \tag{1.65}$$

This transformation changes the time differential operator:

$$\frac{\partial}{\partial t} \to -\frac{\partial}{\partial t'};$$

and substitution in Maxwell equations shows that these are satisfied by the new field

$$\mathbf{e}'(t') = \mp \mathbf{e}(-t), \quad \mathbf{h}'(t') = \pm \mathbf{h}(-t), \tag{1.66}$$

and the new sources

$$\mathbf{j}'_0(t') = \pm \mathbf{j}_0(-t), \quad \rho'_0(t') = \mp \rho_0(-t), \tag{1.67}$$

where the $\mathbf{r}$ dependence has been dropped. A possible time variation for the fields is given in Fig. 1.13.

The initial conditions are time-reversed, too: final values for the unprimed fields become initial values for the primed one. Boundary conditions on perfect

---

27. Joseph Meixner: Percha (Germany), 1908–1976.

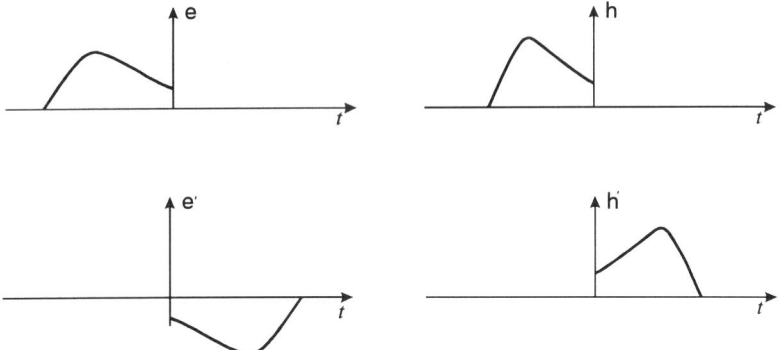

FIGURE 1.13. Time variation for the fields $(\mathbf{e}, \mathbf{h})$ and $(\mathbf{e}', \mathbf{h}')$ at a given identical point in space subject to the upper sign transformation in Eqs. (1.66) and (1.67).

electric (1.60) and magnetic (1.61) conductors remain unchanged, but the primed fields satisfy the *receive condition*

$$\mathbf{e}' = -\zeta \mathbf{h}' \times \hat{\mathbf{n}},$$

rather than the radiation one (1.63). The Poynting vector $\mathbf{s}$ changes sign, as well as the source power density (1.42) $p'_0(t') = -p_0(-t)$, so that we have *sinks* rather than sources. Time reversal is not an allowed transformation if the medium is lossy ($\sigma \neq 0$).

Equation (1.65) defines a time reflection; its spatial dual is a space reflection. We relax the assumption of a lossless medium, consider a z-axis, and decompose fields and sources in their longitudinal and transverse components, namely,

$$\mathbf{a}(\mathbf{r}, t) = \mathbf{a}_\perp(\mathbf{r}, t) + a_z(\mathbf{r}, t)\hat{\mathbf{z}}. \quad (1.68)$$

A little vector algebra shows that

$$\nabla \cdot \mathbf{a} = \nabla_\perp \cdot \mathbf{a}_\perp + \hat{\mathbf{z}} \frac{\partial a_z}{\partial z}, \quad (1.69)$$

$$\nabla \times \mathbf{a} = \frac{\partial}{\partial z}(\hat{\mathbf{z}} \times \mathbf{a}_\perp) - \hat{\mathbf{z}} \times \nabla_\perp a_z + \hat{\mathbf{z}} \nabla_\perp \cdot (\hat{\mathbf{z}} \times \mathbf{a}_\perp), \quad (1.70)$$

where

$$\nabla_\perp = \nabla - \hat{\mathbf{z}} \partial/\partial z$$

is a transverse operator.

FUNDAMENTALS

The *reflection* transformation

$$z' = -z \tag{1.71}$$

changes the space differential operator

$$\frac{\partial}{\partial z} \to -\frac{\partial}{\partial z'};$$

and substitution in Maxwell's equations with the operator decomposition (1.68)–(1.70) shows that they are satisfied by the new fields:

$$\mathbf{e}'_\perp(z') = \mp \mathbf{e}_\perp(-z), \qquad \mathbf{h}'_\perp(z') = \pm \mathbf{h}_\perp(-z), \tag{1.72a}$$

$$e'_z(z') = \pm e_z(-z), \qquad h'_z(z') = \mp h_z(-z), \tag{1.72b}$$

and the new sources:

$$\mathbf{j}'_{0\perp}(z') = \mp \mathbf{j}_{0\perp}(-z), \qquad \mathbf{k}'_{0\perp}(z') = \pm \mathbf{k}_{0\perp}(-z), \tag{1.73a}$$

$$j'_{0z}(z') = \pm j_{0z}(-z), \qquad k'_{0z}(z') = \mp k_{0z}(-z), \tag{1.73b}$$

$$\rho'_0(z') = \mp \rho_0(-z), \qquad \chi'_0(z') = \pm \chi_0(-z), \tag{1.73c}$$

where the (x, y, t) dependence has been dropped. A possible space distribution for the fields and sources is given in Fig. 1.14.

We note that the new fields are located specularly with respect to the plane $z = 0$. The same initial conditions apply to these primed fields. Boundary conditions on perfect electric and magnetic conductors remain unchanged, but these bodies should also be reflected specularly. The radiation condition is also verified.

Another symmetry property of the electromagnetic field is the *duality* transformation:

$$\mathbf{e}' = \zeta \mathbf{h}, \qquad \mathbf{h}' = -\frac{1}{\zeta} \mathbf{e}, \tag{1.74}$$

where the medium is homogeneous and lossless. Substitution in Maxwell's equations shows that the fields $(\mathbf{e}', \mathbf{h}')$ are a solution with the new sources

$$\mathbf{j}'_0 = \frac{1}{\zeta} \mathbf{k}_0, \qquad \mathbf{k}'_0 = -\zeta \mathbf{j}_0, \tag{1.75a}$$

$$\rho'_0 = \frac{1}{\zeta} \chi_0, \qquad \chi'_0 = -\zeta \rho_0. \tag{1.75b}$$

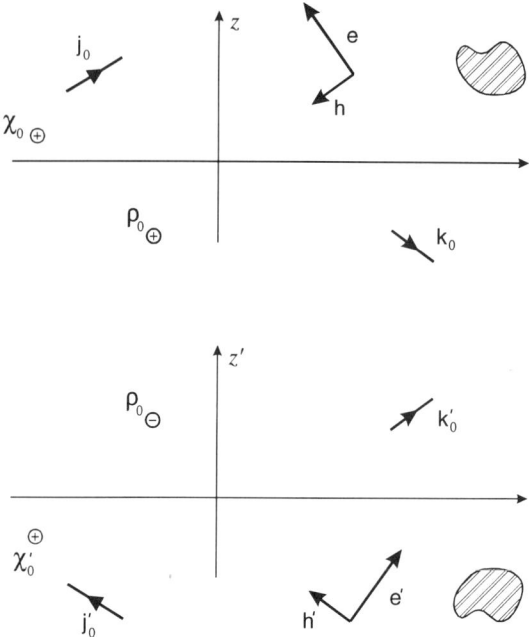

FIGURE 1.14. Space distribution for the fields (**e**, **h**) and (**e**′, **h**′) and also for the sources (j_0, k_0, ρ_0, χ_0) and (j_0', k_0', ρ_0', χ_0'), subject to the upper sign transformation in Eqs. (1.72) and (1.73).

A sketch of possible unprimed and primed fields and sources is given in Fig. 1.15.

Initial and boundary conditions undergo the same duality transformation (1.74). Perfect electric conductors must be replaced by perfect magnetic conductors, and vice versa. The radiation condition is also verified.

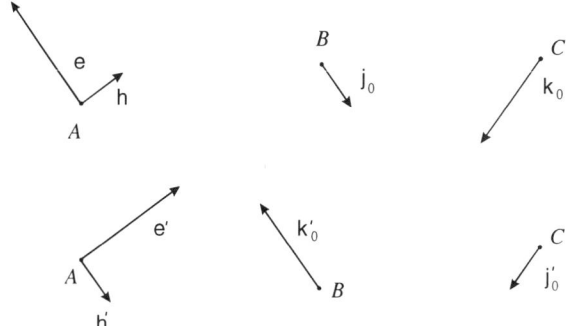

FIGURE 1.15. Unprimed and primed fields and sources with the duality transformation (1.74) and (1.75). Points A, B, C, separated in the figure, should be considered coincident.

1.5.1. Image Theory

Consider the unbounded space and sources $(\mathbf{j}_0, \mathbf{k}_0)$ all located in the half-space $z \geq 0$ and generating the field $(\mathbf{e}, \mathbf{h})$; see Fig. 1.16. By means of the reflection transformation (1.71) we locate new sources $(\mathbf{j}'_0, \mathbf{k}'_0)$ given by Eqs. (1.73) in the half-space $z < 0$, generating the field $(\mathbf{e}', \mathbf{h}')$ according to Eqs. (1.72). Superposition of these primed and unprimed fields is still a solution of Maxwell's equations (the medium is linear).

Consider the superposition field $(\mathbf{e} + \mathbf{e}', \mathbf{h} + \mathbf{h}')$ in the half-space $z \geq 0$. If we choose the upper signs in the transformation (1.72), then on the plane $z = 0$

$$\mathbf{e}_\perp + \mathbf{e}'_\perp = 0,$$

which is the boundary condition corresponding to a perfect electric conductor. The field satisfies the radiation condition at infinity. These boundary conditions assure uniqueness of the solution for $z \geq 0$ and we can state the *image theorem*: the electromagnetic field due to prescribed sources $(\mathbf{j}_0, \mathbf{k}_0)$ in a half-space limited by a perfectly conducting plane can be computed as a superposition of two partial fields, one generated by the given sources and the other by the image sources:

$$\mathbf{j}'_{0\perp}(-z) = -\mathbf{j}_{0\perp}(z), \qquad \mathbf{k}'_{0\perp}(-z) = \mathbf{k}_{0\perp}(z), \tag{1.76a}$$

$$j'_{0z}(-z) = j_{0z}(z), \qquad k'_{0z}(-z) = -k_{0z}(z), \tag{1.76b}$$

both radiating in the same unbounded space.

Clearly, the transformation (1.72) with lower signs would be appropriate for generating a solution of Maxwell's equations in a half-space limited by a perfectly conducting magnetic plane.

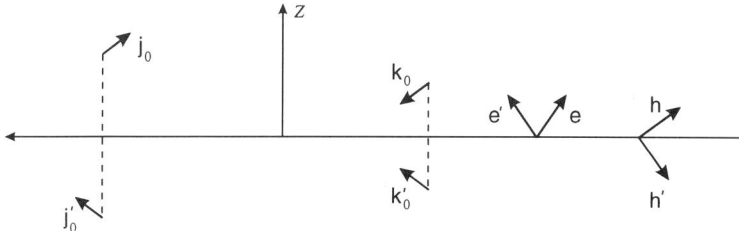

FIGURE 1.16. Real $(\mathbf{j}_0, \mathbf{k}_0)$ and image $(\mathbf{j}'_0, \mathbf{k}'_0)$ sources and fields $(\mathbf{e}, \mathbf{h})$ and $(\mathbf{e}', \mathbf{h}')$, subject to the upper sign transformation in Eqs. (1.72) and (1.73).

1.5.2. The Duality Theorem for Inhomogeneous Media

Let us consider a (space) inhomogeneous lossless medium described by relative permittivity $\varepsilon_r(\mathbf{r}) = \varepsilon(\mathbf{r})/\varepsilon_0$ and permeability $\mu_r(\mathbf{r}) = \mu(\mathbf{r})/\mu_0$. The duality transformation (1.74) is applied with $\zeta = \zeta_0 = \sqrt{\mu_0/\varepsilon_0}$. Substitution in Eqs. (1.1) shows that the primed fields are consistent with Maxwell's equations when the further substitution

$$\varepsilon_r \to \mu_r, \quad \mu_r \to \varepsilon_r$$

is introduced. Continuity conditions (1.33)–(1.37) are also verified at space discontinuities.

1.6. SUMMARY AND SELECTED REFERENCES

This chapter addresses the basic concepts of electromagnetic theory. Relevant equations (Section 1.1), both energy and mechanical (Section 1.3), as well as other theorems (Section 1.5), initial and boundary conditions (Section 1.4) are presented strictly in the time domain. All this material is available in a large number of classical textbooks, although the space given to time-domain analysis is usually rather limited (and sometimes absent) compared to that devoted to steady-state (sinusoidal) fields.

In addition to the original treatise of Maxwell [1.1], several classical books can be cited; a limited number is listed in [1.2–1.13]. A more mathematically oriented approach may be found in [1.4] and [1.14]. Additional references to application-oriented textbooks are given in Chapter 7.

Constitutive relationships (Section 1.2) certainly represent an important issue. Usually, its discussion is limited to the case of local media (see [1.5]), and sometimes even ignored; dispersion of the medium is considered only in the frequency domain. The general approach given in Section 1.2, directly in the time domain, provides a compact formulation, important for wide-band transient analysis in material media. All this is revisited in the frequency domain (Chapter 3) to complete the discussion and provide an alternative approach. Space-dispersive media are also considered in both time (Chapter 1) and transformed (Chapter 3) domains.

More information on chiral media (Section 1.2.6) can be found in [1.15]. A book totally devoted to energy concepts (Section 1.3) is cited in [1.16]. The radiation condition (Section 1.4.2) is included in most textbooks [1.2–1.13]. Its rigorous discussion can be found in [1.14].

The original paper on the edge condition (Section 1.4.3) is listed in [1.17], but it follows a different approach than that presented in Section 1.4.3.

References

[1.1] J. C. Maxwell, *A Treatise on Electricity and Magnetism*, Dover Publications (1974); also published in two volumes by Oxford University Press (1995).
[1.2] R. S. Elliott, *Electromagnetics*, McGraw-Hill, New York (1966).
[1.3] J. D. Jackson, *Classical Electrodynamics*, Wiley, New York (1962).
[1.4] D. S. Jones, *The Theory of Electromagnetism*, Pergamon Press, Oxford (1964).
[1.5] J. A. Kong, *Electromagnetic Wave Theory*, Wiley, New York (1990).
[1.6] J. D. Kraus, *Electromagnetics*, McGraw-Hill, New York (1984).
[1.7] W. K. H. Panofski and M. Phillips, *Classical Electricity and Magnetism*, Addison-Wesley, Reading, Mass., 1962.
[1.8] C. H. Papas, *Theory of Electromagnetic Wave Propagation*, McGraw-Hill, New York (1965).
[1.9] D. T. Paris and G. K. Hurd, *Basic Electromagnetic Theory*, McGraw-Hill, New York (1969).
[1.10] W. P. Smythe, *Static and Dynamic Electricity*, McGraw-Hill, New York (1950).
[1.11] J. A. Stratton, *Electromagnetic Theory*, McGraw-Hill, New York (1941).
[1.12] J. Van Bladel, *Electromagnetic Fields*, Hemisphere, Washington (1985).
[1.13] J. R. Wait, *Electromagnetic Wave Theory*, Harper & Row, New York (1985).
[1.14] C. Müller, *Foundations of the Mathematical Theory of Electromagnetic Waves*, Springer-Verlag, Berlin (1969).
[1.15] D. L. Jaggard, X. Sun, and N. Engheta, "Canonical sources and duality in chiral media," *IEEE Trans. Antennas Propagat.* **AP-35**, 1007–1013 (1988).
[1.16] H. G. Booker, *Energy in Electromagnetism*, Peter Peregrinus, New York (1982).
[1.17] J. Meixner, "Die kanteubedigung in der theorie der Beugung Elektromagnetisher Wellen an Volkommen Lertenden Ebeneu Shirmen," *Ann. Phys.* **6**, 2–6 (1949).

2
Elementary Solutions

2.1. PLANE WAVES

Let us consider the cartesian coordinate system of Fig. 2.1 and a homogeneous, isotropic, lossless, nondispersive medium described by parameters ε and μ. We are examining a possible solution for the fields independent of coordinates (x, y) (*plane waves*), so that $\mathbf{e} = \mathbf{e}(z, t)$ and $\mathbf{h} = \mathbf{h}(z, t)$. We are not concerned with the field excitation mechanism, and expand (the source-free) Maxwell equations (1.1a–b):

$$\frac{\partial e_y}{\partial z} = \mu \frac{\partial h_x}{\partial t}; \quad \frac{\partial h_x}{\partial z} = \varepsilon \frac{\partial e_y}{\partial t}; \qquad (2.1\text{a})$$

$$\frac{\partial e_x}{\partial z} = -\mu \frac{\partial h_y}{\partial t}; \quad \frac{\partial h_y}{\partial z} = -\varepsilon \frac{\partial e_x}{\partial t}; \qquad (2.1\text{b})$$

$$0 = -\mu \frac{\partial h_z}{\partial t}; \quad 0 = \varepsilon \frac{\partial e_z}{\partial t} \qquad (2.1\text{c})$$

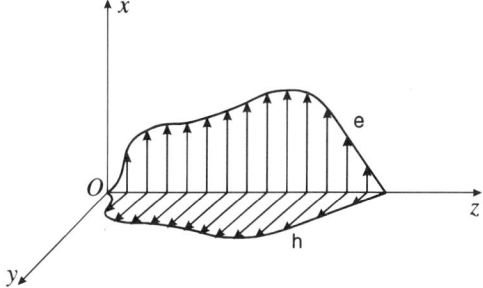

FIGURE 2.1. Plane waves.

and equations (1.1c–d):

$$\frac{\partial e_z}{\partial z} = 0, \quad \frac{\partial h_z}{\partial z} = 0 \tag{2.1d}$$

in component form.

We note from Eqs. (2.1c–d) that e_z and h_z must be constants. If the fields were zero throughout space before a given time (switch-on time of the sources), these constants are zero. In a homogeneous isotropic lossless medium, fields depending only on a single spatial coordinate are orthogonal to this coordinate; they are referred to as *transverse electromagnetic* (or TEM) *fields*.

Examination of Eqs. (2.1a–b) also shows that the field couple (e_x, h_y) is independent of the other field couple (e_y, h_x). We consider only the first one, Eqs. (2.1b), and neglect the subscript in the following. By differentiating the first of Eqs. (2.1b) and substituting the second, we easily obtain

$$\frac{\partial^2 e}{\partial z^2} - \frac{1}{c^2}\frac{\partial^2 e}{\partial t^2} = 0, \quad c = \frac{1}{\sqrt{\varepsilon\mu}}. \tag{2.2}$$

In free space, $c = \sqrt{1/\varepsilon_0\mu_0} \approx 3 \times 10^8$ [m/s] and coincides with the speed of light.

Any twofold differentiable function $e^+(z - ct)$ or $e^-(z + ct)$ is a solution of Eq. (2.2), as follows by substituting in Eq. (2.2) and noting that $\partial e^\pm/\partial t = \mp c\partial e^\pm/\partial z$. The corresponding solution $h^+(z - ct)$ and $h^-(z + ct)$ for the magnetic field is obtained by substitution in the first of Eqs. (2.1b):

$$\frac{\partial}{\partial t}[\zeta h^\pm \mp e^\pm] = 0, \quad \zeta = \sqrt{\frac{\mu}{\varepsilon}}, \tag{2.3}$$

and noting again that $\partial/\partial z = \mp \partial/c\partial t$. The quantity ζ is the *intrinsic resistance* of the medium (see Section 1.4.2); in free space, $\zeta = 377\,[\Omega] \approx 120\pi\,[\Omega]$.

Equation (2.3) states that the bracketed term is independent of time. For fields equal to zero before a given time, this constant value is zero and

$$\zeta h^\pm = \pm e^\pm. \tag{2.4}$$

The most general solution to Eq. (2.2), the *wave equation* in one dimension, is[1]

$$\begin{cases} e(z, t) = e^+(z - ct) + e^-(z + ct), \\ \zeta h(z, t) = e^+(z - ct) - e^-(z + ct). \end{cases} \tag{2.5}$$

---

1. In the following we use also the totally equivalent solutions of type $e^+(ct - z)$; see Section 2.2.

ELEMENTARY SOLUTIONS

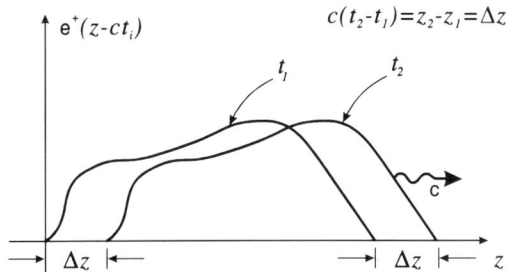

FIGURE 2.2. Propagation of an electromagnetic perturbation.

For the first solution $(\mathbf{e}^+, \mathbf{h}^+)$ the fields are mutually orthogonal, related by the intrinsic resistance ζ, and verify the radiation condition[2] (1.62) with $\hat{\mathbf{n}} = \hat{\mathbf{z}}$ as $z \to +\infty$. They attain equal values for any space–time couple $z - ct = $ const; the electromagnetic perturbation propagates without deformation and with constant speed c along the positive sense of the z-axis (see Fig. 2.2). This perturbation is referred to as an electromagnetic *progressive plane wave*. The Poynting vector,

$$\mathbf{s}^+ = \mathbf{e}^+ \times \mathbf{h}^+ = \frac{(\mathbf{e}^+)^2}{\zeta} \hat{\mathbf{z}} = \zeta(\mathbf{h}^+)^2 \hat{\mathbf{z}},$$

points in the direction of the positive z-axis. The electromagnetic energy density,

$$w^+ = \frac{1}{2}\mu(\mathbf{h}^+)^2 + \frac{1}{2}\varepsilon(\mathbf{e}^+)^2 = \frac{1}{2}\mu(\mathbf{h}^+)^2 + \frac{1}{2}\frac{\zeta}{c}(\mathbf{h}^+)^2 = \frac{1}{c}s^+,$$

is proportional to s^+ and Eq. (1.58b) becomes

$$\frac{\partial s^+}{\partial z} + \frac{1}{c}\frac{\partial s^+}{\partial t} = 0,$$

the solution of which is again of the type $s^+(z - ct)$. As expected, the Poynting vector propagates with speed c, too, along the positive z-axis, and this is the direction of energy propagation.

For the other couple $(\mathbf{e}^-, \mathbf{h}^-)$ the fields are still mutually orthogonal, related by (minus) the intrinsic resistance, and satisfy the radiation condition

---

2. See the footnote on p. 31.

as $z \to -\infty$. The Poynting vector,

$$\mathbf{s}^- = \mathbf{e}^- \times \mathbf{h}^- = -\frac{(\mathbf{e}^-)^2}{\zeta}\hat{\mathbf{z}} = -\zeta(\mathbf{h}^-)^2\hat{\mathbf{z}},$$

as well as the direction of the energy flow, is along the negative z-axis; the perturbation moves without deformation and with speed c along this negative sense, and is referred to as an electromagnetic *regressive plane wave*.

2.1.1. Lossy Media

Let us now relax the assumption $\sigma = 0$. Proceeding as in Section 2.1 we obtain $h_z = 0$, and

$$\varepsilon\frac{\partial e_x}{\partial t} + \sigma e_z = 0, \qquad \frac{\partial e_z}{\partial z} = 0,$$

i.e.,

$$e_z(z, t) = a \exp\left(-\frac{\sigma}{\varepsilon}t\right)$$

with $a = $ const. We are interested in *wave-like solutions*, i.e., solutions whose dominant space–time dependence is of the type $z \mp ct$. For this reason we take $a = 0$ and consider the equations analogous to Eq. (2.1b):

$$\frac{\partial e_x}{\partial z} = -\mu\frac{\partial h_y}{\partial t}, \qquad \frac{\partial h_y}{\partial z} = -\varepsilon\frac{\partial e_x}{\partial t} - \sigma e_x. \tag{2.6}$$

Again neglecting the subscripts and proceeding as in Section 2.1, we obtain by differentiation and substitution

$$\frac{\partial^2 e}{\partial z^2} - \sigma\mu\frac{\partial e}{\partial t} - \frac{1}{c^2}\frac{\partial^2 e}{\partial t^2} = 0, \tag{2.7}$$

which is the counterpart of Eq. (2.2) when $\sigma \neq 0$.

When the conductivity is very small, which is the case of interest, we can obtain the solution of Eq. (2.7) as a perturbation of the solution of Eq. (2.2):

$$e(z, t) = A(z, t)e^+(z - ct) + B(z, t)e^-(z + ct), \tag{2.8}$$

where $A(z, t)$ and $B(z, t)$ are slowly varying functions of z and t, so that their second-order derivatives can be neglected.

ELEMENTARY SOLUTIONS

For the progressive wave appearing in Eq. (2.8), substitution in Eq. (2.7) leads to

$$\frac{\partial e^+}{\partial z}\left[2\frac{\partial A}{\partial z} + \sigma\mu c A\right] + \left[\frac{2}{c}\frac{\partial e^+}{\partial z} - \sigma\mu e^+\right]\frac{\partial A}{\partial t} \approx 0,$$

which is satisfied by the set of equations

$$\frac{\partial A}{\partial t} = 0, \qquad \frac{\partial A}{\partial z} + \frac{\sigma\mu c}{2}A = 0,$$

the solution of which is (but for a multiplicative constant)

$$A = \exp\left[-\frac{\sigma\zeta}{2}z\right].$$

The progressive wave solution is

$$\begin{cases} e^+(z, t) = \exp\left[-\dfrac{\sigma\zeta}{2}z\right]e^+(z - ct), \\ h^+(z, t) = e^+(z, t)/\zeta, \end{cases}$$

and is sketched in Fig. 2.3.

The wave is attenuated as it propagates; the Poynting vector decays as $\exp[-\sigma\zeta z]$ and power is dissipated in the medium. For this case of small losses, it is still possible to define a propagation velocity, and electric and magnetic fields are still related by the intrinsic resistance of the space. Furthermore, the waveform is not deformed but only amplitude-scaled along its propagation.

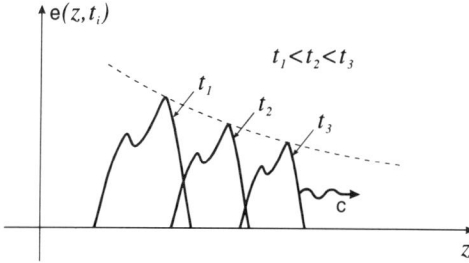

FIGURE 2.3. Propagation of an electromagnetic plane wave in a low-loss dielectric.

Similar results are obtained for the regressive wave; the attenuation factor is given by $\exp[\sigma\zeta z/2]$.

2.1.2. Anisotropic Media

Let us consider a homogeneous, lossless, (electrically) anisotropic uniaxial medium described by the constitutive parameters μ, ε, and ε_r. We take a reference cartesian system (X, Y, Z) parallel to the principal axes of the dielectric (see Fig. 2.4), so that the relative dielectric constant in matrix form is given by (see Section 1.2.3)

$$\varepsilon_r \rightarrow \begin{vmatrix} 1 & 0 & 0 \\ 0 & 1 & 0 \\ 0 & 0 & \varepsilon_r \end{vmatrix}, \qquad (2.9)$$

and is normalized with respect to the dielectric constant ε in the plane orthogonal to the Z-axis.

For plane-wave propagation along the Z-axis, we again derive $h_Z = e_Z = 0$ (TEM wave) and the same Eqs. (2.1)–(2.2). No difference is encountered with respect to the isotropic case: the waves propagate with speed $c = 1/\sqrt{\varepsilon\mu}$ and the fields are related by the intrinsic resistance $\zeta = \sqrt{\mu/\varepsilon}$.

More interesting results are obtained if we consider propagation along a z-axis which makes an angle α with respect to the Z-axis (see Fig. 2.4). The new coordinates (x, y, z) are related to the old ones (X, Y, Z) by the

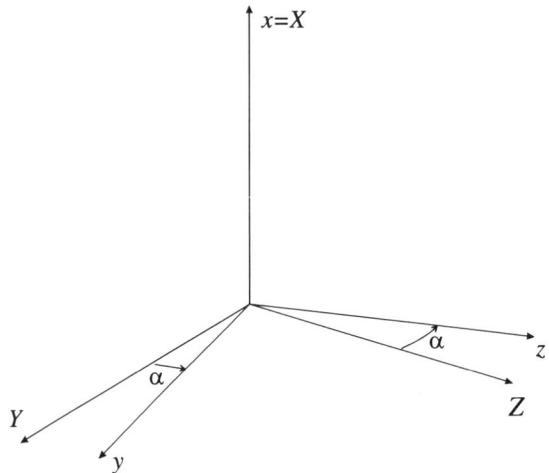

FIGURE 2.4. Propagation in a uniaxial anisotropic medium.

ELEMENTARY SOLUTIONS

rotation matrix

$$R \to \begin{vmatrix} 1 & 0 & 0 \\ 0 & \cos\alpha & \sin\alpha \\ 0 & -\sin\alpha & \cos\alpha \end{vmatrix},$$

and the new representation of the relative dielectric constant is given by $\varepsilon_r \to R \cdot \varepsilon_r \cdot \tilde{R}$:

$$\varepsilon_r \to \begin{vmatrix} 1 & 0 & 0 \\ 0 & a & q \\ 0 & q & b \end{vmatrix},$$

with

$$a = \cos^2\alpha + \varepsilon_r \sin^2\alpha = 1 + (\varepsilon_r - 1)\sin^2\alpha,$$
$$b = \varepsilon_r \cos^2\alpha + \sin^2\alpha = \varepsilon_r - (\varepsilon_r - 1)\sin^2\alpha,$$
$$q = (\varepsilon_r - 1)\sin\alpha\cos\alpha.$$

Source-free Maxwell's equations in component form are readily obtained:

$$\frac{\partial e_y}{\partial z} = \mu \frac{\partial h_x}{\partial t}, \qquad \frac{\partial h_x}{\partial z} = \varepsilon \frac{\partial}{\partial t}[ae_y + qe_z], \tag{2.10a}$$

$$\frac{\partial e_x}{\partial z} = -\mu \frac{\partial h_y}{\partial t}, \qquad \frac{\partial h_y}{\partial z} = -\varepsilon \frac{\partial e_x}{\partial t}, \tag{2.10b}$$

$$0 = -\mu \frac{\partial h_z}{\partial t}, \qquad 0 = \varepsilon \frac{\partial}{\partial t}[qe_y + be_z], \tag{2.10c}$$

and

$$\frac{\partial}{\partial z}(qe_y + be_z) = 0, \qquad \frac{\partial}{\partial z} h_z = 0, \tag{2.10d}$$

which are the analogous form of Eqs. (2.1a–c) and (2.1d), respectively.

A first possible wave solution is described by Eqs. (2.10b), which are decoupled from the others. The fields (e_x, h_y) are TEM (see Fig. 2.5A): progressive and regressive waves propagate with speed $c = 1/\sqrt{\varepsilon\mu}$ and are related by the intrinsic resistance $\zeta = \sqrt{\mu/\varepsilon}$. These waves share the same properties of the electromagnetic perturbation propagating in an isotropic dielectric of constant ε and are referred to as *ordinary waves*.

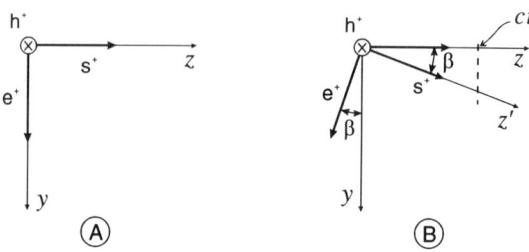

FIGURE 2.5. Ordinary (A) and extraordinary (B) waves in a uniaxial anisotropic dielectric.

A second possible solution is described by Eqs. (2.10a,c,d). From Eqs. (2.10c-d) we obtain

$$h_z = 0, \quad qe_y + be_z = 0,$$

so that this wave exhibits components (e_y, h_x, e_z). The field is *not* TEM, but only *transverse magnetic* (TM) (see Fig. 2.5b), at variance with the vectors **h** and **d**, which are orthogonal to the propagation axis.

The equations governing the transverse components (e_y, h_x) of the fields are easily obtained by eliminating e_z from Eqs. (2.10c) and (2.10a):

$$\frac{\partial e_y}{\partial z} = \mu \frac{\partial h_x}{\partial t}, \quad \frac{\partial h_x}{\partial z} = \varepsilon \frac{ab - q^2}{b} \frac{\partial e_y}{\partial t}, \tag{2.11}$$

and their study is now in order.

Examination of Eqs. (2.11) shows that they describe plane-wave propagation in an equivalent isotropic dielectric of relative permittivity

$$\varepsilon'_r = \frac{ab - q^2}{b} = \frac{\varepsilon_r}{\varepsilon_r - (\varepsilon_r - 1) \sin^2\alpha},$$

which depends on the angle α (see Fig. 2.6). Progressive and regressive waves are TM (see Fig. 2.5b) propagate along the z-axis with speed

$$v = \frac{1}{\sqrt{\varepsilon \varepsilon'_r \mu}} = c \sqrt{1 - \frac{\varepsilon_r - 1}{\varepsilon_r} \sin^2\alpha}, \quad c = \frac{1}{\sqrt{\varepsilon \mu}}$$

which again depends on α, and the transverse components[3] are related by the

---

3. Note that for the progressive wave $e_y^+/(-h_x^+) = Z_0$, while $e_y^-/(-h_x^-) = -Z_0$ for the regressive one.

ELEMENTARY SOLUTIONS

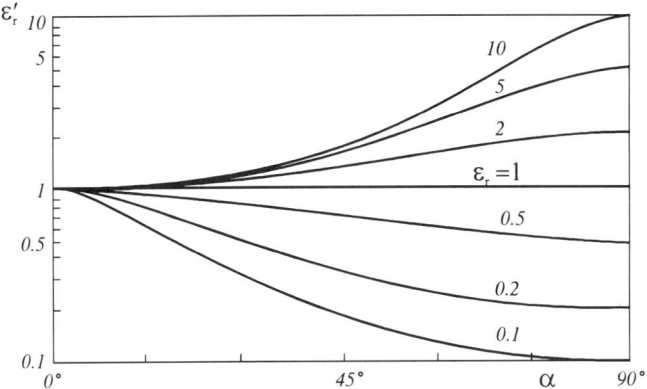

FIGURE 2.6. Plot of the equivalent dielectic constant for the extraordinary wave.

(α-dependent) *wave impedance*:

$$Z_0 = \sqrt{\frac{\mu}{\varepsilon\varepsilon'_r}} = \sqrt{\frac{\mu}{\varepsilon}}\sqrt{1 - \frac{\varepsilon_r - 1}{\varepsilon_r}\sin^2\alpha}.$$

Progressive and regressive electromagnetic perturbations of this type are called *extraordinary waves*.

Let us consider the progressive waves and omit the superscript. We have

$$e_y = -Z_0 h_x, \qquad e_x = -\frac{q}{b}e_y.$$

The Poynting vector,

$$\mathbf{s} = \mathbf{e} \times \mathbf{h} = \frac{e_y^2}{Z_0}\left[\hat{\mathbf{z}} + \frac{q}{b}\hat{\mathbf{y}}\right],$$

is *not* oriented along the assumed direction of propagation z, but is rotated by the angle β, where

$$\tan\beta = -\frac{e_z}{e_y} = \frac{q}{b}, \qquad s = \frac{e_y^2}{Z_0}\frac{1}{\cos\beta};$$

see Fig. 2.5B. Accordingly, fields and energy propagate along different directions.

It should be noted that the definition of the wave speed is related to the assumed direction of propagation; this can be (arbitrarily) chosen in the plane

$x = 0$. It is interesting to compute the speed along the direction of the Poynting vector, because it coincides with the energy velocity. A field displacement $z = ct$ along the z-axis corresponds to a Poynting vector displacement $z' = z/\cos \beta$ along the z'-axis; see Fig. 2.5. Thus

$$s(z - vt) = s(z' \cos \beta - vt) \to s(z' - c't),$$

with

$$c' = \frac{v}{\cos \beta} = c\sqrt{\frac{\varepsilon_r^2 \cos^2\alpha + \sin^2\alpha}{\varepsilon_r^2 \cos^2\alpha + \varepsilon_r \sin^2\alpha}},$$

which is the Poynting vector, i.e., the energy velocity.

Computation of the energy density w and use of Poynting's equations (1.45) with $p_j = p_0 = 0$ leads to

$$\frac{\partial s}{\partial z'} + \frac{1}{c'}\frac{\partial s}{\partial t} = 0,$$

which confirms the above conclusions.

2.2. PLANE WAVES AT DISCONTINUITY BOUNDARIES

Let us consider two semi-infinite, isotropic, homogeneous, lossless, non-dispersive media of parameters (ε_1, μ_1) and (ε_2, μ_2), respectively, separated by a plane surface $z = 0$, as depicted in Fig. 2.7. A plane wave,

$$\mathbf{e}_1^+ = f(c_1 t - z)\hat{\mathbf{x}}, \qquad \zeta_1 \mathbf{h}_1^+ = f(c_1 t - z)\hat{\mathbf{y}}, \qquad z \leq 0, \qquad (2.12)$$

$$c_1 = \frac{1}{\sqrt{\varepsilon_1 \mu_1}}, \qquad \zeta_1 = \sqrt{\frac{\mu_1}{\varepsilon_1}},$$

is incident on the (spatial) discontinuity boundary. Continuity conditions (1.36) and (1.37) at $z = 0$ can be established by introducing *reflected*

$$\mathbf{e}_1^- = g(c_1 t + z)\hat{\mathbf{x}}, \qquad \zeta_1 \mathbf{h}_1^- = -g(c_1 t + z)\hat{\mathbf{y}}, \qquad z \leq 0$$

and *transmitted*

$$\mathbf{e}_2^+ = q(c_2 t - z)\hat{\mathbf{x}}, \qquad \zeta_2 \mathbf{h}_2^+ = q(c_2 t - z)\hat{\mathbf{y}}, \qquad z \geq 0,$$

$$c_2 = \frac{1}{\sqrt{\varepsilon_2 \mu_2}}, \qquad \zeta_2 = \sqrt{\frac{\mu_2}{\varepsilon_2}},$$

ELEMENTARY SOLUTIONS

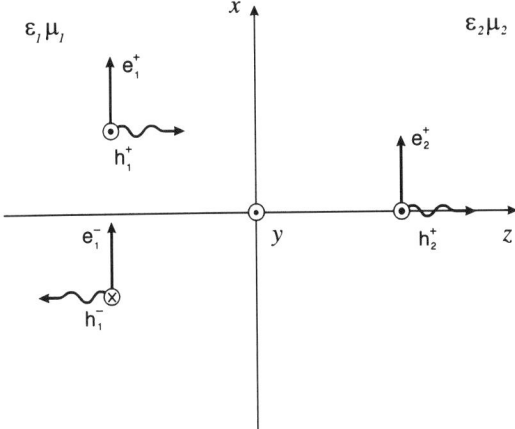

FIGURE 2.7. Electromagnetic plane-wave propagation in the presence of a spatial reflection boundary.

waves in the half-spaces $z \leq 0$ and $z \geq 0$, respectively. Note that the field $(\mathbf{e}_1^-, \mathbf{h}_1^-)$ satisfies the radiation condition when $z \to -\infty$, while the field $(\mathbf{e}_2^+, \mathbf{h}_2^+)$ satisfies the radiation condition when $z = +\infty$.

At the space discontinuity boundary $z = 0$:

$$f(c_1 t) + g(c_1 t) = q(c_2 t) \quad \text{and} \quad \frac{1}{\zeta_1} f(c_1 t) - \frac{1}{\zeta_1} g(c_1 t) = \frac{1}{\zeta_2} q(c_2 t),$$

which must be verified for any t. Letting

$$g(c_1 t) = \Gamma f(c_1 t) \quad \text{and} \quad q(c_2 t) = \mathrm{T} f\left(\frac{c_2 t}{c_2/c_1}\right),$$

where Γ is the (electric field) *reflection* and T the (electric field) *transmission coefficient*, we obtain

$$\Gamma = \frac{\zeta_2 - \zeta_1}{\zeta_2 + \zeta_1} \quad \text{and} \quad \mathrm{T} = \frac{2\zeta_2}{\zeta_1 + \zeta_2}, \qquad (2.13)$$

and the field solution is given by

$$\begin{cases} e_1(z, t) = f(c_1 t - z) + \Gamma f(c_1 t + z), \\ \zeta_1 h_1(z, t) = f(c_1 t - z) - \Gamma f(c_1 t + z) \end{cases} \qquad (2.14)$$

for $z \leqslant 0$, and

$$\begin{cases} e_2(z, t) = Tf\left(\dfrac{c_2 t - z}{c_2/c_1}\right), \\ \zeta_2 h_2(z, t) = Tf\left(\dfrac{c_2 t - z}{c_2/c_1}\right) \end{cases} \quad (2.15)$$

for $z \geqslant 0$.

When $\zeta_2 = 0$ we have $\Gamma = -1$, and the electric field is zero on the plane $z = 0$. This is the boundary condition corresponding to a perfect electric conductor (see also Section 2.2.1 below) and is formally obtained by letting $\varepsilon_2 \to \infty$; the tangential electric field is zero on the plane $z = 0$ and the tangential magnetic field is doubled there. The dual case of a perfect magnetic conductor corresponds to $\Gamma = 1$ and is formally obtained by letting $\mu_2 \to \infty$, i.e., $\zeta_2 \to \infty$. The case of an electromagnetic pulse of spatial width d impinging on a conducting infinite plane is sketched in Fig. 2.8.

If the half-space $z \geqslant 0$ is composed of a permeable medium, the width d of the incident pulse is scaled by the ratio c_2/c_1 [see Eqs. (2.15)]: the pulse is spatially stretched if $c_2 > c_1$, and spatially shrunk if $c_2 < c_1$. The time duration of the pulse remains unchanged.

The reflected wave is zero if $\Gamma = 0$ (*matching condition*), i.e.,

$$\frac{\mu_1}{\mu_2} = \frac{\varepsilon_1}{\varepsilon_2}. \quad (2.16)$$

Relation (2.16) is also referred to as the Heaviside[4] condition.

The Poynting vectors in the two half-spaces $z \leqslant 0$ and $z \geqslant 0$ are given by

$$s_1(z, t) = e_1(z, t) h_1(z, t) = \frac{1}{\zeta_1}[f^2(c_1 t - z) - \Gamma^2 f^2(c_1 t + z)], \quad z \leqslant 0,$$

$$s_2(z, t) = e_2(z, t) h_2(z, t) = \frac{T^2}{\zeta_2} f^2\left(\frac{c_2 t - z}{c_2/c_1}\right), \quad z \leqslant 0.$$

The total flux of energy (per square meter) across any abscissa z is computed as

$$\int_{-\infty}^{+\infty} dt\, s_1(z, t) = \frac{1}{\zeta_1 c_1}(1 - \Gamma^2) \int_{-\infty}^{+\infty} d\xi\, f^2(\xi), \quad z \leqslant 0, \quad (2.17a)$$

$$\int_{-\infty}^{+\infty} dt\, s_2(z, t) = \frac{T^2}{\zeta_2 c_1} \int_{-\infty}^{+\infty} d\xi\, f^2(\xi), \quad z \leqslant 0, \quad (2.17b)$$

---

4. Oliver Heaviside: London (UK), 1850–Torquay, 1925.

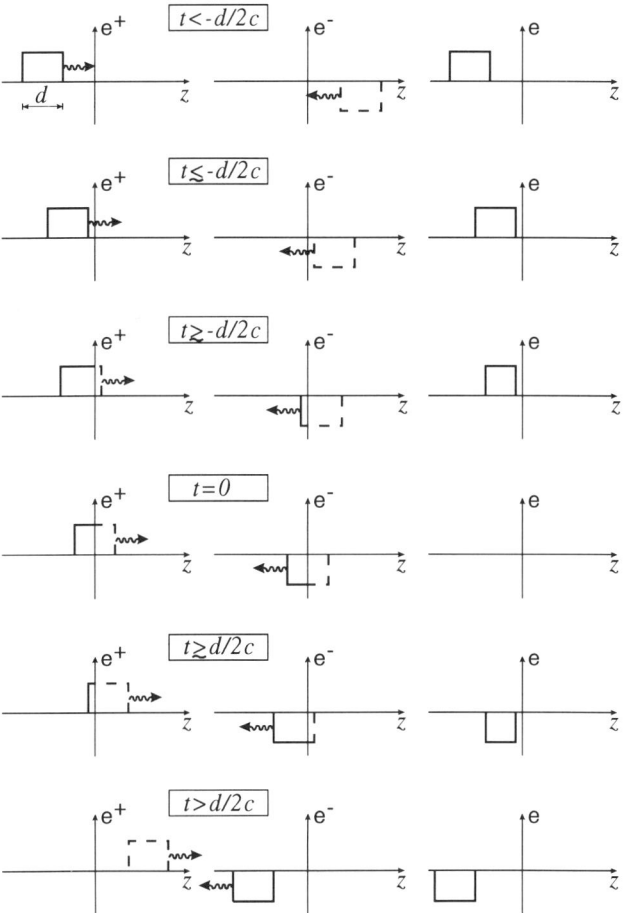

FIGURE 2.8. Reflection of an electromagnetic pulse $\text{rect}[(ct - z)/d]$ by a conducting plane at $z = 0$. The total field, $\mathbf{e}$, is the sum of the incident, $\mathbf{e}^+$, and reflected, $\mathbf{e}^-$, components.

by using the substitution $c_1 t \mp z = \xi$, $(c_2 t - z)/(c_2/c_1) = \xi$. It is immediately seen from Eqs. (2.13) that

$$1 - \Gamma^2 = \frac{\zeta_1}{\zeta_2} T^2, \qquad (2.18)$$

which renders the two integrals (2.17) equal and states energy conservation: the difference between the total incident and reflected energy equals the total transmitted energy.

Let us now consider a plane wave

$$\mathbf{d}_1^+ = f(z - c_1 t)\hat{\mathbf{x}}, \quad \mathbf{b}_1^+ = \zeta_1 f(z - c_1 t)\hat{\mathbf{y}}, \quad t \leqslant 0,$$

propagating in an isotropic, nondispersive, lossless medium described by parameters (ε_1, μ_1). At time $t = 0$ the medium parameters suddenly switch to values (ε_2, μ_2): this is the time-abrupt discontinuity depicted in Fig. 2.9. Continuity conditions (1.38) and (1.39) at $t = 0$ can be established by introducing *backward*

$$\mathbf{d}_2^- = g(z + c_2 t)\hat{\mathbf{x}}, \quad \mathbf{b}_2^- = -\zeta_2 g(z + c_2 t)\hat{\mathbf{y}}$$

and *forward*

$$\mathbf{d}_2^+ = q(z - c_2 t)\hat{\mathbf{x}}, \quad \mathbf{b}_2^+ = \zeta_2 q(z - c_2 t)\hat{\mathbf{y}}$$

waves for $t \geqslant 0$, which satisfy the radiation condition for $z \to -\infty$ and $z \to +\infty$, respectively.

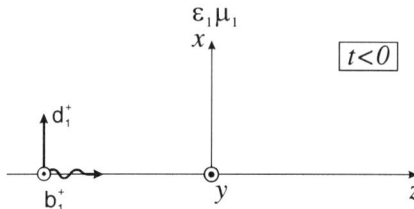

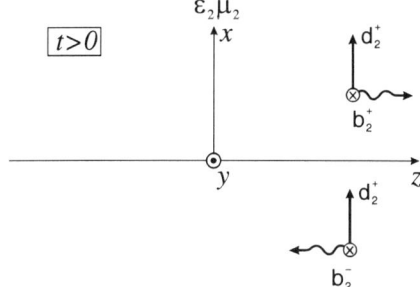

FIGURE 2.9. Electromagnetic wave propagation in the presence of a temporal discontinuity boundary.

ELEMENTARY SOLUTIONS 59

At the time-discontinuity boundary $t = 0$,

$$f(z) = g(z) + q(z),$$
$$\zeta_1 f(z) = -\zeta_2 g(z) + \zeta_2 q(z).$$

If

$$g(z) = Bf(z) \quad \text{and} \quad q(z) = Ff(z),$$

where B is the (electric induction) *backward* and F the (electric induction) *forward coefficient*, we obtain

$$B = \frac{\zeta_2 - \zeta_1}{2\zeta_2}, \quad F = \frac{\zeta_2 + \zeta_1}{2\zeta_2}, \quad (2.19)$$

and the field solution is given by

$$d_1 = f(z - c_1 t), \quad b_1 = \zeta_1 f(z - c_1 t) \quad (2.20)$$

for $t \leq 0$, and

$$\begin{cases} d_2(z, t) = Ff(z - c_2 t) + Bf(z + c_2 t), \\ b_2(z, t) = \zeta_2 [Ff(z - c_2 t) - Bf(z + c_2 t)] \end{cases} \quad (2.21)$$

for $t \geq 0$.

As an example, the case of an electromagnetic pulse of spatial width d before and after the time transition of the medium is depicted in Fig. 2.10. Note that the spatial width of the pulse is preserved. On the other hand, the time duration is scaled by the ratio c_1/c_2 [see Eqs. (2.20) and (2.21)]: the pulse is stretched if $c_2 < c_1$ and shrunk if $c_2 > c_1$.

The backward wave disappears if $\zeta_2 = \zeta_1$, i.e., when the matching condition given by Eq. (2.16) is verified.

Energy densities in the two time intervals $t \leq 0$ and $t \geq 0$ are given by

$$w_1(z, t) = \frac{1}{2\varepsilon_1} d_1^2 + \frac{1}{2\mu_1} b_1^2 = \frac{1}{\varepsilon_1} f^2(z - c_1 t), \quad t \leq 0,$$

$$w_2(z, t) = \frac{1}{2\varepsilon_2} d_2^2 + \frac{1}{2\mu_2} b_2^2$$

$$= \frac{1}{\varepsilon_2} [F^2 f^2(z - c_2 t) + B^2 f^2(z + c_2 t)], \quad t \geq 0,$$

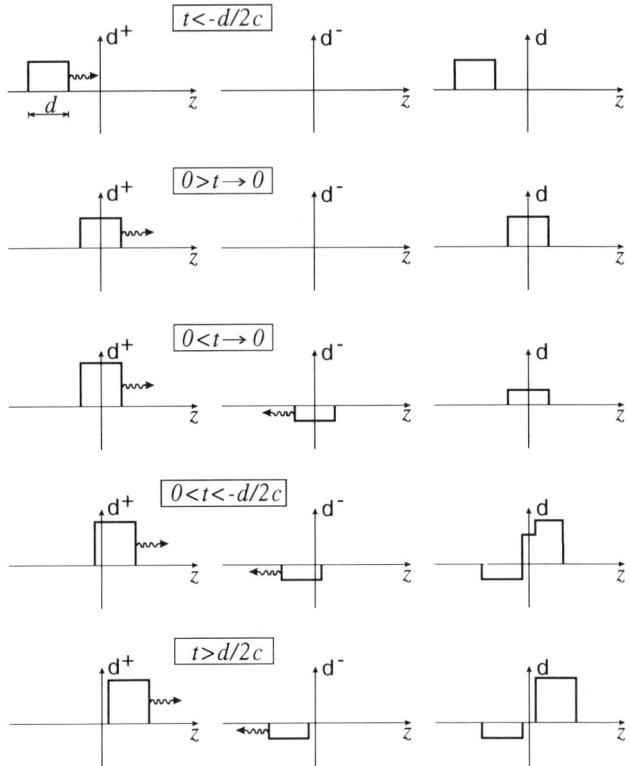

FIGURE 2.10. Reflection and transmission of an electromagnetic pulse $\text{rect}[(ct - z)/d]$ in the presence of a time discontinuity: $\zeta_1 = 2\zeta_2$. The total induction, $\mathbf{d}$, is the sum of the forward, $\mathbf{d}^+$, and backward, $\mathbf{d}^-$, parts.

because $b^2 = \zeta^2 d^2$ for a plane wave. The total energy (within a square-meter tube) at any time t is computed as

$$W_1 = \int_{-\infty}^{+\infty} dz\, w_1(z, t) = \frac{1}{\varepsilon_1} \int_{-\infty}^{+\infty} d\xi\, f^2(\xi), \qquad t \leqslant 0,$$

$$W_2 = \int_{-\infty}^{+\infty} dz\, w_2(z, t) = \frac{1}{\varepsilon_2}(F^2 + B^2) \int_{-\infty}^{+\infty} d\xi\, f^2(\xi), \qquad t \geqslant 0,$$

upon substituting $z - c_1 t = \xi$ and $z \mp c_2 t = \xi$. It is immediately recognized from Eqs. (2.19) that

$$\frac{1}{\varepsilon_1} - \frac{1}{\varepsilon_2}(F^2 + B^2) = \frac{1}{2}\frac{\varepsilon_2 - \varepsilon_1}{\varepsilon_2 \varepsilon_1} + \frac{\zeta_1^2}{2}\frac{\mu_2 - \mu_1}{\mu_2 \mu_1},$$

which renders the two integrals different and shows that the total energy stored in the electromagnetic field is *not* conserved after time $t = 0$:

$$W_2 - W_1 = \frac{1}{2}\frac{\varepsilon_2 - \varepsilon_1}{\varepsilon_2\varepsilon_1}\int_{-\infty}^{+\infty} d\xi d_1^2(\xi) + \frac{1}{2}\frac{\mu_2 - \mu_1}{\mu_2\mu_1}\int_{-\infty}^{+\infty} d\xi b_1^2(\xi).$$

Exchange of energy with external sources is needed to modify the medium parameters, and this energy is either stored or supplied by the field itself. In fact, $W_2 - W_1$ is the difference between the electromagnetic energy before ($t < 0$) and after ($t > 0$) the medium temporal discontinuity.

2.2.1. Conductive Media

Let us consider the same problem depicted in Fig. 2.7, but with the half-space $z > 0$ characterized by constant parameters ε_2, μ_2, and σ_2.

When the conductivity σ_2 is very small, the perturbative analysis of Section 2.1.1 applies and fields in the half-space $z > 0$ are written as

$$e_2(z, t) = \text{T} \exp\left(-\frac{\sigma_2\zeta_2}{2}z\right) f\left(\frac{c_2 t - z}{c_2/c_1}\right)$$

and

$$\zeta_2 h_2(z, t) = \text{T} \exp\left(-\frac{\sigma_2\zeta_2}{2}z\right) f\left(\frac{c_2 t - z}{c_2/c_1}\right)$$

with $c_2 = 1/\sqrt{\varepsilon_2\mu_2}$, $\zeta_2 = \sqrt{\mu_2/\varepsilon_2}$, and the transmission coefficient T given by Eq. (2.13).

The other extreme case occurs when σ_2 is very large, but finite. In Eq. (2.6) we neglect the term $\varepsilon_2 \partial e_2/\partial t$ with respect to $\sigma_2 e_2$ and obtain

$$\frac{\partial e_2}{\partial z} = -\mu_2 \frac{\partial h_2}{\partial t}, \qquad \frac{\partial h_2}{\partial z} = -\sigma_2 e_2. \tag{2.22}$$

We take the z-derivative of the second equation and substitute the electric field from the other:

$$\frac{\partial^2 h_2}{\partial z^2} - \sigma_2\mu_2 \frac{\partial h_2}{\partial t} = 0, \qquad z \geq 0. \tag{2.23}$$

The solution of Eq. (2.23) is obtained by means of a transform technique (see Section 3.2.5). The fields decay in the half-space $z \geq 0$ as z increases: the

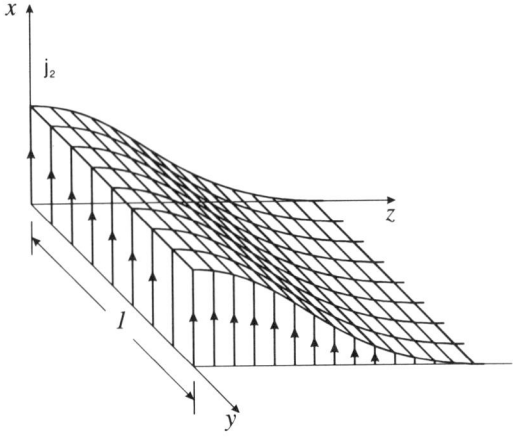

FIGURE 2.11. Plane-wave incidence on a conductive medium: induced current density, for a prescribed given time.

larger σ_2, the faster the spatial attenuation. Also, the induced density current, $\mathbf{j}_2 = \sigma_2 \mathbf{e}_2$, is attenuated; see Fig. 2.11. The total current flux per unit length along y (Fig. 2.11) can be computed by means of Eqs. (2.22),

$$j_s(t) = \int_0^\infty dz \sigma_2 e_2(z, t) = -\int_0^\infty dz \frac{\partial h_2}{\partial z} = h_2(0, t), \qquad (2.24)$$

and depends only on the value of the tangential magnetic field over the discontinuity plane $z = 0$. As the conductivity σ_2 increases, the field, as well as the induced current, is confined within a space interval closest to the boundary. The current density increases accordingly, its total flux remaining constant. For $\sigma_2 \to \infty$ the current density tends to infinity, too, and it is convenient to introduce the current linear density (2.24). The magnetic field is discontinuous at the space boundary, being zero inside the perfect conductor, and its amplitude on the discontinuity boundary is equal to the current linear density; see Eq. (1.36).

2.2.2. Bounded Waves

Let us consider a homogeneous isotropic slab bounded by two perfectly conducting planes, as depicted in Fig. 2.12. The field in the slab is given by the superposition of waves propagating along the positive and negative sense of the z-axis; see Eqs. (2.5). The electric field is oriented along the y-axis and is given by

$$e(z, t) = f(ct - z) + g(ct + z),$$

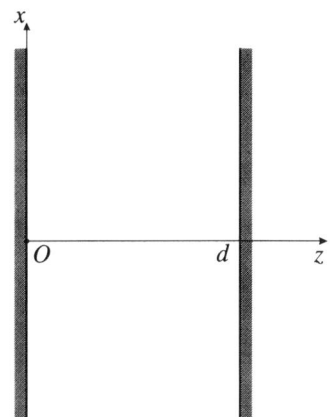

FIGURE 2.12. Geometry of a homogeneous dielectric slab bounded by two perfectly conducting planes.

with boundary conditions

$$f(ct) + g(ct) = 0 \quad \text{and} \quad f(ct - d) + g(ct + d) = 0,$$

at $z = 0$ and d, respectively. The first condition is fulfilled if $g(ct + z) = -f(ct + z)$, while the second, which becomes $f(ct - d) = f(ct + d)$, is verified provided that the function $f(ct - z)$ is periodic with period $2d$. A Fourier[5] series expansion is appropriate:

$$f(ct - z) = -\frac{1}{2}\sum_{1}^{\infty} f_n \sin\left[\frac{n\pi}{d}(ct - z - ct_n)\right],$$

where f_n and t_n are expansion coefficients[6] and the factor $-1/2$ is added for convenience. The field in the slab is given by

$$e(z, t) = f(ct - z) - f(ct + z)$$

$$= \sum_{1}^{\infty} f_n \cos\left[\frac{n\pi}{d}c(t - t_n)\right] \sin\left(\frac{n\pi}{d}z\right), \quad (2.25a)$$

$$\zeta h(z, t) = f(ct - z) + f(ct + z)$$

$$= \sum_{1}^{\infty} f_n \sin\left[\frac{n\pi}{d}c(t - t_n)\right] \cos\left(\frac{n\pi}{d}z\right). \quad (2.25b)$$

---
5. Jean Baptiste-Joseph Fourier: Auxerre (France), 1768 – Paris, 1830.
6. The constant term of the expansion is missing due to the boundary condition.

We note that each term of the series, of amplitude f_n, is a solution of Maxwell equations and is space–time periodic: the space period is $2d/n$, while the time period is $2d/nc$. Each of these terms is referred to as a *mode* of the field. A plot of the space dependence of the first three modes, with $f_n = 1$, is given in Fig. 2.13.

The energy density in the slab is given by

$$w(z, t) = \frac{1}{2}\varepsilon e^2(z, t) + \frac{1}{2}\mu h^2(z, t)$$

$$= \frac{1}{2}\varepsilon \sum_{n}^{1} \sum_{m}^{1} f_n f_m \sin\left(\frac{n\pi}{d}z\right) \sin\left(\frac{m\pi}{d}z\right) \cos\left[\left(\frac{n\pi}{d}c\right)(t - t_n)\right]$$

$$\times \cos\left[\left(\frac{m\pi}{d}c\right)(t - t_m)\right]$$

$$+ \frac{1}{2}\frac{\mu}{\zeta^2} \sum_{n}^{1} \sum_{m}^{1} f_n f_m \cos\left(\frac{n\pi}{d}z\right) \cos\left(\frac{m\pi}{d}z\right) \sin\left[\left(\frac{m\pi}{d}c\right)(t - t_n)\right]$$

$$\times \sin\left[\frac{m\pi}{d}c(t - t_n)\right]$$

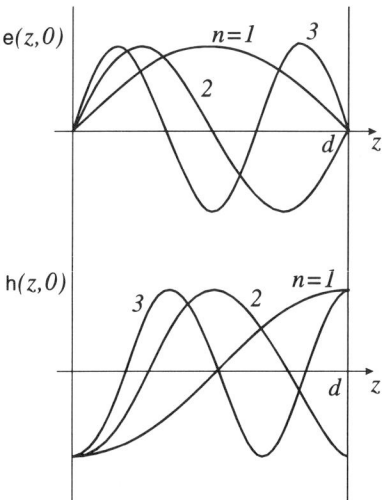

FIGURE 2.13. Plot of the spatial dependence for the first three modes in the slab of Fig. 2.12. The time origin is t_n for the electric field and $t_n - d/(2nc)$ for the magnetic field. The constant f_n is the same for all n.

ELEMENTARY SOLUTIONS

and the total energy per square meter of the slab is obtained by integration,

$$W(t) = \int_0^d dz\, w(z, t)$$

$$= \frac{1}{2}\varepsilon d \sum_n f_n^2 \left\{ \cos^2\left[\frac{n\pi}{d}c(t - t_n)\right] + \sin^2\left[\frac{n\pi}{d}c(t - t_n)\right] \right\} = \frac{1}{2}\varepsilon d \sum_n f_n^2,$$

because

$$\int_0^d \sin\left(\frac{n\pi}{d}z\right) \sin\left(\frac{m\pi}{d}z\right) dz = \frac{d}{2}\delta_{nm}.$$

The total energy is the sum of the individual energies associated with each mode of the field.

2.3. RADIATION FROM PRESCRIBED SOURCES

Radiation from prescribed electric sources in an unbounded, isotropic, homogeneous, lossless, nondispersive medium is governed by Maxwell's equations

$$\nabla \times \mathbf{e} = -\mu \frac{\partial \mathbf{h}}{\partial t}, \tag{2.26a}$$

$$\nabla \times \mathbf{h} = \varepsilon \frac{\partial \mathbf{e}}{\partial t} + \mathbf{j}_0, \tag{2.26b}$$

$$\nabla \cdot \mathbf{e} = \frac{1}{\varepsilon}\rho_0, \tag{2.26c}$$

$$\nabla \cdot \mathbf{h} = 0, \tag{2.26d}$$

where the (prescribed) source current $\mathbf{j}_0$ and charge ρ_0 densities are related by the continuity equation

$$\nabla \cdot \mathbf{j}_0 + \frac{\partial \rho_0}{\partial t} = 0. \tag{2.27}$$

By taking the curl of equation (2.26a) and substituting $\nabla \times \mathbf{h}$ from Eq.

(2.26b), we obtain

$$\nabla \times \nabla \times \mathbf{e} = -\mu\varepsilon\frac{\partial^2 \mathbf{e}}{\partial t^2} - \mu\frac{\partial \mathbf{j}_0}{\partial t}.$$

Equations (A.17) and (2.26c) yield

$$\nabla^2 \mathbf{e} - \frac{1}{c^2}\frac{\partial^2 \mathbf{e}}{\partial t^2} = \mu\frac{\partial \mathbf{j}_0}{\partial t} + \frac{1}{\varepsilon}\nabla\rho_0, \qquad c^2 = 1/\varepsilon\mu. \qquad (2.28)$$

It is convenient to set

$$\mathbf{j}_0 = \frac{\partial \mathbf{p}_0}{\partial t}, \qquad (2.29)$$

which leads to

$$\rho_0 = -\nabla \cdot \mathbf{p}_0 \qquad (2.30)$$

on substituting into Eq. (2.27). Accordingly, Eq. (2.28) becomes

$$\nabla^2 \mathbf{e} - \frac{1}{c^2}\frac{\partial^2 \mathbf{e}}{\partial t^2} = \left[\frac{1}{c^2}\frac{\partial^2}{\partial t^2}\mathbf{I} - \nabla\nabla\right] \cdot \frac{\mathbf{p}_0}{\varepsilon}, \qquad (2.31)$$

and relates the unknown electric field to the source field $\mathbf{p}_0/\varepsilon$.

It is also convenient to write Eq. (2.31) in operator form,

$$\mathbf{\textit{m}}' \cdot \mathbf{e} = \mathbf{\textit{m}} \cdot \frac{\mathbf{p}_0}{\varepsilon}, \qquad (2.32)$$

where the dyadic operators $\mathbf{\textit{m}}$ and $\mathbf{\textit{m}}'$ are given by

$$\mathbf{\textit{m}} = \frac{1}{c^2}\frac{\partial^2}{\partial t^2}\mathbf{I} - \nabla\nabla \qquad \text{and} \qquad \mathbf{\textit{m}}' = \nabla^2 - \frac{1}{c^2}\frac{\partial^2}{\partial t^2}\mathbf{I}$$

with

$$\mathbf{\textit{m}} \cdot \mathbf{a} = \frac{1}{c^2}\frac{\partial^2 \mathbf{a}}{\partial t^2} - \nabla\nabla \cdot \mathbf{a},$$

ELEMENTARY SOLUTIONS

and similarly for m'. In cartesian coordinates

$$m \to \begin{vmatrix} \dfrac{1}{c^2}\dfrac{\partial^2}{\partial t^2} - \dfrac{\partial^2}{\partial x^2} & -\dfrac{\partial^2}{\partial x \partial y} & -\dfrac{\partial^2}{\partial x \partial z} \\ -\dfrac{\partial^2}{\partial y \partial x} & \dfrac{1}{c^2}\dfrac{\partial^2}{\partial t^2} - \dfrac{\partial^2}{\partial y^2} & -\dfrac{\partial^2}{\partial y \partial z} \\ -\dfrac{\partial^2}{\partial z \partial x} & -\dfrac{\partial^2}{\partial z \partial y} & \dfrac{1}{c^2}\dfrac{\partial^2}{\partial t^2} - \dfrac{\partial^2}{\partial z^2} \end{vmatrix}, \qquad (2.33)$$

which shows that the operator m is symmetric: $m = \tilde{m}$. The operator m' is also symmetric because, again in cartesian coordinates,

$$m' \to \begin{vmatrix} \nabla^2 - \dfrac{1}{c^2}\dfrac{\partial^2}{\partial t^2} & 0 & 0 \\ 0 & \nabla^2 - \dfrac{1}{c^2}\dfrac{\partial^2}{\partial t^2} & 0 \\ 0 & 0 & \nabla^2 - \dfrac{1}{c^2}\dfrac{\partial^2}{\partial t^2} \end{vmatrix}, \qquad (2.34)$$

where ∇^2 is the scalar laplacian.

Equation (2.32) may be simplified if we introduce

$$\mathbf{e} = -m \cdot \mathbf{\Pi}. \qquad (2.35)$$

When Eq. (2.35) is substituted into Eq. (2.32) and the symmetric operators m and m' interchanged [see Eq. (B.16)], then we have

$$\nabla^2 \mathbf{\Pi} - \dfrac{1}{c^2}\dfrac{\partial^2 \mathbf{\Pi}}{\partial t^2} = -\dfrac{\mathbf{p}_0}{\varepsilon}. \qquad (2.36)$$

The vector $\mathbf{\Pi}$ is referred to as the Hertz[7] potential. The electric field is given by Eq. (2.35), while the magnetic field is computed by Eq. (2.26a):

$$\zeta \mathbf{h} = \nabla \times \dfrac{1}{c}\dfrac{\partial \mathbf{\Pi}}{\partial t}, \qquad \zeta = \sqrt{\mu/\varepsilon}. \qquad (2.37)$$

7. Heinrich Rudolf Hertz: Hamburg (Germany), 1857–Bonn, 1894.

The solution of Eq. (2.36) requires initial and boundary conditions. If the sources are switched on at time $t = 0$, fields and potentials must be zero for $t < 0$. In the unbounded medium under consideration, boundary conditions are stated in terms of the radiation condition (1.62) at infinity.

2.3.1. Elementary Sources

Let us consider an electric dipole of moment $p = qd$ as depicted in Fig. 2.14. Two charges, of opposite sign and with equal time variation, are aligned along the z-axis and are separated by a distance a. We let this distance approach zero, the moment p remaining constant; this is the elementary dipole discussed in Section 1.1.4. Note that $p(t)$ is a dipole source [coulomb × m], whose corresponding density [coulomb × m/m^3] is $p(t)\delta(\mathbf{r})$.

The source charge density is given by Eq. (1.9),

$$\rho_0 = -p(t)\delta(x)\delta(y)\frac{\partial \delta(z)}{\partial z},$$

and the source current density by Eq. (1.10),

$$\mathbf{j}_0 = \hat{\mathbf{z}}\frac{dp}{dt}\delta(x)\delta(y)\delta(z) = \frac{dp}{dt}\delta(\mathbf{r})\hat{\mathbf{z}}, \qquad (2.38)$$

with $\delta(r) = \delta(x)\delta(y)\delta(z)$. Comparison of Eqs. (2.38) and (2.29) shows that

$$\mathbf{p}_0(\mathbf{r}, t) = p(t)\delta(\mathbf{r})\hat{\mathbf{z}}.$$

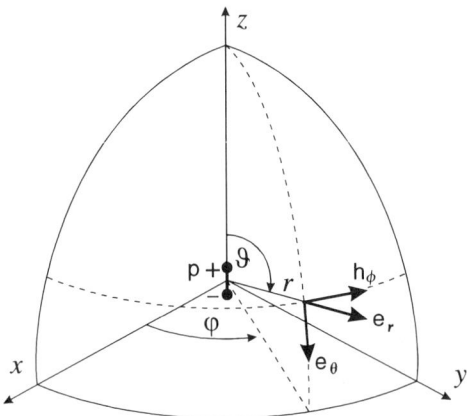

FIGURE 2.14. Fields associated with an electric dipole of moment **p**.

The Hertz potential Π is expressed as the sum of the source-free ($\mathbf{p}_0 = 0$) and the forced solution of Eq. (2.36). The former is taken equal to zero in view of the initial conditions, so that it is reasonable to take $\Pi = \Pi\hat{\mathbf{z}}$, assuming that the boundary conditions can be fulfilled. Equation (2.36) becomes

$$\nabla^2 \Pi - \frac{1}{c^2} \frac{\partial^2 \Pi}{\partial t^2} = -\frac{p(t)}{\varepsilon} \delta(\mathbf{r}), \qquad (2.39)$$

and shows that $\Pi(\mathbf{r}, t) = \Pi(r, t)$, because $\mathbf{p}_0 = p(t)\delta(\mathbf{r})\hat{\mathbf{z}}$ depends spatially only on the radial coordinate. In terms of the spherical system of coordinates in Fig. 2.14, we set

$$\Pi(r, t) = \frac{f(r, t)}{r},$$

thus getting

$$\frac{\partial^2 f}{\partial r^2} - \frac{1}{c^2} \frac{\partial^2 f}{\partial t^2} = 0, \qquad r > 0.$$

The solution fulfilling the radiation condition at infinity is $f(t - r/c)$. We assume this function to be bounded as $r \to 0$ and integrate Eq. (2.39) over a sphere of radius r centered at the origin:

$$\oint dS \frac{\partial}{\partial r}\left[\frac{f(t - r/c)}{r}\right] - \frac{1}{c^2} \frac{\partial^2}{\partial t^2} \iiint dV \frac{f(t - r/c)}{r} = -\frac{p(t)}{\varepsilon},$$

where use has been made of Eq. (A.24) in computing the integral of $\nabla^2 \Pi$. As $r \to 0$ the volume integral tends to zero, while the dominant term $-f(t-r/c)/r^2$ in the surface integral provides

$$-4\pi f(t) = -\frac{p(t)}{\varepsilon},$$

so that

$$\Pi(r, t) = \frac{\mathbf{p}(t - r/c)}{4\pi\varepsilon r}. \qquad (2.40)$$

The electromagnetic field is computed via Eqs. (2.35) and (2.37). In the spherical system of coordinates of Fig. 2.14,

$$\begin{cases} e_r(r, t) = \dfrac{\zeta}{2\pi}\left[\phantom{\dfrac{\ddot{p}(t^*)}{cr}} + \dfrac{\dot{p}(t^*)}{r^2} + \dfrac{cp(t^*)}{r^3}\right]\cos\theta, & (2.41\text{a}) \\[2ex] e_\theta(r, t) = \dfrac{\zeta}{4\pi}\left[\dfrac{\ddot{p}(t^*)}{cr} + \dfrac{\dot{p}(t^*)}{r^2} + \dfrac{cp(t^*)}{r^3}\right]\sin\theta, & (2.41\text{b}) \\[2ex] h_\phi(r, t) = \dfrac{1}{4\pi}\left[\dfrac{\ddot{p}(t^*)}{cr} + \dfrac{\dot{p}(t^*)}{r^2}\phantom{ + \dfrac{cp(t^*)}{r^3}}\right]\sin\theta, & (2.41\text{c}) \end{cases}$$

where

$$\dot{p}(t) = \frac{\partial p}{\partial t}$$

and

$$t^* = t - \frac{r}{c} \qquad (2.42)$$

is the *retarded time*; the signal propagates radially from the origin with finite speed c.

The radial component of the Poynting vector is given by

$$\begin{aligned} s_r = e_\theta h_\phi &= \frac{\zeta\sin^2\theta}{(4\pi)^2}\left[\left(\frac{\ddot{p}}{cr}\right)^2 + \frac{2\ddot{p}\dot{p}}{cr^3} + \frac{\dot{p}^2 + p\ddot{p}}{r^4} + \frac{cp\dot{p}}{r^5}\right] \\ &= \frac{\zeta\sin^2\theta}{(4\pi)^2}\left[\left(\frac{\ddot{p}}{cr}\right)^2 + \frac{\partial}{\partial t}\left(\frac{\dot{p}^2}{cr^3} + \frac{p\dot{p}}{r^4} + \frac{1}{2}\frac{cp^2}{r^5}\right)\right]. \end{aligned}$$

The total radiated energy can be computed in the form

$$2\pi\int_0^\pi d\theta\, r^2\sin\theta\int_{-\infty}^{+\infty}dt\,s_r(r,\theta,t) = \frac{\zeta}{6\pi c^2}\int_{-\infty}^{+\infty}dt\,\ddot{p}^2(t),$$

if $p(\pm\infty) = 0$, i.e., the source is time-limited. Only the fields which decay as $1/r$ are associated with the irreversible radiated energy, while the others are associated with reversible stored energy. The former are referred to as *radiative*, the latter as *local fields*.

ELEMENTARY SOLUTIONS

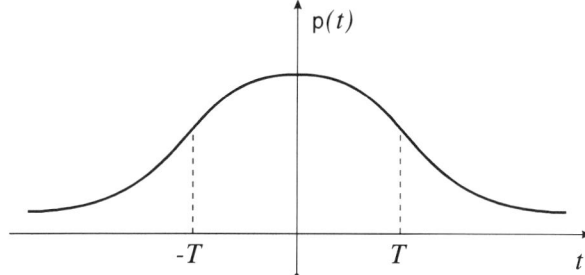

FIGURE 2.15. A gaussian pulse.

For a gaussian pulsed dipole,

$$p(t) = p_0 \exp\left(-\frac{t^2}{2T^2}\right),$$

with $2T$ being the pulse effective time width (see Fig. 2.15), we have

$$\frac{\ddot{p}}{cr} = -\frac{cp}{(cT)^2 r}\left[1 - \left(\frac{t}{T}\right)^2\right], \qquad \frac{\dot{p}}{r^2} = -\frac{cp}{cTr^2}\frac{t}{T},$$

and Eqs. (2.41) show that the radiative fields are dominant provided that $cT \gg r$, i.e., they dominate at a distance from the source large compared to the pulse effective spatial width $2cT$. These fields are also referred to as *far fields*, as opposed to the others, the *near fields*.

When $p = p_0 = $ const, we recover from Eqs. (2.41) the expression for the field of an electric static dipole, and the radiative fields are zero. These are also recognized as the *dynamical fields*, as opposed to the others, the *quasi-static fields*.

In the radiative zone

$$\begin{cases} e_\theta = \dfrac{\zeta}{4\pi cr} \ddot{p}(t^*) \sin\theta, \\ \zeta h_\phi = e_\theta, \end{cases} \qquad (2.43)$$

the fields are mutually orthogonal, orthogonal to the propagation sense $\hat{r}$, and related by the intrinsic impedance of the space; they behave locally as a plane wave. Equations (2.43) describe a simple *spherical wave*.

2.3.2. Magnetic Sources

Let us consider a magnetic dipole of moment **u**, whose associated fields are obtained by duality; see Section 1.5. Hence

$$\begin{cases} h_r = \dfrac{1}{2\pi\zeta}\left[\qquad\quad \dfrac{\dot{u}(t^*)}{r^2} + \dfrac{cu(t^*)}{r^3}\right]\cos\theta, & (2.44\text{a}) \\[2mm] h_\theta = \dfrac{1}{4\pi\zeta}\left[\dfrac{\ddot{u}(t^*)}{cr} + \dfrac{\dot{u}(t^*)}{r^2} + \dfrac{cu(t^*)}{r^3}\right]\sin\theta, & (2.44\text{b}) \\[2mm] e_\phi = -\dfrac{1}{4\pi}\left[\dfrac{\ddot{u}(t^*)}{cr} + \dfrac{\dot{u}(t^*)}{r^2}\qquad\quad\right]\sin\theta, & (2.44\text{c}) \end{cases}$$

by enforcing the transformation (1.74) and (1.75b),

$$\mathbf{e} \to \zeta\mathbf{h}, \qquad \mathbf{h} \to -\frac{1}{\zeta}\mathbf{e}, \qquad \mathbf{p} \to \frac{1}{\zeta}\mathbf{u}.$$

The fields (2.44) are depicted in Fig. 2.16.

We now examine the existence of a physical source which generates the fields (2.44). The magnetic dipole corresponds to a magnetic density current

$$\mathbf{k}_0 = \frac{du}{dt}\delta(\mathbf{r})\hat{\mathbf{z}},$$

which is the magnetic counterpart of Eq. (2.38). From Maxwell's equations we

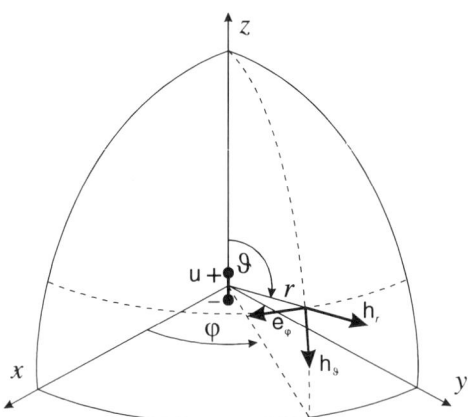

FIGURE 2.16. Fields associated with a magnetic dipole of moment **u**.

obtain

$$\nabla \times \nabla \times \mathbf{e} = -\mu\varepsilon \frac{\partial^2 \mathbf{e}}{\partial t^2} - \nabla \times \mathbf{k}_0,$$

which is equivalent to

$$\nabla^2 \mathbf{e} - \frac{1}{c^2} \frac{\partial^2 \mathbf{e}}{\partial t^2} = \nabla \times \mathbf{k}_0,$$

because $\nabla \cdot \mathbf{e} = 0$ in the absence of electric source charges. Comparison with Eq. (2.28) shows that magnetic and electric sources, $\mathbf{k}_0$ and $\mathbf{j}_0$, are equivalent, i.e., they generate the same field, provided that

$$\nabla \times \mathbf{k}_0 = \mu \frac{\partial \mathbf{j}_0}{\partial t} + \frac{1}{\varepsilon} \nabla \rho_0. \tag{2.45}$$

We now show that the charge density ρ_0 in Eq. (2.45) is equal to zero. The divergence of Eq. (2.45) yields

$$0 = \mu \frac{\partial}{\partial t} \nabla \cdot \mathbf{j} + \frac{1}{\varepsilon} \nabla \cdot \nabla \rho_0 = \frac{1}{\varepsilon} \left[-\frac{1}{c^2} \frac{\partial^2 \rho_0}{\partial t^2} + \nabla^2 \rho_0 \right],$$

where use has been made of the continuity equation (1.7). The charge density is the solution of a source-free wave equation, and can be taken equal to zero. From Eq. (2.45)

$$\nabla \times \mathbf{k}_0 = \frac{\partial}{\partial t} \nabla \times [\hat{\mathbf{z}} u(t) \delta(\mathbf{r})] = \frac{\partial}{\partial t} \mu \mathbf{j}_0,$$

so that

$$\nabla \times [\hat{\mathbf{z}} u \delta(\mathbf{r})] = \mu \mathbf{j}_0. \tag{2.46}$$

Equation (2.46) is the starting point for obtaining a direct relation between the magnetic dipole and the equivalent current density.

Computation of the curl in Eq. (2.46) is in order. The magnetic dipole $\mathbf{u} = u\hat{\mathbf{z}}$ is localized at the origin, $u\hat{\mathbf{z}}\delta(\mathbf{r})$ being its corresponding density. We can spread it over a small area πa^2 (see Fig. 2.17), with $a \to 0$:

$$\mathbf{u} = u\delta(\mathbf{r})\hat{\mathbf{z}} \approx u\delta(z) \frac{1}{\pi a^2} U(a - R)\hat{\mathbf{z}},$$

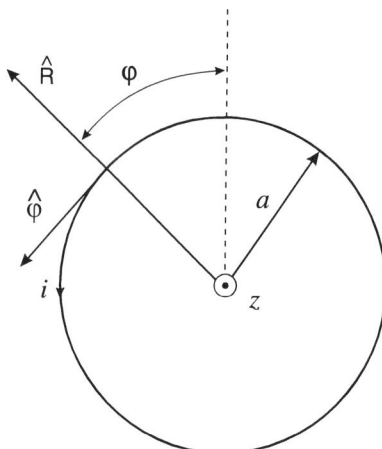

FIGURE 2.17. Equivalence between a current loop and a magnetic dipole.

so that its curl can be easily computed. In a cylindrical system of coordinates (R, ϕ, z) we obtain

$$\nabla \times \left[u\delta(z) \frac{U(a-R)}{\pi a^2} \hat{\mathbf{z}} \right] = u\delta(z) \frac{\delta(R-a)}{\pi a^2} \hat{\boldsymbol{\phi}};$$

see Eq. (A.41). Comparison with Eq. (2.46) states the (Ampére) equivalence between the magnetic dipole $\mathbf{u}(t)$ and the elementary ($a \to 0$) loop of current $i(t)$ whose corresponding density is $i(t)\delta(R-a)\delta(z)$: they generate the same field provided that

$$u(t) = \mu \pi a^2 i(t). \tag{2.47}$$

The sense of the current is counterclockwise with respect to $\hat{\mathbf{z}}$ (see Fig. 2.17), i.e., $\hat{\mathbf{R}} \times \hat{\boldsymbol{\phi}} = \hat{\mathbf{z}}$. It is easily verified that $\nabla \cdot \mathbf{j}_0 = 0$, which checks with the result $\rho_0 = 0$.

2.3.3. Free-Space Green's Functions

In a spherical system of coordinates[8] the (scalar) Green's[9] function is the solution of the (scalar) wave equation

$$\nabla^2 g(\mathbf{r}, t) - \frac{1}{c^2} \frac{\partial^2 g(\mathbf{r}, t)}{\partial t^2} = -\delta(\mathbf{r})\delta(t), \tag{2.48}$$

---

8. Green's functions are defined in other systems of coordinates, too; the appropriate value of the (space) Dirac function should be used.
9. George Green: Sneinton (near Nottingham, UK), 1793–Sneinton, 1841.

subject to the initial condition $g(\mathbf{r}, t < 0) = 0$ and to prescribed boundary conditions. When the latter are the radiation conditions at infinity, the solution of Eq. (2.48) is of the type

$$g(\mathbf{r}, t) = \frac{\delta(t - r/c)}{4\pi r} = \frac{\delta(t^*)}{4\pi r}, \tag{2.49}$$

$t^* = t - r/c$ being the retarded time.

Green's function represents the response of the system described by Eq. (2.48) to a localized unit source of space–time density $-\delta(\mathbf{r})\delta(t)$. It can be used to derive the final expression for the field radiated by an arbitrary source distribution (in general, with arbitrary boundary conditions, too). In the electric case the dipole source density $\mathbf{p}_0(\mathbf{r}, t)$ [see Eqs. (2.29)] can be written as

$$\mathbf{p}_0(\mathbf{r}, t) = \iiint dV' \int dt' \mathbf{p}_0(\mathbf{r}', t') \delta(\mathbf{r} - \mathbf{r}') \delta(t - t'), \tag{2.50}$$

which is a superposition of elementary dipoles $\mathbf{p}_0(\mathbf{r}', t')dV'dt'$ localized at $\mathbf{r} = \mathbf{r}'$ and operating at $t = t'$. The corresponding elementary Hertz potential is obtained by Eq. (2.40) and Fig. 2.18, and the total Hertz potential by superposition:

$$\begin{aligned}\mathbf{\Pi}(\mathbf{r}, t) &= \iiint dV' \int dt' \mathbf{p}_0(\mathbf{r}', t') g\left(t - t' - \frac{|\mathbf{r} - \mathbf{r}'|}{c}\right) \\ &= \frac{1}{4\pi\varepsilon} \iiint dV' \int dt' \mathbf{p}_0(\mathbf{r}', t') \frac{\delta(t - t' - |\mathbf{r} - \mathbf{r}'|/c)}{|\mathbf{r} - \mathbf{r}'|} \\ &= \frac{1}{4\pi\varepsilon} \iiint dV' \frac{\mathbf{p}_0(\mathbf{r}', t^*)}{|\mathbf{r} - \mathbf{r}'|}, \end{aligned} \tag{2.51}$$

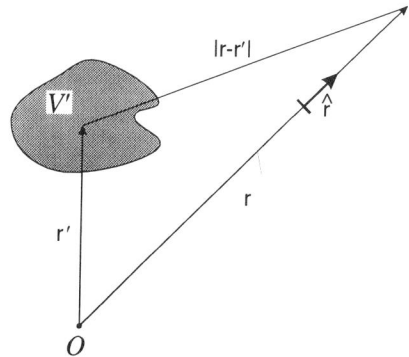

FIGURE 2.18. Use of Green's function.

where $t^* = t - |\mathbf{r} - \mathbf{r}'|/c$ depends on the integration variable $\mathbf{r}'$. The fields $\mathbf{e}(\mathbf{r}, t)$ and $\mathbf{h}(\mathbf{r}, t)$ can be computed via Eqs. (2.35) and (2.37), respectively.

At large distances from the source

$$|\mathbf{r} - \mathbf{r}'| \approx r - \mathbf{r}' \cdot \hat{\mathbf{r}},$$

$$\nabla \cdot \frac{\mathbf{p}_0}{|\mathbf{r} - \mathbf{r}'|} \approx \frac{\nabla \cdot \mathbf{p}_0}{|\mathbf{r} - \mathbf{r}'|} \approx -\frac{1}{|\mathbf{r} - \mathbf{r}'|} \frac{1}{c} \frac{\partial}{\partial t} \mathbf{p}_0 \cdot \hat{\mathbf{r}},$$

as follows by resorting to spherical coordinates. Similarly

$$\nabla \nabla \cdot \frac{\mathbf{p}_0}{|\mathbf{r} - \mathbf{r}'|} \approx \frac{1}{|\mathbf{r} - \mathbf{r}'|} \frac{1}{c^2} \frac{\partial^2}{\partial t^2} \mathbf{p}_0 \cdot \hat{\mathbf{r}}\hat{\mathbf{r}},$$

and the fields are given by

$$\mathbf{e}(\mathbf{r}, t) = -[\mathbf{I} - \hat{\mathbf{r}}\hat{\mathbf{r}}] \cdot \iiint dV' \frac{1}{c^2} \frac{\partial^2}{\partial t^2} \frac{\mathbf{p}_0(\mathbf{r}', t^*)}{4\pi\varepsilon|\mathbf{r} - \mathbf{r}'|}$$

$$= -[\mathbf{I} - \hat{\mathbf{r}}\hat{\mathbf{r}}] \cdot \iiint dV' \frac{\zeta}{c} \frac{\partial \mathbf{j}_0(\mathbf{r}, t^*)/\partial t^*}{4\pi|\mathbf{r} - \mathbf{r}'|}, \qquad (2.52a)$$

$$\zeta \mathbf{h}(\mathbf{r}, t) = \hat{\mathbf{r}} \times \mathbf{e}(\mathbf{r}, t), \qquad (2.52b)$$

where Eq. (2.52b) can be derived simply from the radiation condition. Note that the operator $\mathbf{I} - \hat{\mathbf{r}}\hat{\mathbf{r}}$ renders the far field transverse to the direction of propagation.

2.3.4. Vector and Scalar Potentials

Let us start from Maxwell's equations with electric sources,

$$\begin{cases} \nabla \times \mathbf{e} = -\mu \dfrac{\partial \mathbf{h}}{\partial t}, & (2.53a) \\[6pt] \nabla \times \mathbf{h} = \varepsilon \dfrac{\partial \mathbf{e}}{\partial t} + \mathbf{j}_0, & (2.53b) \\[6pt] \nabla \cdot \mathbf{e} = \dfrac{\rho_0}{\varepsilon}, & (2.53c) \\[6pt] \nabla \cdot \mathbf{b} = 0, & (2.53d) \end{cases}$$

in an isotropic, lossless, homogeneous, nondispersive medium. Equation (2.53d) states that the vector **b** is solenoidal,

$$\mathbf{b} = \mu \mathbf{h} = \nabla \times \mathbf{a}, \tag{2.54}$$

where **a** is referred to as the *vector potential*. Substituting Eq. (2.54) into Eq. (2.53a) yields

$$\nabla \times \left[\mathbf{e} + \frac{\partial \mathbf{a}}{\partial t} \right] = 0,$$

which shows that the vector $\mathbf{e} + \partial \mathbf{a}/\partial t$ is irrotational,

$$\mathbf{e} + \frac{\partial \mathbf{a}}{\partial t} = -\nabla \phi, \tag{2.55}$$

where ϕ is referred to as the *scalar potential*. When we substitute Eqs. (2.54) and (2.55) into Eq. (2.53b) and employ the identity (A.17), we obtain

$$\nabla^2 \mathbf{a} - \frac{1}{c^2} \frac{\partial^2 \mathbf{a}}{\partial t^2} = -\mu \mathbf{j}_0 + \nabla \left[\nabla \cdot \mathbf{a} + \frac{1}{c^2} \frac{\partial \phi}{\partial t} \right]. \tag{2.56}$$

On the other hand, substituting Eq. (2.55) into Eq. (2.53c) and adding the term $-\partial^2 \phi / c^2 \partial t^2$ to both sides leads to

$$\nabla^2 \phi - \frac{1}{c^2} \frac{\partial^2 \phi}{\partial t^2} = -\frac{\rho_0}{\varepsilon} - \frac{\partial}{\partial t} \left[\nabla \cdot \mathbf{a} + \frac{1}{c^2} \frac{\partial \phi}{\partial t} \right]. \tag{2.57}$$

Equations (2.56) and (2.57) are coupled by the term

$$\nabla \cdot \mathbf{a} + \frac{1}{c^2} \frac{\partial \phi}{\partial t} = f(\mathbf{r}, t), \tag{2.58}$$

and the system decouples if $f(\mathbf{r}, t) = 0$. To check if this is possible, let us introduce the transformation

$$\mathbf{a} \to \mathbf{a} + \nabla \psi, \quad \phi \to \phi - \frac{\partial \psi}{\partial t},$$

which, upon substitution into Eqs. (2.54) and (2.55), leaves the fields unchanged. Furthermore, Eq. (2.58) becomes

$$\nabla \cdot \mathbf{a} + \frac{1}{c^2} \frac{\partial \phi}{\partial t} = f(\mathbf{r}, t) - \left[\nabla^2 \psi - \frac{1}{c^2} \frac{\partial^2 \psi}{\partial t^2} \right],$$

which shows that an appropriate choice of function ψ renders $f(\mathbf{r}, t) = 0$. In this case vector and scalar potentials are no longer independent but related by the Lorentz condition

$$\nabla \cdot \mathbf{a} + \frac{1}{c^2} \frac{\partial \phi}{\partial t} = 0,$$

also referred to as *gauge invariance*, and they are solutions of the equations

$$\nabla^2 \mathbf{a} - \frac{1}{c^2} \frac{\partial^2 \mathbf{a}}{\partial t^2} = -\mu \mathbf{j}_0, \qquad (2.59)$$

$$\nabla^2 \phi - \frac{1}{c^2} \frac{\partial^2 \phi}{\partial t^2} = -\frac{\rho_0}{\varepsilon}. \qquad (2.60)$$

Note that

$$\mathbf{a} = \frac{1}{c^2} \frac{\partial \mathbf{\Pi}}{\partial t},$$

as follows from Eqs. (2.36) and (2.59) by inspection.

2.3.5. Radiation from Moving Sources

Let us consider a charge q moving with constant velocity $\mathbf{v} = v\hat{\mathbf{z}}$ (see Fig. 2.19) in a homogeneous, isotropic, nondispersive medium. This source is described in terms of a charge density

$$\rho_0 = q\delta(x)\delta(y)\delta(z - vt),$$

which corresponds to a current density

$$\mathbf{j}_0 = qv\delta(x)\delta(y)\delta(z - vt)\hat{\mathbf{z}} = q\delta(x)\delta(y)\delta(t - z/v)\hat{\mathbf{z}}.$$

This current density is equivalent to an electric dipole moment density

$$\mathbf{p}_0 = q\delta(x)\delta(y)U(t - z/v)\hat{\mathbf{z}},$$

which is consistent with Eq. (2.29).

ELEMENTARY SOLUTIONS

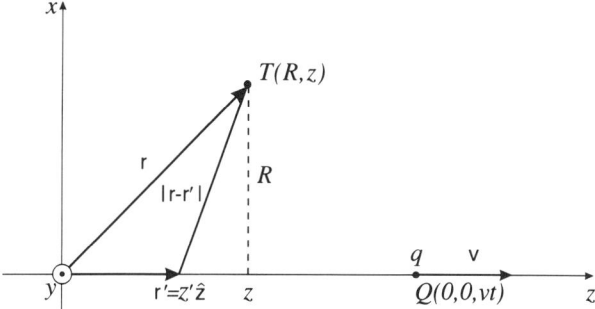

FIGURE 2.19. Radiation from a moving charge.

We compute the Hertz potential $\Pi = \Pi \hat{z}$ via Eq. (2.51):

$$\Pi(\mathbf{r}, t) = \frac{q}{4\pi\varepsilon} \int dz' \frac{U[t - (|\mathbf{r} - z'\hat{z}|/c) - z'/v]}{|\mathbf{r} - z'\hat{z}|}. \quad (2.61)$$

We also have

$$|\mathbf{r} - z'\hat{z}| = \sqrt{R^2 + (z' - z)^2},$$

where R is the radial coordinate of the observation point T, and the integration limits of Eq. (2.61) are set by the condition

$$(vt - z) - \frac{v}{c}\sqrt{R^2 + (z' - z)^2} - (z' - z) \geq 0. \quad (2.62)$$

We consider the *superrelativistic* case $v/c > 1$, i.e., the charge velocity is larger than the speed of a wave propagating in the medium (clearly, this rules out moving charges in vacuo). We first set relation (2.62) equal to zero and solve for $(z' - z)$:

$$z_{1,2} - z = \frac{-(vt - z) \pm (v/c)\sqrt{(vt - z)^2 - [(v/c)^2 - 1]R^2}}{(v/c)^2 - 1}.$$

Condition (2.62) becomes

$$z_2 - z < z' - z < z_1 - z,$$

which can be verified for real values of $z' - z$ provided that

$$(vt - z)^2 > [(v/c)^2 - 1]R^2. \quad (2.63)$$

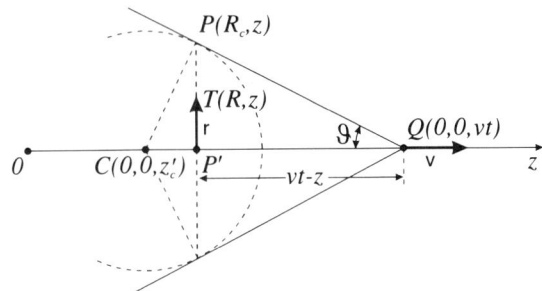

FIGURE 2.20. Čerenkov radiation (only the cone pertinent to the field satisfying the radiation condition at infinity is depicted).

Condition (2.63) sets a constraint on the allowed values of R for a given value of z. The field differs from zero inside a cone (see Fig. 2.20) whose vertex angle θ is given by

$$\cotan \theta = \frac{vt - z}{R_c} = \sqrt{(v/c)^2 - 1} = \sqrt{\frac{1 - (c/v)^2}{c/v}}, \quad \text{i.e., } \sin \theta = \frac{c}{v}.$$

The center of the integration domain of Eq. (2.61) is located at an abscissa z'_c given by

$$z'_c - z = \frac{(z_2 - z) + (z_1 - z)}{2} = -\frac{vt - z}{(v/c)^2 - 1} = -(vt - z)\tan^2\theta,$$

and the width of the integration domain across z'_c depends on the position of the point T inside the cone. If $T \to P$, i.e., close to the conical surface, $z_1 \to z_2$ and a small segment centered at C contributes to the integral.

A model of the process is the following. The moving charge emits spherical waves at each position along its trajectory, and these interfere inside the cone and create the field therein. The envelope of these spherical waves forms a cone — the Cerenkov[10] cone — which is a *caustic* for the field and expands with velocity c (see Fig. 2.21). At large distance from the z-axis the field is transverse to the direction of propagation, and there is a net flux of power along the positive sense of the z-axis (the radiation condition). This power should be provided by external sources which maintain constant the charge speed.

10. Pavel Alekseevič Čerenkov: Novaja Cigla (Russia), 1904–1990.

ELEMENTARY SOLUTIONS

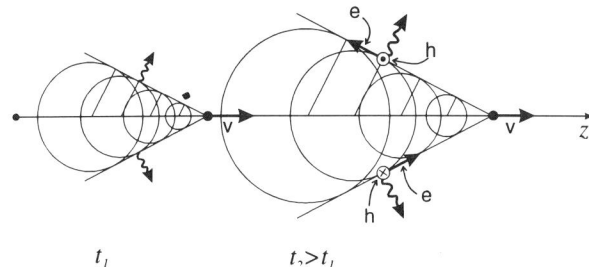

FIGURE 2.21. Formation and movement of the Čerenkov cone.

The Hertz potential is obtained via Eq. (2.61) and Eq. (C.11):

$$\mathbf{\Pi}(\mathbf{r}, t) = \frac{q\hat{\mathbf{z}}}{4\pi\varepsilon} \ln \left[\frac{z_1 - z + \sqrt{R^2 + (z_1 - z)^2}}{z_2 - z + \sqrt{R^2 + (z_2 - z)^2}} \right],$$

and the fields can be computed accordingly.

2.4. SUMMARY AND SELECTED REFERENCES

This chapter presents a number of simple, yet important problems whose solution can be obtained directly in the time domain. It is believed that the use of arbitrary waveforms instead of monochromatic waves (phasors, see Chapter 4) allows a simpler and more intuitive understanding of such fundamental concepts as wave propagation and radiation, finite speed of a signal, relevance of medium characteristics, etc.

Propagation is considered first (Section 2.1) in isotropic as well as anisotropic (Section 2.1.1) media. Both spatial and temporal discontinuities are considered in Section 2.2, showing the inherent symmetry in these changes. Bounded waves, the first step to field quantization, are discussed in Section 2.2.2.

Radiation is considered in Section 2.3, for elementary electric (Section 2.31) and magnetic (Section 2.3.2) sources. For additional information about radiation under transient conditions, a useful reference is [2.1]. Equivalence between electric and magnetic dipoles is discussed directly in the time domain (Section 2.3.2). Its usual derivation in the steady-state case, probably more complicated, is given in Section 4.6.2. It is believed that this systematic comparison between transient and steady-state results is very informative and useful for an improved understanding of all the material.

Green's functions are first introduced in Section 2.3.3 and subsequently expanded in Sections 3.2.9 and 4.6. An exhaustive discussion on this subject can be found in [2.2].
The original paper on radiation from moving sources is [2.3]. Its formal derivation is usually obtained via a frequency-domain approach (see, e.g., [1.4]), as opposed to the very simple and attractive presentation of Section 2.3.5 which, again, remains strictly in the time domain.

References

[2.1] G. Franceschetti and C. H. Papas, "Pulsed antennas," *IEEE Trans. Antennas Propagat.* **AP-22**, 651–661 (1974).

[2.2] C. T. Tai, *Dyadic Green's Functions in Electromagnetic Theory*, Intext, New York (1971).

[2.3] P. A. Čerenkov, "Visible radiation produced by electrons moving in a medium with velocities exceeding that of light," *Phys. Rev.* **52**, 378–379 (1937).

3
Spectral Domains

3.1. PRELIMINARY CONSIDERATIONS

The set of Maxwell differential equations (1.1) can be solved, in principle, when constitutive relationships (Section 1.2) as well as initial and boundary conditions (Section 1.4) are prescribed. The differential operators which appear in Eqs. (1.1) involve either time ($\partial/\partial t$) or space (∇) differential operators. We note that these operators in the case of complex exponential functions transform as follows:

$$\frac{\partial}{\partial t} F(\mathbf{r}, \omega) \exp(i\omega t) = i\omega F(\mathbf{r}, \omega) \exp(i\omega t), \qquad (3.1a)$$

$$\frac{\partial}{\partial t} \mathbf{A}(\mathbf{r}, \omega) \exp(i\omega t) = i\omega \mathbf{A}(\mathbf{r}, \omega) \exp(i\omega t), \qquad (3.1b)$$

$$\nabla F(\mathbf{k}, t) \exp(i\mathbf{k} \cdot \mathbf{r}) = i\mathbf{k} F(\mathbf{k}, t) \exp(i\mathbf{k} \cdot \mathbf{r}) \qquad (3.2a)$$

$$\nabla \cdot \mathbf{A}(\mathbf{k}, t) \exp(i\mathbf{k} \cdot \mathbf{r}) = i\mathbf{k} \cdot \mathbf{A}(\mathbf{k}, t) \exp(i\mathbf{k} \cdot \mathbf{r}), \qquad (3.2b)$$

$$\nabla \times \mathbf{A}(\mathbf{k}, t) \exp(i\mathbf{k} \cdot \mathbf{r}) = i\mathbf{k} \times \mathbf{A}(\mathbf{k}, t) \exp(i\mathbf{k} \cdot \mathbf{r}), \qquad (3.2c)$$

where ω and $\mathbf{k}$ are (scalar and vector) parameters. In other words, the differential relationship becomes an algebraic one.

This transformation for either one or both differential operators may be a convenient tool for seeking the solution of Maxwell equations, provided that a suitable superposition of complex exponential functions represents the electromagnetic field (*expansion completeness*). This is assured by the one-dimensional Fourier Transform (FT)

$$f(t) = \frac{1}{2\pi} \int_{-\infty}^{+\infty} d\omega \, \exp(i\omega t) F(\omega) \qquad (3.3)$$

and the three-dimensional FT

$$f(\mathbf{r}) = \frac{1}{(2\pi)^3} \int_{-\infty}^{+\infty} d\mathbf{k} \exp(i\mathbf{k} \cdot \mathbf{r}) F(\mathbf{k}), \qquad (3.4)$$

with similar expressions for vector functions. The transforms (3.3) and (3.4) can be inverted to yield

$$F(\omega) = \int_{-\infty}^{+\infty} dt \exp(-i\omega t) f(t) \qquad (3.5)$$

and

$$F(\mathbf{k}) = \int_{-\infty}^{+\infty} d\mathbf{k} \exp(-i\mathbf{k} \cdot \mathbf{r}) f(\mathbf{r}). \qquad (3.6)$$

Fourier transforms can be applied to the space L^2 of complex-valued functions which are Lebesque[1]-measurable and square-integrable in a given domain. For electromagnetic fields the latter is defined by initial and boundary conditions. The space L^2 is suitable to represent *physically realizable* sources and fields. However, *idealized* concentrated sources, such as point charges and line currents, are modeled in terms of functions which do not belong to the space L^2. It is therefore necessary to extend the class of considered functions, as shown in Section 3.1.1.

The application of transformations (3.3.6) allows one to solve Maxwell equations (1.1) in new domains of independent variables. According to the transformation employed, we have *space–frequency* $(\mathbf{r}, \omega)$, *wavenumber–time* $(\mathbf{k}, t)$, and *wavenumber–frequency* $(\mathbf{k}, \omega)$ domains, all referred to as *spectral domains*. Solutions in spectral domains, *spectral solutions*, provide the corresponding observable electromagnetic fields by means of the appropriate inverse FT.

3.1.1. Distributions and Dirac Functions

The Dirac[2] function or *delta functions*, $\delta(t)$, is often defined as the limit of a sequence of functions which become increasingly peaked in the immediate vicinity of the origin, their area remaining constant and equal to unity. For

---

1. Henry Leon Lebesgue: Beuvais, Oise (France) 1875–Paris (France), 1941.
2. Paul Adrien Maurice Dirac: Bristol (UK), 1902–Miami (Florida, USA), 1984.

instance (see Fig. 3.1), when $\varepsilon \to 0$,

$$\frac{1}{\pi^{1/2}\varepsilon} \exp\left(-\frac{t^2}{\varepsilon^2}\right) \to \delta(t). \tag{3.7a}$$

Note that, strictly speaking, $\delta(t)$ is not a function but rather a more general mathematical object, usually referred to as a *distribution*. Other useful expressions for the Dirac function are the following, when $a \to \infty$:

$$\frac{\sin at}{\pi t} \to \delta(t), \tag{3.7b}$$

$$\frac{J_1(at)}{t} \to \delta(t). \tag{3.7c}$$

We have

$$\int_{-\infty}^{+\infty} dt f(t)\delta(t - t_0) = \lim_{\varepsilon \to 0} \int_{-\infty}^{+\infty} dt f(t) \frac{1}{\pi^{1/2}\varepsilon} \exp\left[-\frac{(t - t_0)^2}{\varepsilon^2}\right] \to f(t_0), \tag{3.8}$$

which highlights the sifting property of the Dirac function.

Equation (3.8) defines a linear continuous functional on the space of functions $f(t)$. The latter are assumed continuous with their derivatives of any order, and vanish outside some finite domain. Then, the functional is called a *distribution*, and the function $\delta(t)$ is called the *generating function* of the distribution. Application of Eq. (3.7a) and Eq. (C.1) (in Appendix C) with the

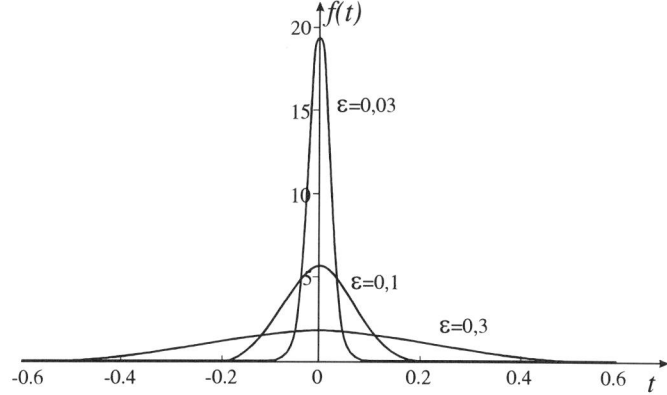

FIGURE 3.1. Definition of the Dirac function.

notation $1/\varepsilon = a$ and $i\omega/2 = b$ yields

$$\int_{-\infty}^{+\infty} dt\, \exp(-i\omega t)\delta(t) = \lim_{\varepsilon \to 0} \frac{1}{\pi^{1/2}\varepsilon} \int_{-\infty}^{+\infty} dt\, \exp\left(-\frac{t^2}{\varepsilon^2} - i\omega t\right)$$

$$= \lim_{\varepsilon \to 0} \exp\left[-\left(\frac{\omega\varepsilon}{2}\right)^2\right] = 1,$$

which allows one to extend the FT to the Dirac function. We have furthermore

$$\frac{1}{2\pi} \int_{-\infty}^{+\infty} d\omega\, \exp(i\omega t) = \delta(t).$$

In three dimensions,

$$\int_{-\infty}^{+\infty} d\mathbf{r}\, \exp(-i\mathbf{k}\cdot\mathbf{r})\delta(\mathbf{r}) = 1$$

and

$$\frac{1}{(2\pi)^3} \int_{-\infty}^{+\infty} d\mathbf{k}\, \exp(i\mathbf{k}\cdot\mathbf{r}) = \delta(\mathbf{r}).$$

3.2. The Frequency Domain

We consider Maxwell equations (1.1) with prescribed (electric) source distributions $\mathbf{j}_0(\mathbf{r}, t)$ and $\rho_0(\mathbf{r}, t)$, and assume the fields to be zero before the sources are switched on (*initial conditions*). Once the sources are given, the corresponding transforms $\mathbf{J}_0(\mathbf{r}, \omega)$ and $\rho_0(\mathbf{r}, \omega)$ are known as well.

We Fourier-expand known sources,

$$\mathbf{j}_0(\mathbf{r}, t) = \frac{1}{2\pi} \int_{-\infty}^{+\infty} d\omega\, \exp(i\omega t) \mathbf{J}_0(\mathbf{r}, \omega), \tag{3.9}$$

$$\cdots \cdots$$

and unknown fields and inductions,

$$\mathbf{e}(\mathbf{r}, t) = \frac{1}{2\pi} \int_{-\infty}^{+\infty} d\omega\, \exp(i\omega t) \mathbf{E}(\mathbf{r}, \omega),$$

$$\cdots \cdots$$

SPECTRAL DOMAINS 87

and substituting into Maxwell equations (1.1). By equating the integrands we obtain

$$\nabla \times \mathbf{E}(\mathbf{r}, \omega) = -i\omega \mathbf{B}(\mathbf{r}, \omega), \tag{3.10a}$$

$$\nabla \times \mathbf{H}(\mathbf{r}, \omega) = i\omega \mathbf{D}(\mathbf{r}, \omega) + \mathbf{J}(\mathbf{r}, \omega) + \mathbf{J}_0(\mathbf{r}, \omega), \tag{3.10b}$$

$$\nabla \cdot \mathbf{D}(\mathbf{r}, \omega) = \rho(\mathbf{r}, \omega) + \rho_0(\mathbf{r}, \omega), \tag{3.10c}$$

$$\nabla \cdot \mathbf{B}(\mathbf{r}, \omega) = 0, \tag{3.10d}$$

which are Maxwell equations in the *space–frequency* spectral domain, or simply the *frequency domain*.

Extension to the case of magnetic sources is immediate; the frequency-domain continuity equation (1.2) can be derived as

$$\nabla \cdot [\mathbf{J}(\mathbf{r}, \omega) + \mathbf{J}_0(\mathbf{r}, \omega)] + i\omega[\rho(\mathbf{r}, \omega) + \rho_0(\mathbf{r}, \omega)] = 0. \tag{3.11}$$

Note that the dimensions of the transformed quantities differ from those of the corresponding time-dependent ones. For instance, $[\mathbf{E}(\mathbf{r}, \omega)] =$ [volt/(m × Hertz)] = [volt × s/m]. Some useful FTs are presented in Table 3.1.

For time-dispersive media which are time-invariant and spatially nondispersive, the constitutive relationships reduce to a convolution, e.g.,

$$\mathbf{d}(\mathbf{r}, t) = \int_{-\infty}^{+\infty} dt' g(\mathbf{r}, t - t') \cdot \mathbf{e}(\mathbf{r}, t'),$$

as shown in Section 1.2. Application of the Borel[3] theorem gives

$$\mathbf{D}(\mathbf{r}, \omega) = \boldsymbol{\varepsilon}(\mathbf{r}, \omega) \cdot \mathbf{E}(\mathbf{r}, \omega), \tag{3.12}$$

where

$$\boldsymbol{\varepsilon}(\mathbf{r}, \omega) = \int_{-\infty}^{+\infty} dt \, \exp(-i\omega t) g(\mathbf{r}, t) \tag{3.13}$$

is the FT of the (electrical) impulse response of the medium. This transform is referred to as the (complex) dielectric constant, or *permittivity*, of the medium. We similarly define a (complex) magnetic constant, or *permeability*, $\boldsymbol{\mu}(\mathbf{r}, \omega)$, and a (complex) *conductivity* $\boldsymbol{\sigma}(\mathbf{r}, \omega)$. For isotropic media we have $\boldsymbol{\varepsilon} = \varepsilon \boldsymbol{I}$,

$$\mathbf{D}(\mathbf{r}, \omega) = \varepsilon(\mathbf{r}, \omega) \boldsymbol{I} \cdot \mathbf{E}(\mathbf{r}, \omega) = \varepsilon(\mathbf{r}, \omega) \mathbf{E}(\mathbf{r}, \omega), \tag{3.14}$$

and similarly for the permittivity $\mu(\mathbf{r}, \omega)$ and conductivity $\sigma(\mathbf{r}, \omega)$.

---

3. Félix Eduard Emile Borel: Saint Affrique, Aveyron (France), 1871–Paris (France), 1956.

TABLE 3.1. Fourier Transform Pairs

$f(t)$	$F(\omega)$
$\delta(t)$	1
1	$2\pi\delta(\omega)$
$\text{sgn}(t)$	$\dfrac{1}{i\omega}$
$U(t)$	$\dfrac{1}{i\omega} + \pi\delta(\omega)$
$\text{rect}\left[\dfrac{t}{T}\right]$	$T\,\text{sinc}\left(\omega\dfrac{T}{2}\right)$
$\dfrac{\Omega}{2\pi}\text{sinc}\left(\dfrac{\Omega}{2}t\right)$	$\text{rect}\left[\dfrac{\omega}{\Omega}\right]$
$\exp\left(-\dfrac{t^2}{2T^2}\right)$	$\sqrt{2\pi}T\exp\left(-\dfrac{\omega^2 T^2}{2}\right)$
$\dfrac{\Omega}{\sqrt{2\pi}}\exp\left(-\dfrac{\Omega^2 t^2}{2}\right)$	$\exp\left(-\dfrac{\omega^2}{2\Omega^2}\right)$
$f(t - t_0)$	$\exp(-i\omega t_0)F(\omega)$
$\cos\omega_0 t$	$\dfrac{1}{2}[\delta(\omega - \omega_0) + \delta(\omega + \omega_0)]$
$\sin\omega_0 t$	$\dfrac{1}{2i}[\delta(\omega - \omega_0) - \delta(\omega - \omega_0)]$
$\cos(\omega_0 t)U(t)$	$\dfrac{i\omega}{\omega_0^2 - \omega^2}$
$\sin(\omega_0 t)U(t)$	$\dfrac{\omega_0}{\omega_0^2 - \omega^2}$

Equations (3.12) and (3.14) are formally similar to the constitutive relationships of a nondispersive medium, but in the latter case ε, μ, and σ are ω-independent, their time-domain transforms being a δ-function. This is usually the case for the conductivity throughout the radio-frequency range.

By using the frequency-domain constitutive relations, Maxwell's equations in a time-invariant spatially nondispersive medium take the form (for the simplest case of an isotropic medium)

$$\nabla \times \mathbf{E} = -i\omega\mu\mathbf{H} - \mathbf{K}_0, \qquad (3.15\text{a})$$

$$\nabla \times \mathbf{H} = i\omega\varepsilon\mathbf{E} + \sigma\mathbf{E} + \mathbf{J}_0, \qquad (3.15\text{b})$$

$$\nabla \cdot (\varepsilon\mathbf{E}) = \rho + \rho_0, \qquad (3.15\text{c})$$

$$\nabla \cdot (\mu\mathbf{H}) = \chi + \chi_0, \qquad (3.15\text{d})$$

where the magnetic current and charge densities have also been included.

3.2.1. Properties of Transformed Fields and Related Quantities

Fields, inductions, as well as charge and current densities are *observables*, macroscopic quantities. They must be represented by real-valued functions, this enforcing some constraint on their FTs. With reference to the electric field, let

$$\mathbf{E}(\omega) = \mathbf{E}_1(\omega) + i\mathbf{E}_2(\omega).$$

By requiring its transform to be a real-valued function, we obtain

$$\mathbf{E}(\omega) = \mathbf{E}^*(-\omega),$$

i.e., the real part of frequency-domain fields is an even function of ω as opposed to the imaginary part, which is an odd function of ω.

Permittivity, permeability, and conductivity are FTs of the corresponding impulse response $g(t)$ of the medium; see Eq. (3.13). For the simplest case of an isotropic medium, $\varepsilon(\omega) = \varepsilon^*(-\omega)$, and similarly for the other constants. With reference to the electric case let

$$\varepsilon(\omega) = \varepsilon_0 + \varepsilon'(\omega), \qquad \varepsilon_0 = \varepsilon(\omega \to \infty), \qquad \varepsilon'(\omega \to \infty) = 0.$$

Then

$$g(t) = \varepsilon_0 \delta(t) + \frac{1}{2\pi} \int_{-\infty}^{+\infty} d\omega \, \exp(i\omega t)\varepsilon'(\omega). \tag{3.16}$$

This equation shows that ε_0 is responsible for the immediate response of the medium, while $\varepsilon'(\omega)$ takes care of the dispersion; see Eq. (1.30) for an example.

By analytical prolongation of $\varepsilon'(\omega)$ in the compex plane $p = \gamma + i\omega$ (see Fig. 3.2), $\varepsilon'(|p| \to \infty) \to 0$, and the integration contour of the integral in Eq. (3.16) can be closed by the circle $R \to \infty$ in the right half-plane for $t < 0$ (Jordan's[4] lemma):

$$g(t < 0) = \frac{1}{2\pi i} \oint_C dp \, \exp(pt)\varepsilon'(p).$$

Causality requires that $g(t < 0) = 0$, so all the singularities of $\varepsilon'(p)$ (as well as those of ε) are confined to the left half p-plane.[5]

---

4. Ernest Pascual Jordan: Hannover (Germany), 1902–1980.
5. Possible singularities located on the imaginary axis should be excluded by means of a small indentation of the integration contour; see Fig. 3.3.

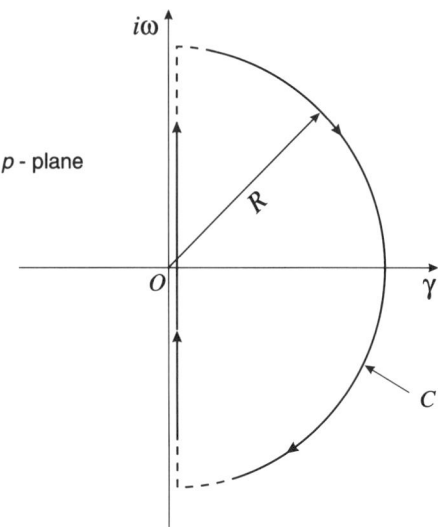

FIGURE 3.2. Integration contour in the complex p-plane relevant to the analytical properties of frequency-domain electromagnetic parameters.

By artificially introducing a pole singularity at $p = i\omega_0$ (see Fig. 3.3), we obtain similarly

$$0 = \oint_C dp \, \frac{\varepsilon'(p)}{p - i\omega_0} = \pi i \varepsilon'(\omega_0) + \int_{-\infty}^{+\infty} d\omega \, \frac{\varepsilon'(\omega)}{\omega - \omega_0}, \qquad (3.17)$$

where the last term is the Cauchy[6] principal value of the integral, $\pi i \varepsilon'(\omega_0)$ is the contribution of the half-circle around the pole, and the contribution along the great circle of the integration path is negligible as $R \to \infty$.

By setting $\varepsilon'(\omega) = \varepsilon_1(\omega) - i\varepsilon_2(\omega)$ and separating real and imaginary parts of Eq. (3.17), we have

$$\varepsilon_1(\omega_0) = \frac{1}{\pi} \int_{-\infty}^{+\infty} d\omega \, \frac{\varepsilon_2(\omega)}{\omega - \omega_0}, \qquad (3.18a)$$

$$\varepsilon_2(\omega_0) = -\frac{1}{\pi} \int_{-\infty}^{+\infty} d\omega \, \frac{\varepsilon_1(\omega)}{\omega - \omega_0}, \qquad (3.18b)$$

which are referred to as Kramers[7]–Kronig[8] relations, and show that in the

---

6. Augustin Louis Cauchy: Paris (France), 1789–Sceaux, 1857.
7. Hendrik Antony Kramers: Rotterdam (Holland), 1894–Leiden, 1952.
8. Ralph Kronig: Dresden (Germany), 1904–1995.

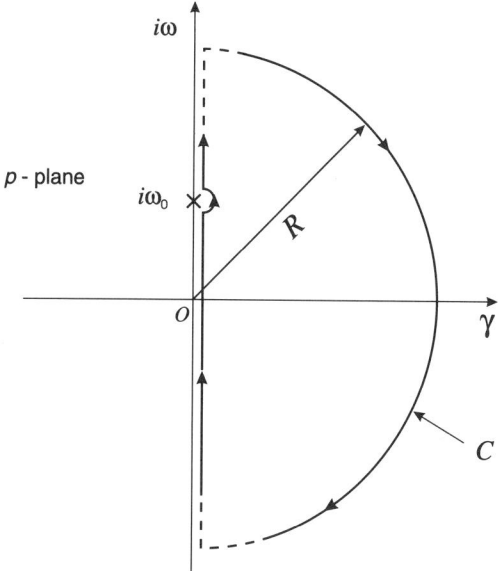

FIGURE 3.3. Derivation of Kramers–Kronig relationships.

frequency domain the real and imaginary parts of medium parameters are related by a Hilbert[9] transform.

We also note that the relationship $\mathbf{D}(\omega) = \varepsilon(\omega)\mathbf{E}(\omega)$ assumes the vector $\mathbf{E}(\omega)$ as input and $\mathbf{D}(\omega)$ as output. However, these roles are immaterial (see Section 1.2) at macroscopic level and can be interchanged. It follows that the singularities of $1/\varepsilon(p)$ are also confined in the left half p-plane, so that $\varepsilon(p)$ is free of poles as well as of zeros in the right half p-plane. If $\varepsilon(p)/\varepsilon_0 = |\varepsilon_r(p)| \exp[-iv(p)]$, then the function

$$\ln \frac{\varepsilon(p)}{\varepsilon_0} = \ln|\varepsilon_r(p)| - iv(p)$$

is singularity-free in the right half p-plane and approaches zero as $|p| \to \infty$. Use of Eqs. (3.18) leads to

$$\ln |\varepsilon_r(\omega_0)| = \frac{1}{\pi} \int_{-\infty}^{+\infty} d\omega \, \frac{v(\omega)}{\omega - \omega_0}, \qquad (3.19a)$$

$$v(\omega_0) = -\frac{1}{\pi} \int_{-\infty}^{+\infty} d\omega \, \frac{\ln |\varepsilon_r(\omega)|}{\omega - \omega_0}. \qquad (3.19b)$$

9. David Hilbert: Königsberg (Germany), 1862–Göttingen, 1943.

Further elaboration of Eq. (3.19b) is possible due to the presence of the slow-varying function $\ln|\varepsilon_r(\omega)|$. Should this be constant, the integral would be equal to zero. Accordingly, its value is essentially determined by the rate of change of $\ln|\varepsilon_r(\omega)|$ around ω_0, and we derive the Bayard[10]–Bode[11] relationship

$$v(\omega_0) \sim \left.\frac{d\,\ln|\varepsilon_r(\omega/\omega_0)|}{d(\omega/\omega_0)}\right|_{\omega_0},$$

where the proportionality constant depends on the assumptions used for the asymptotic evaluation of the integral.

3.2.2. Conductive Media

Consider an isotropic medium characterized in the frequency domain by parameters ε and σ. The source-free Eq. (3.15b) becomes

$$\nabla \times \mathbf{H} = i\omega\left(\varepsilon + \frac{\sigma}{i\omega}\right)\mathbf{E},$$

because $\mathbf{J} = \sigma\mathbf{E}$. Furthermore, Eq. (3.11) yields

$$\nabla \cdot (\sigma\mathbf{E}) + i\omega\rho = 0$$

which, combined with Eq. (3.15c), leads to

$$\nabla \cdot \left[\left(\varepsilon + \frac{\sigma}{i\omega}\right)\mathbf{E}\right] = 0.$$

The conclusion is that a dielectric which exhibits a finite conductivity is equivalent to a dispersive dielectric of permittivity

$$\varepsilon \to \varepsilon - i\frac{\sigma}{\omega}. \tag{3.20}$$

The constitutive relationship becomes $\mathbf{D} = \varepsilon\mathbf{E}$, with $\nabla \cdot \mathbf{D} = 0$ in the absence of sources.

A medium should be regarded as *conductive* or *dielectric* depending on the ratio $\sigma/\omega\varepsilon$, which is frequency-dependent. Metals (see Table 1.2) remain conductors throughout the radio-frequency spectrum.

---

10. Marcel Bayard: 1873–1899–(unk).
11. Hendrik Wade Bode: Madison (Wisconsin, USA), 1905–.

TABLE 3.2. Approximate Values of the Parameters of a Class of Polar Dielectrics[a]

Material	ε_0	ε_p	τ
Monochlorobiphenyl	2.64	1.91	0.39 ns
Dichlorobiphenyl	2.74	3.14	0.79 ns
Trichlorobiphenyl	2.67	2.21	2.14 ns
Tetrachlorobiphenyl	2.68	2.9	6.3 ns
Pentachlorobiphenyl	2.69	2.36	0.14 μs
Exachlorobiphenyl	2.7	2.34	0.20 μs

[a]The constants ε_p and ε_0 are normalized to the free-space permittivity. Note how the relaxation time τ increases with the dimension of the molecule. The temperature is 25 °C.

3.2.3. Polar Dielectrics

Let us consider the polar dielectric of Section 1.2.4, whose dielectric constant in the frequency domain is the FT of Eq. (1.30):

$$\varepsilon(\omega) = \varepsilon_0 + \frac{\varepsilon_p}{1 + i\omega\tau} = \varepsilon_0 + \varepsilon_1(\omega) - i\varepsilon_2(\omega). \quad (3.21)$$

Real $\varepsilon_1(\omega)$ and imaginary $\varepsilon_2(\omega)$ coefficients of the dispersive part of the dielectric constant are depicted in Fig. 3.4. We have also

$$\varepsilon_1^2 + \varepsilon_2^2 = \frac{\varepsilon_p^2}{1 + (\omega\tau)^2} = \varepsilon_p \varepsilon_1.$$

which is the equation of a circle (the Debye[12] circle) in the $(\varepsilon_1, \varepsilon_2)$ plane (see Fig. 3.4).

Values of ε_p and relaxation time τ for some polar dielectrics are given in Table 3.2.

3.2.4. Magnetized Plasma

Let us consider the constitutive relation of the linearized collisionless ($\xi = 0$) plasma of Section 1.2.5, with the addition of the biasing dc magnetic field $\mathbf{h}_0 = H_0 \hat{\mathbf{z}}$. In the frequency domain

$$\begin{cases} \mathbf{J} = -nq\mathbf{V}, \\ i\omega m\mathbf{V} = -q\mathbf{E} - q\mu H_0 \mathbf{V} \times \hat{\mathbf{z}}. \end{cases}$$

12. Petrus Josephus Debye: Mastricht (Holland), 1884–Itacha (NY), 1966.

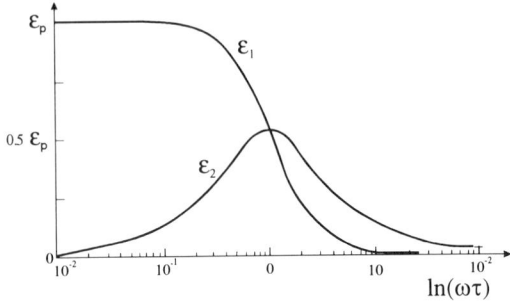

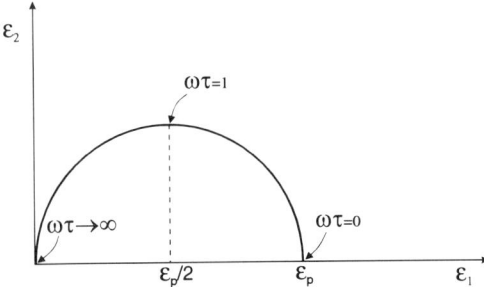

FIGURE 3.4. Dielectric constant of a polar dielectric. Real and imaginary coefficients ε_1 and ε_2, as well as ε_p, are normalized to free-space permittivity.

We make reference to a cartesian coordinate system (x, y, z) and solve the second equation for $\mathbf{V}$, namely, $\mathbf{V} = \mathbf{N} \cdot \mathbf{E}$, where $\mathbf{N}$ is a matrix because $\mathbf{V}$ and $\mathbf{E}$ are not parallel. Substitution into the first equation yields

$$\mathbf{J} = -nq\mathbf{N} \cdot \mathbf{E},$$

which can be inserted in Eq. (3.15b) to give

$$\nabla \times \mathbf{H} = i\omega \left[\varepsilon_0 \mathbf{I} - \frac{nq}{i\omega} \mathbf{N} \right] \cdot \mathbf{E} = i\omega \boldsymbol{\varepsilon} \cdot \mathbf{E},$$

$$\boldsymbol{\varepsilon} = \varepsilon_0 \begin{vmatrix} \varepsilon_1 & i\varepsilon_2 & 0 \\ -i\varepsilon_2 & \varepsilon_1 & 0 \\ 0 & 0 & \varepsilon_3 \end{vmatrix}, \qquad (3.22)$$

with

$$\varepsilon_1 = 1 - \frac{\omega_p^2}{\omega^2 - \omega_c^2}, \quad \varepsilon_2 = \frac{\omega_c \omega_p^2/\omega}{\omega^2 - \omega_c^2}, \quad \varepsilon_3 = 1 - \frac{\omega_p^2}{\omega^2}. \qquad (3.23)$$

In Eq. (3.23)

$$\omega_c = q\mu_0 H_0/m$$

is the *cyclotron angular frequency* (1.26) of the electrons in the magnetic field, and

$$\omega_p^2 = \frac{nq^2}{m\varepsilon_0} \tag{3.24}$$

is the *plasma angular frequency*.

The Earth is surrounded by a plasma region due to ionization of air molecules by sunrays. This region, called the ionosphere, extends in several layers from about 90 to 300 km above the Earth's surface. The plasma is magnetized by the Earth's magnetic field. Parameters relevant to the ionosphere are presented in Table 3.3.

The magnetized plasma is equivalent to a dispersive anisotropic dielectric. The matrix $\boldsymbol{\varepsilon}$ is hermitian, $\boldsymbol{\varepsilon} = \boldsymbol{\tilde{\varepsilon}}^*$; this type of medium is called *gyrotropic*. When $H_0 = 0$, then $\omega_c = 0$, $\varepsilon_1 = \varepsilon_3$, $\varepsilon_2 = 0$, and the plasma behaves as an isotropic dispersive dielectric with relative dielectric constant

$$\varepsilon_r = 1 - \frac{\omega_p^2}{\omega^2}. \tag{3.25}$$

We have

$$\nabla \times \mathbf{H} = i\omega\varepsilon_0\varepsilon_r\mathbf{E} = i\omega\varepsilon_0\mathbf{E} - i\frac{\omega_p^2\varepsilon_0}{\omega}\mathbf{E},$$

which models also the plasma in terms of an equivalent conductivity $\omega_p^2\varepsilon_0/i\omega$, the FT of Eq. (1.32) with $\xi = 0$.

When $H_0 \to \infty$, then $\omega_c \to \infty$, $\varepsilon_1 \to 1$, $\varepsilon_2 \to 0$, and the plasma becomes uniaxial (see Section 1.2.3).

TABLE 3.3. Plasma Parameters in the Ionosphere

Parameters	Values
Electron charge q	1.6×10^{-19} coulomb
Electron mass m	9.1×10^{-31} kg
Electron density n	$10^7 - 10^{12}$ m^{-3}
Biasing magnetic field H_0	40 A/m
Cyclotron frequency $f_c = \omega_c/2\pi$	1.4 MHz
Plasma frequency $f_p = \omega_p/2\pi$	28 KHz + 9 MHz

3.2.5. Plane-Wave Propagation in Dispersive Media

Plane-wave propagation in isotropic homogeneous dispersive media can be studied following same approach as in Section 2.1. The only difference is that we use Eq. (3.15) with $\mathbf{K}_0 = \mathbf{J}_0 = 0$ instead of Eqs (2.1). Setting

$$\mathbf{E} = E(z,\omega)\hat{\mathbf{x}} \quad \text{and} \quad \mathbf{H} = H(z,\omega)\hat{\mathbf{y}},$$

we easily obtain

$$\begin{cases} \dfrac{dE}{dz} = -i\omega\mu H, \\ \dfrac{dH}{dz} = -i\omega\varepsilon E, \end{cases}$$

whose solution,

$$E(z,\omega) = E^+(\omega)\exp(-ikz) + E^-(\omega)\exp(ikz), \quad (3.26a)$$

$$\zeta H(z,\omega) = E^+(\omega)\exp(-ikz) - E^-(\omega)\exp(ikz), \quad (3.26b)$$

is obtained by taking the derivative of the first equation with respect to z, substituting dH/dz from the second, and then solving the differential equation

$$\frac{d^2E}{dz^2} + k^2 E = 0.$$

In Eqs. (3.26)

$$k = \omega\sqrt{\varepsilon\mu} = \beta - i\alpha \quad (3.27)$$

is the (complex) *propagation constant*, and

$$\zeta = \sqrt{\mu/\varepsilon} \quad (3.28)$$

is the *intrinsic impedance* of the medium.

For a nondispersive medium, ε and μ are frequency-independent real constants. If

$$k = \beta = \omega/c \quad \text{and} \quad c = 1/\sqrt{\varepsilon\mu},$$

we have from Eq. (3.26a)

$$e(z, t) = \frac{1}{2\pi} \int_{-\infty}^{+\infty} d\omega \, \exp(i\omega t) E(z, \omega)$$
$$= e^+(t - z/c) + e^-(t + z/c),$$

where

$$e^\pm(t) = \frac{1}{2\pi} \int_{-\infty}^{+\infty} d\omega \, \exp(i\omega t) \, E^\pm(\omega). \tag{3.29a}$$

We recognize that the $E^+(\omega)$ solution corresponds to a wave propagating along the positive-z sense (*progressing wave*), while the other term $E^-(\omega)$ corresponds to a wave moving in the opposite sense along the z-axis (*regressive wave*).

In an unbounded medium only one wave, either progressive or regressive, is present, depending on the assumed position of the sources. It these are located in the half-space $z \leq 0$, we have for $z > 0$ only the progressive wave, and the value of $E^+(\omega)$ is determined by the boundary conditions, i.e., the values of the field at $z = 0$:

$$E^+(\omega) = \int_{-\infty}^{+\infty} d\omega \, \exp(-i\omega t) \, e^+(t). \tag{3.29b}$$

For an inhomogeneous stratified medium, both types of wave are present (see Section 3.2.8).

In the dispersive case the frequency dependence of the propagation constant becomes more involved than the simple linear one, and the intrinsic impedance becomes frequency-dependent, too. Assume that only the progressive wave is present. Then

$$e(z, t) = \frac{1}{2\pi} \int_{-\infty}^{+\infty} d\omega \, \exp(i\omega t) \, E^+(\omega) \, \exp(-ikz). \tag{3.30}$$

Let

$$\varepsilon_r(\omega) = \frac{\varepsilon(\omega)}{\varepsilon_0} \quad \text{and} \quad \mu_r(\omega) = \frac{\mu(\omega)}{\mu_0},$$

where ε_0 and μ_0 are the values of the electromagnetic parameters for $|\omega| \to \infty$. Note that ε_0 and μ_0 are real constants, due to a general property of materials to become nondispersive in the high-frequency range (at least at radio frequencies). With the further definitions

$$c = \frac{1}{\sqrt{\varepsilon_0 \mu_0}}, \qquad n^2(\omega) = \varepsilon_r(\omega)\mu_r(\omega), \qquad t^* = t - \frac{z}{c},$$

Eq. (3.30) becomes

$$e(z, t) = \frac{1}{2\pi} \int_{-\infty}^{+\infty} d\omega \, \exp(i\omega t^*) E^+(\omega) \, \exp\left[-i\omega(n-1)\frac{z}{c}\right]$$
$$= e_1(t - z/c) - e_2(z, t), \qquad (3.31)$$

where

$$e_1(t - z/c) = \frac{1}{2\pi} \int_{-\infty}^{+\infty} d\omega \, \exp(i\omega t^*) E^+(\omega), \qquad (3.32a)$$

$$e_2(z, t) = \frac{1}{2\pi} \int_{-\infty}^{+\infty} d\omega \, \exp(i\omega t^*) E^+(\omega) \left\{1 - \exp\left[-i\omega(n-1)\frac{z}{c}\right]\right\}. \qquad (3.32b)$$

In Eq. (3.31), $n(\omega)$ is the (relative) *refractive index* of the medium and t^* the retarded time; see Eq. (2.42).

Equation (3.31) shows that in a dispersive medium the field can be formally split into the two parts e_1 and e_2. The former corresponds to the nondispersive component and propagates with speed c. The latter accounts for the medium dispersion and is the convolution at the retarded time between the same $e_1(t^*)$ and the function

$$f(z, t) = \frac{1}{2\pi} \int_{-\infty}^{+\infty} d\omega \, \exp(i\omega t^*) \left\{1 - \exp\left[-i\omega(n-1)\frac{z}{c}\right]\right\}. \qquad (3.33)$$

The unit impulse response of the medium, i.e., the response to the cause $e(t) \to \delta(t)$, is given by $\delta(t^*) - f(z, t)$. The splitting described by Eq. (3.32) may be convenient to highlight the effect of the dispersion.

In order to study the effect of the dispersion, it may be convenient to refer to the complex frequency $p = \gamma + i\omega$ as in Section 3.2.1, thus considering the analytical continuation in the complex p-plane of the integrand in Eq. (3.33). Let $\varepsilon_r(p)$, $\mu_r(p)$, and $n(p)$ be the analytical continuation of $\varepsilon_r(\omega)$, $\mu_r(\omega)$, and

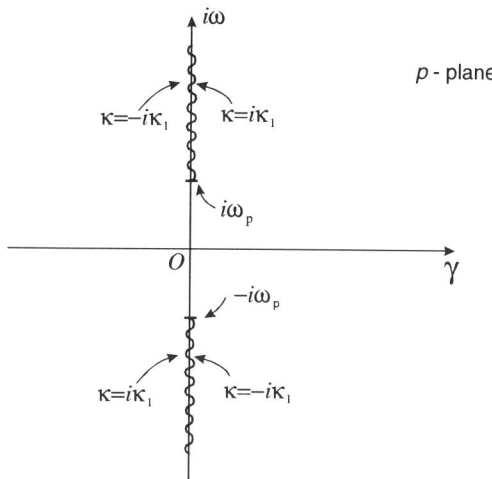

FIGURE 3.5. Branch cuts in the complex p-plane for an isotropic lossless plasma. Values of κ along the two sides of the branches are reported: $\kappa_1 = \sqrt{\omega^2 - \omega_p^2}$, $|\omega| \geqslant \omega_p$, with the positive sign for the square root.

$n(\omega)$, respectively. Equation (3.33) then assumes the form

$$f(z,t) = \frac{1}{2\pi i} \int_{-i\infty}^{+i\infty} dp \, \exp(pt^*) \left\{ 1 - exp\left[-p(n-1)\frac{z}{c} \right] \right\}, \qquad (3.34)$$

which corresponds to the use of Laplace[13] instead of Fourier transforms.

The presence of the square roots in expressions (3.27) for k and (3.28) for ζ calls for appropriate cuts in the p-plane to render these functions single-valued. It is convenient to make the cut along the (real) negative values of $\kappa^2 = (ikc)^2 = (pnc)^2$, i.e.,

$$\kappa^2 = (\alpha + i\beta)^2 c^2 = -\xi^2, \qquad 0 \leqslant \xi \leqslant \infty,$$

so that $\alpha \geqslant 0$ over all the (used) Riemann[14] sheet of the p-plane. For instance, the nonmagnetized ($H_0 = 0$) lossless plasma of Section 3.2.4 is characterized by the relative dielectric constant

$$\varepsilon_r(p) = 1 + \frac{\omega_p^2}{p^2}$$

and the cuts are depicted in Fig. 3.5. Similarly, the cuts appropriate to the

---

13. Pierre Simon de Laplace: Beaumont en Auge (France), 1749–Paris (France), 1827.
14. Bernhard Riemann: Reselenz (Hanover, Germany), 1826–Selasca (Novara, Italy), 1866.

conductive medium of Section 3.2.2 with

$$\varepsilon_r(p) = 1 + \frac{\sigma}{p\varepsilon_0}$$

are depicted in Fig. 3.6.

It is noted that κ^2 exhibits neither poles nor zeros in the right half p-plane (see section 3.2.1): κ is free of singularities (poles and branch points) there. When $|p| \to \infty$, the integrand in Eq. (3.34) approaches zero and the integration contour can be closed for $t^* < 0$ with a circle of radius $R \to \infty$ in the right half p-plane. Use of Cauchy's theorem and lack of singularities there shows that

$$f(z, t^* < 0) = 0,$$

and causality is fulfilled.

For $t^* > 0$, the integration coutour can be closed to the left half p-plane for computing either $f(z, t)$ from Eq. (3.33), or computing directly $e(z, t)$ via the analytical continuation of Eq. (3.31). The field $e(z, t)$ is represented by the singularity contributions: poles, usually present in the (analytical continuation of the) spectrum $E^*(\omega)$, and branch cuts appearing in the integrand of Eq. (3.33), accounting for the medium dispersion. Examples of plane-wave propagation are given in Section 3.2.6 and 3.2.7.

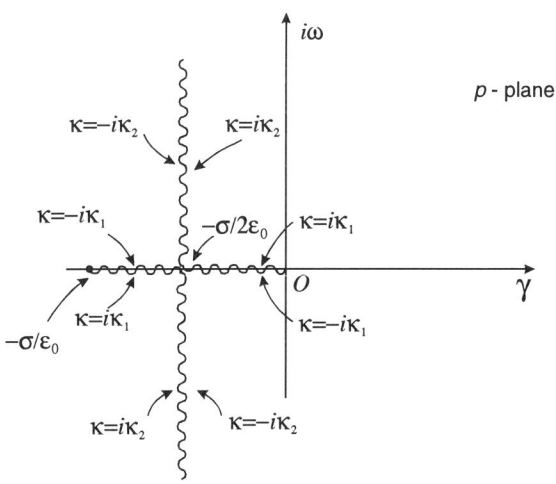

FIGURE 3.6. Branch cuts in the complex p-plane for a conductive medium. Values of κ along the two sides of the branches are reported: $\kappa_1 = \sqrt{-\gamma(\gamma + \sigma/\varepsilon_0)}$, $-\sigma/\varepsilon_0 < \gamma < 0$; $\kappa_2 = \sqrt{(\sigma/2\varepsilon_0)^2 + \omega^2}$, with the positive sign for the square root.

SPECTRAL DOMAINS

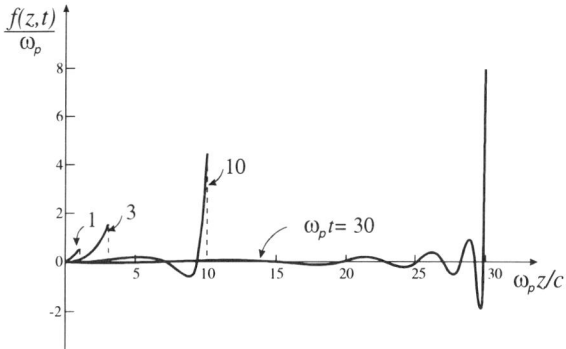

FIGURE 3.7. Dispersive component of the unit response for plane-wave propagation in a collisionless plasma.

3.2.6. Plane-Wave Propagation in a Plasma Medium

In some (a few) cases the transform $f(t, z)$ is known. For the collisionless isotropic plasma we have

$$f(z, t) = \frac{1}{2\pi i} \int_{-i\infty}^{+i\infty} dp \, \exp(pt^*) \, \{1 - \exp[-\sqrt{p^2 + \omega_p^2} z/c + pz/c]\}.$$

We use the result (C.2) with the substitution

$$b = z/c, \quad a = \omega_p, \quad t \Rightarrow t^*,$$

to derive

$$f(z, t) = \omega_p^2 \frac{z}{c} \frac{J_1(\omega_p \sqrt{t^*(t^* + 2z/c)})}{\omega_p \sqrt{t^*(t^* + 2z/c)}} U(t^*), \tag{3.35}$$

$J_1(\cdot)$ being the Bessel[15] function. It is immediately seen that the solution (3.35) no longer depends on t^* alone, but on

$$t^*(t^* + 2z/c) = t^2 - (z/c)^2,$$

so that the wave changes its shape as it propagates. A wave speed independent of wave location cannot be defined.

Eq. (3.35), divided by ω_p, is presented in Fig. 3.7 as a function of $\omega_p z/c$ for several values of $\omega_p t$. The wave becomes increasingly peaked for increasing

---

15. Friederich Wilhelm Bessel: Minden (Germany), 1784–Königsberg, 1846.

values of time (and distance), thus resembling a Dirac function. As a matter of fact,

$$\omega_p \frac{z}{c} \frac{J_1(\omega_p\sqrt{t^2 - (z/c)^2})}{\sqrt{t^2 - (z/c)^2}} = \frac{\partial}{\partial(z/c)} J_0(\omega_p\sqrt{t^2 - (z/c)^2}), \qquad t > z/c;$$

see Eq. (D.7). The area spanned by the function in the interval $0 \leqslant z/c < t$ is given by

$$1 - J_0(\omega_p t) \approx 1, \qquad \omega_p t \gg 1;$$

see Eqs. (D.17a) and (D.20a). Accordingly, the field $e_2(z, t)$ given by Eq. (3.32b) tends to cancel out the field $e_1(z, t)$ given by Eq. (3.32a) on the wavefront $\omega_p t^* = 0$; for $\omega_p t^* \sim 0$, the total signal varies rapidly, is small, and propagates with the speed of light (*precursors*). On the contrary, when $t^2 \gg (z/c)^2$, use of the asymptotic expansion (D.20a) shows that

$$f(z, t) \sim \omega_p \frac{z}{ct} \sqrt{\frac{2}{\pi \omega_p t}} \sin\left(\omega_p t - \frac{\pi}{4}\right);$$

the solution ceases to exhibit a wave-like behavior, the signal is again small (*tail*), and it consists of oscillations at plasma frequency ω_p. The intermediate values of $\omega_p t^*$ yield the most important part of the signal, which propagates with a velocity whose value is a matter of definition and is ambiguous at a certain extent. As a matter of fact, the displacement of each point of the waveform is not only due to its propagation, but also to its shape change.

3.2.7. Plane-Wave Propagation in a Conductive Medium

For the conductive medium Eq. (3.33) becomes

$$f(z, t) = \frac{1}{2\pi i} \int_{-i\infty}^{+i\infty} dp \, \exp(pt^*) \left\{ 1 - \exp\left[-\sqrt{p^2 + p\sigma/\varepsilon_0} \frac{z}{c} + p\frac{z}{c}\right]\right\}.$$

We deform the original integration contour along the imaginary $p = i\omega$ axis onto the new integration contour (solid line) depicted in Fig. 3.8 (for the branch cuts, see Fig. 3.6). Integration of the first term in parentheses equals $\delta(t^*)$, and we have

$$f(z, t) = \delta(t^*) - \frac{1}{\pi} \int_0^{\sigma/2\varepsilon_0} d\gamma \, \exp(-\gamma t) \sin\left(\sqrt{\gamma\left(\frac{\sigma}{\varepsilon_0} - \gamma\right)} \frac{z}{c}\right)$$

$$- \frac{1}{\pi} \exp(-\sigma t/2\varepsilon_0) \int_0^{+\infty} d\omega \, \cos(\omega t - \sqrt{\omega^2 + (\sigma/2\varepsilon_0)^2} z/c), \qquad t^* \geqslant 0.$$

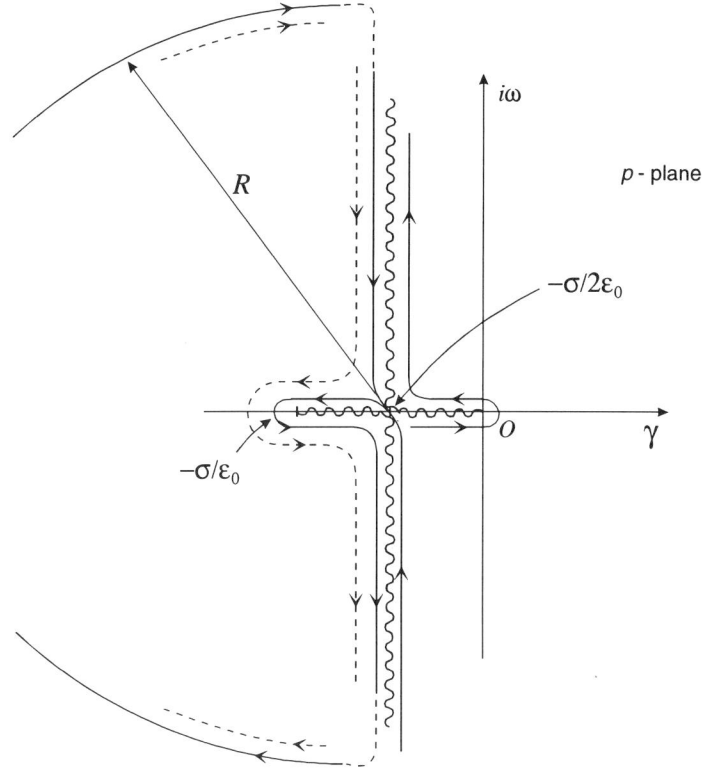

FIGURE 3.8. Deformed integration contours relevant to plane-wave propagation in a conductive medium. Solid line: new integration contour obtained by deforming the original one. Dotted line: part of the deformed integration contour which provides no contribution to the integral.

The Dirac function $\delta(t^*)$ just compensates the analogous term appearing in the response e_1; see Eq. (3.31). The overall unit response, $g(t) = \delta(t^*) - f(z, t)$, depends on the parameter

$$\tau = \frac{2\varepsilon_0}{\sigma},$$

and can be conveniently cast in the form

$$g(z, t) = \frac{1}{\pi t} \int_0^{t/\tau} d\gamma \, \exp(-\gamma) \sin\left(\sqrt{\gamma\left(2\frac{t}{\tau} - \gamma\right)} \frac{z}{ct}\right)$$
$$+ \frac{1}{\pi t} \exp\left(-\frac{t}{\tau}\right) \int_0^{+\infty} d\omega \, \cos\left(\omega - \sqrt{\omega^2 + (t/\tau)^2}\, \frac{z}{ct}\right), \quad t^* \geqslant 0,$$

with the substitution of variables $\omega \to \omega t$ and $\gamma \to \gamma t$.

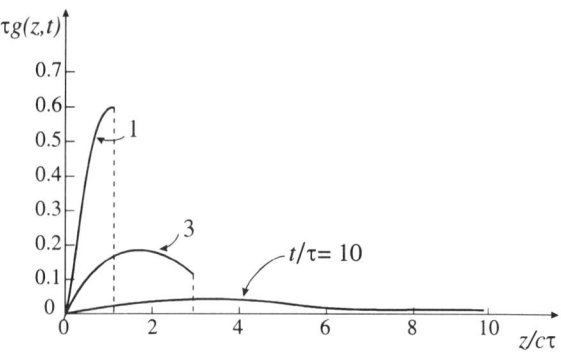

FIGURE 3.9. Unit response (times $\tau = 2\varepsilon_0/\sigma$) for plane-wave propagation in a lossy medium for $t/\tau \gg 1$. The limiting case $t/\tau = 1$ is also included.

When $t/\tau \ll 1$, the first integral is negligible compared to the second. We put $\omega^2 + (t/\tau)^2 \approx \omega^2$, $\omega \to \omega t$ in the secont, thus recovering the result

$$g(z,t) \approx \frac{1}{2\pi} \exp(-t/\tau) \int_0^{+\infty} d\omega [\exp(i\omega t^*) + \exp(-i\omega t^*)]$$

$$= \frac{1}{2\pi} \exp(-t/\tau) \int_{-\infty}^{+\infty} d\omega \, \exp(i\omega t^*) = \exp(-t/\tau)\delta(t^*), \qquad t/\tau \ll 1$$

of Section 2.1.1. The wave propagates with speed c and is amplitude-scaled as time elapses. On the other hand, when $t/\tau \gg 1$ the second integral is negligible. In order to compute the first integral, we first set $2t/\tau - \gamma \approx 2t/\tau$ and extend to infinity its upper integration limit. Then, we make the change of variables $\sqrt{\gamma} \to u$, integrate by parts, and employ the result (C.3) with the values $a = 1$ and $b = \sqrt{2t/\tau}(z/ct)$ to obtain

$$g(z,t) \approx \sqrt{\frac{1}{2\pi} \frac{1}{\tau} \frac{z}{c\tau}} \left(\frac{\tau}{t}\right)^{3/2} \exp\left[-\frac{1}{2}\left(\frac{z}{c\tau}\right)^2 \frac{\tau}{t}\right], \qquad t > z/c, \quad t/\tau \gg 1.$$

The wave-like characteristic of the signal is lost, substituted by a diffusion-like behavior (see Fig. 3.9).

3.2.8. Plane Wave at a (Space) Discontinuity Boundary

We consider the case of two media separated by the plane surface $z = 0$, as in Section 2.2. The medium filling the half-space $z > 0$ is dispersive, and characterized in the frequency domain by ω-dependent electromagnetic parameters $\varepsilon_2\varepsilon_r(\omega)$ and $\mu_2\mu_r(\omega)$; the other medium is nondispersive and characterized by constant (real) parameters ε_1 and μ_1.

We proceed as in Section 2.2, assuming that the electric field is linearly polarized along $\hat{x}$, the magnetic field is oriented along $\hat{y}$, and the sources of the field are remotely located in the half-space $z < 0$. We have, for $z \leq 0$,

$$\begin{cases} E_1(z,\omega) = E_1^+(\omega)\exp(-ik_1 z) + E_1^-(\omega)\exp(ik_1 z), \\ \zeta_1 H_1(z,\omega) = E_1^+(\omega)\exp(-ik_1 z) - E_1^-(\omega)\exp(ik_1 z), \end{cases}$$

with

$$k_1 = \omega\sqrt{\varepsilon_1\mu_1} \quad \text{and} \quad \zeta_1 = \sqrt{\mu_1/\varepsilon_1};$$

and for $z \geq 0$,

$$\begin{cases} E_2(z,\omega) = E_2^+(\omega)\exp(-ik_2 z), \\ \zeta_2 H_2(z,\omega) = E_2^+(\omega)\exp(-ik_2 z), \end{cases}$$

with

$$k_2 = \omega\sqrt{\varepsilon_2\varepsilon_r(\omega)\mu_2\mu_r(\omega)} \quad \text{and} \quad \zeta_2 = \sqrt{\mu_2\mu_r(\omega)/\varepsilon_2\varepsilon_r(\omega)}.$$

We define (electric field) reflection

$$\Gamma(\omega) = \frac{E_1^-(\omega)}{E_1^+(\omega)} \tag{3.36}$$

and transmission

$$T(\omega) = \frac{E_2^+(\omega)}{E_1^+(\omega)} \tag{3.37}$$

coefficients, which are the frequency-domain counterparts of the analogous coefficients (2.13) for time-dependent fields.

The field at $z = 0$ is everywhere tangent to the discontinuity plane. By enforcing its continuity we obtain

$$\Gamma(\omega) = \frac{\zeta_2 - \zeta_1}{\zeta_2 + \zeta_1} \quad \text{and} \quad T(\omega) = \frac{2\zeta_2}{\zeta_2 + \zeta_1}. \tag{3.38}$$

which are formally identical to Eqs. (2.13) but for their frequency dependence.

We use Eq. (3.29b) for $E^+(\omega)$ and obtain in time domain

$$e_1^-(z,t) = \frac{1}{2\pi}\int_{-\infty}^{+\infty} dt' f(z, t-t') e^+(0, t'), \qquad z < 0, \qquad (3.39)$$

$$e_2^+(z,t) = \frac{1}{2\pi}\int_{-\infty}^{+\infty} dt' g(z, t-t') e^+(0, t'), \qquad z > 0, \qquad (3.40)$$

where

$$f(t,z) = \int_{-\infty}^{+\infty} d\omega \, \exp(i\omega t - i k_1 z) \Gamma(\omega), \qquad z < 0, \qquad (3.41)$$

$$g(t,z) = \int_{-\infty}^{+\infty} d\omega \, \exp(i\omega t - i k_2 z) T(\omega). \qquad z > 0. \qquad (3.42)$$

For the collisionless isotropic plasma

$$\Gamma(\omega) = \frac{1-\sqrt{\varepsilon_r}}{1+\sqrt{\varepsilon_r}} = \frac{1+\varepsilon_r - 2\sqrt{\varepsilon_r}}{1-\varepsilon_r} = \frac{2\omega^2 - \omega_p^2 - 2\omega\sqrt{\omega^2 - \omega_p^2}}{\omega_p^2}$$

and

$$\Gamma(p) = \frac{2p}{\omega_p^2}(\sqrt{p^2 + \omega_p^2} - p) - 1.$$

We analytically continue Eq. (3.41) in the complex p-plane,

$$f(t,z) = \frac{1}{2\pi i}\int_{-i\infty}^{+i\infty} dp \, \exp(pt^*) \Gamma(p)$$

$$= \frac{2}{\omega_p^2}\frac{\partial}{\partial t^*}\frac{1}{2\pi i}\left\{\int_{-i\infty}^{+i\infty} dp \, \exp(pt^*)[\sqrt{p^2+\omega_p^2} - p]\right\} - \delta(t^*),$$

with $t^* = t - z/c$, $c = 1/\sqrt{\varepsilon_1\mu_1}$, $\partial/\partial t^* \to p$, and make use of the result (C.4) with $a = \omega_p$ and $t \to t^*$ to yield

$$f(t,z) = \frac{1}{\omega_p}\frac{\partial}{\partial t^*}\left\{\frac{2J_1(\omega_p t^*)}{\omega_p t^*} U(t^*)\right\} - \delta(t^*). \qquad (3.43)$$

Note that $2J_1(\omega_p t^*)/\omega_p t^* \to 1$ as $\omega_p t^* \to 0$, as follows from Eq. (D.17b), and that $-\delta(t^*)$ exactly compensates for the derivative discontinuity as $\omega_p t^* \to 0$, so that $f(t,z)$ is zero for $t^* = 0$. This function is depicted in Fig. 3.10.

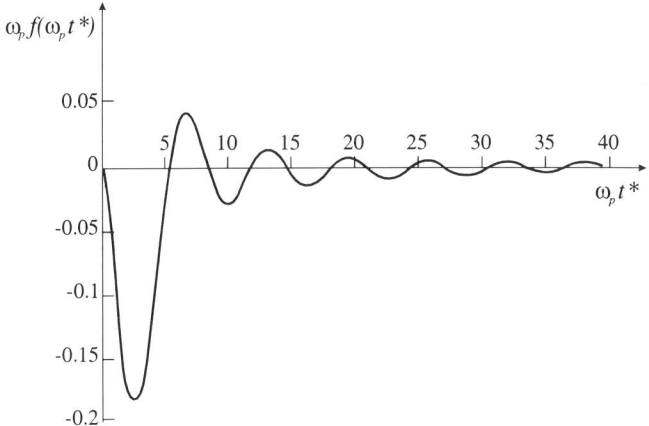

FIGURE 3.10. Reflection of an impulsive plane wave from a half-space isotropic plasma.

The transmitted field can be computed similarly, noting that

$$T(p) = \frac{2}{\omega_p^2} p(\sqrt{p^2 + \omega_p^2} - p)$$

and using the functional identity

$$p(\sqrt{p^2 + \omega_p^2} - p) \to -\frac{\partial}{\partial t}\left(\frac{\partial}{\partial t} + \frac{\partial}{\partial(z/c)}\right)$$

in the analytical continuation of Eq. (3.42).

3.2.9. Radiation in Dispersive Media

Let us consider a homogeneous isotropic time-invariant medium and a prescribed source distribution $\mathbf{j}_0(\mathbf{r}, t)$. From Maxwell equations (3.15a–b) we have

$$\nabla \times \nabla \times \mathbf{E} = -i\omega\mu(i\omega\varepsilon\mathbf{E} + \mathbf{J}_0).$$

The use of vector identity (A.17) and Eq. (3.15c) yields

$$\frac{1}{\varepsilon}\nabla\rho_0 - \nabla^2\mathbf{E} = k^2\mathbf{E} - i\omega\mu\mathbf{J}_0,$$

which is modified to the form

$$\nabla^2 \mathbf{E} + k^2 \mathbf{E} = i\omega\mu \left[\mathbf{J}_0 + \frac{1}{k^2} \nabla\nabla \cdot \mathbf{J}_0 \right]$$

by means of Eq. (3.11). We proceed as in Section 2.3 and introduce the operator

$$\mathbf{M} = i\omega\mu \left[\mathbf{I} + \frac{\nabla\nabla}{k^2} \right], \quad k^2 = \omega^2 \varepsilon \mu,$$

thus obtaining

$$\nabla^2 \mathbf{E} + k^2 \mathbf{E} = \mathbf{M} \cdot \mathbf{J}_0. \tag{3.44}$$

Equation (3.44) relates electric field $\mathbf{E}(\mathbf{r}, \omega)$ and source $\mathbf{J}_0(\mathbf{r}, \omega)$ spectra. Note that $\mathbf{M}$ is the frequency-domain counterpart of the operator $\mathbf{m}$ appearing in Eq. (2.32), as follows by computing the FT of $\mathbf{m} \cdot \mathbf{p}_0/\varepsilon$ with the aid of the FT of Eq. (2.29).

If

$$\mathbf{J}_0(\mathbf{r}, \omega) = \iiint_{-\infty}^{+\infty} d\mathbf{V}' \delta(\mathbf{r} - \mathbf{r}') \mathbf{I} \cdot \mathbf{J}_0(\mathbf{r}', \omega) \tag{3.45}$$

and

$$\mathbf{E}(\mathbf{r}, \omega) = \iiint_{-\infty}^{+\infty} d\mathbf{V}' \mathbf{G}(\mathbf{r} - \mathbf{r}', \omega) \cdot \mathbf{J}_0(\mathbf{r}', \omega), \tag{3.46}$$

then

$$\nabla^2 \mathbf{G}(\mathbf{r} - \mathbf{r}', \omega) + k^2 \mathbf{G}(\mathbf{r} - \mathbf{r}', \omega) = \mathbf{M}\delta(\mathbf{r} - \mathbf{r}') \tag{3.47}$$

upon substituting Eqs. (3.45) and (3.46) into Eq. (3.44). This equation must be valid for any source spectra $\mathbf{J}_0(\mathbf{r}', \omega)$.

The matrix $\mathbf{G}(\mathbf{r} - \mathbf{r}', \omega)$ is referred to as the (frequency-domain) *dyadic Green's function*. It is defined by Eq. (3.47) together with appropriate boundary conditions and relates (electric) source and field spectra via Eq. (3.46). Proceeding again as in Section 2.3, we set

$$\mathbf{G}(\mathbf{r} - \mathbf{r}', \omega) = -\mathbf{M} \cdot \mathbf{I} G(\mathbf{r} - \mathbf{r}', \omega), \tag{3.48}$$

thus deriving

$$\nabla^2 G(\mathbf{r} - \mathbf{r}', \omega) + k^2 G(\mathbf{r} - \mathbf{r}', \omega) = -\delta(\mathbf{r} - \mathbf{r}') \quad (3.49)$$

upon substitution in Eq. (3.47). The function $G(\mathbf{r} - \mathbf{r}', \omega)$ defined by Eq. (3.49) is the (frequency-domain) *scalar Green's function* and is the frequency-domain counterpart of the time-domain Green's function (2.48). Its knowledge allows computation of the field spectra via Eqs. (3.48), (3.46), and (3.15a).

3.2.10. Scalar Green's Function Evaluation

The scalar Green's function in an unbounded homogeneous isotropic medium is the solution of Eq. (3.49) with appropriate radiation conditions at infinity. In a spherical system of coordinates centered at $\mathbf{r}'$,

$$\nabla^2 G(r, \omega) + k^2 G(r, \omega) = -\delta(r), \quad (3.50)$$

i.e.,

$$G(r, \omega) = A \frac{\exp(-ikr)}{r} + B \frac{\exp(ikr)}{r}, \quad (3.51)$$

following the same derivation as in Section 2.3.1.

It was shown in Section 3.2.5 that $k = \beta - i\alpha$; see Eq. (3.27). It is clear that in lossy media ($\alpha > 0$) the exponential term $\exp(ikr)$ diverges for $r \to \infty$, so that we set the constant B equal to zero in the solution (3.51). We can argue that this is also true for lossless media, but it is instructive to also consider this case in some detail.

First, we note that the source of Eq. (3.50), $-\delta(r)$, exhibits a constant ω-spectrum, so that its time-dependent counterpart is $-2\pi\delta(t)\delta(r)$ (see Table 3.1). A causal Green's function $G(r, \omega)$ requires that its FT is identically equal to zero for $t^* = t - r/c < 0$. Then, we proceed as in Section 3.2.5 and separate the nondispersive part of the response, $\delta(t^*)$, from the dispersive part. The latter is now of interest. For the first term in the solution (3.51) we are lead to study the Fourier integral:

$$\frac{1}{2\pi} \int_{-\infty}^{+\infty} d\omega \exp(i\omega t^*) \left\{ 1 - \exp\left[-i\omega \frac{r}{c}(n-1) \right] \right\}$$

$$= \frac{1}{2\pi} \int_{-\infty}^{+\infty} d\omega \exp(i\omega t^*) F(r, \omega), \quad n = \sqrt{\varepsilon_r(\omega)\mu_r(\omega)}.$$

In the right half complex p-plane the function $F(r, p)$ is analytical; it approaches zero as $|p| \to \infty$, and lack of poles and zeros for $n^2(p)$ (see Section

3.2.1) implies lack of poles and branch points for $n(p)$. The FT integration contour can be closed to the right for $t^* < 0$ (see Section 3.2.1) and causality is verified.

For the second term in the solution (3.51) we have analogously

$$\frac{1}{2\pi}\int_{-\infty}^{+\infty} d\omega \, \exp\left[i\omega\left(t + \frac{r}{c}\right)\right]\left\{1 - \exp\left[i\omega\frac{r}{c}(n-1)\right]\right\}.$$

For $t + r/c > 0$ the integration contour can be closed to the left, thus encircling singularities of the integrand. This solution is not causal, because a response may appear for $t < -r/c < 0$.

We conclude that a causal bounded solution involves setting $B = 0$ in the solution (3.51). We apply the same procedure used in connection with Eq. (2.39) to obtain

$$G(r, \omega) = \frac{\exp(-ikr)}{4\pi r}. \tag{3.52}$$

The spectral Green's function (3.52) can be Fourier-transformed to provide its time-dependent counterpart. We must multiply Eq. (3.52) by 2π, the spectral amplitude of the time-dependent Green's function, to yield

$$g(r, t^*) = \frac{1}{4\pi r}\int_{-\infty}^{+\infty} d\omega \, \exp(i\omega t^*) \exp\left[-i\frac{\omega}{c}r(n-1)\right]. \tag{3.53}$$

For a nondispersive medium $n = 1$ and $g(r, t) = \delta(t^*)/4\pi r$, see also Eq. (2.49). For a dispersive medium the analysis of this section shows that the signal cannot propagate with a speed larger than $c = 1/\sqrt{\varepsilon_0 \mu_0}$.

3.3. THE WAVENUMBER DOMAIN

We consider Maxwell equations (1.1) with prescribed (electric) source distributions $j_0(\mathbf{r}, t)$ and $\rho_0(\mathbf{r}, t)$, which are now expanded in a triple Fourier integral:

$$\mathbf{j}_0(\mathbf{r}, t) = \frac{1}{(2\pi)^3}\int_{-\infty}^{+\infty} d\mathbf{k} \, \exp(i\mathbf{k}\cdot\mathbf{r})\mathbf{J}_0(\mathbf{k}, t), \tag{3.54}$$

... ...

SPECTRAL DOMAINS

where

$$\mathbf{k} = u\hat{\mathbf{x}} + v\hat{\mathbf{y}} + w\hat{\mathbf{z}} \qquad (3.55)$$

is the (vector) wavenumber. We similarly expand fields and inductions,

$$\mathbf{e}(\mathbf{r}, t) = \frac{1}{(2\pi)^3} \int_{-\infty}^{+\infty} d\mathbf{k} \, \exp(i\mathbf{k} \cdot \mathbf{r}) \mathbf{E}(\mathbf{k}, t), \qquad (3.56)$$

$$\cdots \cdots$$

and substitute in Maxwell equation (1.1). By equating the integrals we obtain

$$\begin{cases} -\dfrac{d\mathbf{B}(\mathbf{k}, t)}{dt} = i\mathbf{k} \times \mathbf{E}(\mathbf{k}, t), \\[6pt] -\dfrac{d\mathbf{D}(\mathbf{k}, t)}{dt} = i\mathbf{H}(\mathbf{k}, t) \times \mathbf{k} + \mathbf{J}(\mathbf{k}, t) + \mathbf{J}_0(\mathbf{k}, t), \\[6pt] i\mathbf{k} \cdot \mathbf{D}(\mathbf{k}, t) = \rho(\mathbf{k}, t) + \rho_0(\mathbf{k}, t), \\[6pt] i\mathbf{k} \cdot \mathbf{B}(k, t) = 0. \end{cases}$$

Note that the dimensions of the transformed quantities differ from those of the corresponding space-dependent ones. For instance, $[\mathbf{E}(\mathbf{k}, t)] = [(\text{volt}/\text{m})\text{m}^3] = [\text{volt} \times \text{m}^2]$.

To proceed as in Section 3.2, we should consider space-dispersive media which are space-invariant and nondispersive in time. But this is not possible, because spatially dispersive media are time-dispersive, too (see Section 1.2). Therefore, we limit ourselves to nondispersive space-invariant media and derive (for the simplest case of an isotropic medium)

$$\begin{cases} -\varepsilon \dfrac{d\mathbf{B}}{dt} = i\mathbf{k} \times \mathbf{D}, & (3.57\text{a}) \\[6pt] -\mu \dfrac{d\mathbf{D}}{dt} = i\mathbf{B} \times \mathbf{k} + \mu\mathbf{J} + \mu\mathbf{J}_0, & (3.57\text{b}) \\[6pt] i\mathbf{k} \cdot \mathbf{D} = \rho + \rho_0, & (3.57\text{c}) \\[6pt] i\mathbf{k} \cdot \mathbf{B} = 0, & (3.57\text{d}) \end{cases}$$

where ε and μ are real (possibly time-dependent) constants.

Equations (3.57) are referred to as Maxwell equations in the *wavenumber–time* spectral domain, or simply the *wavenumber domain*.

3.3.1. Radiation from Prescribed Sources

Let us consider a nonconductive medium ($\sigma = 0$). Taking the time derivative of Eq. (3.57a) and substituting from Eq. (3.57b) leads to

$$-\mu \frac{d}{dt}\varepsilon\frac{d\mathbf{B}}{dt} = \mathbf{k} \times \mathbf{B} \times \mathbf{k} - i\mu\mathbf{k} \times \mathbf{J}_0. \tag{3.58}$$

On the other hand, Eq. (3.57d) shows that $\mathbf{k}$ and $\mathbf{B}$ are orthogonal:

$$\mathbf{k} \times \mathbf{B} \times \mathbf{k} = \mathbf{k} \cdot \mathbf{k}\mathbf{B} = k^2\mathbf{B}, \qquad k^2 = \mathbf{k} \cdot \mathbf{k}, \tag{3.59}$$

which is used to modify Eq. (3.58) as follows:

$$\mu\frac{d}{dt}\varepsilon\frac{d\mathbf{B}}{dt} + k^2\mathbf{B} = i\mu\mathbf{k} \times \mathbf{J}_0. \tag{3.60}$$

In order to solve Eq. (3.60), it may be convenient to introduce a Green's function as in Section 2.3.3. We set

$$\mathbf{J}_0(\mathbf{k}, t) = \int_{-\infty}^{+\infty} dt'\delta(t - t')\mathbf{J}_0(\mathbf{k}, t')$$

and

$$\mathbf{B}(\mathbf{k}, t) = i\mu\mathbf{k} \times \int_{-\infty}^{+\infty} dt' G(\mathbf{k}, t - t')\mathbf{J}_0(\mathbf{k}, t').$$

Substitution into Eq. (3.60) yields

$$\mu\frac{d}{dt}\varepsilon\frac{d}{dt}G(\mathbf{k}, t - t') + k^2 G(\mathbf{k}, t - t') = \delta(t - t'), \tag{3.61}$$

which shows that the Green's function depends only on $k^2 = \mathbf{k} \cdot \mathbf{k}$, i.e., $G(\mathbf{k}, t) \to G(k, t)$. Furthermore, comparison with Eq. (3.60) shows that $i\mu\mathbf{k} \times \hat{\mathbf{a}}G(k, t)$ is the magnetic induction response (in the wavenumber domain) to the unit electric source $\hat{\mathbf{a}}\delta(t)$.

When

$$\mathbf{j}_0(\mathbf{r}, t) = i_s(t)\delta(z)\hat{\mathbf{x}},$$

the source is a current linear density $i_s(t)$ [ampere/m] located on the plane

$z = 0$, and is independent of the spatial coordinates (x, y) (see Fig. 3.11). Then

$$\mathbf{J}_0(\mathbf{k}, t) = \hat{\mathbf{x}} i_s(t) \int_{-\infty}^{+\infty} dx \int_{-\infty}^{+\infty} dy \int_{-\infty}^{+\infty} dz \exp(-iux - ivy - iwz)\delta(z)$$
$$= \hat{\mathbf{x}}(2\pi)^2 \delta(u)\delta(v) i_s(t),$$

which shows that the u, v components of $\mathbf{k}$ can be set equal to zero and that we can take $\mathbf{k} = w\hat{\mathbf{z}}$. Field and induction notation may be simplified, e.g.,

$$\mathbf{b}(z, t) = \frac{1}{2\pi} \int_{-\infty}^{+\infty} dw \exp(iwz)\mathbf{B}(w, t).$$

For a time-invariant medium ε and μ are constant, $\varepsilon\mu = 1/c^2$, and Eq. (3.61) reduces to

$$\frac{d^2 G}{dt^2} + c^2 w^2 G = c^2 \delta(t). \tag{3.62}$$

For $t < 0$ we take $G(w, t) = 0$ (causality). For $t > 0$ the forcing term disappears and we derive the solution

$$G(w, t) = A^+(w) \exp(-iwct) + A^-(w) \exp(iwct).$$

The radiation condition at infinity for $z = 0$ (see Section 1.4.2 and the footnote at p. 31) is satisfied by the first term for $z > 0$, and by the second term for $z < 0$. Accordingly, we set

$$\begin{cases} G(w, t) = G^+(w, t) = A^+(w) \exp(-iwct)U(t), & z > 0, \quad (3.63a) \\ G(w, t) = G^-(w, t) = A^-(w) \exp(iwct)U(t), & z < 0. \quad (3.63b) \end{cases}$$

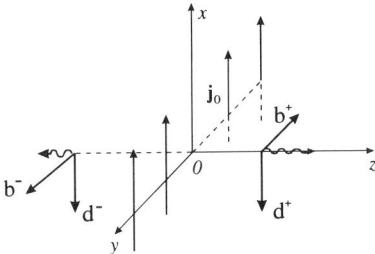

FIGURE 3.11. Excitation of a plane wave by a current sheet.

The correct behavior of the solution for $t \to 0$ is obtained by integrating Eq. (3.62) over a small time interval $2\Delta t \to 0$ across $t = 0$:

$$\lim_{\Delta t \to 0} \left\{ \left. \frac{dG^\pm}{dt} \right|_{-\Delta t}^{\Delta t} + c^2 w^2 \int_{-\Delta t}^{\Delta t} dt\, G^\pm(w,t) \right\} = \mp iwcA^\pm = c^2,$$

$$A^\pm = \mp \frac{c}{iw}.$$

The solution for the magnetic induction (in the wavenumber domain) is readily obtained:

$$\mathbf{B}^\pm(\mathbf{k},t) = \mp i\mu w \hat{\mathbf{z}} \times \hat{\mathbf{x}} \int_{-\infty}^{+\infty} dt'\, i_s(t') \frac{c\,\exp[\mp icw(t-t')]}{iw} U(t-t')$$

for $z > 0$ (upper sign) and $z < 0$ (lower sign), respectively. The electric induction follows from Eq. (3.57a):

$$\pm icw\varepsilon \mathbf{B}^\pm = iw\hat{\mathbf{z}} \times \mathbf{D}^\pm,$$

$$\mathbf{D}^\pm = \pm c\varepsilon \mathbf{B}^\pm \times \hat{\mathbf{z}}.$$

To proceed to the time domain we set $t - t' = \tau$, which yields

$$\mathbf{b}^\pm(z,t) = \mp \mu c \hat{\mathbf{y}} \int_{-\infty}^{+\infty} d\tau\, i_s(t-\tau) U(\tau) \frac{1}{2\pi} \int_{-\infty}^{+\infty} dw\, \exp[iw(z \mp c\tau)]$$

$$= \mp \mu \hat{\mathbf{y}} \int_{-\infty}^{+\infty} d\tau\, i_s(t-\tau) U(\tau) \delta(z \mp c\tau) d(c\tau)$$

$$= \mp \frac{1}{2} \hat{\mathbf{y}} \mu i_s(t \mp z/c), \tag{3.64a}$$

$$\mathbf{d}^\pm(z,t) = -\frac{1}{2c} \hat{\mathbf{x}} i_s(t \mp z/c). \tag{3.64b}$$

These represent two plane waves (see Fig. 3.11) moving in opposite senses $\pm \hat{\mathbf{z}}$; the plane uniform current sheet is the appropriate (though physically not realizable) source for plane-wave excitation. For computing $\mathbf{b}^\pm$ and $\mathbf{d}^\pm$ we exchanged integrations over τ and w; the factor $1/2$ is due to the presence of the unit steep function $U(\tau)$ that halves the integration domain.

It is further noted that

$$\lim_{z \to 0} \frac{1}{\mu} \hat{\mathbf{z}} \times (\mathbf{b}^+ - \mathbf{b}^-) = \lim_{z \to 0} \hat{\mathbf{z}} \times (\mathbf{h}^+ - \mathbf{h}^-) = \mathbf{j}_0,$$

as expected. The magnetic field is discontinuous at $z = 0$ due to the presence of the linear current density (see Section 1.2.7).

3.4. THE WAVENUMBER–FREQUENCY DOMAIN

We consider Maxwell equations (1.1) with prescribed (electric) source distribution $\mathbf{j}_0(\mathbf{r}, t)$ and $\rho_0(\mathbf{r}, t)$, as in Section 3.2. The latter functions are now expanded in a quadruple Fourier integral:

$$\mathbf{j}_0(\mathbf{r}, t) = \frac{1}{(2\pi)^4} \int_{-\infty}^{+\infty} d\omega \int_{-\infty}^{+\infty} d\mathbf{k} \ \exp(i\omega t + i\mathbf{k}\cdot\mathbf{r})\mathbf{J}_0(\mathbf{k}, \omega),$$

and similarly for $\rho_0(\mathbf{r}, t)$. Fields and inductions are also expanded. Substitution into Maxwell equations leads to

$$\mathbf{k} \times \mathbf{E}(\mathbf{k}, \omega) = -\omega \mathbf{B}(\mathbf{k}, \omega), \tag{3.65a}$$

$$\mathbf{H}(\mathbf{k}, \omega) \times \mathbf{k} = -\omega \mathbf{D}(\mathbf{k}, \omega) + i\mathbf{J}(\mathbf{k}, \omega) + i\mathbf{J}_0(\mathbf{k}, \omega), \tag{3.65b}$$

$$\mathbf{k} \cdot \mathbf{D}(\mathbf{k}, \omega) = -i\rho(\mathbf{k}, \omega) - i\rho_0(\mathbf{k}, \omega), \tag{3.65c}$$

$$\mathbf{k} \cdot \mathbf{B}(\mathbf{k}, \omega) = 0. \tag{3.65d}$$

Note that the dimensions of the transformed quantities differ from those of the corresponding space–time dependent ones. For instance, $[\mathbf{E}(\mathbf{k}, \omega)] = [(\text{volt}/\text{m})(\text{m}^3 \times \text{s})] = [\text{volt} \times \text{m}^2 \times \text{s}]$.

For space–time homogeneous media which may be space–time dispersive, the constitutive relations take the form

$$\mathbf{d}(\mathbf{r}, t) = \int d\mathbf{r}' g(\mathbf{r} - \mathbf{r}', t - t') \cdot \mathbf{e}(\mathbf{r}', t'),$$

$$\ldots \ \ldots$$

[see Eq. (1.23)]. Transformation into the $(\mathbf{k}, \omega)$ domain and use of Borel's theorem (see Section 3.2) yield

$$\mathbf{D}(\mathbf{k}, \omega) = \varepsilon(\mathbf{k}, \omega) \cdot \mathbf{E}(\mathbf{k}, \omega),$$

where

$$\varepsilon(\mathbf{k}, \omega) = \int_{-\infty}^{+\infty} dt \int_{-\infty}^{+\infty} d\mathbf{r} \ \exp(i\omega t + i\mathbf{k}\cdot\mathbf{r})g(r, t)$$

is the medium permittivity in the wavenumber–frequency domain. In this domain the medium is fully characterized by the permittivity $\varepsilon(\mathbf{k}, \omega)$ in addition to the permeability $\mu(\mathbf{k}, \omega)$ and the conductivity $\sigma(\mathbf{k}, \omega)$. The latter can be included in the permittivity,

$$\varepsilon(\mathbf{k}, \omega) + \frac{1}{i\omega} \sigma(\mathbf{k}, \omega) \to \varepsilon(\mathbf{k}, \omega); \qquad (3.66)$$

see section 3.2.2.
Substitution into Eqs. (3.65) leads to

$$\begin{cases} \mathbf{k} \times \mathbf{E} = -\omega\mu \cdot \mathbf{H}, & (3.67a) \\ \mathbf{H} \times \mathbf{k} = -\omega\varepsilon \cdot \mathbf{E} + i\mathbf{J}_0, & (3.67b) \\ \mathbf{k} \cdot \varepsilon \cdot \mathbf{E} = -i\rho_0, & (3.67c) \\ \mathbf{k} \cdot \mu \cdot \mathbf{H} = 0, & (3.67d) \end{cases}$$

which are referred to as Maxwell equations in the *wavenumber–frequency spectral domain* or simply the *spectral domain*.

3.4.1. Space Dispersive Media. Compressible Plasma

Constitutive relations for a plasma medium are considered in Sections 1.2.5 and 3.2.4. In the former case the plasma is isotopic and collisions are included, as opposed to the latter case which refers to a collisionless anisotropic plasma. The common (linearized) model is that of an electron fluid of average particle density n, subject to the Lorenz force density (1.2) and described by a space–time varying velocity $\mathbf{v}(\mathbf{r}, t)$.

In both cases we implicitly assume that changes of the electron density in the electron fluid are negligible. If this assumption is relaxed, the model includes a space–time varying pressure $p(\mathbf{r}, t)$, which is simply related to the charge density,

$$p = mu^2 n, \qquad (3.68)$$

when linearization is enforced. In Eq. (3.68), m is the electron mass and u [m/s] is a proportionality coefficient, subsequently identified as the acoustic velocity in the fluid.

The constitutive equations are

$$\begin{cases} \mathbf{j} = -qn\mathbf{v}, & (3.69a) \\ m\dfrac{\partial n\mathbf{v}}{\partial t} = -qn\mathbf{e} - \nabla p, & (3.69b) \\ -q\nabla \cdot n\mathbf{v} - \dfrac{q}{mu^2}\dfrac{\partial p}{\partial t} = 0, & (3.69c) \end{cases}$$

where the last one is the current density continuity equation (1.7), with charge density $-qn$ expressed in terms of the pressure, given in Eq. (3.68). In the absence of an electric field, we take the divergence of Eq. (3.69b) and substitute from Eq. (3.69c) to obtain

$$\nabla^2 p - \frac{1}{u^2}\frac{\partial^2 p}{\partial t^2} = 0,$$

which is the scalar wave equation (2.2) in three dimensions and identifies u^2 as the speed of a pressure wave in the fluid.

In the wavenumber–frequency domain the constitutive relationships (3.69) transform as follows:

$$\begin{cases} \omega m \mathbf{J} - q\mathbf{k}P = -iq^2 n\mathbf{E}, \\ \mathbf{k} \cdot \mathbf{J} - \dfrac{\omega q}{mu^2}P = 0, \end{cases}$$

where n is now the *average* electron density (linearization). Hence

$$\left(\mathbf{I} - \frac{u^2}{\omega^2}\mathbf{k}\mathbf{k}\right) \cdot \mathbf{J} = -i\omega\varepsilon_0 \frac{\omega_p^2}{\omega^2}\mathbf{E},$$

where ω_p is the plasma angular frequency (3.24) and $\mathbf{k}\mathbf{k}$ the symmetric dyadic of the wavenumber vector (see Appendix B, Section B.1). Accordingly

$$\mathbf{J} = -i\omega\varepsilon_0 \frac{\omega_p^2}{\omega^2}\left(\mathbf{I} - \frac{u^2}{\omega^2}\mathbf{k}\mathbf{k}\right)^{-1} \cdot \mathbf{E}$$

and, upon substitution into Eq. (3.67b),

$$\mathbf{H} \times \mathbf{k} = -\omega\varepsilon_0 \mathbf{E} + \omega\varepsilon_0 \frac{\omega_p^2}{\omega^2}\left(\mathbf{I} - \frac{u^2}{\omega^2}\mathbf{k}\mathbf{k}\right)^{-1} \cdot \mathbf{E} = -\omega\varepsilon_0 \boldsymbol{\varepsilon}_r \cdot \mathbf{E},$$

with

$$\varepsilon_r = I - \frac{\omega_p^2}{\omega^2}\left(I - \frac{u^2}{\omega^2}\mathbf{kk}\right)^{-1} = \left(1 - \frac{\omega_p^2}{\omega^2}\right)I - \frac{\omega_p^2}{\omega^2}\frac{u^2\mathbf{kk}}{\omega^2 - u^2k^2}; \quad (3.70)$$

the last result is obtained by formally expanding the dyadic operator:

$$\left(I - \frac{u^2}{\omega^2}\mathbf{kk}\right)^{-1} = I + \left(\frac{u}{\omega}\right)^2\mathbf{kk} + \left(\frac{u}{\omega}\right)^4 k^2\mathbf{kk} + \cdots$$
$$= I + \left(\frac{u}{\omega}\right)^2\mathbf{kk}\frac{1}{1 - (u/\omega)^2k^2}.$$

Examination of Eq. (3.70) shows that time dispersion is relevant at temporal frequencies close to the plasma frequency, and space dispersion at spatial frequencies close to ω/u.

Note that the medium is isotropic, although its properties are described by the symmetric matrix ε_r.

3.4.2. Radiation in Isotropic Homogeneous Media

We start from Eqs. (3.67) with $\boldsymbol{\mu} \to \mu$ and $\boldsymbol{\varepsilon} \to \varepsilon$ and obtain

$$\mathbf{k} \times \mathbf{E} \times \mathbf{k} = -\omega\mu\mathbf{H} \times \mathbf{k} = \omega^2\varepsilon\mu\mathbf{E} - i\omega\mu\mathbf{J}_0. \quad (3.71)$$

We note that $\mathbf{k} \times \mathbf{E} \times \mathbf{k}$ is $\mathbf{k} \cdot \mathbf{k} = k^2$ times the component of the electric field (spectrum) orthogonal to $\mathbf{k}$:

$$\mathbf{k} \times \mathbf{K} \times \mathbf{k} = k^2\mathbf{E} - \mathbf{k}(\mathbf{k} \cdot \mathbf{E}).$$

We now use Eq. (3.67c) to express $\mathbf{k} \cdot \mathbf{E}$ in terms of the source charge density ρ_0, and the transform of the current density continuity equation,

$$\mathbf{k} \cdot \mathbf{J}_0 + \omega\rho_0 = 0, \quad (3.72)$$

to eliminate ρ_0. We finally arrive at

$$\mathbf{k} \times \mathbf{E} \times \mathbf{k} = k^2\mathbf{E} + \frac{1}{i\omega\varepsilon}\mathbf{kk} \cdot \mathbf{J}_0.$$

Substitution into Eq. (3.71) and subsequent use of Eq. (3.67a) yield

$$\begin{cases} \mathbf{E} = \dfrac{-i\omega\mu}{k^2 - \omega^2\varepsilon\mu}\left[\mathbf{I} - \dfrac{\mathbf{kk}}{\omega^2\varepsilon\mu}\right]\cdot\mathbf{J}_0, & (3.73a) \\[2mm] \mathbf{H} = \dfrac{i\mathbf{k}\times\mathbf{J}_0}{k^2 - \omega^2\varepsilon\mu}. & (3.73b) \end{cases}$$

In the space–time domain

$$\begin{cases} \mathbf{h}(\mathbf{r}, t) = \dfrac{1}{(2\pi)^4}\int_{-\infty}^{+\infty} d\mathbf{k}\int_{-\infty}^{+\infty} d\omega\, \exp(i\mathbf{k}\cdot\mathbf{r})\exp(i\omega t)\cdot\left[\dfrac{-i\mathbf{k}\times\mathbf{J}_0(\mathbf{k},\omega)}{\omega^2\varepsilon\mu - k^2}\right] \\[2mm] \quad = \nabla\times\left\{\dfrac{1}{(2\pi)^4}\int_{-\infty}^{+\infty} d\mathbf{k}\int_{-\infty}^{+\infty} d\omega\, \exp(i\mathbf{k}\cdot\mathbf{r})\exp(i\omega t)\mathbf{A}(\mathbf{k},\omega)\right\}, \\[2mm] \mathbf{A}(\mathbf{k},\omega) = -\dfrac{\mathbf{J}_0(\mathbf{k},\omega)}{\omega^2\varepsilon\mu - k^2}, & (3.74) \end{cases}$$

and similarly for $\mathbf{e}(\mathbf{r}, t)$. We have also

$$\mathbf{a}(\mathbf{r}, t) = \frac{1}{(2\pi)^4}\int_{-\infty}^{+\infty} d\mathbf{k}\int_{-\infty}^{+\infty} d\omega\, \exp(i\mathbf{k}\cdot\mathbf{r})\exp(i\omega t)\mathbf{A}(\mathbf{k},\omega). \quad (3.75)$$

3.4.3. The Resonant Wave Solution

Computation of the (quadruple) FT (3.75) is necessary to derive the field solution in the $(\mathbf{r}, t)$ domain. However, intermediate expressions, resulting from partial inversion of the FT (3.75), are of interest. One of these intermediate results is obtained when the ω-integration is performed. Accordingly, computation of the integral

$$\mathbf{R}(\mathbf{k}, t) = \frac{1}{2\pi}\int_{-\infty}^{+\infty} d\omega\, \exp(i\omega t)\mathbf{A}(\mathbf{k},\omega) \quad (3.76)$$

is of interest and provides the (generalized) *resonant mode representation* for the field.

When $\mathbf{J}_0(\mathbf{k}, \omega)$ is bounded for $|\omega| \to \infty$, the ω-integration contour in Eq. (3.76) can be closed in the upper half $\Omega = \omega + i\omega'$ plane for $t > 0$ [see Eq. (3.74) and Fig. 3.12], and $\mathbf{R}(\mathbf{k}, t)$ is given by the contributions of the singularities of the integrand. These can be categorized as *source singularities*, i.e., those appearing in the transform $\mathbf{J}_0(\mathbf{k}, \omega)$, and *system singularities*, i.e., the zeros of

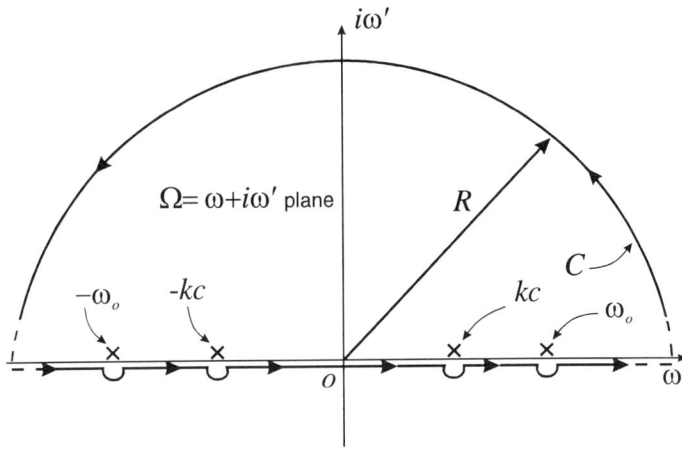

FIGURE 3.12. Resonant wave representation: integration contour in the complex $\Omega = \omega + i\omega'$ plane.

the dispersion equation

$$\omega^2 \varepsilon \mu - k^2 = 0 \tag{3.77}$$

[see Eq. (3.74)], together with the (possible) branch cuts of ε and μ in the Ω-plane.

As an example, let us consider a nondispersive medium described by parameters ε_0 and μ_0 (free space), with the (artificial) introduction of a vanishingly small conductivity $\sigma \to 0$, which helps to locate the position of singularities. Hence

$$\mu = \mu_0, \qquad \varepsilon = \varepsilon_0 - i\frac{\sigma}{\omega}, \qquad \sigma \to 0.$$

The system singularities are readily obtained by Eq. (3.77) and are given by the poles

$$\omega_d = \pm kc + i\frac{\sigma}{2\varepsilon_0}, \qquad \sigma \to 0, \qquad c = \frac{1}{\sqrt{\varepsilon_0 \mu_0}},$$

located closest to the real axis in the upper half Ω-plane (see Fig. 3.12). When

$$\mathbf{j}_0(\mathbf{r}, t) = \mathbf{f}(\mathbf{r}) \, \sin(\omega_0 t) U(t) \quad \text{and} \quad \mathbf{J}_0(\mathbf{k}, \omega) = \mathbf{F}(\mathbf{k}) \, \frac{\omega_0}{\omega_0^2 - \omega^2}$$

(see Table 3.1), the source singularities are again poles at the angular frequencies

$$\omega_s = \pm \omega_0.$$

These poles are also slightly displaced in the upper half-plane (see Fig. 3.12), in order to generate a source identically equal to zero for $t < 0$.

We can now carry out the ω-integration in Eq. (3.76). For $t > 0$ we can close the integration contour in the upper Ω-plane (see Fig. 3.12) and compute the integral in terms of the residues at the poles ω_d and ω_s:

$$\begin{aligned}\mathbf{R}(\mathbf{k}, t) &= -\frac{1}{2\pi} \oint d\Omega \, \exp(i\Omega t) \frac{\omega_0 c^2 \mathbf{F}(\mathbf{k})}{(\omega_0 - \Omega)(\omega_0 + \Omega)(\Omega - kc)(\Omega + kc)} \\ &= -\mathbf{F}(\mathbf{k}) \frac{\omega_0 c^2}{(kc)^2 - \omega_0^2} \left(\frac{\sin kct}{kc} - \frac{\sin \omega_0 t}{\omega_0} \right) U(t),\end{aligned}$$

which is the *resonant solution* in the **k** domain.

It is clear that the resonant wave solution eventually leads to the final solution: the **k** transform is in order. As an example, we consider the case

$$\mathbf{f}(\mathbf{r}) = \hat{\mathbf{x}} i_0 \delta(z), \qquad [i_0] = [\text{ampere/m}],$$

which corresponds to the current sheet of Section 3.3.1. We have

$$\begin{cases} \mathbf{F}(\mathbf{k}) = \hat{\mathbf{x}}(2\pi)^2 i_0 \delta(u)\delta(v), \\ \mathbf{a}(z, t) = \frac{i_0}{2\pi} \hat{\mathbf{x}} \int_{-\infty}^{+\infty} dw \, \exp(iwz) \cdot \frac{\sin(\omega_0 t) - (\omega_0/wc)\sin(wct)}{w^2 - (\omega_0/c)^2} U(t). \end{cases} \quad (3.78)$$

The singularities of the integrand in Eq. (3.78) are again poles:

$$w_s = \pm \frac{\omega_0}{c}.$$

To correctly locate them in the complex W-plane (see Fig. 3.13), we remember that ω_0 has a vanishingly small positive imaginary part. Accordingly, the two poles are closest to the real W-axis, one in the upper and one in the lower half-plane, as depicted in Fig. 3.13.

It is convenient to individually integrate the four factors which are

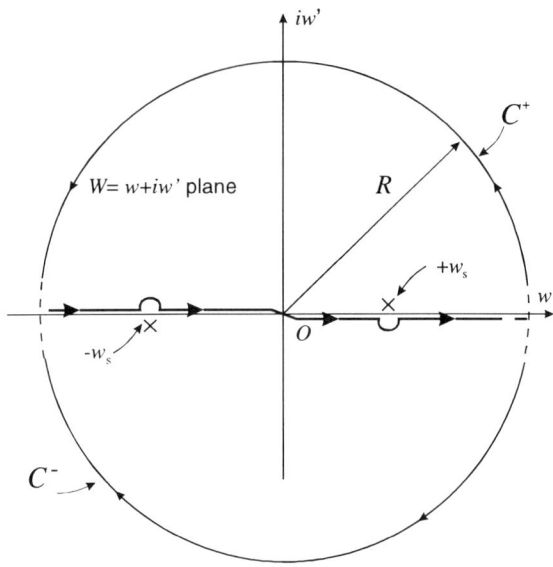

FIGURE 3.13. Resonant wave representation: integration contour in the complex $W = w + iw'$ plane.

obtained in Eq. (3.78) by expanding the circular functions in complex exponentials:

$$\frac{\exp[i(\omega_0 t + wz)] - \exp[i(-\omega_0 t + wz)] - (\omega_0/wc)\exp[iw(ct+z)] + (\omega_0/wc)\exp[-iw(ct-z)]}{2i(w + \omega_0/c)(w - \omega_0/c)}.$$

For $z > 0$, $ct - z < 0$, the integration contour can be closed along the path C^+ of Fig. 3.13 for all the four terms, thus obtaining

$$\mathbf{a}(z, t) = \hat{\mathbf{x}} \frac{i_0}{4} \frac{c}{\omega_0} \{\exp[i\omega_0(t + z/c)] - \exp[-i\omega_0(t - z/c)]$$

$$- \exp[i\omega_0(t + z/c)] + \exp[-i\omega_0(t - z/c)]\} U(t) = 0,$$

and causality is verified. For $z > 0$, $ct - z > 0$, the integration contour is again closed along C^+ for the first three terms, but along C^- for the last, with the final result

$$\mathbf{a}(z, t) = -\hat{\mathbf{x}} \frac{i_0}{2} \frac{c}{\omega_0} \cos[\omega_0(t - z/c)] U(t - z/c), \quad z \geq 0.$$

SPECTRAL DOMAINS

It is easy to check that this result is consistent with the solution (3.64): the curl of $\mathbf{a}(z,t)$ times μ provides Eq. (3.64a), $z > 0$. Similar results hold for $z < 0$.

3.4.4. Guided-Wave Representation

An intermediate field representation alternative to that of Section 3.4.3 can be obtained by starting from Eq. (3.75) and focusing attention on the w-integration:

$$\mathbf{S}(\mathbf{k}_t, z, \omega) = \frac{1}{2\pi} \int_{-\infty}^{+\infty} dw \, \exp(iwz) \mathbf{A}(\mathbf{k}_t, w, \omega), \quad \mathbf{k} = \mathbf{k}_t + w\hat{\mathbf{z}}. \quad (3.79)$$

For space nondispersive media, ε and μ do not depend on $\mathbf{k}$ and the integrand appearing in Eq. (3.79) exhibits poles in the complex $W = w + iw'$ plane located at

$$\pm w_p = \pm \sqrt{\omega^2 \varepsilon \mu - k_t^2}, \quad k_t^2 = u^2 + v^2 = \mathbf{k}_t \cdot \mathbf{k}_t,$$

as depicted in Fig. 3.14. The proper location of the poles is chosen by assuming a vanishingly small imaginary part for ε, as in Section 3.4.3. We assumed in the figure that $\omega > 0$; in the opposite case the position of the poles is reversed with respect to the ω-axis. In addition, $k_t^2 < \omega^2 \varepsilon \mu$; should this not be the case, the poles are located along the imaginary w'-axis.

Assume now that the sources are located on the plane $z = 0$: $\mathbf{j}_0(\mathbf{r}, t) \to \mathbf{j}_0(x, y, t)\delta(z)$. Then $\mathbf{J}_0(\mathbf{k}, \omega) \to \mathbf{J}_0(\mathbf{k}_t, \omega)$ and no additional singularity may appear in the complex W-plane. For $z > 0$, the integration contour in

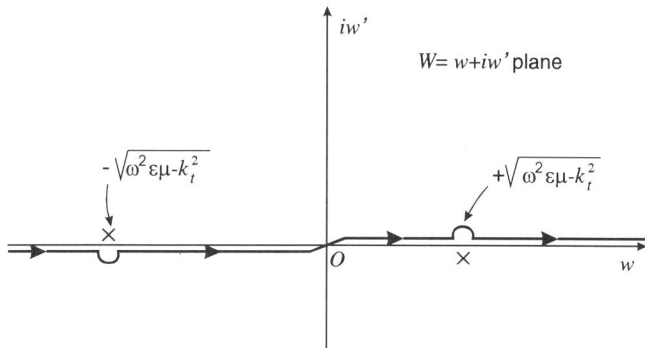

FIGURE 3.14. Guided-wave representation: integration contour in the complex $W = w + iw'$ plane.

Fig. 3.14 can be closed in the upper half-plane to yield

$$S(\mathbf{k}_t, z, \omega) = \frac{1}{2\pi} \oint dW \exp(iWz) \frac{\mathbf{J}_0(\mathbf{k}_t, \omega)}{W^2 + k_t^2 - \omega^2 \varepsilon \mu} U(z)$$

$$= \frac{-i\mathbf{J}_0(\mathbf{k}_t, \omega)}{2\sqrt{\omega^2 \varepsilon \mu - k_t^2}} \exp(-i\sqrt{\omega^2 \varepsilon \mu - k_t^2} z) U(z), \quad (3.80a)$$

which is the *guided-wave solution* in the $(\mathbf{k}_t, \omega)$ domain.

It has been noted that solution (3.80a) is valid for $\omega > 0$. In the opposite case the pole position is reversed with respect to the ω-axis, so we obtain

$$S(\mathbf{k}_t, z, \omega) = \frac{i\mathbf{J}_0(\mathbf{k}_t, \omega)}{2\sqrt{\omega^2 \varepsilon \mu - k_t^2}} \exp(i\sqrt{\omega^2 \varepsilon \mu - k_t^2} z) U(z). \quad (3.80b)$$

Since $\mathbf{J}_0(\mathbf{k}_t, -\omega) = \mathbf{J}_0^*(\mathbf{k}_t, \omega)$ (see Section 3.2.1), we have $S(\mathbf{k}_t, z, \omega) = S^*(\mathbf{k}_t, z, -\omega)$.

Also in this case, to proceed to the final solution, the $\mathbf{k}_t$ and ω transforms are in order. As an example, we consider the case

$$\mathbf{j}_0(\mathbf{r}, t) = \hat{\mathbf{x}} i_0(t) \delta(z) \sum_{n=-\infty}^{+\infty} \text{rect}\left[\frac{y - 2nY}{Y}\right],$$

as sketched in Fig. 3.15. The source is an $\hat{\mathbf{x}}$-oriented current source on the plane $z = 0$, x-independent, and spatially pulsed along the y-axis with spatial period $2Y$. This (spatially) periodic source distribution in an unbounded space is equivalent to a single (spatial) period of the source in the bounded space limited at $y = \pm Y$ by two magnetic walls, as follows from image theory (see Section 1.5.1).

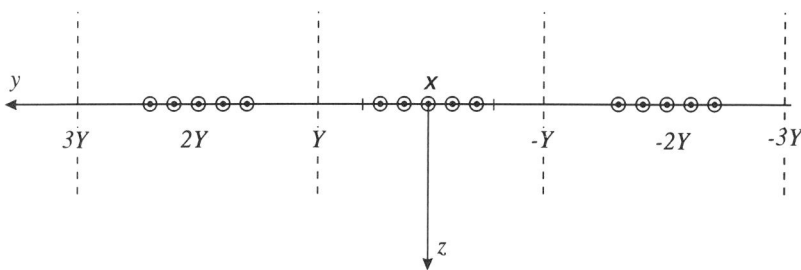

FIGURE 3.15. Periodic current source distribution.

We expand the source in a Fourier series,

$$\mathbf{j}_0(\mathbf{r}, t) = \hat{\mathbf{x}} i_0(t) \delta(z) \left\{ \frac{1}{2} + \frac{1}{\pi} \sum_{-\infty}^{+\infty} \frac{(-)^n}{2n+1} \exp\left[i \frac{(2n+1)\pi}{Y} y \right] \right\},$$

thus easily computing the associated spectrum:

$$\mathbf{J}_0(\mathbf{k}, \omega) = \hat{\mathbf{x}} I_0(\omega)(2\pi)^2 \delta(u) \left\{ \frac{1}{2}\delta(v) + \frac{1}{\pi} \sum_{-\infty}^{+\infty} \frac{(-)^n}{2n+1} \delta\left(v - \frac{(2n+1)\pi}{Y} \right) \right\}$$

$$= \mathbf{J}_0(\mathbf{k}_t, \omega),$$

$I_0(\omega)$ being the FT of $i_0(t)$.

The first term inside the large parentheses coincides with that of Section 3.4.3, but for the factor 1/2, and can be dealt with similarly; so it will not be considered further. Upon substituting the remaining terms in Eq. (3.80a), assuming $\varepsilon = \varepsilon_0$ and $\mu = \mu_0$, and performing the $\mathbf{k}_t$-integration, we obtain

$$\hat{\mathbf{S}}(\mathbf{r}, \omega) = \frac{1}{(2\pi)^2} \int_{-\infty}^{+\infty} d\mathbf{k}_t \exp(i\mathbf{k}_t \cdot \mathbf{r}) \mathbf{S}(\mathbf{k}_t, z, \omega)$$

$$= -i I_0(\omega) \frac{c^2}{2Y} \hat{\mathbf{x}} \sum_{-\infty}^{+\infty} (-)^n \frac{\exp(i\omega_n y/c - i\sqrt{\omega^2 - \omega_n^2} z/c)}{\omega_n \sqrt{\omega^2 - \omega_n^2}},$$

with

$$\omega_n = \frac{(2n+1)\pi c}{Y}.$$

The above expression is valid for $\omega > 0$. For $\omega < 0$, $\hat{\mathbf{S}}(\mathbf{r}, \omega)$ changes to its complex conjugate: $\hat{\mathbf{S}}(\mathbf{r}, -\omega) = \hat{\mathbf{S}}^*(\mathbf{r}, \omega)$.

When

$$i_0(t) = I_0 \cos \omega_0 t, \quad [I_0] = [\text{ampere/m}],$$

$$I_0(\omega) = \frac{I_0}{2}[\delta(\omega - \omega_0) + \delta(\omega + \omega_0)]$$

(see Table 3.1), we easily derive for $\omega_0 > \omega_n$,

$$\mathbf{a}(\mathbf{r}, t) = I_0 \frac{c^2}{Y} \hat{\mathbf{x}} \sum_0^{+\infty} (-)^n \frac{\cos(\omega_n y/c)}{\omega_n} \frac{\sin[\omega_0 t - \sqrt{\omega_0^2 - \omega_n^2} z/c]}{\sqrt{\omega_0^2 - \omega_n^2}}.$$

In the opposite case, $\omega_0 < \omega_n$,

$$\mathbf{a}(\mathbf{r}, t) = 2I_0 \frac{c^2}{Y} \hat{\mathbf{x}} \sum_{n}^{+\infty} (-)^n \frac{\cos(\omega_n y/c)}{\omega_n} \frac{\exp[-\sqrt{\omega_n^2 - \omega_0^2}\, z/c]}{\sqrt{\omega_n^2 - \omega_0^2}},$$

and propagation is impaired due to the exponentially decaying term.

3.5. THE FIELD REPRESENTATION

In Section 3.4 the electromagnetic fields **e** and **h** are given in terms of the spectrum $\mathbf{J}_0(\mathbf{k}, \omega)$ of the sources. As an alternative representation, it may be convenient to represent the field in terms of its spectrum over a conveniently prescribed surface.

We skip the ω-integration, which is of no concern here (and in the following), and make reference to $\hat{\mathbf{A}}(\mathbf{r}, \omega)$, the FT of $\mathbf{a}(\mathbf{r}, t)$. From Eqs. (3.80) we have

$$\hat{\mathbf{A}}(\mathbf{r}, \omega) = \frac{1}{(2\pi)^2} \int d\mathbf{k}_t \, \exp(i\mathbf{k}_t \cdot \mathbf{r}) \mathbf{S}(\mathbf{k}_t, 0, \omega) \exp(-iw_p z),$$

$$w_p = \sqrt{\omega^2 \varepsilon\mu - k^2},$$

where the guided-wave spectrum has been factorized into its z-independent and z-dependent parts:

$$\mathbf{S}(\mathbf{k}_t, z, \omega) = \mathbf{S}(\mathbf{k}_t, 0, \omega) \exp(-iw_p z).$$

It is evident that $\mathbf{S}(\mathbf{k}_t, 0, \omega)$ is the $\mathbf{k}_t$ transform of $\hat{\mathbf{A}}(\mathbf{r}, \omega)$ evaluated over the plane $z = 0$:

$$\mathbf{S}(\mathbf{k}_t, 0, \omega) = \iint dx\, dy \, \exp(-iux - ivy) \hat{\mathbf{A}}(x, y, 0, \omega),$$

$$\mathbf{k}_t = u\hat{\mathbf{x}} + v\hat{\mathbf{y}},$$

and we conclude that $\hat{\mathbf{A}}(\mathbf{r}, \omega)$ can be computed if we know the spectrum $\mathbf{S}(\mathbf{k}_t, 0, \omega)$ of its determination over the plane $z = 0$[16].

A similar representation is valid for the spectra **E** and **H**, which are related to **A** by means of simple operations in the $(\mathbf{k}, \omega)$ domain; see Eqs. (3.73). For

---

16. Any other convenient plane may be used.

SPECTRAL DOMAINS

$z \geq 0$,

$$\mathbf{E}(\mathbf{r}, \omega) = \frac{1}{(2\pi)^2} \int\!\!\!\int_{-\infty}^{+\infty} du\, dv \, \exp(iux + ivy) \cdot \exp(-iw_p z) \mathbf{E}_0(u, v, \omega), \qquad (3.81)$$

with

$$\mathbf{E}_0(u, v, \omega) = \int\!\!\!\int_{-\infty}^{+\infty} dx\, dy \, \exp(-iux - ivy) \mathbf{E}(x, y, 0, \omega).$$

Similar expressions are valid for the magnetic field $\mathbf{H}$. We remind the reader that $\mathbf{E}(\mathbf{r}, \omega)$ is just the ω-transform of $\mathbf{e}(\mathbf{r}, t)$.

The three components of $\mathbf{E}_0$ are not independent, as follows from Eq. (3.67c):

$$u E_{0x} + v E_{0y} - w_p E_{0z} = 0.$$

Accordingly, knowledge of the tangential component $\mathbf{E}_t(x, y, 0, \omega)$ of the field over the plane $z = 0$ is the information necessary to compute the field everywhere for $z \geq 0$:

$$\mathbf{E}_0(u, v, \omega) = \left[\mathbf{I}_t + \frac{\hat{\mathbf{z}} \mathbf{k}_t}{w_p} \right] \cdot \int\!\!\!\int_{-\infty}^{+\infty} dx\, dy \cdot \exp(-iux - ivy) \mathbf{E}_t(x, y, 0, \omega), \qquad (3.82)$$

($w = w_p$, see Section 3.4.4), and the field for $z \geq 0$ can be computed by using Eq. (3.81).

Equation (3.81) expresses the field as a superposition of monochromatic plane waves, $\exp(iux + ivy - iw_p z)$, whose propagation constant satisfies the dispersion equation (3.77) because

$$u^2 + v^2 + w_p^2 = \omega^2 \varepsilon \mu.$$

These elementary plane waves are source-free solutions of Maxwell equations, as follows from Eq. (3.73): when $\mathbf{J}_0 \to 0$, the spectra $\mathbf{E}$ and $\mathbf{H}$ can differ from zero only if $k^2 \to \omega^2 \varepsilon \mu$. Equation (3.82) is referred to as the *plane-wave representation* for the field, and $\mathbf{E}_0(u, v, \omega)$ as the *plane-wave spectrum*.

3.5.1. Fields and the Plane-Wave Spectrum Relationship

Let us consider the (frequency-domain) field distribution over the plane $z = 0$,

$$\mathbf{E}(y) = \hat{\mathbf{x}} f(y),$$

as depicted in Fig. 3.16 (the ω-dependence is understood and suppressed). We use Eq. (3.82) to compute the associated plane wave spectrum:

$$\mathbf{E}_0(u, v) = \left[\mathbf{I}_t + \frac{\hat{\mathbf{z}} \mathbf{k}_t}{w_p}\right] \cdot \hat{\mathbf{x}} \int\!\!\!\int_{-\infty}^{+\infty} dx\,dy\, \exp(-iux - ivy) f(y)$$

$$= \left(\hat{\mathbf{x}} + \hat{\mathbf{z}} \frac{u}{w_p}\right) 2\pi \delta(u) F(v),$$

$F(v)$ being the (one-dimensional) FT of $f(y)$.

For $w \neq 0$ we can set $u = 0$, thus ignoring the u-dependence of the spectrum. Hence

$$\mathbf{E}_0(v) = \hat{\mathbf{x}} F(v),$$

$$\mathbf{E}(\mathbf{r}) = \frac{\hat{\mathbf{x}}}{2\pi} \int_{-\infty}^{+\infty} dv\, \exp(ivy)\, \exp(-i\sqrt{\omega^2 \varepsilon \mu - v^2}\, z) F(v).$$

It is convenient to define *effective widths*, Y and V, of the field and its associated plane-wave spectrum. A possible definition is the following[17]:

$$\int_{-\infty}^{+\infty} dy |f(y)| = |f(y_0)| Y \quad \text{and} \quad \int_{-\infty}^{+\infty} dv |F(v)| = |F(v_0)| V,$$

where we assume that $|f(y)|$ and $|F(v)|$ attain their maximum values for $y = y_0$ and $v = v_0$, respectively ($y_0 = v_0 = 0$ in Fig. 3.16). Therefore

$$f(y_0) = \frac{1}{2\pi} \int_{-\infty}^{+\infty} dv\, \exp(ivy_0) F(v) \quad \text{and} \quad F(v_0) = \int_{-\infty}^{+\infty} dy\, \exp(-iv_0 y) f(y),$$

---

17. This definition is by no means unique. Alternative, sometimes more convenient, definitions are possible.

SPECTRAL DOMAINS

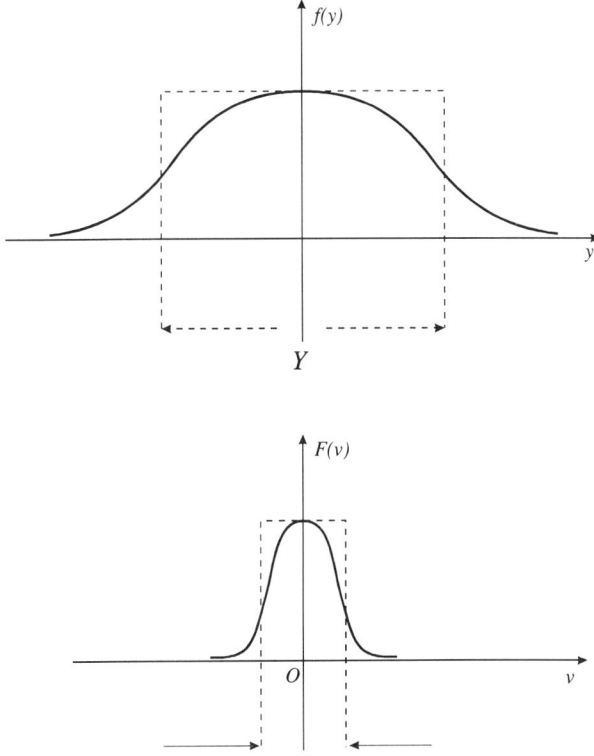

FIGURE 3.16. Field distribution over the plane $z = 0$ and its associated plane-wave spectrum.

so that

$$YV = 2\pi \frac{\int_{-\infty}^{+\infty} dy |f(y)|}{\left|\int_{-\infty}^{+\infty} dy \exp(-iv_0 y) f(y)\right|} \frac{\int_{-\infty}^{+\infty} dv |F(v)|}{\left|\int_{-\infty}^{+\infty} dv \exp(ivy_0) F(v)\right|} \geqslant 2\pi,$$

because

$$\left|\int_{-\infty}^{+\infty} dy \exp(-iv_0 y) f(y)\right| \leqslant \int_{-\infty}^{+\infty} dy |\exp(-iv_0 y) f(y)| = \int_{-\infty}^{+\infty} dy |f(y)|,$$

and similarly for the other integral.

We conclude that

$$YV \geqslant 2\pi. \tag{3.83}$$

Equality is attained when $f(y)$ is gaussian:

$$f(y) = f_0 \exp\left(-\frac{y^2}{2a^2}\right),$$

because it is always real positive, as well as its transform $F(v)$ (see Table 3.1). Application of Eq. (C.3) with $b = 0$ yields

$$f_0 \sqrt{2\pi a^2} = f_0 Y,$$

so that effective widths of a gaussian field distribution and its associated plane-wave spectrum are given by

$$Y = \sqrt{2\pi} a, \qquad V = \frac{\sqrt{2\pi}}{a}.$$

Relation (3.83) states that a narrow plane-wave spectrum corresponds to a wide field distribution, and vice versa. The implications of this statement are discussed in Section 3.5.3.

3.5.2. Asymptotic Evaluation of the Far Field

Equation (3.81) provides the field for $z \geqslant 0$ when its plane-wave spectrum is prescribed over the plane $z = 0$. We wish to evaluate the integral at large distances from the origin of coordinates.

Let

$$x = ar, \qquad y = br, \qquad z = cr, \qquad a^2 + b^2 + c^2 = 1,$$
$$u = \alpha k, \qquad v = \beta k, \qquad w_p = \gamma k, \qquad \alpha^2 + \beta^2 + \gamma^2 = 1,$$

with $k^2 = \omega^2 \varepsilon \mu$. Hence Eq. (3.81) can be expressed as

$$\mathbf{E}(\mathbf{r}, \omega) = \frac{1}{(2\pi)^2} \int\!\!\!\int_{-\infty}^{+\infty} du\, dv\, \exp[ikr(a\alpha + b\beta - c\gamma)] \mathbf{E}_0(u, v, \omega)$$

$$= \frac{1}{(2\pi)^2} k^2 \int\!\!\!\int_{-\infty}^{+\infty} d\alpha\, d\beta\, \exp[ikr(a\alpha + b\beta - c\gamma)] \mathbf{E}_0(k\alpha, k\beta, \omega).$$

We evaluate the integral in the limit $kr \to \infty$ (the *far field*). Due to the fast-varying phase term, the integral is negligible unless

$$\psi(\alpha, \beta) = a\alpha + b\beta - c\gamma$$

is stationary. This happens for the solution α_s and β_s of the system

$$\begin{cases} \dfrac{\partial \psi}{\partial \alpha} = a + c\dfrac{\alpha}{\sqrt{1 - \alpha^2 - \beta^2}} = 0, \\ \dfrac{\partial \psi}{\partial \beta} = b + c\dfrac{\beta}{\sqrt{1 - \alpha^2 - \beta^2}} = 0, \end{cases}$$

i.e.,

$$\begin{cases} a\gamma_s + c\alpha_s = 0, \\ b\gamma_s + c\beta_s = 0, \end{cases}$$

namely,

$$\alpha_s = -a, \qquad \beta_s = -b, \qquad \gamma_s = c.$$

The phase term can be expanded in the neighborhood of the *stationary phase point* (α_s, β_s):

$$\begin{aligned} \psi(\alpha, \beta) &\approx \psi(\alpha_s, \beta_s) + \frac{1}{2}\frac{\partial^2 \psi}{\partial \alpha^2}(\alpha - \alpha_s)^2 + \frac{1}{2}\frac{\partial^2 \psi}{\partial \beta^2}(\beta - \beta_s)^2 + \frac{\partial^2 \psi}{\partial \alpha \partial \beta}(\alpha - \alpha_s)(\beta - \beta_s) \\ &= -1 + \frac{1 - b^2}{2c^2}(\alpha - \alpha_s)^2 + \frac{1 - a^2}{2c^2}(\beta - \beta_s)^2 + \frac{ab}{c^2}(\alpha - \alpha_s)(\beta - \beta_s) \\ &= -1 + \phi(\alpha, \beta). \end{aligned}$$

Asymptotically, for $kr \to \infty$,

$$\begin{aligned} \mathbf{E}(r, \omega) &\sim \frac{1}{(2\pi)^2} k^2 \exp(-ikr) \iint_{-\infty}^{+\infty} d\alpha d\beta \, \exp[ikr\phi(\alpha, \beta)]\mathbf{E}_0(k\alpha, k\beta, \omega) \\ &\sim \frac{1}{(2\pi)^2} k^2 \exp(-ikr)\mathbf{E}_0(-ka, -kb, \omega) \iint_{-\infty}^{+\infty} d\alpha d\beta \, \exp[ikr\phi(\alpha, \beta)], \quad (3.84) \end{aligned}$$

because only a very limited portion of the (α, β) plane around (α_s, β_s) contributes significantly to the integral in the limit $kr \to \infty$. This asymptotic procedure is referred to as the *stationary-phase evaluation* of integrals.

Some comments are in order before completing the evaluation of the far field. The asymptotic condition $kr \to \infty$, which is the basis of the procedure, requires that $k = \omega\sqrt{\varepsilon\mu} \neq 0$. In practice, this is assured by fields whose frequency-domain spectrum is negligible in the low-frequency limit. Also, the plane-wave spectrum $\mathbf{E}_0(k\alpha, k\beta)$ should be singularity-free near the stationary-phase point, so that it can be evaluated at (α_s, β_s) and taken outside the integral. Under these hypotheses Eq. (3.84) states that the far field along the direction $\hat{\mathbf{r}}$ is related to the value of the spectrum along the same direction $\hat{\mathbf{k}} = \hat{\mathbf{r}}$. Only plane waves moving close to $\hat{\mathbf{r}}$ contribute significantly to the far field; the others interfere destructively and cancel out.

To evaluate the remaining integral in Eq. (3.84), we translate the origin of coordinates to the point (α_s, β_s) and introduce a rotation of coordinates:

$$\begin{cases} \alpha = \alpha' \cos \delta + \beta' \sin \delta, \\ \beta = -\alpha' \sin \delta + \beta' \cos \delta. \end{cases}$$

These values are substituted into the expression for $\phi(\alpha, \beta)$ and the angle δ chosen so that the coefficient of the mixed term $\alpha'\beta'$ is equal to zero. Hence

$$\tan 2\delta = \frac{2ab}{b^2 - a^2},$$

and

$$\phi(\alpha', \beta') = \frac{\alpha'^2}{2} + \frac{\beta'^2}{2c^2}.$$

Now the double integral in Eq. (3.84) factorizes as the product of two integrals with integration variables α' and β'. Application of Eq. (C.5) leads to

$$\mathbf{E}(\mathbf{r}, \omega) = ik \cos \theta \, \frac{\mathbf{E}_0(-k \sin \theta \cos \phi, \, -k \sin \theta \sin \phi, \omega)}{2\pi r} \exp(-ikr), \quad (3.85)$$

where reference is made to Fig. 3.17 and[18]

$$\mathbf{r} = (ar, \, br, \, cr) = (r \sin \theta \cos \phi, \, r \sin \theta \sin \phi, \, r \cos \theta).$$

---

18. We note that the stationary-phase point coordinates (u_s, v_s) appear with a minus sign, as well as the corresponding value of $w_s = k \cos \theta$; see Eq. (3.82). This is due to the (arbitrary) choice of the FT representation [see Eqs. (3.2)] and corresponds to plane waves moving from the plane $z = 0$ toward the half-space $z > 0$.

SPECTRAL DOMAINS

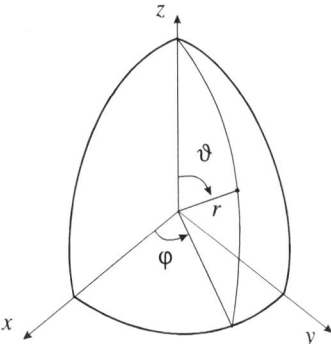

FIGURE 3.17. Coordinate system for the asymptotic evaluation of the field. The plane-wave spectrum is known over the plane $z = 0$.

3.5.3. Radiation from Apertures

Let us consider an aperture in a ground plane with a prescribed tangential component of the electric field (see Fig. 3.18):

$$\mathbf{E}_t(x, y, 0, \omega) = F(\omega)\hat{\mathbf{x}} \cos\left(\frac{\pi}{2Y}y\right) \text{rect}\left(\frac{y}{2Y}\right) \text{rect}\left(\frac{x}{2X}\right).$$

Note that the tangential electric field is different from zero over the aperture, and equal to zero over the remaining part of the plane $z = 0$.
Equation (3.82) is used to compute the plane-wave spectrum:

$$\mathbf{E}_0(u, v, \omega) = F(\omega)\left[\mathbf{I}_t + \frac{\hat{\mathbf{z}}\mathbf{k}_t}{w_p}\right] \cdot \hat{\mathbf{x}} \int_{-X}^{X} dx \exp(-iux) \int_{-Y}^{Y} dy \exp(-ivy) \cos\left(\frac{\pi}{2Y}y\right)$$

$$= F(\omega)\left(\hat{\mathbf{x}} + \hat{\mathbf{z}}\frac{u}{w_p}\right)\frac{2}{\pi} 4XY \text{sinc}(uX) \frac{\cos(vY)}{1 - (2vY/\pi)^2}.$$

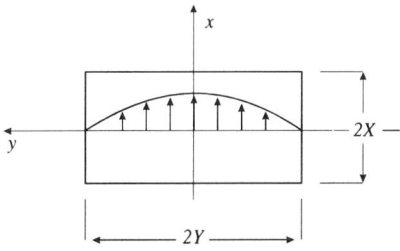

FIGURE 3.18. Field distribution of a rectangular aperture.

At large distances from the aperture we immediately obtain from Eq. (3.85)

$$\mathbf{E}(\mathbf{r}, \omega) = F(\omega) \frac{ik \exp(-ikr)}{2\pi r} (\hat{\mathbf{x}} \cos\theta - \hat{\mathbf{z}} \sin\theta \cos\phi) \cdot \frac{2}{\pi} 4XYD(u, v), \qquad (3.86)$$

where

$$D(u, v) = \text{sinc}(uX) \frac{\cos(vY)}{1 - (2vY/\pi)^2}$$

with $u = k \sin\theta \cos\phi$ and $v = k \sin\theta \sin\phi$. The reference coordinate system is centered in the aperture with the z-axis orthogonal to it.

The function $D(u, v)$ s plotted in Fig. 3.19 for the two cuts $v = 0$ and $u = 0$,

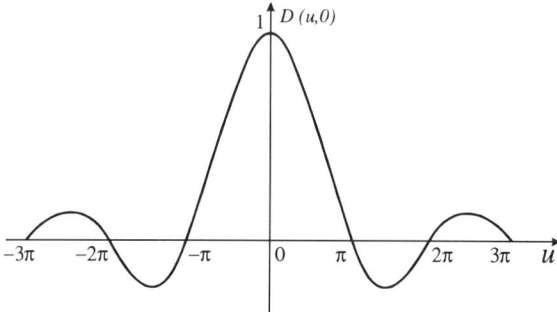

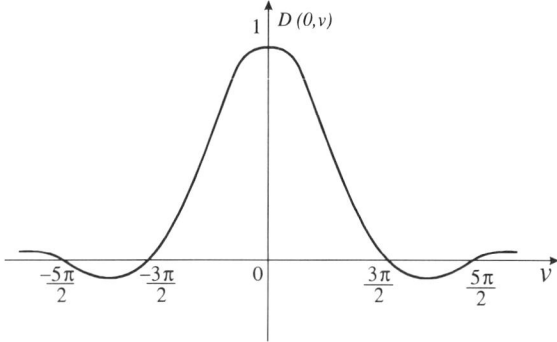

FIGURE 3.19. Radiation diagram of a rectangular aperture in the two principal planes $\phi = 0, \pi$ and $\phi = \pi/2, 3\pi/2$.

SPECTRAL DOMAINS

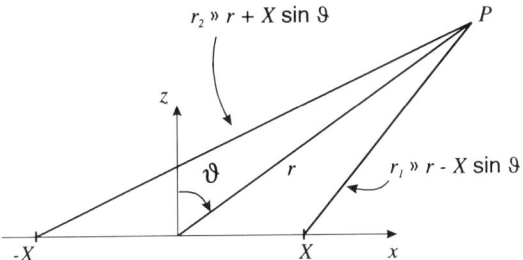

FIGURE 3.20. Model of radiation from a rectangular aperture. Principal plane $\phi = 0, \pi$.

and describes the radiation properties of the aperture. It is evident that, in general, the larger the aperture, the narrower is the radiation beam. Furthermore, the beam tends to spread in the plane where the aperture field is smoother; for the aperture under consideration, this is the plane $x = 0$. Additional discussion relevant to radiation from apertures is deferred to Section 8.5.

In the time domain, the far field $\mathbf{e}(\mathbf{r}, t)$ is the FT of Eq. (3.86), i.e., it is proportional to the convolution[19] of the FTs of $F(\omega)$ and $kD(u, v)\exp(-ikr)$. For instance, in the plane $\phi = 0, \pi$ we have

$$\frac{1}{2\pi} \int_{-\infty}^{+\infty} d\omega \, \exp(i\omega t) ikD(u, 0) \exp(-ikr)$$

$$= \frac{1}{2\pi} \int_{-\infty}^{+\infty} d\omega \, \exp(i\omega t) \frac{i\omega}{c} \operatorname{sinc}\left(\omega \frac{X \sin \theta}{c}\right) \exp(-i\omega r/c)$$

$$= \frac{\delta(t - [r - X \sin \theta]/c) - \delta(t - [r + X \sin \theta]/c)}{2X \sin \theta i}$$

This solution suggests the model depicted in Fig. 3.20; the endpoints of the aperture are responsible for the radiated field. On the z-axis, $\theta \to 0$ and

$$f(t) \to \frac{1}{c} \delta'\left(t - \frac{r}{c}\right).$$

3.5.4. Gaussian Beams

Let us consider the transverse component of the plane-wave spectrum

$$\mathbf{E}_{0t} = \mathbf{F}(\omega) \exp\left(-\frac{u^2 X^2}{2} - \frac{v^2 Y^2}{2}\right),$$

---

[19]. As noted in Section 3.5.2, use of the asymptotic expression (3.86) implies that $F(\omega)$ is negligible in the low-frequency limit.

where $\mathbf{F}(\omega)$ is a complex vector in the plane $z = 0$. The transverse component of the frequency-domain field $\mathbf{E}_t(\mathbf{r}, \omega)$ is computed via Eq. (3.81):

$$\mathbf{E}_t(r, \omega) = \frac{1}{(2\pi)^2} \mathbf{F}(\omega) \int\int_{-\infty}^{+\infty} du\,dv \exp(iux + ivy)$$

$$\cdot \exp\left(-\frac{u^2 X^2}{2} - \frac{v^2 Y^2}{2} - i\sqrt{\omega^2 \varepsilon\mu - u^2 - v^2}\, z\right).$$

Note that the integrand becomes negligible when $uX > 1$ and $vY > 1$. If

$$kX \gg 1, \qquad kY \gg 1, \qquad k = \omega\sqrt{\varepsilon\mu}, \tag{3.87}$$

then we can safely set

$$\sqrt{k^2 - u^2 - v^2} \approx k - \frac{u^2 + v^2}{2k},$$

and

$$\mathbf{E}_t(\mathbf{r}, \omega) \approx \frac{1}{(2\pi)^2} \mathbf{F}(\omega) \exp(-ikz) \int\int_{-\infty}^{+\infty} du\,dv \exp(iux + ivy)$$

$$\cdot \exp\left(-\frac{u^2 X^2}{2} - \frac{v^2 Y^2}{2} + i\frac{u^2 + v^2}{k} z\right).$$

This double integral factorizes into the product of two one-dimensional integrals in the variables u and v, respectively:

$$\mathbf{E}_t(\mathbf{r}, \omega) = \mathbf{F}(\omega) \exp(-ikz) \cdot \frac{1}{2\pi} \int_{-\infty}^{+\infty} du \exp(iux) \exp\left(-\frac{u^2 R_x^2}{2}\right)$$

$$\cdot \frac{1}{2\pi} \int_{-\infty}^{+\infty} dv \exp(ivy) \exp\left(-\frac{v^2 R_y^2}{2}\right),$$

with

$$R_x^2 = X^2 - i\frac{z}{k} \quad \text{and} \quad R_y^2 = Y^2 - i\frac{z}{k}.$$

Each integral is readily recognized as the FT of a Gaussian function of complex width R_x and R_y, respectively. From Table 3.1, with the substitution $T \to R_x$, R_y we obtain

$$E_t(\mathbf{r}, \omega) = \mathbf{F}(\omega) \frac{1}{2\pi R_x R_y} \exp(-ikz) \exp\left(-\frac{x^2}{2R_x^2} - \frac{y^2}{2R_x^2}\right).$$

When $X = Y = R_0$ the field exhibits cylindrical symmetry with respect to the z-axis. We have $x^2 + y^2 = r^2$,

$$R_x^2 = R_y^2 = R_x R_y = R_0^2 - i\frac{z}{k},$$

$$\frac{1}{R_x^2} = \frac{1}{R_y^2} = \frac{1}{R^2} + i\frac{kz}{\rho^2},$$

with

$$R^2 = R^2(z) = R_0^2 + \left(\frac{z}{kR_0}\right)^2, \qquad (3.88)$$

$$\rho^2 = \rho^2(z) = \frac{k^2 R_0^4 + z^2}{kz}. \qquad (3.89)$$

Substitution into the expression for the frequency-domain field yields

$$E_t(r, \omega) = \mathbf{F}(\omega) \frac{1}{2\pi R_0 R} \exp\left[-\frac{1}{2}\left(\frac{r}{R}\right)^2\right]$$
$$\cdot \exp\left[-i\left(kz + \frac{r^2}{2\rho^2}\right) + i\tan^{-1}\frac{kz}{(kR_0)^2}\right]. \qquad (3.90)$$

Equation (3.90) represents a *gaussian beam*.

The r-dependent exponentially decaying term $\exp[-1/2(r/R)^2]$ describes the width of the beam vs. z. The function $R(z)$, the *beam waist*, is an estimate of the beam width: when $r > R$, the intensity of the field exhibits a sharp decay with r. Along the curves $r = \alpha \cdot R$, $\alpha = $ const, the field decays as $1/R = \alpha/r$, because the exponential term remains constant. These curves are depicted in Fig. 3.21. When $\alpha = \sqrt{2}$, we derive the plot of the beam waist: the beam opens up as z increases.

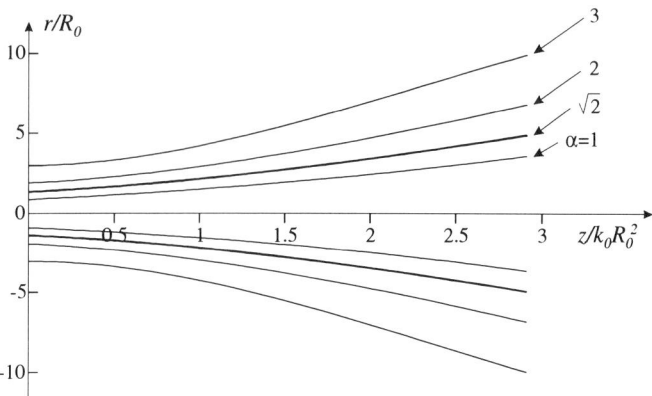

FIGURE 3.21. Curves of constant amplitude for the exponential factor of a gaussian beam. The radial distance is normalized to R_0 and the longitudinal distance to kR_0^2, $1/R_0$ being the plane-wave spectrum width.

The r-dependent phase term $-kz - r^2/2\rho^2 + \tan^{-1}(z/kR_0^2)$ describes the phase front. The phase is constant along the surfaces

$$-kz - \frac{r^2}{2\rho^2(z)} + \tan^{-1}\frac{z}{kR_0^2} = -kz_0 + \tan^{-1}\frac{z_0}{kR_0^2},$$

where z_0 is the value of z for $r = 0$. Since $kR_0 \gg 1$ [see Eq. (3.87)], we neglect the term $\tan^{-1}(z/kR_0^2)$ and have

$$k(z_0 - z) = \frac{r^2}{2\rho^2} \tag{3.91}$$

for the phase front. Equiphase surfaces inside the beam waist are depicted in Fig. 3.22. Note that $r^2/2\rho^2$ ranges from 0 to z/kR_0^2 within the beam waist, so that the change of z within the phase font is very small, which implies also that the change of ρ is negligible there. If we take $\rho = $ const, then it is identified with the curvature of the phase front, i.e., the second derivative of Eq. (3.91) with respect to r.

3.6. SUMMARY AND SELECTED REFERENCES

In this chapter the full spectral domain of electromagnetic fields is employed: frequency (Section 3.2), wavenumber (Section 3.3), and frequency–wavenumber (Section 3.4). A large number of case studies is presented, with systematic use of contour integration in the complex plane [3.1–3.4].

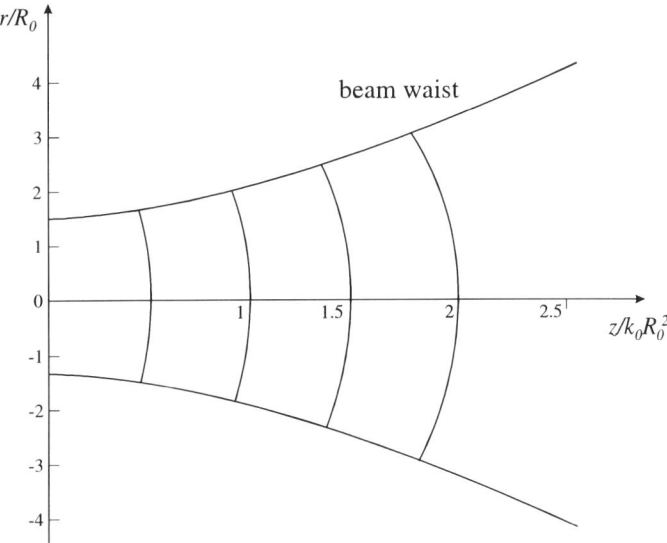

FIGURE 3.22. Curves of constant phase inside the beam waist. Normalization is the same as in Fig. 3.21.

In Section 3.2, constitutive relations (Sections 3.2.2–3.2.4) are derived in the frequency domain, a complementary approach to the time-domain derivation in Sections 1.2.2, 1.2.4, and 1.2.5. In addition, their properties in the complex frequency plane are fully exploited, leading to the dispersion Kramers–Kronig relationships. A cold magnetized plasma is considered in Section 3.2.4 [3.5]; when the pressure is accounted for [3.6], derivation in the frequency-wavenumber domain is necessary (Section 3.4.1). Transient plane-wave propagation in dispersive media is considered in Section 3.2.5. Case studies (Section 3.2.6–3.2.8) include unbounded and bounded plasma as well as conductive media. Transient radiation in dispersive media is presented in Section 3.2.9, where scalar and dyadic Green's functions [2.2] are systematically used, showing also their connection.

The wavenumber domain is examined in Section 3.3. It is important not only for studying propagation and radiation in time-varying media, but also as a counterpart to the frequency domain, and as a preparation for the fully spectral frequency–wavenumber domain. The latter is introduced in Section 3.4, where two important representations of the field are introduced: the resonant field solution in Section 3.4.3 [3.7], and the guided-wave solution in Section 3.4.4 [3.7]. Examples with extensive use of contour integration in the complex plane are included.

The guided-wave representation is the starting point for obtaining the important and useful plane-wave representation of the field [3.8] (Section

3.5.1). Its asymptotic evaluation [3.7] is provided in Section 3.5.2, including a full derivation of the stationary-phase method [3.9] in two dimensions. Examples relevant to plane-wave expansion are given in Section 3.5.3 and 3.5.4, including radiation from apertures and gaussian beams.

References

[3.1] R. Constant and D. Hilbert, *Methods for Mathematical Physics*, Interscience, New York (1953).
[3.2] B. Friedman, *Principles and Techniques of Applied Mathematics*, Wiley, New York (1961).
[3.3] H. Jeffreys and B. S. Jeffreys, *Methods of Mathematical Physics*, Cambridge University Press (1956).
[3.4] E. T. Whittaker and G. N. Watson, *A Course of Modern Analysis*, Cambridge University Press (1940).
[3.5] V. L. Ginzburg, *Propagation of Electromagnetic Waves in Plasmas*, Addison-Wesley, Reading Mass. (1964).
[3.6] J. R. Wait, *Electromagnetics and Plasmas*, Holt, Rinehart & Winston, New York (1968).
[3.7] L. B. Felsen and N. Marcuvite, *Radiation and Scattering of Waves*, IEEE Press, New York (1994).
[3.8] P. C. Clemmow, *The Plane Wave Spectrum Representation of Electromagnetic Fields*, Pergamon Press, New York (1966).
[3.9] N. G. De Bruijn, *Asymptotic Methods in Analysis*, North-Holland, Amsterdam (1958).

4
Narrowband Signals and Phasor Fields

4.1. NARROWBAND SIGNALS

Consider a signal $a(t)$ whose FT, $A(\omega)$, is of the type depicted in Fig. 4.1. The spectrum is concentrated near $\omega = 0$. In the figure $A(\omega)$ is real (and even, see Section 3.2.1), but complex spectra can be considered, the only condition being tht $|A(\omega)|^2$ is essentially *localized* around $\omega = 0$.

Effective widths of space and wavenumber-dependent functions have already been defined in Section 3.5.1, showing that their product cannot be less than a given constant; the latter depends on the assumed (and to some extent arbitrary) definition of the widths. Similar constraints exist in time and frequency domains.

If the definition of Section 3.5.1 is employed, then we have an *effective timewidth* Δt:

$$\int_{-\infty}^{+\infty} dt |a(t)| = |a_0| \Delta t, \tag{4.1a}$$

and an *effective bandwidth*:

$$\int_{-\infty}^{+\infty} d\omega |A(\omega)| = |A_0| \Delta\omega, \tag{4.1b}$$

with

$$\Delta\omega \cdot \Delta t \geqslant 2\pi, \tag{4.2}$$

$|a_0|$ and $|A_0|$ being the maximum values of $|a(t)|$ and $|A(\omega)|$, respectively.

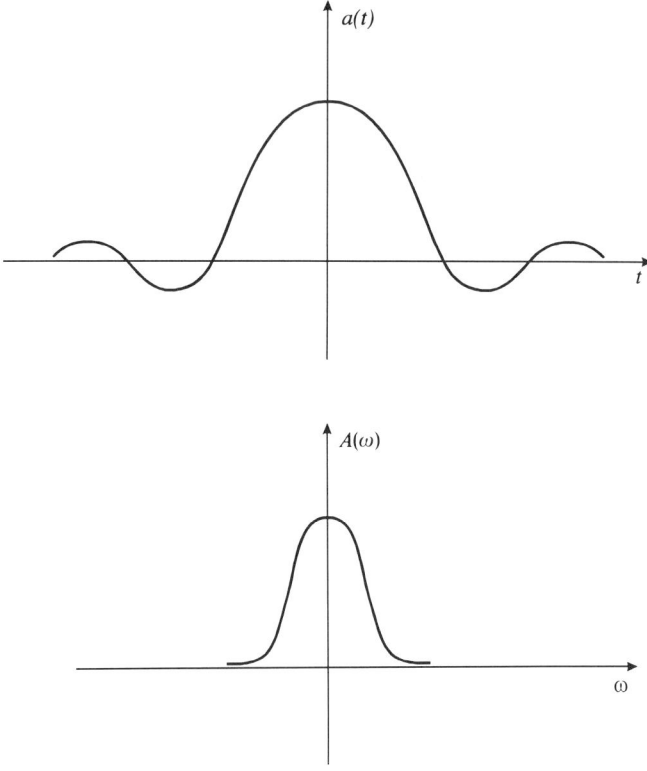

FIGURE 4.1. A narrowband signal $a(t)$ and its associated spectrum $A(\omega)$.

Inequality (4.2) shows that a peaked spectrum implies a spread time waveform (and vice versa), and is usually referred to as the *uncertainty principle* (in quantum mechanics) or *time–frequency product* (in communications theory).

As already noted, alternative definitions of the widths are possible; in any case, the product of the two widths is always larger than a given constant.

Let us now consider the function

$$a(t) = a_e(t) \cos \omega_0 t,$$

where $\omega_0 = 2\pi/T$ is a prescribed angular frequency and T the corresponding (time) period. We have a sinusoidal signal (or *carrier*) whose amplitude is modulated by the function (or *envelope*) $a_e(t)$, as depicted in Fig. 4.2. The associated spectrum,

$$A(\omega) = \tfrac{1}{2} A_e(\omega + \omega_0) + \tfrac{1}{2} A_e(\omega - \omega_0), \tag{4.3}$$

NARROWBAND SIGNALS AND PHASOR FIELDS

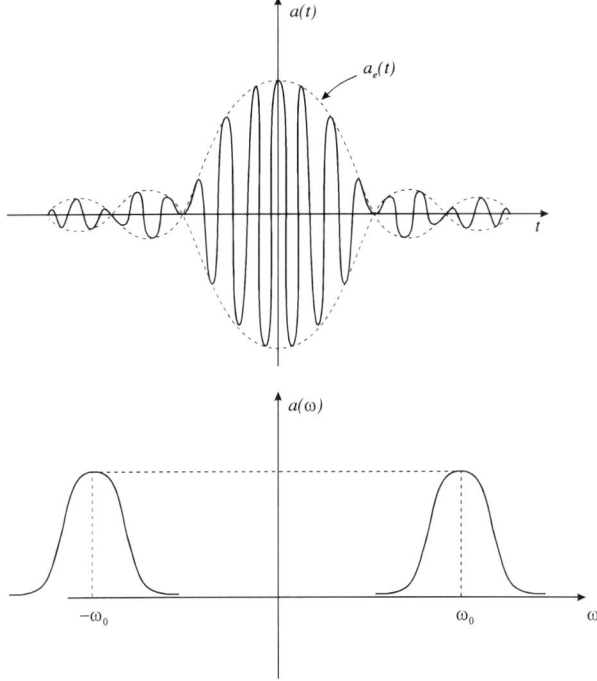

FIGURE 4.2. A modulated signal and its associated spectrum. The dotted line represents the envelope signal $a_e(t)$.

is easily obtained by expanding $\cos\omega_0 t$ in its complex exponential form, and consists of two scaled replicas of $A_e(\omega)$, the FT of $a_e(t)$, translated $\pm\omega_0$ along the ω-axis (see Fig. 4.2).

Most generally,

$$a(t) = a_e(t) \cos[\omega_0 t + \alpha(t)]$$

and

$$\begin{aligned}A(\omega) &= \tfrac{1}{2}\int_{-\infty}^{+\infty} dt \exp[-i(\omega-\omega_0)t]a_e(t)\exp[i\alpha(t)] \\ &+ \tfrac{1}{2}\int_{-\infty}^{+\infty} dt \exp[-i(\omega+\omega_0)t]a_e(t)\exp[-i\alpha(t)] \\ &= \tfrac{1}{2}A_e(\omega-\omega_0) + \tfrac{1}{2}A_e^*(-\omega-\omega_0) \\ &= \tfrac{1}{2}A_e(\omega-\omega_0) + \tfrac{1}{2}A_e(\omega+\omega_0),\end{aligned}$$

with

$$A_e(\omega) = \int_{-\infty}^{+\infty} dt \, \exp(-i\omega t) a_e(t) \exp[i\alpha(t)].$$

Note that we have derived the same formal expression (4.3) because $A^*(-\omega - \omega_0) = A(\omega + \omega_0)$ (see Section 3.2.1).

The function $a(t)$ represents a *narrowband* signal if the inequality

$$\frac{\Delta\omega}{\omega_0} \ll 1 \tag{4.4a}$$

is verified, where $\Delta\omega$ is the effective bandwidth of the envelope signal. Equation (4.2) implies that

$$\Delta t \geq \frac{2\pi}{\Delta\omega} = \frac{2\pi}{\omega_0} \frac{\omega_0}{\Delta\omega} \gg \frac{2\pi}{\omega_0} = T \tag{4.4b}$$

upon use of Eq. (4.4a), i.e., the time rate of the envelope variation is small compared to the corresponding rate of the carrier, namely, the envelope remains essentially constant over several periods of the carrier. This condition is met exactly when the signal bandwidth shrinks to zero: $a_e(t) \to a = \text{const}$, $\alpha(t) \to \alpha = \text{const}$, and $a(t)$ becomes purely sinusoidal (*monochromatic signal*).

In conclusion, narrowband signals are characterized by the product of *two-scale factors*, a (sinusoidal) *fast-varying* and a *slow-varying* one. These signals play a dominant role in many applications, e.g., in the telecommunication area, where the transmitted bandwidths are almost always very small and often negligible compared to the carrier (ω_0) angular frequency.

A convenient way to represent narrowband signals is the following:

$$a(t) = \text{Re}[A(t) \exp(i\omega_0 t)], \tag{4.5a}$$

where

$$A(t) = \frac{2}{T} \int_t^{t+T} d\tau \, \exp(-i\omega_0 \tau) a(\tau) \approx a_e(t) \exp[i\alpha(t)] \tag{4.5b}$$

is a slow-varying complex function. For monochromatic signals α_e and α are constants, and $A = a_e \exp(i\alpha)$ is the *phasor* associated with the sinusoidal signal $a(t) = \alpha_e \cos(\omega_0 t + \alpha)$. For a narrowband signal, $A(t)$ is the generalized phasor associated with the signal $a(t)$.

NARROWBAND SIGNALS AND PHASOR FIELDS

Equations (4.5) are the simplified counterparts for narrowband signals of FTs for wideband signals: Eq. (4.5b) plays the role of FT and Eq. (4.5a) that of its inverse. Note, however, that dimensions of phasors are coincident with dimensions of their corresponding time-domain quantities, unlike true FTs.

To simplify notation, the symbol ω is introduced instead of ω_0 in the absence of any confusion. Also, we usually do not explicitly indicate the (slow) time variation of the phasors.

4.1.1. Phasor Evaluation

Phasors can be more conveniently evaluated by inspection rather than by using Eq. (4.5b). If

$$a(t) = a_e(t) \cos[\omega t + \alpha(t)], \qquad (4.6a)$$

then

$$A = a_e \exp(i\alpha). \qquad (4.6b)$$

When

$$a(t) = a_e(t) \sin[\omega t + \alpha(t)] = a_e(t) \cos\left[\omega t - \frac{\pi}{2} + \alpha(t)\right], \qquad (4.7a)$$

then

$$A = a_e \exp\left(i\alpha - i\frac{\pi}{2}\right) = -i a_e \exp(i\alpha). \qquad (4.7b)$$

4.1.2. Linear Operations on Phasors

To any narrowband signal $a(t)$ we associate the phasor A such that

$$a(t) = \text{Re}[A \exp(i\omega t)];$$

see Eq. (4.5a). We now compute the phasor associated with the time derivative of $a(t)$:

$$y(t) = \frac{da}{dt} = \text{Re}\left\{\frac{d}{dt} A \exp(i\omega t)\right\}$$

$$= \text{Re}\left\{\left[i\omega A + \frac{dA}{dt}\right] \exp(i\omega t)\right\}. \qquad (4.8a)$$

Comparison with Eq. (4.5a) shows that

$$Y = i\omega A + \frac{dA}{dt} \approx i\omega A, \qquad (4.8b)$$

where dA/dt is exactly zero for true phasors ($A =$ const) and negligible otherwise, in view of the narrowband condition (4.4b).

Let the narrowband signal $a(t)$ drive a linear time-invariant circuit of unit response $g(t)$, as depicted in Fig. 4.3, where $u(t)$ is the output of the circuit. The relation between output and input is essentially that examined in Section 1.2, and we have

$$\begin{aligned} u(t) &= \int_{-\infty}^{+\infty} d\tau\, g(t-\tau) a(\tau) \\ &= \operatorname{Re}\left\{ \int_{-\infty}^{+\infty} d\tau\, g(t-\tau) A(\tau) \exp(i\omega\tau) \right\}, \end{aligned}$$

where use has been made of Eq. (4.5a). We have further

$$\int_{-\infty}^{+\infty} d\tau\, g(t-\tau) A(\tau) \exp(i\omega\tau) = \exp(i\omega t) \int_{-\infty}^{+\infty} d\xi\, g(\xi) A(t-\xi) \exp(-i\omega\xi),$$

with the substitution $t - \tau = \xi$. We expand $A(t - \xi)$ in the form

$$A(t - \xi) = A(t) - \frac{dA}{dt} \xi + \cdots$$

and note that

$$\xi \exp(-i\omega\xi) \to \frac{1}{(-i)} \frac{\partial}{\partial \omega} \exp(-i\omega\xi).$$

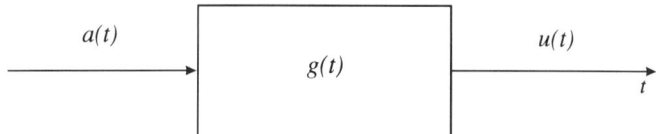

FIGURE 4.3. Response of a linear time-invariant filter to a narrowband signal.

Hence

$$u(t) = \text{Re}\left\{\exp(i\omega t)\left[A - i\frac{dA}{dt}\frac{\partial}{\partial\omega} + \cdots\right]\int_{-\infty}^{+\infty}d\xi\exp(-i\omega\xi)g(\xi)\right\}$$

$$= \text{Re}\left\{\exp(i\omega t)\left[AG(\omega) - i\frac{dA}{dt}\frac{dG}{d\omega} + \cdots\right]\right\}, \quad (4.9a)$$

where $G(\omega)$ is the FT of $g(t)$, i.e., the transfer function of the system.
Comparison with Eq. (4.5a) shows that

$$U \approx G(\omega)A - i\frac{dA}{dt}\frac{dG}{d\omega} \approx AG(\omega), \quad (4.9b)$$

where again the relation is exact for a true phasor (A = const) and usually met unless $dG/d\omega$ is not too large.

For the signal du/dt we have

$$\frac{du}{dt} = \text{Re}\left\{\int_{-\infty}^{+\infty}d\tau\,\frac{dg(t-\tau)}{dt}A(\tau)\exp(i\omega\tau)\right\}, \quad (4.10a)$$

and the first two terms of the expansion of the associated phasor are

$$i\omega G(\omega)A + \frac{dA}{dt}\frac{d(\omega G)}{d\omega}, \quad (4.10b)$$

by using the result (4.9b) and noting that the FT of dg/dt is just $i\omega G(\omega)$. It is again noted that the second term appearing in Eq. (4.10b) is exactly zero for true phasors (A = const) and usually (but not always, see Section 4.3.2) negligible.

4.1.3. Response of a Linear Time-Invariant Circuit to a Narrowband Signal

Let us consider the linear time-invariant circuit of Fig. 4.3. We have

$$U = G(\omega)A,$$

with $U = |U|\exp(i\psi)$ and $A = |A|\exp(i\alpha)$ the phasors associated with $u(t)$ and $a(t)$, respectively.

In the time domain

$$u(t) = \text{Re}[U\exp(i\omega t)] = |A||G|\cos(\omega t + \alpha + \chi),$$

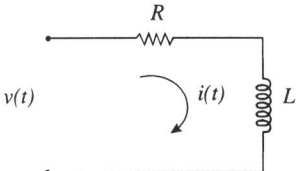

FIGURE 4.4. A linear circuit driven by the narrowband voltage $v(t)$.

with

$$G = |G| \exp(i\chi).$$

For the simple circuit of Fig. 4.4 we identify the input $a(t)$ with the applied voltage $v(t) = V \cos \omega t$, and the output $u(t)$ with the current $i(t)$, where

$$G = \frac{1}{R + i\omega L}, \quad \chi = -\tan^{-1} \frac{\omega L}{R}, \quad \text{and} \quad i(t) = \frac{|V|}{\sqrt{R^2 + \omega^2 L^2}} \cos(\omega t + \chi).$$

4.1.4. Quadratic Averages

Time averages of products of narrowband signals over the period T of the carrier signals are of interest. Clearly, the average of the single narrowband signal is very close to zero, and exactly zero for true phasors.

For the square of the signal we have

$$\overline{a^2(t)} = \frac{1}{T} \int_t^{t+T} d\tau\, a^2(\tau) \approx \frac{a_e^2(t)}{T} \int_t^{t+T} d\tau \cos^2[\omega\tau + \alpha(\tau)]$$
$$= \tfrac{1}{2} a_e^2(t) \cos \alpha(t) = \operatorname{Re}[\tfrac{1}{2} A A^*], \tag{4.11}$$

in view of the narrowband condition (4.4b). Most generally, set

$$a(t) = a_e(t) \cos[\omega t + \alpha(t)] \quad \text{and} \quad b(t) = b_e(t) \cos[\omega t + \beta(t)],$$

in which case

$$\frac{1}{T} \int_t^{t+T} d\tau\, a(\tau) b(\tau) = \tfrac{1}{2} a_e(t) b_e(t) \cos[\alpha(t) - \beta(t)]$$
$$= \operatorname{Re}[\tfrac{1}{2} A B^*], \tag{4.12}$$
$$A = a_e \exp(i\alpha), \quad \text{and} \quad B = b_e \exp(i\beta).$$

NARROWBAND SIGNALS AND PHASOR FIELDS 149

The integrals (4.11) and (4.12) are slowly varying functions over the time scale T, and are constant for monochromatic signals.

Similarly, we obtain

$$\frac{1}{T} \int_t^{t+T} d\tau\, a(\tau) \frac{db}{d\tau} = \frac{\omega}{2} a_e(t) b_e(t) \sin[\alpha(t) - \beta(t)]$$
$$= \omega \operatorname{Im}[\tfrac{1}{2} AB^*]. \quad (4.13)$$

Note that Eq. (4.13) can also be written as

$$\omega \operatorname{Im}\left[\frac{1}{2} AB^*\right] = \omega \operatorname{Re}\left[-\frac{i}{2} AB^*\right] = \operatorname{Re}\left[\frac{1}{2} A(i\omega B)^*\right],$$

which is consistent with rule (4.8b) for computing the phasor of the derivative of a narrowband signal.

4.1.5. Power and Phasors

Consider the simple circuit of Fig. 4.5, $v(t) = \operatorname{Re}[V \exp(i\omega t)]$ and $i(t) = \operatorname{Re}[I \exp(i\omega t)]$ being the narrowband voltages and currents at its input terminals. We define

$$P = \tfrac{1}{2} VI^* = P_1 + iP_2, \quad (4.14)$$

the *complex power* at the circuit input terminals.

Equation (4.12) shows that the real part P_1 of Eq. (4.14) equals the time average, over the period T, of the circuit input power,

$$P_1 = \overline{vi} = \overline{v_R i} + \overline{v_L i} + \overline{v_C i}$$
$$= \overline{Ri^2} + \overline{L \frac{di}{dt} i} + \overline{v_C C \frac{dv_C}{dt}} = \overline{Ri^2}$$

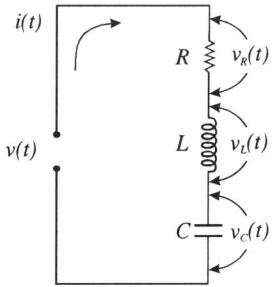

FIGURE 4.5. A simple circuit driven by narrowband signals.

because

$$\overline{L\frac{di}{dt}i} = \frac{1}{2}\overline{L\frac{di^2}{dt}} = 0,$$

and similarly for $\overline{v_C i}$. We have also

$$P_1 = \tfrac{1}{2}R|I|^2,$$

which corresponds to the average power delivered to the circuit and dissipated over the resistor R. For this reason P_1 is referred to as the *real power*.

Equation (4.13) shows that, for the imaginary coefficient P_2 of Eq. (4.14),

$$\omega P_2 = \overline{v\frac{di}{dt}} = \overline{v_R\frac{di}{dt}} + \overline{v_L\frac{di}{dt}} + \overline{v_C\frac{di}{dt}}$$

$$= \overline{Ri\frac{di}{dt}} + \overline{L\frac{di}{dt}\frac{di}{dt}} - \overline{\frac{dv_C}{dt}C\frac{dv_C}{dt}}$$

$$= \omega^2 \tfrac{1}{2}L|I|^2 - \omega^2 \tfrac{1}{2}C|V_C|^2,$$

upon integrating by parts the term $v_C di/dt$. Accordingly

$$P_2 = 2\omega(\tfrac{1}{4}L|I|^2 - \tfrac{1}{4}C|V_C|^2),$$

and is equal to 2ω times the difference between the average energies stored in the inductor and the capacitor, respectively. Furthermore,

$$P_2 = \frac{1}{2}\left(\omega L - \frac{1}{\omega C}\right)|I|^2 = \frac{1}{2}X|I|^2,$$

and is referred to as the *reactive power*.

The real power P_1 represents the average power flux from the sources to the circuit (*irreversible power*). The reactive power P_2 represents also an average power flux (*reversible power*), from the sources to the circuit and vice versa, to compensate for the unbalance of energy storage in the inductors (magnetic energy) and condensers (electric energy) of the circuit. The average magnetic energy prevails in the circuit if $P_1 > 0$, and vice versa if $P_2 < 0$. When $P_2 = 0$ the average magnetic energy equals the average electric energy.

4.1.6. Bandlimited Signals

Consider the *bandlimited signal* characterized by the envelope FT spectrum,

$$A_e(\omega) = \frac{2\pi}{2\Omega} a_0 S(\omega) \text{ rect}\left(\frac{\omega}{2\Omega}\right),$$

exactly zero outside the (angular) frequency band $(-\Omega, \Omega)$. Bandlimited signals are important, because real signals are always bandlimited due to the essentially low-pass response of any (physically realizable) transfer device.

When $S(\omega) = 1$ the spectrum is constant within the bandwidth, and we obtain from Table 3.1

$$a_e(t) = a_0 \text{ sinc}(\Omega t). \tag{4.15}$$

Equation (4.15) represents the envelope (and also the generalized phasor) of the narrowband signal

$$a(t) = a_0 \text{ sinc}(\Omega t) \cos \omega t, \quad \Omega \ll \omega,$$

presented in Fig. 4.6.

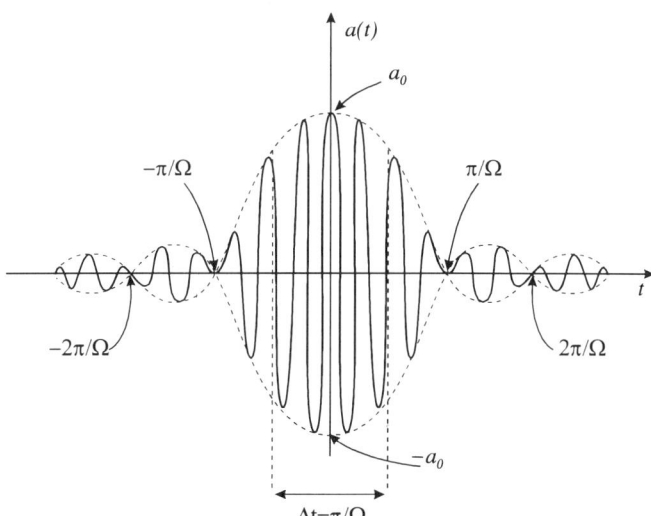

FIGURE 4.6. Time-domain representation (solid line) of a bandlimited signal whose envelope (dotted line) is given by Eq. (4.15).

An interesting representation of the signal envelope $a_e(t)$ is possible even when $S(\omega) \neq$ const. With reference to Fig. 4.7, we generate a periodic spectrum $S'(\omega)$ and note that

$$\frac{\pi}{\Omega} a_0 S(\omega) \operatorname{rect}\left(\frac{\omega}{2\Omega}\right) = \frac{\pi}{\Omega} a_0 S'(\omega) \operatorname{rect}\left(\frac{\omega}{2\Omega'}\right), \tag{4.16}$$

provided that $\Omega' \geqslant \Omega$. The FT of Eq. (4.16), i.e., $a_e(t)$, can be computed as the convolution of the FTs of the $S'(\omega)$ and of the window $(\pi a_0/\Omega)\operatorname{rect}(\omega/2\Omega')$. We first expand $S'(\omega)$ in a Fourier series,

$$S'(\omega) = \sum_{n=-\infty}^{+\infty} S_n \exp\left(-i\frac{2\pi n}{2\Omega'}\omega\right),$$

whose coefficients need not be evaluated for subsequent analysis. Then, its FT is simply given by

$$s'(t) = \sum_{n=-\infty}^{+\infty} S_n \delta\left(t - \frac{n\pi}{\Omega'}\right).$$

The FT of the window is $a_0(\Omega'/\Omega)\operatorname{sinc}(\Omega' t)$; see Table 3.1. Hence

$$a_e(t) = a_0 \frac{\Omega'}{\Omega} \int_{-\infty}^{+\infty} d\tau \operatorname{sinc}(\Omega'\tau) \sum_{n=-\infty}^{+\infty} S_n \delta\left(t - \tau - \frac{n\pi}{\Omega'}\right)$$

$$= a_0 \frac{\Omega'}{\Omega} \sum_{n=-\infty}^{+\infty} S_n \operatorname{sinc}[\Omega'(t - t_n)], \qquad t_n = \frac{n\pi}{\Omega'}.$$

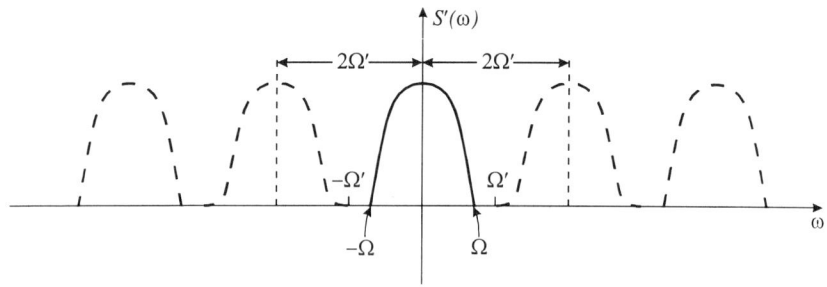

FIGURE 4.7. Periodic spectrum $S'(\omega)$ obtained by successive replicas of the bandlimited spectrum $S(\omega)$, $\Omega' \geqslant \Omega$.

NARROWBAND SIGNALS AND PHASOR FIELDS 153

For $t = t_m$, only the term $n = m$ of the series is different from zero because

$$\Omega'(t_m - t_n) = (m - n)\pi,$$

and we have $a(t_m) = a_0 \Omega' S_m / \Omega$. Accordingly

$$a_e(t) = \sum_{n=-\infty}^{+\infty} a_n \operatorname{sinc}[\Omega'(t - t_n)], \qquad a_n = a_e(t_n), \qquad (4.17)$$

which is usually referred to as the *sampling expansion*, or the Whittaker–Shannon–Kotelnikov[1] series for the bandlimited function. The latter is *exactly* reconstructed by means of its *samples* $a_n = a_e(t_n)$ taken at intervals $\Delta t = \pi/\Omega'$ $\leqslant \pi/\Omega$, 2Ω being the bandwidth of the function. The required minimum number of samples Ω/π per unit time is usually referred to as the *Nyquist*[2] *rate*. The ratio

$$\chi = \frac{\Omega'}{\Omega}, \qquad \Omega' \geqslant \Omega, \qquad (4.18)$$

is the *oversampling factor*.

In all practical situations the infinite series (4.17) must be truncated to finite order, and this results in a *truncation error*:

$$\delta_N(t) = a_e(t) - \sum_{n=-N}^{N} a_n \operatorname{sinc}[\Omega'(t - t_n)].$$

An overall (normalized) estimate of this error is the following:

$$\Delta_N^2 = \frac{\int_{-\infty}^{+\infty} dt \, \delta_N^2(t)}{\int_{-\infty}^{+\infty} dt \, a^2(t)} = \frac{\sum_{|n|=N+1}^{\infty} a_n^2}{\sum_{-\infty}^{+\infty} a_n^2},$$

where the integrals have been evaluated by means of Eqs. (C.6) and (C.7). Since a bandlimited function decays at least as $1/t$ for large values of time, Δ_N^2 decays at least as $1/N$ for large values of N.

1. Edmund Taylor Whittaker: Birkdale, Lancashire (UK), 1873–Edinburgh (Scotland), 1956.
   Claude Elwood Shannon: Gaylord (Michigan, USA), 1916.
2. Alexander Petrovich Kotelnikov: Kazan (Russia), 1865–Moscow, 1944.
   Herry Nyquist: Nilsby (Sweden), 1889–Harligen (Texas), 1976.

4.1.7. Almost Bandlimited Signals

Functions that exhibit a theoretical infinite bandwidth, but concentrated within a limited region (the effective bandwidth) are referred to as *almost bandlimited*. Use of the sampling expansion to represent (or *reconstruct*) an almost bandlimited function requires truncation of the latter to a finite bandwidth $|\omega| \leq \Omega$. This generates an error—the *aliasing error*—because successive replicas of the spectrum $S(\omega)$ overlap along the ω-axis; see Fig. 4.8. The tails of the spectrum $S(\omega)$ are wrapped inside its truncated version $S_\Omega(\omega)$. This happens also for truly bandlimited functions if $\Omega' < \Omega$, i.e., if the sampling rate is smaller than the Nyquist one.

Let $a_e(t)$ be the almost bandlimited function and $a_\Omega(t)$ its bandlimited version, different from $a_e(t)$ by the amount

$$\delta_\Omega(t) = a_e(t) - a_\Omega(t).$$

An overall (normalized) estimate of this aliasing error is the following:

$$\Delta_\Omega^2 = \frac{\int_{-\infty}^{+\infty} dt [a_e(t) - a_\Omega(t)]^2}{\int_{-\infty}^{+\infty} dt \, a_e^2(t)} = \frac{\int_{-\infty}^{-\infty} d\omega |A_e(\omega) - A_\Omega(\omega)|^2}{\int_{-\infty}^{+\infty} d\omega |A_e(\omega)|^2},$$

where $A_e(\omega)$ and $A_\Omega(\omega)$ are the FT spectra of $a_e(t)$ and $a_\Omega(t)$, respectively, and use has been made of Parseval[3] theorem (D.32).

For almost bandlimited functions only the first two nearby replicas are important, so

$$\int_{-\infty}^{+\infty} d\omega |A_e(\omega) - A_\Omega(\omega)|^2 \approx 4 \int_\Omega^\infty d\omega |A_e(\omega)|^2.$$

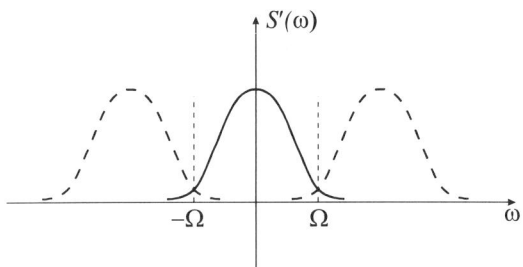

FIGURE 4.8. The aliasing error.

3. Marc-Antoine Parseval des Chênes: Rosière-aux-Salines (France), 1755–Paris, 1836.

For the gaussian signal

$$a_e(t) = a_0 \exp\left(-\frac{t^2}{2T^2}\right),$$

we have from Table 3.1

$$A_e(\omega) = a_0\sqrt{2\pi}\,T \exp\left(-\frac{\omega^2 T^2}{2}\right),$$

and the effective bandwidth is readily computed to be $\sqrt{2\pi}/T$; see also Section 3.5.1. We integrate by parts the numerator of Δ_Ω^2 and obtain

$$\Delta_\Omega^2 \approx \frac{2\exp(-\Omega^2 T^2)}{\sqrt{\pi}\,\Omega T}.$$

When $2\Omega = \sqrt{2\pi/T}$ we have $\Delta_\Omega^2 = 0.187$. Should 2Ω be slightly increased by a factor $\sqrt{2}$, we would have $\Delta_\Omega^2 \approx 0.027$.

4.2. COMPLEX VECTORS

Narrowband signals considered in Section 4.1 are represented by scalar functions. Similarly, we have narrowband vectors,

$$\mathbf{a}(\mathbf{r}, t) = a_1(\mathbf{r}, t)\hat{\mathbf{x}}_1 + a_2(\mathbf{r}, t)\hat{\mathbf{x}}_2 + a_3(\mathbf{r}, t)\hat{\mathbf{x}}_3, \tag{4.19}$$

where the components of the vector along the coordinate system (x_1, x_2, x_3) are narrowband functions of time.

As we associate phasors with narrowband signals, we can similarly associate a *complex vector* with the narrowband vector (4.19),

$$\mathbf{A} = A_1\hat{\mathbf{x}}_1 + A_2\hat{\mathbf{x}}_2 + A_3\hat{\mathbf{x}}_3, \tag{4.20}$$

where A_1, A_2, and A_3 are the phasors associated with a_1, a_2, and a_3, respectively.

Equation (4.20) is a first possible representation of a complex vector, different from the usual real ones because the vector components are complex functions. A second representation is obtained by summing all the real components, thus obtaining a real vector $\mathbf{A}_p$, and then all the imaginary components, thus generating a purely imaginary vector $i\mathbf{A}_q$. Hence

$$\mathbf{A} = \mathbf{A}_p + i\mathbf{A}_q = A_p\hat{\mathbf{p}} + iA_q\hat{\mathbf{q}}, \tag{4.21}$$

where the vector $\mathbf{A}$ is represented as the *complex* combination of two real vectors, which need not be aligned and define the *polarization plane* (indeterminate when the two vectors are aligned). Two real vectors are necessary, in general, to specify a complex vector (see Fig. 4.9); addition of the vectors must be done separately for the real and imaginary parts, respectively. Accordingly, the usual graphical representation of $\mathbf{A}$ as a single vector is just conventional: two vectors (perhaps of different colors) should be used.

A third representation of a complex vector is possible. We first define the module (see also Section 4.2.1)

$$|\mathbf{A}| = \sqrt{\mathbf{A} \cdot \mathbf{A}^*} \tag{4.22}$$

and the phase

$$\tan \phi = \frac{A_q}{A_p} \tag{4.23}$$

of the vector $\mathbf{A}$. We have (see Fig. 4.9)

$$\begin{aligned}\mathbf{A} &= |\mathbf{A}| \cos \phi \hat{\mathbf{p}} + i|\mathbf{A}| \sin \phi \hat{\mathbf{q}} \\ &= |\mathbf{A}| \exp(-i\phi) \frac{\hat{\mathbf{p}} - \hat{\mathbf{q}}}{2} + |\mathbf{A}| \exp(i\phi) \frac{\hat{\mathbf{p}} + \hat{\mathbf{q}}}{2} \\ &= A^* \sin \frac{\theta}{2} \hat{\mathbf{u}} + A \cos \frac{\theta}{2} \hat{\mathbf{v}},\end{aligned} \tag{4.24}$$

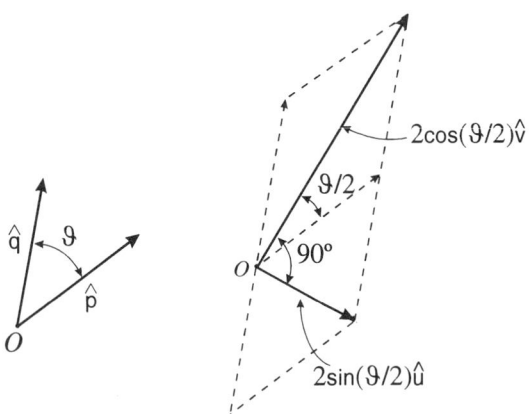

FIGURE 4.9. Representation of a complex vector.

where $(\hat{\mathbf{u}}, \hat{\mathbf{v}})$ are two orthogonal unit vectors with

$$\hat{\mathbf{u}}\sin\frac{\theta}{2} = \frac{\hat{\mathbf{p}} - \hat{\mathbf{q}}}{2}, \quad \hat{\mathbf{v}}\cos\frac{\theta}{2} = \frac{\hat{\mathbf{p}} + \hat{\mathbf{q}}}{2}, \quad \hat{\mathbf{u}} \times \hat{\mathbf{v}} = \hat{\mathbf{p}} \times \hat{\mathbf{q}},$$

and $A = |A|\exp(i\phi)$ is a phasor. Unlike Eq. (4.21), Eq. (4.24) represents **A** as a *real* combination of two (spatially) orthogonal complex vectors. The real and imaginary parts of Eq. (4.24) just provide the vectors $\mathbf{A}_p$ and $\mathbf{A}_q$ of Eq. (4.21), as expected. The connections between the three types of representation are listed in Table 4.1.

The relation linking vectors in the time domain to the associated complex vectors is the generalization of Eq. (4.5a):

$$\mathbf{a}(t) = \mathrm{Re}[\mathbf{A}\exp(i\omega t)].$$

In terms of representation (4.21),

$$\mathbf{a}(t) = A_p \cos\omega t\,\hat{\mathbf{p}} - A_q \sin\omega t\,\hat{\mathbf{q}}, \tag{4.25a}$$

while representation (4.24) leads to the equivalent form:

$$\mathbf{a}(t) = |A|\sin\frac{\theta}{2}\cos(\omega t - \phi)\hat{\mathbf{u}} + |A|\cos\frac{\theta}{2}\cos(\omega t + \phi)\hat{\mathbf{v}}. \tag{4.25b}$$

Both representations show that the vector $\mathbf{a}(t)$ lies in the polarization plane $(\hat{\mathbf{p}}, \hat{\mathbf{q}})$, coincident with the plane $(\hat{\mathbf{u}}, \hat{\mathbf{v}})$, and that it changes in general both its amplitude and its direction as time elapses. Some cases are of interest.

When $\theta = 0$ Eq. (4.25b) yields

$$\mathbf{a}(t) = |A|\cos(\omega t + \phi)\hat{\mathbf{v}}. \tag{4.26a}$$

TABLE 4.1. Connection between the Representations of a Complex vector

$\mathbf{A} = A_1\hat{\mathbf{x}}_1 + A_2\hat{\mathbf{x}}_2 + A_3\hat{\mathbf{x}}_3$	$A_1 = A'_1 + iA''_1$ $A_2 = A'_2 + iA''_2$	$A_3 = A'_3 + iA''_3$
$\mathbf{A} = A_p\hat{\mathbf{p}} + iA_q\hat{\mathbf{q}}$ $= \mathbf{A}_p + i\mathbf{A}_q$	$\mathbf{A}_p = A'_1\hat{\mathbf{x}}_1 + A'_2\hat{\mathbf{x}}_2 + A'_3\hat{\mathbf{x}}_3$ $\mathbf{A}_q = A''_1\hat{\mathbf{x}}_1 + A''_2\hat{\mathbf{x}}_2 + A''_3\hat{\mathbf{x}}_3$	$\hat{\mathbf{p}} = \hat{\mathbf{u}}\sin(\theta/2) + \hat{\mathbf{v}}\cos(\theta/2)$ $\hat{\mathbf{q}} = -\hat{\mathbf{u}}\sin(\theta/2) + \hat{\mathbf{v}}\cos(\theta/2)$
$\mathbf{A} = A^*\sin(\theta/2)\hat{\mathbf{u}} + A\cos(\theta/2)\hat{\mathbf{v}}$	$A = A\exp(i\phi)$ $A^2 = A_p^2 + A_q^2$ $\tan^{-1}\phi = A_q/A_p$	$2\hat{\mathbf{u}}\sin(\theta/2) = \hat{\mathbf{p}} - \hat{\mathbf{q}}$ $2\hat{\mathbf{v}}\cos(\theta/2) = \hat{\mathbf{p}} + \hat{\mathbf{q}}$

The vector **a**(t) does not change its direction and its tip moves along a straight line; see Fig. 4.10. This is the case of *linear polarization*: the vectors $\mathbf{A}_p$ and $\mathbf{A}_q$ [see Eqs. (4.21) and (4.24)] are aligned along the same $\hat{\mathbf{v}}$-axis ($\hat{\mathbf{p}} = \hat{\mathbf{q}} = \hat{\mathbf{v}}$):

$$\mathbf{A} = A\hat{\mathbf{v}} = |A|\cos\phi\hat{\mathbf{v}} + i|A|\sin\phi\hat{\mathbf{v}}. \tag{4.26b}$$

When $\theta° = 90°$ and $\phi = \pm\pi/4$ we have, again from Eq. (4.25b),

$$\mathbf{a}(t) = \frac{|A|}{\sqrt{2}}\left[\cos\left(\omega t \mp \frac{\pi}{4}\right)\hat{\mathbf{u}} + \cos\left(\omega t \pm \frac{\pi}{4}\right)\hat{\mathbf{v}}\right]$$

$$= \frac{|A|}{\sqrt{2}}\left[\cos\left(\omega t \mp \frac{\pi}{4}\right)\hat{\mathbf{u}} \mp \sin\left(\omega t \mp \frac{\pi}{4}\right)\hat{\mathbf{v}}\right]. \tag{4.27a}$$

The vector **a**(t) maintains a constant modulus $|\mathbf{A}|/\sqrt{2}$ and its tip moves along a circle with angular velocity ω (see Fig. 4.10). This is the case of *circular*

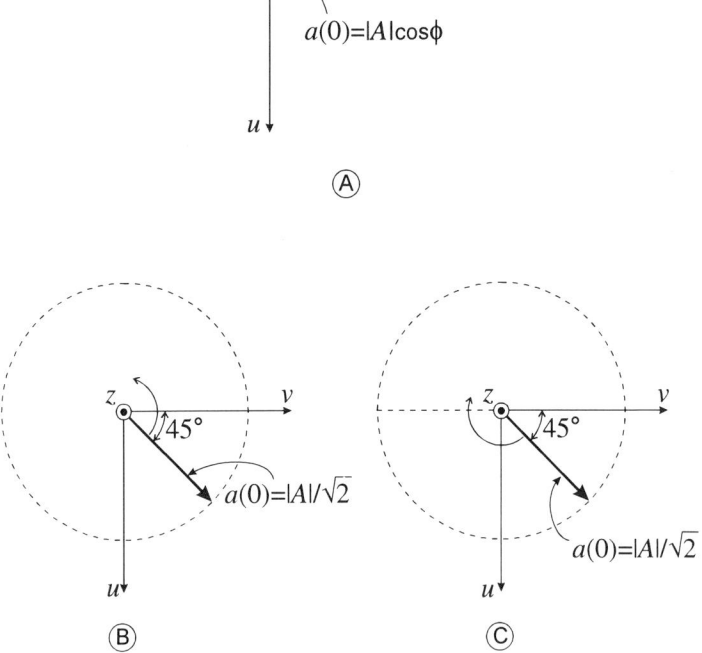

FIGURE 4.10. (A) Linear, (B) positive, and (C) negative circular polarization.

positive ($\phi = -\pi/4$) or *negative* ($\phi = \pi/4$) *polarization*. In the former case, rotation is the same as a screwdriver progressing along the $+z$-axis; in the latter case, the screwdriver is regressing along the $-z$-axis. As seen by an observer oriented along $\hat{z} = \hat{u} \times \hat{v}$, the vector rotates in a counterclockwise sense for $\phi = -\pi/4$ and in a clockwise sense for $\phi = \pi/2$. The vectors $\mathbf{A}_p$ and $\mathbf{A}_q$ [see Eqs. (4.24) and (4.21)] are orthogonal with the same modulus:

$$\mathbf{A} = \frac{A}{\sqrt{2}} \hat{u} + \frac{A}{\sqrt{2}} \hat{v} = \frac{|A|}{\sqrt{2}} \frac{\hat{u}+\hat{v}}{\sqrt{2}} \pm i \frac{|A|}{\sqrt{2}} \frac{\hat{v}-\hat{u}}{\sqrt{2}}$$
$$= \frac{|A|}{\sqrt{2}} \hat{p} \pm i \frac{|A|}{\sqrt{2}} \hat{q}. \qquad (4.27b)$$

In the general case the tip of the vector $\mathbf{a}(t)$ moves along an ellipse in the polarization plane; see Fig. 4.11. We compute $|a(t)|^2$, and set its derivative with respect to ωt equal to zero, to determine the major (L) and minor (l) axes of the ellipse. Lengthy computations show that these are given by

$$\frac{L}{l} = \frac{|A|}{\sqrt{2}} [1 \pm (1 - \sin^2\theta \sin^2 2\phi)^{1/2}]^{1/2}, \qquad (4.28a)$$

where the major axis is rotated through an angle α from the $\hat{v}$-axis and is given by

$$\tan 2\alpha = \tan \theta \cos 2\phi. \qquad (4.28b)$$

This is the case of *elliptic negative* ($0 < \phi < \pi/2$) or *positive* ($-\pi/2 < \phi < 0$) *polarization*.

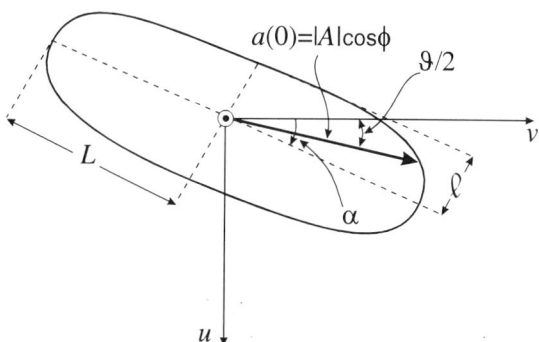

FIGURE 4.11. Elliptic polarization.

4.2.1. Scalar Product for Complex Vectors and Some of Their Properties

Given two complex vectors **A** and **B** (also simply referred to as vectors, if there is no ambiguity), it is usually convenient to define the scalar product of the first vector times the complex conjugate of the second one: $\mathbf{A} \cdot \mathbf{B}^*$. This is the natural extension to complex vectors of the phasor product defined in Section 4.1.4 and leading to the narrowband-signal quadratic averages. It is also a convenient way to extend geometrical properties of real vectors to complex vectors.

We start from the Schwartz[4] inequality (D.34) applied to vectors:

$$|\mathbf{A} \cdot \mathbf{B}^*|^2 \leqslant |\mathbf{A}|^2 |\mathbf{B}|^2, \qquad (4.29)$$

the maximum being obtained for $\mathbf{B} = \alpha \mathbf{A}$, with α a constant. In the latter case, the real as well as imaginary parts of the two vectors are mutually parallel.

Two vectors are said to be *parallel*[5] if

$$|\mathbf{A} \cdot \mathbf{B}^*|^2 = |\mathbf{A}|^2 |\mathbf{B}|^2 \qquad (4.30)$$

and *orthogonal* if

$$\mathbf{A} \cdot \mathbf{B}^* = 0. \qquad (4.31)$$

In the former case $\mathbf{B} = \alpha \mathbf{A}$, as noted.

The cases of linear and circular polarization are considered in detail as examples.

For linear polarization we have, from Eq. (4.26b),

$$\mathbf{A} = A\hat{\mathbf{a}} = |A|\exp(i\alpha)\hat{\mathbf{a}} \quad \text{and} \quad \mathbf{B} = B\hat{\mathbf{b}} = |B|\exp(i\beta)\hat{\mathbf{b}},$$

so

$$\mathbf{A} \cdot \mathbf{B}^* = |A||B|\exp[i(\alpha - \beta)]\hat{\mathbf{a}} \cdot \hat{\mathbf{b}}$$

The vectors are parallel[6] if $|\hat{\mathbf{a}} \cdot \hat{\mathbf{b}}| = 1$ (same direction); they are orthogonal if $|\hat{\mathbf{a}} \cdot \hat{\mathbf{b}}| = 0$ (spatially orthogonal).

---

4. Hermann Schwartz: Hermsdorf (Germany), 1787–Berlin, 1826.
5. Definition (4.30) is related to inequality (4.29) and does not necessarily render the product $\mathbf{A} \cdot \mathbf{B}^*$ real. This may be convenient if the scalar product is related to power averages; see Section 4.1.4. In this case Eq. (4.30) should be substituted by $\mathbf{A} \cdot \mathbf{B}^* = |A||B|$ and the constant α must be real.
6. Maximization of $\mathrm{Re}[\mathbf{A} \cdot \mathbf{B}^*]$ requires $\alpha = \beta$.

For circular polarization of the same (say positive) sign, we have from Eq. (4.27b)

$$\mathbf{A} = \frac{|A|}{\sqrt{2}} [\hat{\mathbf{a}} + i\hat{\mathbf{a}} \times \hat{\mathbf{z}}] \quad \text{and} \quad \mathbf{B} = \frac{|B|}{\sqrt{2}} [\hat{\mathbf{b}} + i\hat{\mathbf{b}} \times \hat{\mathbf{z}}],$$

the vectors lying in the plane $z = 0$, so

$$\mathbf{A} \cdot \mathbf{B}^* = |A||B| [\hat{\mathbf{a}} \cdot \hat{\mathbf{b}} - i\hat{\mathbf{z}} \cdot (\hat{\mathbf{a}} \times \hat{\mathbf{b}})].$$

The vectors are always[7] parallel because $|\mathbf{A} \cdot \mathbf{B}^*|^2 = |\mathbf{A}|^2 |\mathbf{B}|^2$.
For circular polarization of opposite sign

$$\mathbf{A} = \frac{|A|}{\sqrt{2}} (\hat{\mathbf{a}} + i\hat{\mathbf{z}} \times \hat{\mathbf{a}}) \quad \text{and} \quad \mathbf{B} = \frac{|B|}{\sqrt{2}} (\hat{\mathbf{b}} - i\hat{\mathbf{z}} \times \hat{\mathbf{b}}),$$

so

$$\mathbf{A} \cdot \mathbf{B}^* = 0.$$

The vectors are always orthogonal.

Coordinate systems appropriate for complex vectors are defined by mutually orthogonal unit vectors:

$$\hat{\mathbf{x}}_n \cdot \hat{\mathbf{x}}_m^* = \delta_{nm}.$$

Accordingly, a vector can be represented in component form as

$$\mathbf{A} = (\mathbf{A} \cdot \hat{\mathbf{x}}_1^*)\hat{\mathbf{x}}_1 + (\mathbf{A} \cdot \hat{\mathbf{x}}_2^*)\hat{\mathbf{x}}_2 + (\mathbf{A} \cdot \hat{\mathbf{x}}_3^*)\hat{\mathbf{x}}_3, \qquad (4.32)$$

which is the generalization of Eq. (4.20) because the unit vectors need not be real. For instance, let

$$\hat{\mathbf{x}}_1 = \frac{1}{\sqrt{2}} (\hat{\mathbf{x}} + i\hat{\mathbf{y}}), \qquad \hat{\mathbf{x}}_2 = \frac{1}{\sqrt{2}} (\hat{\mathbf{x}} - i\hat{\mathbf{y}}), \qquad \hat{\mathbf{x}}_3 = i\hat{\mathbf{z}},$$

so that

$$\hat{\mathbf{x}}_n \cdot \hat{\mathbf{x}}_m^* = \delta_{mn} \quad \text{and} \quad \hat{\mathbf{x}}_n \times \hat{\mathbf{x}}_m = \hat{\mathbf{x}}_s^*.$$

---

7. Maximization of $\text{Re}[\mathbf{A} \cdot \mathbf{B}^*]$ requires $|\hat{\mathbf{a}} \cdot \hat{\mathbf{b}}| = 1$.

In this coordinate system circularly negative and positive polarized vectors are represented by

$$\mathbf{A} = |A|\hat{\mathbf{x}}_1 \quad \text{and} \quad \mathbf{B} = |B|\hat{\mathbf{x}}_2,$$

respectively.

Quadratic averages (see Section 4.1.4) are immediately extended to complex vectors:

$$\overline{\mathbf{a}(t) \cdot \mathbf{b}(t)} = \operatorname{Re}\left[\frac{1}{2}\mathbf{A} \cdot \mathbf{B}^*\right], \tag{4.33}$$

$$\overline{\mathbf{a}(t) \cdot \frac{\partial \mathbf{b}}{\partial t}} = \omega \operatorname{Im}\left[\frac{1}{2}\mathbf{A} \cdot \mathbf{B}^*\right], \tag{4.34}$$

which are the generalization to vectors of Eqs. (4.12) and (4.13), respectively.

4.2.2. Polarization States

A complex vector can be referred to an arbitrary coordinate system. Sometimes, a particular choice is convenient.

Any vector **A** can be represented in its polarization plane by means of two unit vectors, $\hat{\mathbf{p}}$ and $\hat{\mathbf{q}}$ (these vectors are not those considered in Section 4.2):

$$\mathbf{A} = (\mathbf{A} \cdot \hat{\mathbf{p}}^*)\hat{\mathbf{p}} + (\mathbf{A} \cdot \hat{\mathbf{q}}^*)\hat{\mathbf{q}} = A_p \hat{\mathbf{p}} + A_q \hat{\mathbf{q}},$$

with

$$\hat{\mathbf{p}} \cdot \hat{\mathbf{p}}^* = \hat{\mathbf{q}} \cdot \hat{\mathbf{q}}^* = 1 \quad \text{and} \quad \hat{\mathbf{p}} \cdot \hat{\mathbf{q}}^* = \hat{\mathbf{p}}^* \cdot \hat{\mathbf{q}} = 0;$$

see Eq. (4.32). Each of the two vectors, $\hat{\mathbf{p}}$ and $\hat{\mathbf{q}}$, describes a single *polarization state*, and form a *basis* for representing (in the plane) any vector, whose polarization is given by the (weighted) superposition of these two polarization states. Often, only one polarization state is desired, and is referred to as the *copolar component* of the vector field, while the other, undesired, one is the *crosspolar component*.

If (x, y) is the polarization plane, then a possible choice for the polarization basis is

$$\hat{\mathbf{p}} = \hat{\mathbf{x}} \quad \text{and} \quad \hat{\mathbf{q}} = \hat{\mathbf{y}}, \tag{4.35a}$$

which implies that use is made of two orthogonal, linearly polarized states. Alternatively,

$$\hat{\mathbf{p}} = \frac{1}{\sqrt{2}}(\hat{\mathbf{x}} + i\hat{\mathbf{y}}) \quad \text{and} \quad \hat{\mathbf{q}} = \frac{1}{\sqrt{2}}(\hat{\mathbf{x}} - i\hat{\mathbf{y}}), \tag{4.35b}$$

and the polarization states are the circular negative and positive ones.

TABLE 4.2. Some Choices for Polarization Unit Vectors

$\hat{\mathbf{p}}$	$\hat{\mathbf{q}}$
$\dfrac{\cos\theta\cos\phi\hat{\boldsymbol{\theta}} - \sin\phi\hat{\boldsymbol{\phi}}}{\sqrt{\cos^2\theta\cos^2\phi + \sin^2\phi}}$	$\dfrac{\sin\phi\hat{\boldsymbol{\theta}} + \cos\theta\cos\phi\hat{\boldsymbol{\phi}}}{\sqrt{\cos^2\theta\cos^2\phi + \sin^2\phi}}$
$\dfrac{\cos\phi\hat{\boldsymbol{\theta}} - \cos\theta\sin\phi\hat{\boldsymbol{\phi}}}{\sqrt{\cos^2\theta\sin^2\phi + \cos^2\phi}}$	$\dfrac{\cos\theta\sin\phi\hat{\boldsymbol{\theta}} + \cos\phi\hat{\boldsymbol{\phi}}}{\sqrt{\cos^2\theta\sin^2\phi + \cos^2\phi}}$
$\cos\phi\hat{\boldsymbol{\theta}} - \sin\phi\hat{\boldsymbol{\phi}}$	$\sin\phi\hat{\boldsymbol{\theta}} + \cos\phi\hat{\boldsymbol{\phi}}$

Convenient choices for polarization bases are presented in Table 4.2. The vector $\hat{\mathbf{p}}$ in the top row describes the polarization state of the electric far field radiated by an x-oriented electric dipole[8]; see Eq. (4.97a). The corresponding polarization lines are given in Fig. 4.12. The orthogonal unit vector $\hat{\mathbf{q}}$ in Table 4.2 describes the polarization state of the $-x$-oriented magnetic dipole; see Fig. 4.12. Should the dipoles be oriented along the y-axis, we obtain the polarization basis in the second row of Table 4.2: $\hat{\mathbf{p}}$ describes the polarization state of the electric far field of the magnetic dipole [see Eq. (4.97b)] and $\hat{\mathbf{q}}$ the polarization state of the electric dipole (see Fig. 4.13).

A third possible basis choice is given in the third row of Table 4.2, where both the vectors $\hat{\mathbf{p}}$ and $\hat{\mathbf{q}}$ are associated with polarization states of another elementary source (see section 4.6.3): $\hat{\mathbf{p}}$ describes the polarization state of a Huygens[9] source with the electric dipole oriented along the x-axis [see Eq. (4.97c)], while the electric dipole is oriented along the y-axis for the vector $\hat{\mathbf{q}}$; see Fig. 4.14.

For all three bases, $\hat{\mathbf{p}} \times \hat{\mathbf{q}} = \hat{\mathbf{r}}$. Furthermore, the three bases coincide around the z-axis, i.e., for $\theta° \approx 0°$ and $180°$.

4.2.3. Stokes Parameters and the Poincaré Sphere

Let $\mathbf{A} = \mathbf{A}_p + i\mathbf{A}_q = A_x\hat{\mathbf{x}} + A_y\hat{\mathbf{y}}$ be a complex vector in the polarization plane $z = 0$. We define the *polarization matrix*:

$$S = \begin{vmatrix} A_y A_y^* & A_y A_x^* \\ A_y^* A_x & A_x A_x^* \end{vmatrix} = \begin{vmatrix} S_{11} & S_{12} \\ S_{21} & S_{22} \end{vmatrix}, \quad (4.36)$$

---
8. The far field of the dipole is given by $-i\zeta(I\Delta z/\lambda r)\exp(-ikr)\hat{\mathbf{p}}$.
9. Christiaan Huygens: Den Haag (The Netherlands), 1629–Den Haag, 1695.

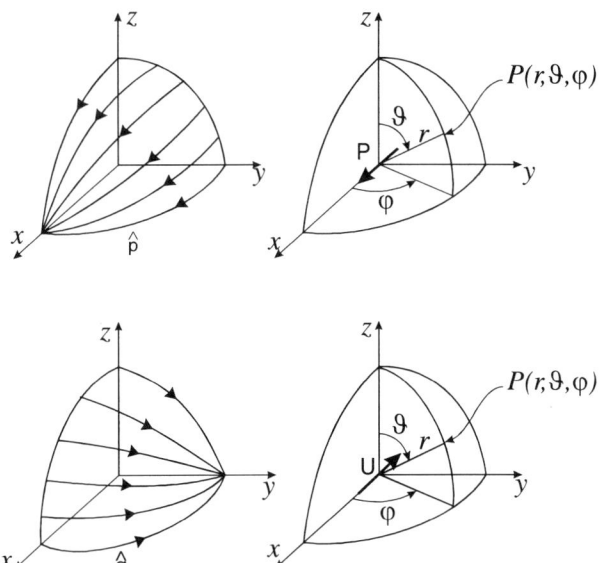

FIGURE 4.12. Polarization lines of an electric dipole oriented along the x-axis and of its crosspolar mate.

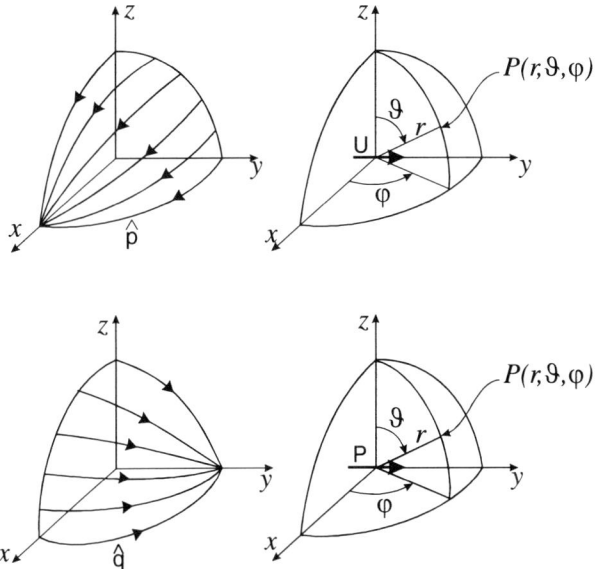

FIGURE 4.13. Polarization lines of a magnetic dipole oriented along the $-x$-axis and of its crosspolar mate.

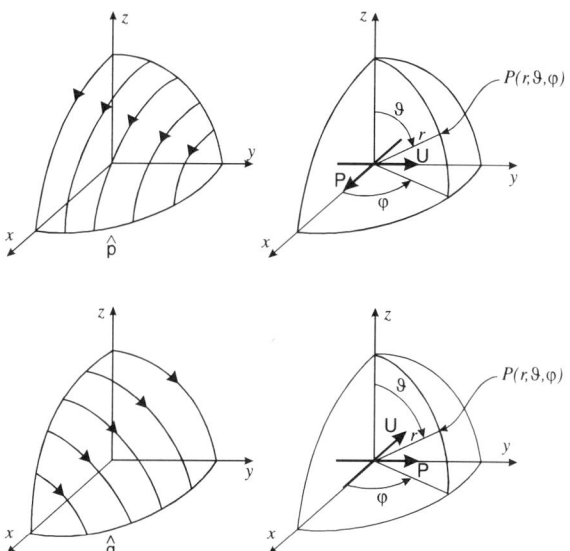

FIGURE 4.14. Polarization lines of two Huygens sources.

and the *Stokes parameters:*

$$\begin{cases} S_0 = A_y A_y^* + A_x A_x^* = S_{11} + S_{22}, & (4.37a) \\ S_1 = A_y A_y^* - A_x A_x^* = S_{11} - S_{22}, & (4.37b) \\ S_2 = A_x^* A_y + A_x A_y^* = S_{12} + S_{21}, & (4.37c) \\ iS_3 = A_x^* A_y - A_x A_y^* = S_{12} - S_{21}, & (4.37d) \end{cases}$$

with respect to cartesian axes (x, y).

It is clear that values of both S entries and Stokes parameters depend on the choice of coordinate axes. A natural choice are the axes[10] related to the complex vector representation (4.24) and depicted in Fig. 4.15. Hence

$$A_x = A^* \sin \frac{\theta}{2}, \quad A_y = A \cos \frac{\theta}{2}, \quad A = |A| \exp(i\phi),$$

$$\begin{cases} S_0 = |A|^2, \\ S_1 = |A|^2 \cos \theta \\ S_2 = |A|^2 \sin \theta \cos 2\phi, \\ S_3 = |A|^2 \sin \theta \sin 2\phi, \end{cases}$$

10. These coordinates undergo rotations if the vector **A** changes (slowly) in time, unless the bisector of the angle θ remains fixed (see Fig. 4.15).

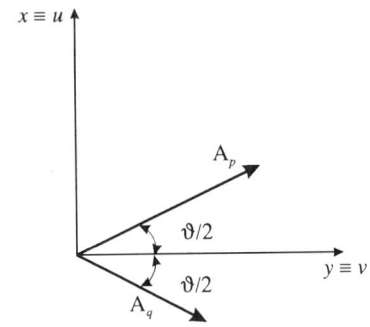

FIGURE 4.15. Definition of Stokes parameters.

and

$$S = \frac{1}{2}|A|^2 \begin{vmatrix} 1 + \cos\theta & \sin\theta\exp(2i\phi) \\ \sin\theta\exp(-2i\phi) & 1 - \cos\theta \end{vmatrix}.$$

Stokes parameters in their usual normalized form

$$s_0 = 1, \quad s_1 = \cos\theta, \quad s_2 = \sin\theta\cos 2\phi, \quad s_3 = \sin\theta\sin 2\phi, \quad (4.38)$$

with

$$s_1^2 + s_2^2 + s_3^2 = 1, \quad (4.39)$$

are cartesian coordinates of the unitary *Poincaré*[11] *sphere* (see Fig. 4.16), where

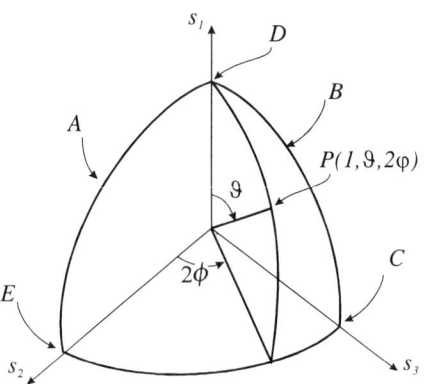

FIGURE 4.16. The Poincaré sphere.

11. Jules Henri Poincaré: Nancy (France), 1854–Paris, 1912.

NARROWBAND SIGNALS AND PHASOR FIELDS 167

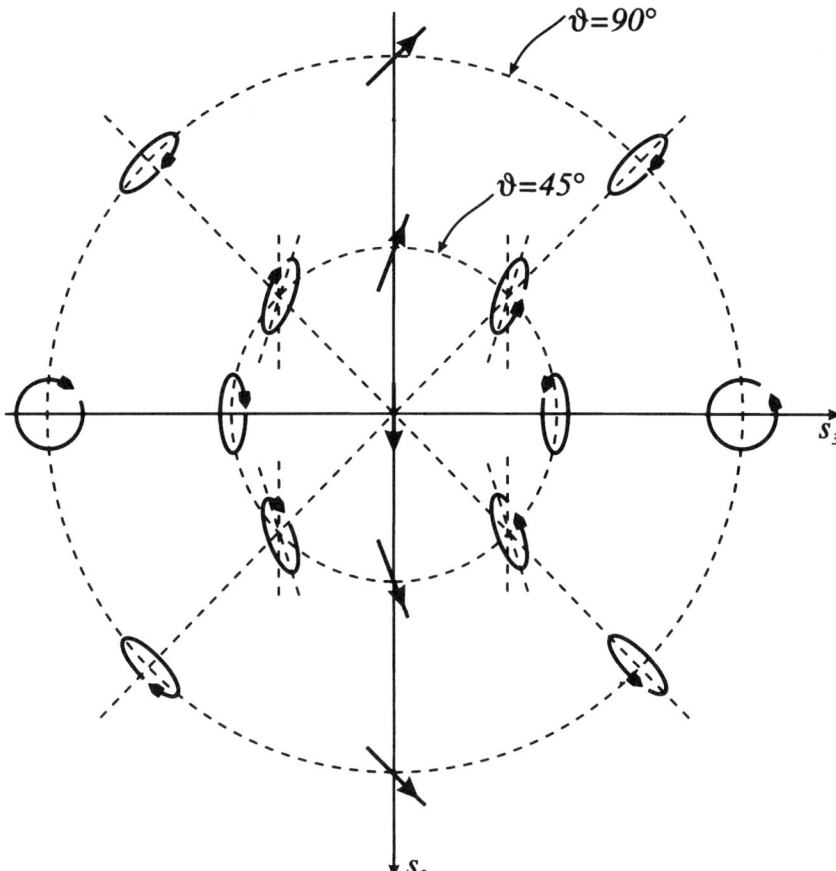

FIGURE 4.17. Top view of the Poincaré sphere with corresponding polarization states.

$\hat{\mathbf{s}}_2 = \hat{\mathbf{y}}$ and $\hat{\mathbf{s}}_3 = \hat{\mathbf{x}}$. Each point of the sphere is representative of the polarization state of a vector with its polarization plane parallel to the $s_1 = 0$ plane. In Fig. 4.16, the circle A ($\phi = 0$) corresponds to linear polarization states; Eq. (4.28a) provides $L = |A|$ and $l = 0$. At the north pole (point D) the polarization is along the y-axis, and at the equator (point E) it is rotated at 45° degrees; see Eq. (4.28b). The circle B ($\phi = \pi/4$) corresponds to elliptic polarization states; Eq. (4.28a) provides $L = |A|\cos(\theta/2)$ and $l = |A|\sin(\theta/2)$. The principal axes of the ellipse are parallel to the y- and x-axes, respectively [see Eq. (4.28b)]; at the equator (point C) the polarization is (negative) circular. Polarization states on the half-sphere are depicted in Fig. 4.17; obviously, all points of the remaining ($\theta \geqslant 90°$) sphere are representative of appropriate polarization states.

4.2.4. Field Coherence

Consider a vector field $\mathbf{A}(\mathbf{r}, t)$ whose time-domain counterpart is given by the narrowband vector $\mathbf{a}(\mathbf{r}, t)$. Loosely speaking, coherence of the field is its ability to maintain stable relationships linking amplitude and polarization with other reference fields as time elapses.

It is convenient to refer to the averaged polarization matrix (4.36),

$$\mathbf{C} = \begin{vmatrix} C_{11} & C_{12} \\ C_{21} & C_{22} \end{vmatrix}, \quad C_{12} = C_{21}^*, \tag{4.40}$$

each entry being the time average of the corresponding entry of $\mathbf{S}$. In the case of narrowband fields, this average can be extended to their timewidth only. If the field at any point is a (narrowband) random process, the coherence matrix $\mathbf{C}$ can be considered as the ensemble average of $\mathbf{S}$. The matrix $\mathbf{C}$ is referred to as the *coherence matrix*.

The entries of the matrix $\mathbf{C}$, as well as those of $\mathbf{S}$, depend on the chosen reference system. We define the *degree of coherence* γ,

$$\gamma^2 = \frac{|C_{12}|^2}{C_{11} C_{22}}, \tag{4.41}$$

in the reference system where it attains its maximum value. Accordingly, we study the change of $\mathbf{C}$ under a rotation α of the reference system. Hence

$$\begin{cases} A'_x = A_x \cos \alpha + A_y \sin \alpha, \\ A'_y = -A_x \sin \alpha + A_y \cos \alpha, \end{cases}$$

and

$$C'_{11} = \tfrac{1}{2}[(C_{11} + C_{22}) + (C_{11} - C_{22}) \cos 2\alpha + (C_{12} + C_{21}) \sin 2\alpha],$$
$$C'_{22} = \tfrac{1}{2}[(C_{11} + C_{22}) - (C_{11} - C_{22}) \cos 2\alpha - (C_{12} + C_{21}) \sin 2\alpha],$$
$$C'_{12} = \tfrac{1}{2}[(C_{12} - C_{21}) + (C_{12} + C_{21}) \cos 2\alpha - (C_{11} - C_{22}) \sin 2\alpha] = C'^*_{21}.$$

Substitution into Eq. (4.41) yields a function of α which attains its maximum value,

$$\gamma^2 = \frac{(C_{11} - C_{22})^2 + (C_{12} + C_{21})^2 - (C_{12} - C_{21})^2}{(C_{11} + C_{22})^2}, \tag{4.42}$$

for

$$\tan 2\alpha = -\frac{C_{11} - C_{22}}{C_{12} + C_{21}}.$$

A field is referred to as *completely coherent*, or simply *coherent*, if $\gamma = 1$. In this case Eq. (4.42) yields

$$\frac{(C_{11} - C_{22})^2}{(C_{11} + C_{22})^2} + \frac{(C_{12} + C_{21})^2}{(C_{11} + C_{22})^2} - \frac{(C_{12} - C_{21})^2}{(C_{11} + C_{22})^2} = 1.$$

Comparison with Eqs. (4.37) shows that we can define averaged Stokes parameters satisfying relation (4.39) and identify a definite polarization for the field, which is *completely polarized*. Monochromatic fields are always completely polarized.

The opposite situation occurs when $\gamma = 0$, in the system where it attains its maximum value, and therefore in any reference system. We obtain from Eq. (4.42)

$$(C_{11} - C_{22})^2 + 4C_{12}C_{21} = (C_{11} - C_{22})^2 + 4|C_{12}|^2 = 0,$$

$$C_{11} = C_{22}, \quad C_{12} = C_{21} = 0,$$

and the (averaged) Stokes parameters are all zero except for the first one; see Eq. (4.37). In this case no polarization can be identified for the field, which is *completely unpolarized*. Statistically independent fields A_x and A_y are an example of lack of polarization.

In intermediate cases the field is referred to as *partly polarized*. For example, if

$$A_x = \exp\left(-\frac{t^2}{2a}\right), \quad A_y = \exp\left(-\frac{t^2}{2a^*}\right),$$

$$a = a_1 + ia_2, \quad a_1 > 0,$$

then repeated use of Eq. (C.3) for computing the time averages shows that

$$C_{11} = C_{22} = \sqrt{2\pi}\,\frac{|a|}{\sqrt{2a_1}}, \quad C_{12} = C_{21}^* = \sqrt{2\pi}\,\sqrt{\frac{a}{2}},$$

and

$$\gamma = \sqrt{\frac{a_1}{|a|}}.$$

Any field can be split into its polarized

$$\gamma(C_{11} + C_{22}) = C_{11}^P + C_{22}^P \tag{4.43}$$

and unpolarized

$$\sqrt{1-\gamma^2}\,(C_{11} + C_{22}) = C_{11}^U + C_{22}^U \tag{4.44}$$

parts.

For the polarized part we take $C_{12}^P = C_{12}$ and $C_{21}^P = C_{21}$, and require the entries of the coherence matrix to fulfil Eq. (4.42) with $\gamma = 1$:

$$(C_{11}^P - C_{21}^P)^2 + 4C_{12}^P C_{21}^P = (C_{11}^P + C_{22}^P)^2.$$

Accordingly, we derive the system

$$\begin{cases} C_{11}^P + C_{22}^P = \gamma(C_{11} + C_{22}), \\ (C_{11}^P - C_{22}^P)^2 = \gamma^2(C_{11} + C_{22})^2 - 4C_{12}C_{21}, \end{cases}$$

which determines the other two entries C_{11}^P and C_{22}^P.

For the unpolarized part we take

$$C_{11}^U = C_{22}^U = \frac{\sqrt{1-\gamma^2}}{2}(C_{11} + C_{22}) \quad \text{and} \quad C_{12}^U = C_{21}^U = 0.$$

It is readily verified that the (averaged) Stokes parameters of the field are the sum of their polarized and unpolarized parts.

4.3. MAXWELL EQUATIONS IN PHASOR FORM

Maxwell equations for narrowband signals are easily obtained by setting

$$\mathbf{e}(\mathbf{r}, t) = \mathrm{Re}[\mathbf{E}\exp(i\omega t)],$$

and similarly for the other field quantities. On substituting into Eqs. (1.1) and equating the complex vectors, we obtain

$$\begin{cases} \nabla \times \mathbf{E}(\mathbf{r}) = -i\omega \mathbf{B}(\mathbf{r}), & (4.45\mathrm{a}) \\ \nabla \times \mathbf{H}(\mathbf{r}) = i\omega \mathbf{D}(\mathbf{r}) + \mathbf{J}(\mathbf{r}) + \mathbf{J}_0(\mathbf{r}), & (4.45\mathrm{b}) \\ \nabla \cdot \mathbf{D}(\mathbf{r}) = \rho(\mathbf{r}) + \rho_0(\mathbf{r}), & (4.45\mathrm{c}) \\ \nabla \cdot \mathbf{B}(\mathbf{r}) = 0, & (4.45\mathrm{d}) \end{cases}$$

which are formally equivalent to Eqs. (3.10). However, phasors exhibit a noticeable difference when compared to spectral fields: the former have the same dimensions as their time-domain counterpart and may possibly change (slowly) with time. Furthermore, the (carrier) angular frequency ω is fixed, not being an independent variable. Also, Eqs. (4.45) do not require initial conditions, unlike their time-domain counterparts (1.1). Only boundary conditions are required, to be derived from a new formulation of the uniqueness theorem; see Section 4.3.3.

Inclusion of magnetic sources in Eq. (4.45) is straightforward [see Eqs. (3.15)], as well as derivation of the current-density continuity equation [see Eq. (3.11)].

For time-invariant isotropic media

$$\begin{cases} \nabla \times \mathbf{E} = -i\omega\mu \mathbf{H}, & (4.46\mathrm{a}) \\ \nabla \times \mathbf{H} = i\omega\varepsilon \mathbf{E} + \sigma \mathbf{E} + \mathbf{J}_0, & (4.46\mathrm{b}) \\ \nabla \cdot \varepsilon \mathbf{E} = \rho + \rho_0, & (4.46\mathrm{c}) \\ \nabla \cdot \mu \mathbf{H} = 0, & (4.46\mathrm{d}) \end{cases}$$

where the complex permeability ε and permittivity μ describe the response of the medium under sinusoidal excitation, e.g.,

$$d(\mathbf{r}, t) = \mathrm{Re}[\varepsilon(\mathbf{r})\mathbf{E}(\mathbf{r}) \exp(i\omega t)].$$

Their values coincide with those of the FTs of the (electric and magnetic) unit response of the medium (see Section 3.2), evaluated at the carrier angular frequency ω. Note that we have neglected in Eqs. (4.45) and (4.46) the additional contributions $(\partial \mathbf{E}/\partial t)[\partial(\omega\varepsilon)/\partial\omega], \ldots$ (see Section 4.1.2).

Extension of Eqs. (4.46) to anisotropic media is quite evident, as well as use of the equivalent dielectric constant (3.20).

4.3.1. Poynting's Theorem

Poynting's theorem for phasor fields is easily obtained by time-averaging Eq. (1.41) over the carrier period T, or directly from Eqs. (4.46) following the same procedure in Section 1.3. If

$$\mathbf{S} = \tfrac{1}{2}\mathbf{E} \times \mathbf{H}^* = \mathbf{S}_1 + i\mathbf{S}_2 \tag{4.47}$$

is the complex Poynting vector, we have for isotropic time-invariant media

$$\nabla \cdot \mathbf{S} - \tfrac{1}{2}i\omega\varepsilon^* \mathbf{E} \cdot \mathbf{E}^* + \tfrac{1}{2}i\omega\mu \mathbf{H} \cdot \mathbf{H}^* + \tfrac{1}{2}\sigma \mathbf{E} \cdot \mathbf{E}^*$$
$$= -\tfrac{1}{2}\mathbf{E} \cdot \mathbf{J}_0^*. \tag{4.48}$$

Integration of the real part of Eq. (4.48) over a volume bounded by a closed surface (see Fig. 4.18) leads to

$$\oiint_A dA\, \mathbf{S}_1 \cdot \hat{\mathbf{n}} + \iiint_V dV \left[\frac{\omega\varepsilon_2}{2} |\mathbf{E}|^2 + \frac{\omega\mu_2}{2} |\mathbf{H}|^2 + \frac{\sigma}{2} |\mathbf{E}|^2 \right]$$

$$= \iiint_V dV\, \text{Re}[-\tfrac{1}{2}\mathbf{E} \cdot \mathbf{J}_0^*], \tag{4.49a}$$

with $\varepsilon = \varepsilon_1 - i\varepsilon_2$ and $\mu = \mu_1 - i\mu_2$. This is a relation involving time-averaged quantities: the real power delivered by the sources within the volume V (last term) equals the flux of the field's real power out of the surface A (first term) plus the dissipated power within the volume. The latter is a nonnegative quantity, so that not only σ (see Section 1.3.4) but also ε_2 and μ_2 are nonnegative quantities (as well as ω, the carrier angular frequency). The losses are divided into *ohmic* (σ term), *dielectric* (ε_2 term), and *magnetic* (μ_2 term) components.

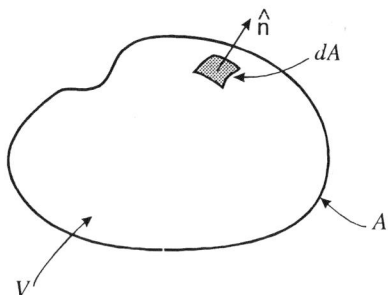

FIGURE 4.18. Poynting's theorem.

NARROWBAND SIGNALS AND PHASOR FIELDS

Integration of the imaginary part of Eq. (4.48) over the same volume leads to

$$\oiint_A dA\, \mathbf{S}_2 \cdot \hat{\mathbf{n}} + 2\omega \iiint_V dV[\tfrac{1}{4}\mu_1|\mathbf{H}|^2 - \tfrac{1}{4}\varepsilon_1|\mathbf{E}|^2]$$

$$= \iiint_V dV\, \mathrm{Im}[-\tfrac{1}{2}\mathbf{E}\cdot\mathbf{J}^*]. \qquad (4.49\mathrm{b})$$

The last term is the reactive power (see Section 4.1.5) delivered by the sources within the volume V, and the first term is the flux of the same reactive power out of the volume V. For nondispersive media, the remaining term is 2ω times the difference between the (time-averaged) magnetic and electric energies stored in the volume V (see also Section 4.1.5). In other words, the reactive power of the source divided by 2ω equals the unbalance of these two energies, computed inside (volume integral) and outside (surface integral) the volume V. If dispersion is present, the reactive power can still be defined, but the two terms $\mu_1|\mathbf{H}|^2/4$ and $\varepsilon_1|\mathbf{E}|^2/4$ cannot be identified with true energies (see Section 4.3.2 below).

4.3.2. The Energy Theorem

Formulation of an alternative energy theorem is possible in a lossless medium. We make use of Eqs. (4.10b) to obtain

$$\frac{\partial \mathbf{D}}{\partial t} = i\omega\varepsilon\mathbf{E} + \frac{\partial(\omega\varepsilon)}{\partial\omega}\frac{\partial\mathbf{E}}{\partial t} \quad \text{and} \quad \frac{\partial\mathbf{B}}{\partial t} = i\omega\mu\mathbf{H} + \frac{\partial(\omega\mu)}{\partial\omega}\frac{\partial\mathbf{H}}{\partial t}$$

with ε and μ real, thus including the extra terms of Section 4.1.2 that are missing in Eq. (4.46). We compute the real part of the Poynting vector,

$$\mathbf{S}_1 = \tfrac{1}{2}(\mathbf{S} + \mathbf{S}^*),$$

and proceed as in Section 4.3.1 to yield

$$\nabla\cdot\mathbf{S}_1 + \frac{\partial}{\partial t}\left[\frac{1}{4}\frac{\partial(\omega\varepsilon)}{(\partial\omega)}|\mathbf{E}|^2 + \frac{1}{4}\frac{\partial(\omega\mu)}{\partial\omega}|\mathbf{H}|^2\right] = \mathrm{Re}\left[-\frac{1}{2}\mathbf{E}\cdot\mathbf{J}_0^*\right], \qquad (4.50)$$

because

$$\mathbf{E}\cdot\frac{\partial\mathbf{E}^*}{\partial t} + \mathbf{E}^*\cdot\frac{\partial\mathbf{E}}{\partial t} = \frac{\partial}{\partial t}|\mathbf{E}|^2,$$

and similarly for the magnetic term. Equation (4.50) is the counterpart of Eq. (4.48) and provides the alternative form of the energy theorem, being valid also for dispersive media.

Integration of Eq. (4.50) over the volume V of Fig. 4.18 leads to the relation

$$\oint_A dA\, \mathbf{S}_1 \cdot \hat{\mathbf{n}} + \frac{\partial}{\partial t} \iiint_V dV \left[\frac{1}{4} \frac{\partial(\omega\varepsilon)}{\partial \omega} |\mathbf{E}|^2 + \frac{1}{4} \frac{\partial(\omega\mu)}{\partial \omega} |\mathbf{H}|^2 \right] = \iiint_V dV\, \mathrm{Re}\left[-\frac{1}{2} \mathbf{E} \cdot \mathbf{J}_0^* \right],$$

which is a conservation equation valid for narrowband signals and equivalent to Eq. (1.46). The quantity

$$W = \frac{1}{4} \frac{\partial(\omega\varepsilon)}{\partial \omega} |\mathbf{E}|^2 + \frac{1}{4} \frac{\partial(\omega\mu)}{\partial \omega} |\mathbf{H}|^2 = W_e + W_m \qquad (4.51)$$

is identified as the (time-averaged) energy density of the narrowband field.

For an isotropic collisionless plasma we have

$$\mu = \mu_0 \quad \text{and} \quad \varepsilon = \varepsilon_0 \left(1 - \frac{\omega_p^2}{\omega^2} \right)$$

[see Eq. (3.25)], as well as

$$\frac{\partial \omega\mu}{\partial \omega} = \mu_0 \quad \text{and} \quad \frac{\partial \omega\varepsilon}{\partial \omega} = \varepsilon_0 + \frac{Nq^2}{m\omega^2}.$$

Accordingly

$$W = \frac{1}{4} \mu_0 |\mathbf{H}|^2 + \frac{1}{4} \varepsilon_0 |\mathbf{E}|^2 + \frac{1}{4} Nm \left(\frac{q}{\omega} \right)^2 |\mathbf{E}|^2,$$

where the last term,

$$\frac{1}{4} Nm \left(\frac{q}{\omega} \right)^2 |\mathbf{E}|^2 = \frac{1}{4} Nm |\mathbf{V}|^2,$$

is recognized as the (time-averaged) kinetic energy density of the electrons in the collisionless plasma fluid.

4.3.3. Uniqueness

As noted, uniqueness of the solution of Eqs. (4.46) requires only an appropriate statement of the boundary conditions, because the initial conditions are of no concern since the field spans the entire time interval. The appropriate uniqueness theorem can be stated along the same lines as in Section 1.4.

We consider a linear time-invariant medium with the electromagnetic field generated by prescribed sources $\mathbf{J}_0(\mathbf{r})$. Lack of uniqueness would imply the existence of at least two solutions; their difference, $(\mathbf{E}, \mathbf{H})$, is the solution of source-free Maxwell equations with null boundary conditions.

We consider first the exterior problem (see Section 1.4), where the volume V lies outside the surface A of Fig. 4.18, and we use Eq. (4.49a) with $\mathbf{J}_0 = 0$. Boundary conditions on A are stated for the tangential components of *either* the electric *or* the magnetic field. For the field difference, either $\hat{\mathbf{n}} \times \mathbf{E}$ or $\hat{\mathbf{n}} \times \mathbf{H}$ is zero on A,

$$\mathbf{S} \cdot (-\hat{\mathbf{n}}) = \tfrac{1}{2}\mathbf{E} \cdot (\hat{\mathbf{n}} \times \mathbf{H}^*) = \tfrac{1}{2}(\mathbf{E} \times \hat{\mathbf{n}}) \cdot \mathbf{H}^* = 0,$$

and the corresonding flux in Eq. (4.49a) is zero as well. On the sphere at infinity we enforce the radiation condition [see Eq. (1.63)], which reads

$$\mathbf{E} = \zeta \mathbf{H} \times \hat{\mathbf{n}} \qquad (4.52)$$

for the field difference as well. In Eq. (4.52), ζ is the (assumed real) value of the intrinsic impedance of the medium at infinity. We have there

$$\mathbf{S} \cdot \hat{\mathbf{n}} = \tfrac{1}{2}\zeta |\mathbf{H} \times \hat{\mathbf{n}}|^2,$$

in view of Eq. (4.52), and the outgoing flux at infinity is a *positive* quantity if the medium is lossless (see Section 1.4.2). The left-hand side of Eq. (4.49a) is now the sum of nonnegative terms, with at least a positive one: the first in the lossless case, and the others in the lossy case. Accordingly $\mathbf{E} = \mathbf{H} = 0$, and uniqueness is assured.

For the interior problem we proceed similarly. The flux of $\mathbf{S}$ out of A is zero, and Eq. (4.49a) assures again uniqueness unless $\varepsilon_2 = \mu_2 = \sigma = 0$ (lossless medium). For the latter case we resort to Eq. (4.49b),

$$\iiint_V dV \frac{1}{4}\mu|\mathbf{H}|^2 = \iiint_V dV \frac{1}{4}\varepsilon|\mathbf{E}|^2, \qquad (4.53)$$

which may be verified for fields different from zero. Equation (4.53) states the condition that forbids uniqueness to be assured and corresponds to the existence of a source-free solution ($J_0 = 0$) in the volume V with null boundary conditions.

Equation (4.53) states a balance between electric and magnetic quadratic averages and is identified as a *resonant condition*; the corresponding fields are a *resonant* solution ($J_0 = 0$). In the nondispersive case the two integrals appearing in Eq. (4.53) are the magnetic and electric (time-averaged) energies stored in the volume, and are equal at resonance.

4.3.4. Image Theory

Image theory as presented in Section 1.5.1 is immediately extended to phasor fields without any modification. As a matter of fact, all derivations make use of spatial operators only.

4.3.5. Duality Transformation

As in Section 4.3.4, duality transformation developed for the time-dependent fields (see Section 1.5) is immediately extended to phasor fields.

4.3.6. Reciprocity

Consider two source distributions J_1 and J_2 with their associated fields (E_1, H_1) and (E_2, H_2), respectively, as depicted in Fig. 4.19. We define the mixed Poynting-like vector

$$S_{12} = E_1 \times H_2 - E_2 \times H_1,$$

and obtain for an isotropic medium

$$\nabla \cdot S_{12} = E_1 \cdot J_2 - E_2 \cdot J_1,$$

after expanding the divergence operator via Eq. (A.11) and then using Maxwell's equations (4.46). Integration over the volume V depicted in Fig. 4.19 leads to

$$\iint_A dA\, \hat{n} \cdot S_{12} = \iiint_V dV\, [E_1 \cdot J_2 - E_2 \cdot J_1], \tag{4.54}$$

which is the most general form of the *reciprocity theorem*.

NARROWBAND SIGNALS AND PHASOR FIELDS 177

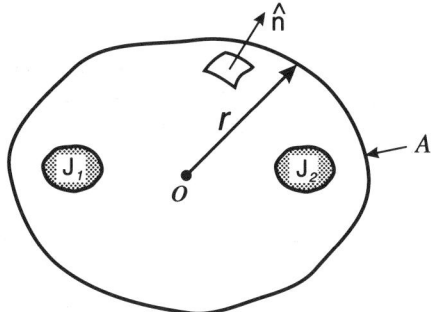

FIGURE 4.19. The reciprocity theorem.

There are two interesting cases that render the surface integral equal to zero, so that the volume integral is equal to zero, too.

The first case occurs when the surface material is a perfect (electric or magnetic) conductor, because $\mathbf{S}_{12} \cdot \hat{\mathbf{n}} = 0$ everywhere on A. The second case is obtained if the volume encompasses all the space, so that A can recede to infinity and the radiation condition (1.63) applies. In this case

$$\mathbf{E} \sim O(1/r), \qquad \mathbf{H} \sim O(1/r), \qquad \mathbf{E} - \zeta \mathbf{H} \times \hat{\mathbf{n}} \sim o(1/r), \qquad \text{as } r \to \infty,$$

and

$$\mathbf{S}_{12} = \zeta[\mathbf{H}_1 \times \hat{\mathbf{n}} + o(1/r)] \times \mathbf{H}_2 - \zeta[\mathbf{H}_2 \times \hat{\mathbf{n}} + o(1/r)] \times \mathbf{H}_1,$$
$$\mathbf{S}_{12} \cdot \hat{\mathbf{n}} \sim o(1/r^2), \qquad \text{as } r \to \infty,$$

so that the flux integral in Eq. (4.54) approaches zero as $r \to \infty$.

In both cases, i.e., the volume is either bounded by a perfect material or encompasses all space, we have

$$\iiint_V dV\, \mathbf{E}_1 \cdot \mathbf{J}_2 = \iiint_V dV\, \mathbf{E}_2 \cdot \mathbf{J}_1, \tag{4.55}$$

which is the most useful form of the reciprocity theorem. Equation (4.55) continues to be valid for anisotropic media with symmetrical matrixes (*reciprocal media*); see Section 1.2.3. We can use the conjugate scalar product if the medium is lossless.

4.3.7. The Equivalence Theorem

Consider a source distribution $\mathbf{J}_0$ with its associated electromagnetic field $(\mathbf{E}, \mathbf{H})$. A (smooth) surface S with an (everywhere defined) unit normal $\hat{\mathbf{n}}$ surrounds the sources. The equivalence theorem states that the original sources $\mathbf{J}_0$ can be removed and substituted by *equivalent sources*,

$$\mathbf{J}_S = \hat{\mathbf{n}} \times \mathbf{H}_2 \quad \text{and} \quad \mathbf{K}_S = -\hat{\mathbf{n}} \times \mathbf{E}_S, \tag{4.56}$$

i.e., electric $\mathbf{J}_S$ and magnetic $\mathbf{K}_S$ current (linear) densities (see Section 1.2.7) distributed over the surface S; see Fig. 4.20A. In Eq. (4.56) the fields $\mathbf{E}_S$ and $\mathbf{H}_S$ are those associated with the sources $\mathbf{J}_0$ and evaluated over S.

We prove that the equivalent sources (4.56) generate a field $(\mathbf{E}', \mathbf{H}')$ coincident with $(\mathbf{E}, \mathbf{H})$ outside S and identically equal to zero inside S.

The field $(\mathbf{E}', \mathbf{H}')$ is a solution of Maxwell equations both outside S, being coincident with $(\mathbf{E}, \mathbf{H})$, and inside S, too, because a null field is a solution of the homogeneous Maxwell equations, the sources $\mathbf{J}_0$ having been removed. The field $(\mathbf{E}', \mathbf{H}')$ also satisfies the radiation condition at infinity because this was verified by the field $(\mathbf{E}, \mathbf{H})$; it is discontinuous over S, and this is properly accounted for by the current linear densities (4.56); see Eqs. (1.36) and (1.37). The uniqueness theorem (see Section 4.3.3) states that this *possible* field is the *only* field associated with the new sources $(\mathbf{J}_S, \mathbf{K}_S)$, and the equivalence theorem is proved. In particular, the surface S separating the two regions, with and without the real sources, can coincide with an infinite planar surface, and we derive the cartesian geometry of Fig. 4.20B, where the equivalent sources are located on the plane.

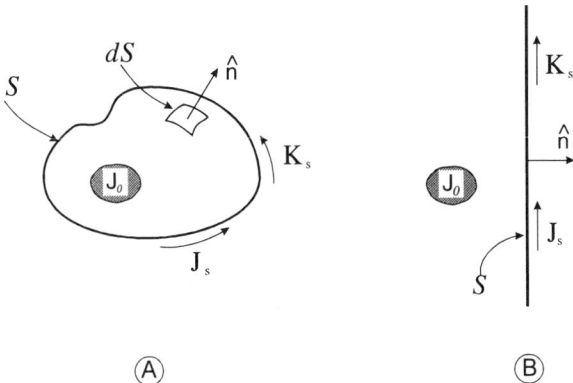

FIGURE 4.20. Two geometries relevant to the equivalence theorem: (A) basic geometry, (B) planar geometry.

The equivalence theorem can be regarded as a rigorous statement of the *Huygens principle*: each point over S becomes a source of radiation for the outside space, while the internal field is zero. This does *not* imply that the equivalent sources do not individually generate any field inside S, only that their collective result is zero.

An additional comment is in order. If the original sources are embedded (and radiate) in free space, the equivalent sources $\mathbf{J}_S$ and $\mathbf{K}_S$ radiate in free space, too. We can always include dielectric or conductive materials inside S (Fig. 4.20A) because the total field there is zero. This inclusion does not modify the field outside S, although the partial contributions from $\mathbf{J}_S$ and $\mathbf{K}_S$ change, because these sources now radiate in an inhomogeneous space. If the volume inside S is filled with, say, a perfect electric conductor, only the magnetic sources $\mathbf{K}_S$ need be applied. These radiate in the presence of a perfect conductor, and the linear density $\mathbf{J}_S$ of the electric current is automatically set in virtue of the boundary conditions on S (see Section 1.4.1). For the geometry of Fig. 4.20B the equivalent sources are

$$\mathbf{K}_S = -2\hat{\mathbf{n}} \times \mathbf{E}_S,$$

as an application of image theory (see Section 1.5.1).

4.4. PLANE-WAVE PROPAGATION

Plane-wave propagation is studied as in Section 3.2.5. For an isotropic homogeneous medium and a monochromatic wave we set

$$\mathbf{E} = E(z)\hat{\mathbf{x}} \quad \text{and} \quad \mathbf{H} = H(z)\hat{\mathbf{y}},$$

and substitute into Maxwell's equations (4.46) with $\mathbf{J}_0 = 0$. The solution is

$$\begin{cases} E(z) = E^+ \exp(-ikz) + E^- \exp(ikz), \\ \zeta H(z) = E^+ \exp(-ikz) - E^- \exp(ikz) \end{cases}$$

[see Eq. (3.26)] with

$$k = \omega\sqrt{\varepsilon\mu} \quad \text{and} \quad \zeta = \sqrt{\mu/\varepsilon}.$$

For the first solution we have

$$\begin{aligned} e(z, t) &= \text{Re}[E(z)\exp(i\omega t)] \\ &= \text{Re}[E^+ \exp[i(\omega t - \beta z)] = E^+ \cos(\omega t - \beta z), \end{aligned} \quad (4.57)$$

where we have assumed E^+ real and $k = \beta$ real as well. At any given point $z = z_0$ the field amplitude $E^+ \cos(\omega t - \beta z_0)$ attains the same value after the minimum time interval T such that

$$\omega(t + T) - \beta z_0 = \omega t - \beta z_0 + 2\pi,$$

which leads to

$$T = 2\pi/\omega \qquad (4.58)$$

and is referred to as the *wave period* (see Fig. 4.21). Similarly, at any time instant t_0 the field amplitude attains the same value past the minimum space interval λ such that

$$\omega t_0 - \beta(z + \lambda) = \omega t_0 - \beta z - 2\pi,$$

which leads to

$$\lambda = 2\pi/\beta \qquad (4.59)$$

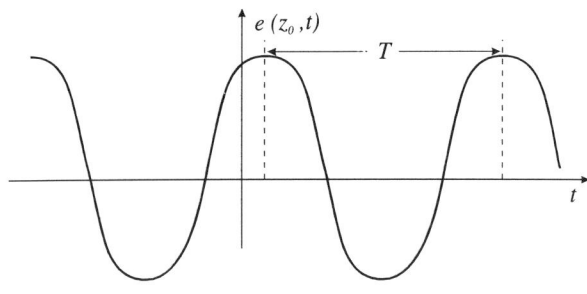

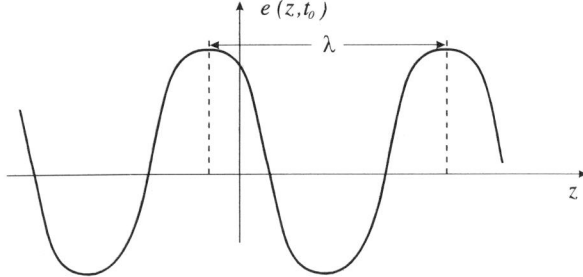

FIGURE 4.21. Definitions of wave period and wavelength.

NARROWBAND SIGNALS AND PHASOR FIELDS 181

and is referred to as the *wavelength*. Furthermore, the wave amplitude remains unchanged for time variation Δt and space displacement Δz such that

$$\omega(t + \Delta t) - \beta(z + \Delta z) = \omega t - \beta z,$$

$$v_P = \frac{\Delta z}{\Delta t} = \frac{\omega}{\beta} = \frac{2\pi}{T} \cdot \frac{\lambda}{2\pi} = \frac{\lambda}{T}. \tag{4.60}$$

The wave pattern *appears to move* along the positive z-direction with velocity v_P, thus explaining the terminology *progressive* for this wave component, as in Section 3.2.5, Eq. (3.29). We stress, however, that this is only an apparent movement, because this monochromatic wave is neither space nor time bounded and definition of displacement as well as its rate is meaningless. For this reason v_P is referred to as the *phase velocity*.

We turn now to narrowband plane-wave propagation and consider the progressive wave alone. As an intermediate step we need to employ the full spectral representation (no phasors). We start from the plane-wave representation (3.81) with $E_0(u, v, \omega) \to (2\pi)^2 \delta(u) \delta(v) E(\omega)$ and $w_P = \beta$:

$$e(z, t) = \frac{1}{2\pi} \int_{-\infty}^{+\infty} d\omega \exp(i\omega t) E(\omega) \exp[-i\beta(\omega)z], \tag{4.61a}$$

where

$$\omega^2 \varepsilon(\omega) \mu(\omega) = \beta^2 \tag{4.62}$$

is the dispersion equation (3.77). In Eq. (4.61a)

$$E(\omega) = \tfrac{1}{2} E_e(\omega + \omega_0) + \tfrac{1}{2} E_e(\omega - \omega_0)$$

is the frequency spectrum of the field at $z = 0$, i.e., the FT of $e(0, t)$, while $E_e(\omega)$ is the same spectrum for the envelope signal; see Section 4.1 and Eq. (4.3).

Equation (4.61a) represents the field as a superposition of plane waves, all with the same polarization and all progressing along the positive z-axis. It may be convenient to change the integration variable from ω to β, thus obtaining

$$e(z, t) = \int_{-\infty}^{+\infty} d\beta \exp(-i\beta z) E(\beta) \exp[i\omega(\beta)t], \tag{4.61b}$$

where

$$\frac{1}{2\pi} E[\omega(\beta)] \frac{d\omega}{d\beta} \to E(\beta)$$

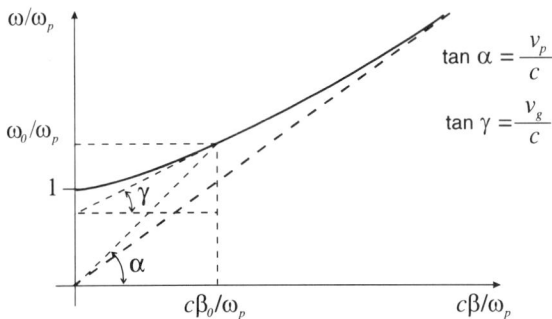

FIGURE 4.22. The Brillouin diagram for an isotropic collisionless plasma medium of plasma angular frequency ω_p. The dotted line corresponds to the free-space medium.

is the wavenumber spectrum of the field at $t = 0$, i.e., the FT of $e(z, 0)$. Hence

$$E(\beta) = \tfrac{1}{2} E_e(\beta + \beta_0) + \tfrac{1}{2} E_e(\beta - \beta_0),$$

$E_e(\beta)$ being the same wavenumber spectrum for the envelope signal. Note that

$$\omega_0^2 \varepsilon(\omega_0) \mu(\omega_0) = \beta_0^2,$$

so that β_0 is the wavenumber of the carrier frequency. We expect a signal with narrow frequency band $2\Delta\omega$ centered at $\pm\omega_0$ to exhibit a similar narrow wavenumber band $2\Delta\beta$ centered at $\pm\beta_0$, provided that the function $\omega(\beta)$ is regular near $\beta = \pm\beta_0$.

We make use of Eq. (4.61a) and solve Eq. (4.62) for ω,

$$\omega = \omega(\beta), \qquad (4.63)$$

whose graphical representation is referred to as the *Brillouin*[12] *diagram* (for the case of a plasma medium, see Section 4.4.1 and Fig. 4.22). In order to evaluate the integral (4.61b), we take advantage of the narrowband condition, expand $\omega(\beta)$[13] around $\pm\beta_0$ in the form

$$\omega(\beta) \approx \pm\omega_0 + \omega'(\beta_0)(\beta \mp \beta_0) \pm \tfrac{1}{2}\omega''(\beta_0)(\beta \mp \beta_0)^2,$$

and approximate the function $E(\beta)$ by $E_e(\beta - \beta_0)$ near β_0 and by $E_e(\beta + \beta_0)$ near $-\beta_0$. On substituting into Eq. (4.61a) with change of integration variables

$$\beta - \beta_0 \to \xi \qquad \text{and} \qquad -(\beta + \beta_0) \to \xi,$$

---

12. Marcel Louis Brillouin: Melle, Deux-Sévres (France), 1854–Paris, 1948.
13. Note that $\omega(\beta)$ is an odd function of β.

and recalling that $E(-\beta) = E^*(\beta)$, we obtain

$$e(z,t) \approx \text{Re}\left\{\exp[i(\omega_0 t - \beta_0 z)] \cdot \int_{-\infty}^{+\infty} d\xi E_e(\xi) \exp\left[-i(z - v_g t)\xi + i\frac{\omega_0'' t}{2}\xi^2\right]\right\}, \quad (4.64)$$

where $\omega_0'' = \omega''(\beta_0)$ and the derivative of the dispersion Eq. (4.63),

$$\omega'(\beta_0) = v_g = \left.\frac{d\omega}{d\beta}\right|_0, \quad (4.65)$$

is referred to as the *group velocity*.

Equation (4.64) is the generalization of Eq. (4.57) to narrowband signals; the phasor E^+ is replaced by the FT of the product of the envelope spectrum $E_e(\beta)$ with an additional quadratic phase function. The role of this latter term is irrelevant within the effective integration domain $(-\Delta\beta, \Delta\beta)$ provided that

$$\frac{\omega_0'' t}{2}\Delta\beta^2 \ll 1.$$

In such a case this extra term can be set equal to unity, in which case

$$e(z,t) = E(z - v_g t)\cos[\beta_0(z - v_P t)],$$

where $E(z)$ is the envelope space pattern at $t = 0$. The latter moves with unchanged shape at a speed equal to the group velocity, unlike the carrier frequency which apparently moves with the phase velocity (see also Section 4.4.1 and Fig. 4.23).

The considered signal is the result of superposition of elementary plane waves moving individually with their own phase velocities, and is usually referred to as a *wavepacket*. For a nondispersive medium $\omega = \beta/c$, and $v_P = v_g = c$. When dispersion is present, phase and group velocities differ. Noting that

$$\frac{\partial}{\partial t} \to -v_g \frac{\partial}{\partial z}$$

as far as the envelope is concerned, Eqs. (4.51), and (4.50) with $\mathbf{J}_0 = 0$, yield

$$\frac{\partial}{\partial z}[S_1 - v_g W] = 0, \quad (4.66)$$

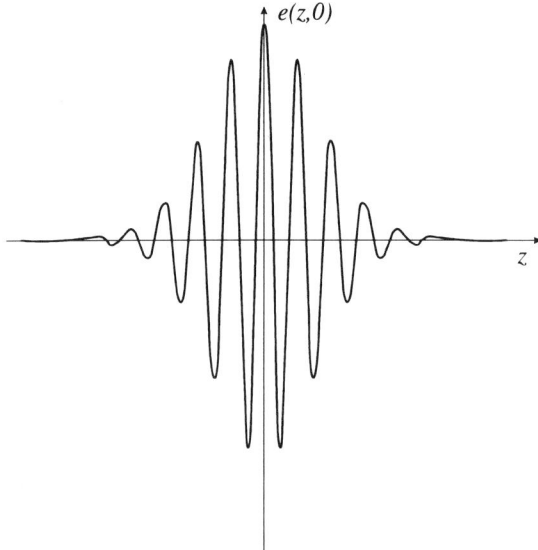

FIGURE 4.23. Gaussian wavepacket pattern at $t = 0$.

which suggests that the (time-averaged) electromagnetic energy of a narrow-band signal in a dispersive environment propagates with the group velocity.

On the contrary, when

$$\sqrt{\frac{2}{\omega_0'' t}} < \Delta\beta,$$

the effect of the medium dispersion cannot be neglected. The integral (4.64) tends to be averaged out by the fast varying phase term in the integration domain $|\beta| > \sqrt{2/\omega_0'' t}$. This results in a reduction of the effective (wavenumber) bandwidth of the envelope signal, and its space width spreads accordingly (see Section 4.1). An example is given in Section 4.4.1.

4.4.1. Propagation of a Gaussian Wavepacket in a Plasma Medium

Consider an unbounded collisionless isotropic plasma medium whose relative dielectric constant in the frequency domain is given by Eq. (3.25), and a wavepacket of envelope spectrum

$$E_e(\beta) = \frac{E_0}{\sqrt{2\pi}B} \exp\left(-\frac{\beta^2}{2B^2}\right).$$

From Eq. (4.62)

$$\omega^2 = \omega_P^2 + \beta^2 c^2,$$

and the corresponding Brillouin diagram is presented in Fig. 4.22. We have also

$$v_P = \frac{\omega_0}{\beta_0}, \qquad v_g = c^2 \frac{\beta_0}{\omega_0}, \qquad \omega_0'' = c^2 \frac{\omega_P^2}{\omega_0^3}.$$

At time $t = 0$ the wavepacket space pattern $e(z, 0)$ is given by Eq. (4.64) with $t = 0$:

$$e(z, 0) = \operatorname{Re}\left\{\exp[-i\beta_0 z)] \int_{-\infty}^{+\infty} d\beta \exp(-i\beta z) E_e(\beta)\right\}$$
$$= E_0 \exp\left(\frac{-B^2 z^2}{2}\right) \cos(\beta_0 z)$$

(see Table 3.1), and is presented in Fig. 4.23. For $t > 0$, we have from Eqs. (4.64) and (C.1)

$$e(z, t) = \operatorname{Re}\left\{\exp[i(\omega_0 t - \beta_0 z)] \frac{E_0}{\sqrt{1 - i\alpha}} \exp\left[-\frac{(z - v_g t)^2 B^2}{2(1 - i\alpha)}\right]\right\},$$

where

$$\alpha = \left(\frac{B}{\beta_0}\right)^2 \frac{\omega_P}{\omega_0} \omega_P t.$$

The envelope of the wavepacket,

$$a_e(z, t) = \frac{E_0}{\sqrt[4]{1 + \alpha^2}} \exp\left[-\frac{(z - v_g t)^2 B^2}{2(1 + \alpha^2)}\right],$$

undergoes deformation (see Fig. 4.24), which becomes noticeable at distances Z from the undeformed pattern situation such that

$$\alpha = \left(\frac{B}{\beta_0}\right)^2 \frac{\omega_P}{\omega_0} \omega_P \frac{Z}{v_g} = \left(\frac{B}{\beta_0}\right)^2 \left(\frac{\omega_P}{c\beta_0}\right)^2 \beta_0 Z \geqslant 1.$$

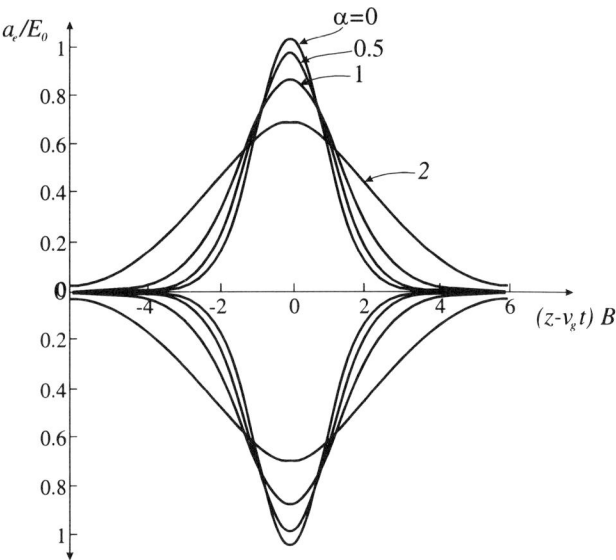

FIGURE 4.24. Envelope pattern of a gaussian wavepacket propagating in a plasma medium, for increasing values of the spreading factor α.

4.4.2. Information Scrambling through a Dispersive Channel

A bandlimited signal can be fully recovered by knowledge of its values sampled at proper discrete time intervals, see Eq. (4.17). If these samples are quantified in a number of discrete amplitude levels, and each level is represented by a binary sequence, signal transmission can be accomplished by sending and receiving a sequence of pulses. These pulses represent numbers in a binary code and can attain values equal either to one or to zero. The *transmission rate* equals the number of transmitted pulses, or *bits*, per unit time (bit/s). In an idealized situation (*gedankten experiment*) we can transmit the successive bits by means of a train of time-delayed waveforms,

$$F(t) = \Sigma_z a_n \operatorname{sinc}[\Delta\omega(t - t_n)], \qquad \Delta\omega t_n = n\pi,$$

where $a_n = a(t_n)$ is the value of the transmitted bit: one or zero. This signal is bandlimited to $2\Delta\omega$ and is used to modulate the carrier frequency ω before being transmitted to the receiving point at a distance z.

Recovering the envelope signal is possible if the channel dispersion factor $\omega_0'' t \Delta\beta^2/2$ ($c\Delta\beta = \Delta\omega$) is negligible. In this hypothesis the envelope signal at the

NARROWBAND SIGNALS AND PHASOR FIELDS   187

receiving point is $F(z - v_g t)$ (see Section 4.4) and we have

$$F(t - z/v_g) = \Sigma a_n \operatorname{sinc}[\Delta\omega(t - t_n - z/v_g)],$$

but for an (amplitude) scaling due to the attenuation of the transmission channel. Then each signal sample, say a_m, transmitted at time t_m, is recovered by measuring the received signal at the retarded time[14]:

$$t^* = t_m + z/v_g.$$

The *channel capacity*, i.e., the number of samples that can be transmitted per unit time (bit/s), is given by

$$C = \frac{1}{t_{n+1} - t_n} = \frac{v_g \Delta\beta}{\pi} = \frac{2\Delta\omega}{2\pi},$$

and equals the transmitted bandwidth.

The situation is different if the propagation time t^* and the propagation channel dispersion are such that $\omega_0'' t^* \Delta\beta^2/2$ cannot be neglected. We noted in Section 4.1 that the main effect on the signal is its (spatial) bandwidth reduction from $\Delta\beta$ to a smaller value $\Delta\beta_1$, approximately equal to $\sqrt{2/\omega_0'' t^*}$. In a first approximation the received signal is roughly given by

$$F(t - z/v_g) \approx \Sigma a_n \operatorname{sinc}[\Delta\omega_1(t - t_n - z/v_g)], \quad \text{where } \Delta\omega_1 = \Delta\omega \Delta\beta_1/\Delta\beta < \Delta\omega.$$

If we implement the same recovery procedure and evaluate the signal samples at the retarded times, the signal recovery is scrambled because values of samples near a_m contribute to its estimate. As an example let $\Delta\beta_1 = \Delta\beta/2$, so that

$$\Delta\omega_1(t - t_n - z/v_g) = \Delta\omega_1(t - z/v_g) - n\pi/2.$$

If $a_m = 0$ and $a_{m+1} = a_{m-1} = 1$ at the retarded time $t_m^* = t_m - z/v_g$, the contribution of the two samples near a_m is given by $2\operatorname{sinc}(\pi/2) = 4/\pi$, which exceeds unity, and the estimate of $a_m = 0$ is impaired. With this simple model the capacity of a dispersive channel turns out to be inversely proportional to the square root of the propagation time, i.e., to the square root of the transmission length.

---

14. Distorsions due to the coding and modulation system and also to thermal noise are ignored.

4.4.3. Plane-Wave Propagation in a Magnetized Plasma

The generalization of Maxwell equations (4.46) to anisotropic media is the following:

$$\begin{cases} \nabla \times \mathbf{E} = -i\omega\boldsymbol{\mu} \cdot \mathbf{H}, & (4.67\text{a}) \\ \nabla \times \mathbf{H} = i\omega\boldsymbol{\varepsilon} \cdot \mathbf{E} + \mathbf{J}_0, & (4.67\text{b}) \\ \nabla \cdot \boldsymbol{\varepsilon} \cdot \mathbf{E} = \rho_0, & (4.67\text{c}) \\ \nabla \cdot \boldsymbol{\varepsilon} \cdot \mathbf{H} = 0, & (4.67\text{d}) \end{cases}$$

ω being the carrier frequency. For a collisionless plasma magnetized along the z-axis we have $\boldsymbol{\mu} = \mu_0 \mathbf{I}$, and $\boldsymbol{\varepsilon}$ is given by Eqs. (3.22). Source-free solutions are obtained by setting $\mathbf{J}_0 = \rho_0 = 0$.

We consider a homogeneous plasma magnetized along z, with plane-wave propagation along the x-axis, so that

$$\partial/\partial y = \partial/\partial z = 0 \quad \text{and} \quad \partial/\partial x = -i\beta,$$

β being the plane-wave propagation constant.

From Eqs. (4.67c–d)

$$H_x = 0 \quad \text{and} \quad \varepsilon_1 E_x + i\varepsilon_2 E_y = 0,$$

so that the field is transverse magnetic (TM) with respect to the direction of propagation. Equations (4.67a–b) yield

$$\begin{vmatrix} \beta & -\omega\mu_0 & 0 & 0 \\ -\omega\varepsilon_0 \dfrac{\varepsilon_1^2 - \varepsilon_2^2}{\varepsilon_1} & \beta & 0 & 0 \\ 0 & 0 & \beta & \omega\mu_0 \\ 0 & 0 & \omega\varepsilon_0\varepsilon_3 & \beta \end{vmatrix} \begin{vmatrix} E_y \\ H_z \\ E_z \\ H_y \end{vmatrix} = 0, \quad (4.68)$$

which shows that the fields (E_y, H_z) and (E_z, H_y) are decoupled from each other, so two plane-wave modes can propagate in the plasma.

One mode exhibits components (E_z, H_y), thus being not only TM, but TEM with respect to x. Nonzero values for the field components require the determinant of the matrix of the coefficients to be zero, i.e.,

$$c^2\beta^2 = \omega^2\varepsilon_3 = \omega^2 - \omega_p^2.$$

The field is linearly polarized, unaffected by the dc magnetization, and is

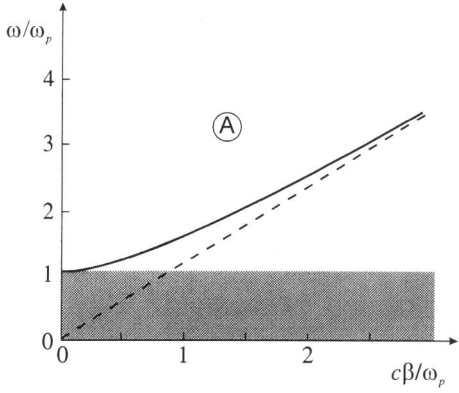

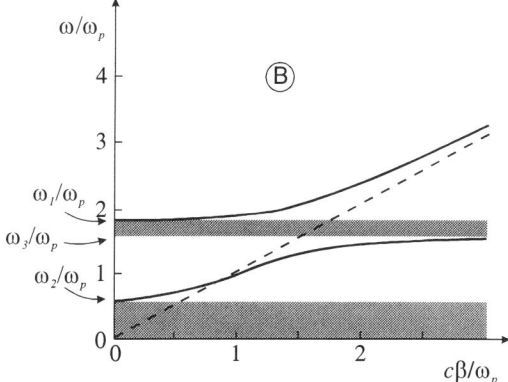

FIGURE 4.25. Brillouin diagram for a plasma medium. Propagation is perpendicular to the biasing magnetic field. The forbidden bands are darkened. (A) Ordinary wave, (B) extraordinary wave. $\omega_c/\omega_p = \sqrt{2}$. The dotted line corresponds to the free-space medium.

referred to as the *ordinary wave*. The corresponding Brillouin diagram is given in Fig. 4.25A. No propagation is possible in the *forbidden band* $0 \leqslant \omega < \omega_p$.

The other mode exhibits components $(E_x = -i\varepsilon_2 E_y/\varepsilon_1, E_y, H_z)$, is TM with respect to x, and

$$c^2\beta^2 = \omega^2 \frac{\varepsilon_1^2 - \varepsilon_2^2}{\varepsilon_1}.$$

The field has its electric components polarized elliptically in the (x, y) plane, its propagation constant depends upon the dc magnetization, and is referred to as the *extraordinary wave*. The corresponding Brillouin diagram is given in

Fig. 4.25B, and is split into two branches with cutoff (angular) frequencies[15]

$$\omega_{1,2} = \sqrt{\frac{(2\omega_p^2 + \omega_c^2) \pm \sqrt{(2\omega_p^2 + \omega_c^2)^2 - 4\omega_p^4}}{2}}$$

respectively, and two forbidden bands:

$$0 \leqslant \omega < \omega_2 \quad \text{and} \quad \omega_3 < \omega < \omega_1,$$

where

$$\omega_3 = \sqrt{\omega_p^2 + \omega_c^2}.$$

We turn now to the propagation along the z-axis with

$$\partial/\partial x = \partial/\partial y = 0 \quad \text{and} \quad \partial/\partial z = -i\beta,$$

β being the pertinent propagation constant. From Eqs. (4.67c–d) we have

$$E_z = H_z = 0,$$

so that the field is TEM with respect to z. From Eqs. (4.67a–b) we obtain

$$\begin{cases} \beta E_y = -\omega\mu_0 H_x, \\ \beta E_x = \omega\mu_0 H_y, \\ \beta H_y = \omega\varepsilon_0[\varepsilon_1 E_x + i\varepsilon_2 E_y], \\ \beta H_x = \omega\varepsilon_0[i\varepsilon_2 E_x - \varepsilon_1 E_y]. \end{cases}$$

A linear combination of the first two equations with coefficients $(i, 1)$, $(-i, 1)$, and the last two with coefficients $(1, i)$, $(1, -i)$, yields

$$\begin{vmatrix} \beta & -\omega\mu_0 & 0 & 0 \\ -\omega\varepsilon_0(\varepsilon_1 + \varepsilon_2) & \beta & 0 & 0 \\ 0 & 0 & \beta & -\omega\mu_0 \\ 0 & 0 & -\omega\varepsilon_0(\varepsilon_1 - \varepsilon_2) & \beta \end{vmatrix} \begin{vmatrix} E^+ \\ H^+ \\ E^- \\ H^- \end{vmatrix} = 0$$

where

$$E^+ = E_x + iE_y, \quad H^+ = (H_y - iH_x),$$
$$E^- = E_x - iE_y, \quad H^- = (H_y + iH_x),$$

---

15. It is convenient to first set $\beta = 0$ and then solve the dispersion equation with respect to $\omega^2 - \omega_c^2$.

which shows that the fields (E^+, H^+) and (E^-, H^-) decouple from each other. This, however, does not imply that solution (E^+, H^+) can exist independently of solution (E^-, H^-), and vice versa, because the same phasor components (e.g., E_x and E_y) of the fields appear in both of them. Additional constraints should be enforced in order to derive two *independent* solutions.

Suppose

$$E_x = \pm iE_y = \frac{E}{\sqrt{2}} \quad \text{and} \quad H_y = \mp iH_x = \frac{H}{\sqrt{2}}.$$

With the upper sign we have

$$E^- = H^- = 0,$$

$$\mathbf{E}^+ = E_x\hat{\mathbf{x}} + E_y\hat{\mathbf{y}} = E\frac{\hat{\mathbf{x}} + i\hat{\mathbf{y}}}{\sqrt{2}} = E\hat{\mathbf{p}},$$

$$\mathbf{H}^+ = H_x\hat{\mathbf{x}} + H_y\hat{\mathbf{y}} = H\frac{\hat{\mathbf{x}} + i\hat{\mathbf{y}}}{\sqrt{2}} = H\hat{\mathbf{p}},$$

where $\hat{\mathbf{p}}$ is the (negative) polarization vector of Eq. (4.35b). Similarly, the lower sign leads to

$$E^+ = H^+ = 0,$$

$$\mathbf{E}^- = E_x\hat{\mathbf{x}} + E_y\hat{\mathbf{y}} = E\frac{\hat{\mathbf{x}} - i\hat{\mathbf{y}}}{\sqrt{2}} = E\hat{\mathbf{q}},$$

$$\mathbf{H}^- = H_x\hat{\mathbf{x}} + H_y\hat{\mathbf{y}} = H\frac{\hat{\mathbf{x}} - i\hat{\mathbf{y}}}{\sqrt{2}} = H\hat{\mathbf{q}},$$

where $\hat{\mathbf{q}}$ is the (positive) polarization vector of Eq. (4.35b). We conclude that two circularly polarized plane-wave modes can propagate in the plasma.

The dispersion equation of the mode (E^+, H^+) is again obtained by setting to zero the determinant of the matrix of coefficients:

$$c^2\beta^2 = \omega^2(\varepsilon_1 + \varepsilon_2).$$

Its Brillouin diagram is depicted in Fig. 4.26. The forbidden band[16] is defined by $0 \leqslant \omega < \omega_1$ with

$$\omega_1 = \frac{\sqrt{4\omega_p^2 + \omega_c^2} - \omega_c}{2}.$$

16. Also in this case, it is convenient to first set $\beta = 0$ and then to solve the dispersion equation with respect to $\omega - \omega_c$ for the (E^+, H^+) and to $\omega + \omega_c$ for the (E^-, H^-) modes.

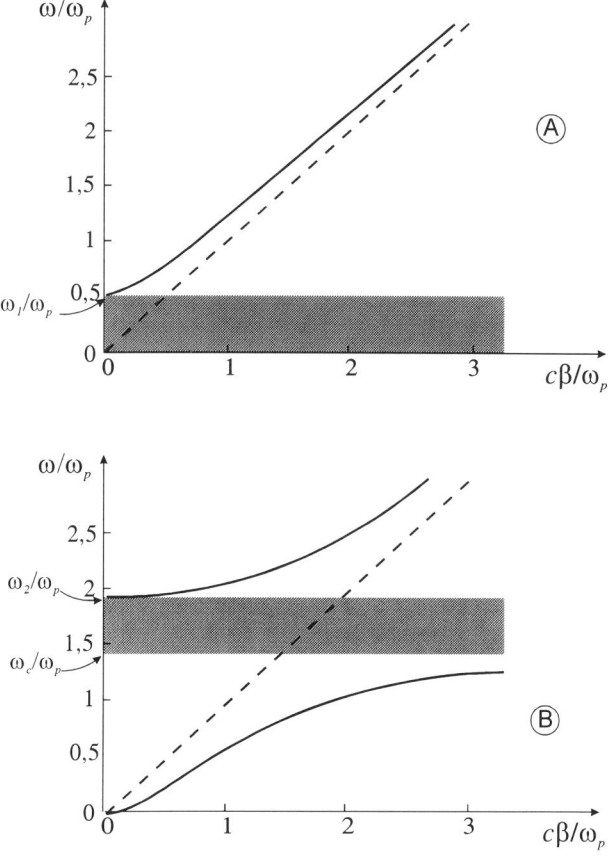

FIGURE 4.26. Brillouin diagram for a plasma medium. Propagation is parallel to the biasing magnetic field. The forbidden bands are darkened. Circularly polarized (A) negative and (B) positive waves. $\omega_c/\omega_p = \sqrt{2}$. The dotted line corresponds to the free-space medium.

Similarly, the equation of the mode (E^-, H^-) is

$$c^2\beta^2 = \omega^2(\varepsilon_1 - \varepsilon_2).$$

The corresponding Brillouin diagram is given in Fig. 4.26 and is split into two branches, $\omega \leqslant \omega_c$ and $\omega \geqslant \omega_2$, with

$$\omega_2 = \frac{\sqrt{4\omega_p^2 + \omega_c^2} + \omega_c}{2}.$$

The forbidden band is $\omega_c \leqslant \omega \leqslant \omega_2$.

4.5. GUIDED PROPAGATION

We consider electromagnetic propagation along a structure *uniform* in the direction of the z-axis, i.e., with a transverse cross section which does not change with z. A few examples are given in Fig. 4.27. We assume the structure to be composed of regions with homogeneous isotropic materials characterized by electromagnetic parameters ε and μ.

We start from Maxwell's equation (4.46) with $\mathbf{J}_0 = \rho_0 = 0$ (source-free solution), and use them within each homogeneous region of the structure. We take the curl of Eq. (4.46a) and substitute from Eq. (4.46b) to obtain

$$\nabla \times \nabla \times \mathbf{E} = k^2 \mathbf{E} \quad \text{with} \quad k^2 = \omega^2 \varepsilon \mu.$$

Use of Eqs. (A.17) and (4.46c) leads to

$$\nabla^2 \mathbf{E} + k^2 \mathbf{E} = \nabla \nabla \cdot \mathbf{E} = 0,$$

because ε and μ are locally constant. Similarly

$$\nabla^2 \mathbf{H} + k^2 \mathbf{H} = 0.$$

For uniform structure a convenient z-dependence of the fields is $\exp(-i\beta z)$, with the same value of β in each region. This is a necessary condition for enforcing boundary conditions at interfaces among different regions within the cross section of the structure. If

$$\mathbf{E}(x, y, z) \to \mathbf{E}(x, y) \exp(-i\beta z) \quad \text{and} \quad \mathbf{H}(x, y, z) \to \mathbf{H}(x, y) \exp(-i\beta z),$$

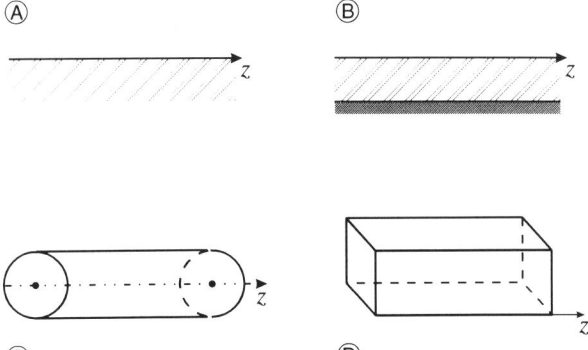

FIGURE 4.27. Examples of uniform guiding structures: (A) Two dielectric half-spaces, (B) a dielectric slab, (C) a circular pipe, (D) a rectangular guide.

and we project onto the z-axis, then

$$\nabla_t^2 E_z + k_t^2 E_z = 0, \quad (4.69\text{a})$$

$$\nabla_t^2 H_z + k_t^2 H_z = 0, \quad (4.69\text{b})$$

with

$$k_t^2 = k^2 - \beta^2, \quad (4.70)$$

because

$$\nabla^2 \to \nabla_t^2 + \frac{\partial^2}{\partial z^2} = \nabla_t^2 - \beta^2.$$

In Eqs. (4.69), k_t is the transverse component of the propagation constant $\mathbf{k}$. In cartesian coordinates the x and y components of Eq. (4.46a–b) are

$$\begin{vmatrix} \beta & -\omega\mu & 0 & 0 \\ -\omega\varepsilon & \beta & 0 & 0 \\ 0 & 0 & \beta & -\omega\mu \\ 0 & 0 & -\omega\varepsilon & \beta \end{vmatrix} \begin{vmatrix} E_x \\ H_y \\ E_y \\ -H_x \end{vmatrix} = i \begin{vmatrix} \partial E_z/\partial x \\ \partial H_z/\partial y \\ \partial E_z/\partial y \\ -\partial H_z/\partial x \end{vmatrix},$$

which shows that the electromagnetic field can be derived from its *longitudinal* components (E_z, H_z), and that the *transverse* components (E_x, H_y) can be computed independently from the other components $(E_y, -H_x)$. The solution for these two sets of field components yields

$$\begin{vmatrix} E_x \\ H_y \end{vmatrix} = \frac{1}{ik_t^2} \mathbf{T} \cdot \begin{vmatrix} \partial E_z/\partial x \\ \partial H_z/\partial y \end{vmatrix}, \quad (4.71\text{a})$$

$$\begin{vmatrix} E_y \\ -H_x \end{vmatrix} = \frac{1}{ik_t^2} \mathbf{T} \cdot \begin{vmatrix} \partial E_z/\partial y \\ -\partial H_z/\partial x \end{vmatrix}, \quad (4.71\text{b})$$

with

$$\mathbf{T} = \begin{vmatrix} \beta & \omega\mu \\ \omega\varepsilon & \beta \end{vmatrix}. \quad (4.71\text{c})$$

Formulation (4.71) is by no means limited to cartesian geometries and can be easily generalized to cylindrical-type coordinates systems; see Section 4.5.4.

The solution of Eqs. (4.69) provides four integration constants for each homogeneous region of the structure. These constants are determined by imposing continuity of the four tangential components of the fields at media discontinuities (see Section 1.2.7); only one single constant remains undetermined because the sources have not been specified. Accordingly, both the electric E_z and magnetic H_z fields are necessary and present in the solution, which is composed of *hybrid modes* with respect to the z-axis. In some cases, enforcement of boundary conditions requires only two constants per homogeneous section (see Sections 4.5.1–4.5.3); in these cases, TE ($E_z = 0$) or TM ($H_z = 0$) *modes* are possible with respect to the propagation direction z.

4.5.1. Oblique Incidence on a Dielectric Half-Space

In Fig. 4.28, we consider y-independent fields (so that $\nabla_t^2 \to d^2/dx^2$) and plane-wave propagation.

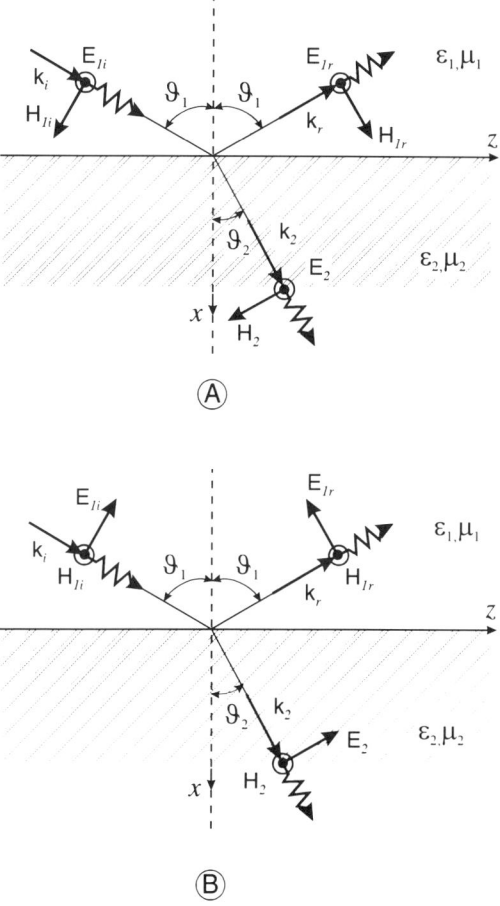

FIGURE 4.28. Oblique incidence of a plane wave over a dielectric half-space: (A) perpendicular and (B) parallel polarization.

When the sources are (remotely) located in the half-space $x \leq 0$, the appropriate solution of Eqs. (4.69) in this region is the following:

$$E_{1z} = A \exp(-iu_1 x) + B \exp(iu_1 x)$$

and

$$H_{1z} = C \exp(-iu_1 x) + D \exp(iu_1 x)$$

with

$$k_1^2 = \omega^2 \varepsilon_1 \mu_1 = \beta^2 + u_1^2. \quad (4.72a)$$

The components with constants A and C are recognized to be the incoming fields generated by the sources, while the other components with constant B and D correspond to the fields reflected at the discontinuity boundary $x = 0$; the latter verify the radiation condition at $x \to -\infty$.

In the region $x \geq 0$, the appropriate solution fulfilling the radiation conditions at $x \to +\infty$ is

$$E_{2z} = F \exp(-iu_2 x) \quad \text{and} \quad H_{2z} = G \exp(-iu_2 x)$$

with

$$k_2^2 = \omega^2 \varepsilon_2 \mu_2 = \beta^2 + u_2^2. \quad (4.72b)$$

Equations (4.71) show that two independent solutions are possible for the field, if the boundary conditions can be verified: one solution is obtained by setting $H_z = 0$ and exhibits components (E_x, H_y, E_z), see Eq. (4.71a), while the other is obtained by setting $E_z = 0$ and exhibits components $(E_y, -H_x, H_z)$, see Eq. (4.71b).

For the second solution, continuity conditions at $x = 0$ involve the components (E_y, H_z); these conditions can be enforced with the aid of the available constants C, D, and G. This solution is valid in its own right, is TE with respect to z, and is usually referred to as *perpendicular polarization* (see Fig. 4.28A); the electric field is perpendicular to the *plane of incidence* $y = 0$. Similarly, for the other independent solution the continuity conditions involve the components (H_y, H_z); the solution is TM with respect to z and is usually referred to as *parallel polarization*, with the electric field in the plane of incidence (see Fig. 4.28B).

Equations (4.72) suggest the positions:

$$\beta = k_1 \sin \theta_1, \quad u_1 = k_1 \cos \theta_1, \quad (4.73a)$$

$$\beta = k_2 \sin \theta_2, \quad u_2 = k_2 \cos \theta_2, \quad (4.73b)$$

whose physical model is depicted in Fig. 4.28. The plane wave is incident with propagation vector $\mathbf{k}_i = u_1\hat{\mathbf{x}} + \beta\hat{\mathbf{z}}$, is reflected with propagation vector $\mathbf{k}_r = -u_1\hat{\mathbf{x}} + \beta\hat{\mathbf{z}}$, and is transmitted with propagation vector $\mathbf{k}_2 = u_2\hat{\mathbf{x}} + \beta\hat{\mathbf{z}}$. The angles of *incidence* θ_i and *reflection* θ_r are equal,

$$\theta_i = \theta_r = \theta_1, \tag{4.74a}$$

and

$$\sqrt{\varepsilon_1\mu_1}\sin\theta_1 = \sqrt{\varepsilon_2\mu_2}\sin\theta_2, \tag{4.74b}$$

which are *Snell's*[17] (also referred to as Snell–Cartesio[18]) *laws* in optics. They are a consequence of the continuity conditions, which require β to be the same in the two media.

For perpendicular polarization, the continuity conditions at $x = 0$ can be enforced by computing E_y via Eq. (4.71b) with $k_t = u_1$ for $x \leqslant 0$ and $k_t = u_2$ for $x \geqslant 0$. This procedure yields

$$\begin{cases} C + D = G, & (4.75a) \\ \dfrac{\omega\mu_1}{u_1}(C - D) = \dfrac{\omega\mu_2}{u_2}G, & (4.75b) \end{cases}$$

where Eq. (4.75a) refers to the magnetic field H_z and Eq. (4.75b) to the electric field E_y, respectively. We can define reflection and transmission coefficients as in Section 3.2.8 if reference is made to the field components (E_y, H_z) transverse to the x-axis, which may now be considered as the direction of propagation. The reflection coefficient Γ is the ratio between the y-components of the reflected and incident electric field:

$$\Gamma = \frac{-\omega\mu_1 D/u_1}{\omega\mu_1 C/u_1},$$

and the transmission coefficient T is the ratio between the same transmitted and incident components:

$$T = \frac{\omega\mu_2 G/u_2}{\omega\mu_1 C/u_1}.$$

If we set

$$Z_{01} = \frac{\omega\mu_1}{u_1} \quad \text{and} \quad Z_{02} = \frac{\omega\mu_2}{u_2} \tag{4.76a}$$

---

17. Willebrordus Snell van Royen (Snellius): Leiden (The Netherlands), 1581–Leiden, 1626.
18. René Descartes (Cartesio): La Haye, Turenna (France), 1596–Stockholm, 1650.

and solve Eqs. (4.75) for D and G, we obtain

$$\Gamma = \frac{Z_{02} - Z_{01}}{Z_{02} + Z_{01}} \tag{4.77a}$$

and

$$T = \frac{2Z_{02}}{Z_{02} + Z_{01}}, \tag{4.77b}$$

which are formally identical to Eqs. (3.38) provided that the intrinsic impedances of Eq. (3.28) are substituted by the *characteristic impedances* of Eqs. (4.76a). Equations (4.73b) and (4.74) lead to

$$u_2^2 = k_2^2(1 - \sin^2\theta_2) = k_2^2 - k_1^2 \sin^2\theta_1,$$

and Eq. (4.77a) takes the form

$$\Gamma \to \Gamma_\perp = \frac{\cos\theta_1 - (\mu_1/\mu_2)\sqrt{(k_2/k_1)^2 - \sin^2\theta_1}}{\cos\theta_1 + (\mu_1/\mu_2)\sqrt{(k_2/k_1)^2 - \sin^2\theta_1}}, \tag{4.78a}$$

which is referred to as the *Fresnel*[19] *reflection coefficient* for perpendicular polarization; see Fig. 4.29.

Parallel polarization can be dealt with similarly. Expressions (4.77) are still valid provided that the characteristic impedances are defined as

$$Z_{01} = \frac{u_1}{\omega\varepsilon_1} \quad \text{and} \quad Z_{02} = \frac{u_2}{\omega\varepsilon_2}. \tag{4.76b}$$

The Fresnel reflection coefficient (see Fig. 4.29) assumes the following form:

$$\Gamma_\parallel = -\frac{\cos\theta_1 - (\varepsilon_1/\varepsilon_2)\sqrt{(k_2/k_1)^2 - \sin^2\theta_1}}{\cos\theta_1 + (\varepsilon_1/\varepsilon_2)\sqrt{(k_2/k_1)^2 - \sin^2\theta_1}}. \tag{4.78b}$$

When $k_2/k_1 < 1$, at the angle of incidence

$$\sin\bar{\theta}_1 = k_2/k_1$$

we have $\theta_2 = 90°$ for both polarizations [see Eq. (4.74)], and the wave emerges

19. Augustin Jean Fresnel: Broglie (France), 1788–Ville d'Avray, Paris, 1827.

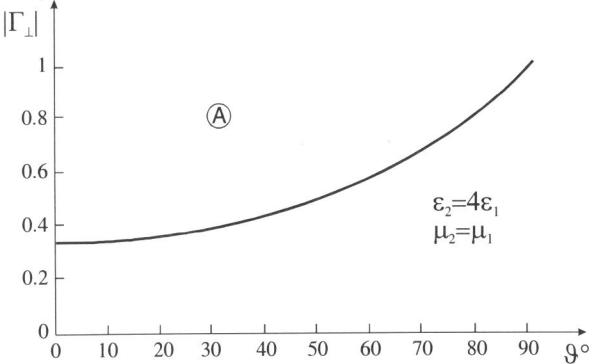

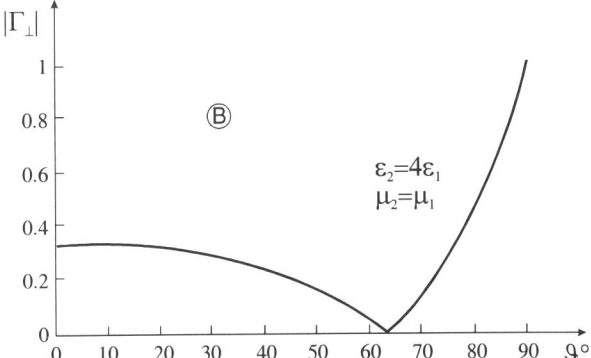

FIGURE 4.29. Modulus of the Fresnel reflection coefficient vs. the angle of incidence θ with $\mu_1 = \mu_2$ and $\varepsilon_2 = 4\varepsilon_1$: (A) perpendicular and (B) parallel polarization.

in the half-space $x \geq 0$ at a grazing angle. For $\theta_1 \geq \bar{\theta}_1$, $|\Gamma_\perp| = |\Gamma_\parallel| = 1$ and there is no flux of real power in the half-space $x \geq 0$. We also have

$$u_2 = \sqrt{k_2^2 - k_1^2 \sin^2\theta_1} = -i\sqrt{k_1^2 \sin^2\theta_1 - k_2^2} = -i\alpha_2,$$

and the field is attenuated exponentially for $x > 0$, where propagation is along the z-axis only with

$$\beta = \sqrt{k_2^2 + \alpha_2^2} > k_2.$$

The field differs appreciably from zero only close to the boundary, and is referred to as the *surface wave*. The phase velocity ω/β is larger (*fast* wave) than that (ω/k_2) of a plane wave in the same unbounded medium. The angle $\bar{\theta}_1$ is referred to as the *limit angle* in optics.

It is interesting to explore the possible existence of an angle of incidence such that there is no reflected wave, even if the two media are different. This condition is met if

$$Z_{01} = Z_{02},$$

which generalizes to oblique incidence in the Heaviside matching condition (2.16). In the simplest, yet important, nonmagnetic case, $\mu_1 = \mu_2$, no solution exists for perpendicular polarization if $\varepsilon_1 \neq \varepsilon_2$. For parallel polarization, use of expressions (4.76b) for the characteristic impedances leads to the condition

$$1 - \sin^2\theta_B = \frac{\varepsilon_1}{\varepsilon_2} - \left(\frac{\varepsilon_1}{\varepsilon_2}\right)^2 \sin^2\theta_B,$$

i.e.,

$$\sin^2\theta_B = \frac{\varepsilon_2}{\varepsilon_1 + \varepsilon_2},$$

and the angle θ_B is referred to as the *Brewster*[20] *angle*. An unpolarized plane wave incident at angle θ_B is reflected with perpendicular polarization.

If the half-space is highly conducting, namely,

$$\frac{\sigma_2}{\omega\varepsilon_2} \gg 1,$$

we have for normal incidence ($\theta_1 = 0°$) and nonmagnetic materials ($\mu_1 = \mu_2 = \mu_0$)

$$Z_{02} = \sqrt{\frac{\mu_0}{\varepsilon_2 + \sigma_2/i\omega}} \approx \sqrt{\frac{i\omega\mu_0}{\sigma_2}} \ll 1, \quad T \approx \frac{2Z_{02}}{Z_{01}} \approx 2\sqrt{\frac{i\omega\varepsilon_1}{\sigma_2}}$$

$$u_2 = k_2 \approx \sqrt{-i\omega\mu_0\sigma_2}.$$

It is convenient to define the *skin depth*

$$\delta = \sqrt{\frac{2}{\omega\mu_0\sigma_2}}, \tag{4.79}$$

---

20. Davis Brewster: Jedburg, Roxburshire (UK), 1781–Allerby on Tweed, Melrose, 1868.

so that

$$T \approx 2\frac{1+i}{\zeta_1\sigma_2\delta}, \quad u_2 \approx \frac{1-i}{\delta}, \quad \text{and} \quad \zeta_1 = \sqrt{\frac{\mu_0}{\varepsilon_1}}.$$

If E_0 is the incident electric field at $x = 0$, the transmitted field is given by

$$E_2 = TE_0 \exp\left(-\frac{x}{\delta}\right) \exp\left(-i\frac{x}{\delta}\right),$$

and is exponentially damped away from the boundary surface. The good conductor *screens* the incident radiation, which becomes negligible past a few skin depths. A current density

$$J_2 = \sigma_2 E_2 = 2\frac{1+i}{\zeta_1\delta} E_0 \exp\left(-\frac{1+i}{\delta}x\right)$$

is induced within the conducting medium. Integration of the entire current flux per unit length along y, from $x = 0$ to $x \to \infty$ (see Fig. 2.11), provides the current linear density

$$I = \frac{2E_0}{\zeta_1} = 2H_0,$$

twice as large as the incident magnetic field H_0. The induced current shrinks close to the boundary surface as $\sigma_2 \to \infty$, and a current linear density is set up on the surface [see Eqs. (2.24) and (1.36)].

4.5.2. Guided Propagation along a Dielectric Slab

In Fig. 4.30, we consider y-independent plane-wave propagation along a grounded dielectric slab. As in Section 4.5.1, the appropriate solution of Eqs. (4.69) in the half-space $x \geq 0$ is

$$E_{0z} = A \exp(iu_0 x) + B \exp(-iu_0 x)$$

and

$$H_{0z} = C \exp(iu_0 x) + D \exp(-iu_0 x)$$

with

$$k_0^2 = \omega^2 \varepsilon_0 \mu_0 = \beta^2 + u_0^2.$$

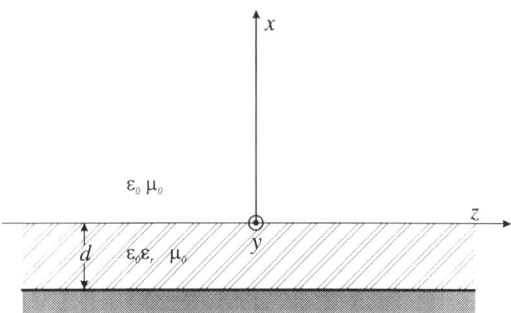

FIGURE 4.30. Plane-wave propagation along a dielectric grounded slab.

The fields with constants A and C are recognized to be the incoming plane waves associated with the sources, while the fields with constants B and D are those reflected by the discontinuity boundary at $x = 0$.

The solution in the dielectric slab $-d \leqslant x \leqslant 0$ fulfilling boundary conditions at $x = -d$ is

$$E_z = F \sin[(u(x + d)] \quad \text{and} \quad H_z = G \cos[(u(x + d)]$$

with

$$k^2 = \omega^2 \varepsilon_0 \mu_0 \varepsilon_r = \beta^2 + u^2.$$

As a matter of fact, on the perfectly conducting plane $x = -d$ we have $E_z(x = -d) = 0$ and also $E_y(x = -d) = 0$, as follows from Eq. (4.71b).

As in Section 4.5.1, Eqs. (4.71) show that two possible solutions exist, characterized by field components (E_x, H_y, E_z) and $(E_y, -H_x, H_z)$. These solutions satisfy individually all the boundary conditions and are independent of each other.

The second solution is TE with respect to z. Continuity of the tangential components of the field (E_y, H_z) at $x = 0$ requires that

$$\begin{cases} C + D = G \cos ud, \\ \dfrac{1}{u_0} C - \dfrac{1}{u_0} D = iG \dfrac{\sin ud}{u}, \end{cases}$$

where E_y has been computed with the aid of Eq. (4.71b). This system can be recast as follows:

$$\begin{vmatrix} -1 & \cos ud \\ 1 & \dfrac{iu_0}{u}\sin ud \end{vmatrix} \begin{vmatrix} D \\ G \end{vmatrix} = C \begin{vmatrix} 1 \\ 1 \end{vmatrix}, \qquad (4.80)$$

and the constants (D, G) can be computed in terms of C whose value remains unspecified, depending on the sources. If $C \to 0$ (source-free solution) we can still derive a nonzero solution for (D, G) provided that the determinant of the system (4.80) is equal to zero:

$$\frac{iu_0}{u}\sin ud + \cos ud = 0,$$

where u and u_0 are related via Eqs. (4.72):

$$u^2 - u_0^2 = k_0^2(\varepsilon_r - 1).$$

The corresponding fields are recognized to be the *resonant wave patterns* guided along the structure. Letting $u_0 \to -i\alpha$, $\alpha \geqslant 0$, the pertinent values of u and α are solutions of the system

$$\begin{cases} \alpha d = -(ud)\cotan(ud), \\ (ud)^2 + (\alpha d)^2 = (k_0 d)^2(\varepsilon_r - 1), \qquad \alpha d \geqslant 0. \end{cases} \qquad (4.81)$$

A graphical solution of system (4.81) is given in Fig. 4.31. We note that the frequency should exceed the minimum value

$$k_0 d\sqrt{\varepsilon_r - 1} = \frac{2\pi f_c}{c}d\sqrt{\varepsilon_r - 1} = \frac{\pi}{2}$$

for the existence of guided modes. This value f_c is referred to as the *cutoff frequency* of the *fundamental mode*:

$$f_c = \frac{c}{4d\sqrt{\varepsilon_r - 1}}.$$

An alternative equivalent parameter, the *cutoff wavelength* λ_c, can be used:

$$\frac{2\pi}{\lambda_c} = \frac{2\pi f_c}{c}, \qquad \text{i.e., } \lambda_c = 4d\sqrt{\varepsilon_r - 1}.$$

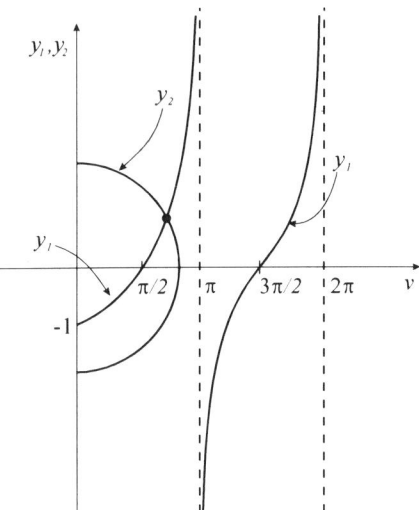

FIGURE 4.31. Plot of the functions $y_1 = -v \cotan v$ and $y_2^2 + v^2 = (k_0 d)^2 (\varepsilon_r - 1) y_1$, $y_2 = \alpha d$; $v = ud$.

As the frequency increases, *higher-order modes* can propagate.

The field exhibits a stationary pattern inside the dielectric slab, and is exponentially damped outside. Again, we recognize a surface-wave solution as in Section 4.5.1, with a phase velocity larger that the speed of light in the vacuum. Plots of the field are given in Fig. 4.32.

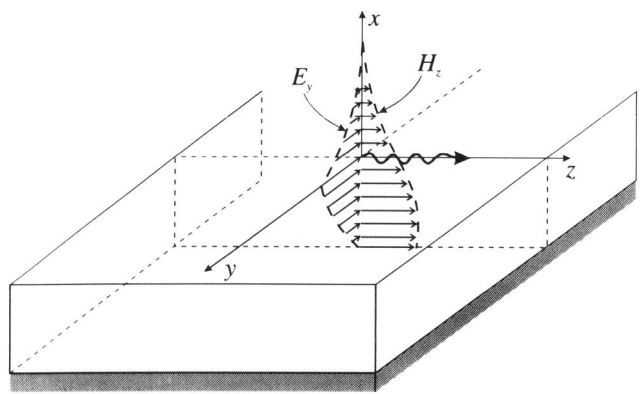

FIGURE 4.32. Plot of the field components (E_y, H_z) guided by a grounded dielectric slab. The additional component H_x, not represented in the figure, is equal to $-(\beta/\omega\mu_0)E_y$.

The (independent) TM modes can be studied similarly. Additional properties of guided wave modes are given in Section 7.3.3.

4.5.3. Propagation inside a Parallel-Plate Guide

In Fig. 4.33, we consider y-independent propagation along a parallel-plate guide, the two planes being perfectly conducting.

The solution of Eqs. (4.69) is as follows:

$$E_z = A \cos ux + B \sin ux \quad \text{and} \quad H_z = C \cos ux + D \sin ux.$$

Equations (4.71) are used to generate the other field components. Boundary conditions require E_z and E_y to be zero on the metal plates at $x = 0, a$.

The longitudinal field component H_z generates the fields $(E_y, -H_x)$; see Eq. (4.71b). Accordingly, the set of fields $(E_y, -H_x, H_z)$ is a possible independent solution if the electric component,

$$E_y = -\frac{1}{iu}(C \sin ux - D \cos ux),$$

satisfies the boundary conditions at $x = 0, a$, namely, $E_y(0) = E_y(a) = 0$. This is possible if $D = 0$ and

$$ua = n\pi, \quad n = 1, 2, \ldots.$$

The value $n = 0$ is excluded because $E_y = $ const does not satisfy the boundary conditions. The field is TE with respect to z, and consists of an infinite number of modes with propagation constants

$$\beta_n^2 = k_0^2 - (n\pi/a)^2, \quad \text{where } k_0^2 = \omega^2 \varepsilon_0 \mu_0.$$

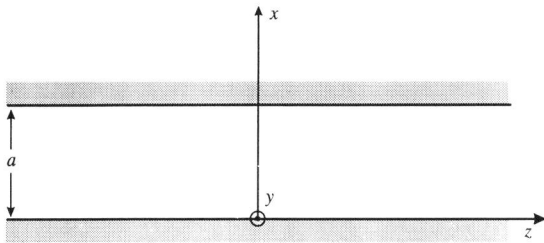

FIGURE 4.33. A parallel-plate guide structure.

These modes are numbered (and can be classified) according to the number n. The fundamental mode is obtained by setting $n = 1$ and its cutoff wavelength is given by

$$\lambda_{c1} = 2a.$$

The longitudinal field E_z generates the components (E_x, H_y); see Eq. (4.71a). The set (E_x, H_y, E_z) is an independent solution if boundary conditions $E_z(0) = E_z(a) = 0$ are satisfied. This is possible if $A = 0$ and

$$u = \frac{n\pi}{a}, \quad n = 0, 1, 2, \ldots.$$

The field is TM with respect to z. The fundamental mode is that corresponding to $n = 0$. We obtain $E_z = 0$, and the field becomes TEM with respect to z with propagation constant $\beta = \omega\sqrt{\varepsilon_0\mu_0}$ equal to that of free space.

4.5.4. Guided Propagation along Cylindrical Structures

In Fig. 4.34 we consider propagation along cylindrical structures of circular cross section. We extend the procedures presented in Sections 4.5.1–4.5.3 from cartesian to cylindrical geometries.

Equations (4.69) are still valid, the transverse scalar laplacian being given by

$$\nabla_t^2 \to \frac{1}{r}\frac{\partial}{\partial r} r \frac{\partial}{\partial r} + \frac{1}{r^2}\frac{\partial^2}{\partial \phi^2}$$

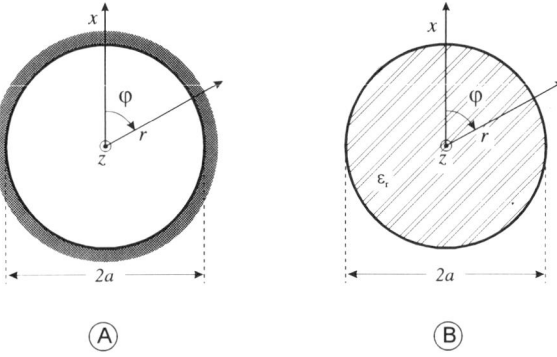

FIGURE 4.34. Cross section of (A) a hollow pipe of circular cross section, and of (B) a dielectric fiber of circular cross section.

[see Eq. (A.42)], and can be solved by standard application of separation of variables techniques. For the longitudinal electric field we set

$$E_z(r, \phi) = f(r)g(\phi),$$

thus obtaining

$$\frac{1}{f} r \frac{d}{dr} r \frac{df}{dr} + \frac{1}{g} \frac{d^2 g}{d\phi^2} + k_t^2 r^2 = 0,$$

which is the sum of three factors, the first and last depending only on r, and the second only on ϕ. Accordingly, each of the two groups is taken to be constant, the sum of the two constants being equal to zero. We have

$$\frac{d^2 g}{d\phi^2} + n^2 g(\phi) = 0,$$

in which case

$$g(\phi) = A \exp(in\phi),$$

with integer $n = \ldots, -2, -1, 0, 1, 2, \ldots$, to yield a single-valued field for any $\phi + 2m\pi \to \phi$ in the guide. In addition

$$\frac{1}{r} \frac{d}{dr} r \frac{df}{dr} + \left(k_t^2 - \frac{n^2}{r^2}\right) f = 0,$$

which is the differential equation governing Bessel functions; see Section D.1 in Appendix D.

In the region $r \leqslant a$ containing the origin, a finite solution for $E_z(r, \phi)$ is given by

$$E_z(r, \phi) = A J_n(k_t r) \exp(in\phi), \qquad 0 \leqslant r \leqslant a; \tag{4.82a}$$

see Eqs. (D.17). In the unbounded region $r \geqslant a$ excluding the origin, a solution fulfilling the radiation condition at infinity is given by

$$E_z(r, \phi) = F H_n^{(2)}(k_t r) \exp(in\phi), \qquad a \leqslant r; \tag{4.82b}$$

see Eqs. (D.21). For the longitudinal magnetic field we obtain similarly

$$H_z(r, \phi) = C J_n(k_t r) \exp(in\phi), \qquad 0 \leqslant r \leqslant a, \tag{4.83a}$$

$$H_z(r, \phi) = G H_n^{(2)}(k_t r) \exp(in\phi), \qquad a \leqslant r. \tag{4.83b}$$

The other field components are derived by computing the r and ϕ components of Maxwell equations (4.46a–b) and, as in Section 4.5, we obtain

$$\begin{vmatrix} E_r \\ H_\phi \end{vmatrix} = \frac{1}{ik_t^2} \, \boldsymbol{T} \cdot \begin{vmatrix} \partial E_z/\partial r \\ \partial H_z/r\partial\phi \end{vmatrix}, \tag{4.84a}$$

$$\begin{vmatrix} E_\phi \\ -H_r \end{vmatrix} = \frac{1}{ik_t^2} \, \boldsymbol{T} \cdot \begin{vmatrix} \partial E_z/r\partial\phi \\ -\partial H_z/\partial r \end{vmatrix}, \tag{4.84b}$$

where $\boldsymbol{T}$ is given by Eq. (4.71c). Equations (4.84) are analogous for the cylindrical coordinates of Eqs. (4.71). We note that Eqs. (4.84) can be generalized to cross sections other than cylindrical, in which case r and ϕ differentiation on the right side is replaced by $\partial/h_1\partial x_1$ and $\partial/h_2\partial x_2$, respectively, (x_1, x_2) being the appropriate transverse orthogonal coordinates and (h_1, h_2) the corresponding scale factors.

For the circular pipe of Fig. 4.34 with perfectly conducting metal boundaries, solutions (4.82a) and (4.83a) are appropriate.

The longitudinal field E_z generates *all* transverse components, unlike the cases examined in Section 4.5.1–4.5.3. The components E_z and E_ϕ are involved in the boundary conditions, but

$$E_\phi = \frac{\beta}{ik_t^2} \frac{\partial E_z}{r\partial\phi} = \frac{\beta n}{k_t^2 r} E_z$$

turns out to be just proportional to E_z. If

$$J_n(k_t a) = 0, \tag{4.85a}$$

then the boundary conditions are verified and we obtain an independent field solution TM with respect to z. Allowed values of $k_t a$ are those listed in Table 4.3, the successive zeros p_{nm} of Bessel functions. A double set of TM modes can propagate along the pipe with

$$\beta^2 = k_0^2 - \frac{p_{nm}^2}{a^2}.$$

Each mode is characterized by two indexes: the first, n, corresponds to the order of the Bessel function and is responsible for the azimuthal variation in the field; the second, m, corresponds to the order of the zeros of the Bessel function and is responsible for the radial variation in the field. Note that the value $k_t = 0$ is a solution of Eq. (4.85a) for $n \geqslant 1$, but the corresponding fields are identically equal to zero. Each mode exhibits a cutoff and the dominant one is TM_{01} with $\lambda_c = 0.83\pi a$.

TABLE 4.3. Successive Zeros of Bessel Functions of first kind, $J_n(p_{nm}) = 0^a$

n \ m	1	2	3
0	2.40	5.52	8.65
1	3.83	7.01	10.17
2	5.13	8.41	11.62

[a]The zeros $p_{n0} = 0$, $n \geq 1$ are not listed.

Similar reasoning applies for TE modes, the dispersion equation becoming

$$J'_n(k_t a) = 0, \tag{4.85b}$$

where the prime implies the derivative with respect to the argument. Solutions to Eq. (4.85b) are given in Table 4.4. All modes exhibit a cutoff and the dominant one is TE_{11} with $\lambda_c = 1.07\pi a$.

Additional properties of propagating modes are given in Section 7.4.2.

For the dielectric fiber of Fig. 4.34, the solutions (4.82a) in the dielectric ($r \leq a$) and (4.82b) in vacuo ($r \geq a$) are appropriate. For azimuthally independent modes, $n = 0$, the longitudinal field E_z generates the components (E_r, H_ϕ); in particular

$$H_\phi = \frac{\omega \varepsilon}{i k_t^2} \frac{\partial E_z}{\partial r}.$$

Continuity of tangential components of the field at $r = a$ requires that

$$\begin{vmatrix} J_0(ua) & H_0^{(2)}(u_0 a) \\ \varepsilon_r \dfrac{J'_0(ua)}{ua} & \dfrac{H_0^{(2)\prime}(u_0 a)}{u_0 a} \end{vmatrix} \begin{vmatrix} A \\ F \end{vmatrix} = 0, \tag{4.86a}$$

TABLE 4.4. Successive Zeros of the Derivative of Bessel Functions, $J'_n(q_{nm}) = 0$

n \ m	1	2	3
0	0	3.83	7.01
1	1.84	5.33	8.53
2	3.05	6.70	9.97

with $u = k_{t1}$, $u_0 = k_{t2}$, and

$$\varepsilon_r k_0^2 = \beta^2 + u^2 \quad \text{and} \quad k_0^2 = \beta^2 + u_0^2. \tag{4.86b}$$

We now require the modes to decay exponentially away from the fiber (*guided modes*), and this is obtained by setting

$$u_0 = -i\alpha;$$

see Eq. (D.21). It may be convenient to use Kelvin[21] instead of Hankel functions of complex argument. Equations (D.2), (D.3b), (D.7), and (D.11) yield

$$H_0^{(2)}(-i\alpha a) = \frac{2i}{\pi} K_0(\alpha a),$$

$$H_0^{(2)\prime}(-i\alpha a) = -H_1^{(2)}(-i\alpha a) = -\frac{2i}{\pi} K_1(\alpha a) = \frac{2i}{\pi} i K_0'(\alpha a),$$

and Eq. (4.86a) becomes

$$\begin{vmatrix} J_0(ua) & K_0(\alpha a) \\ -\varepsilon_r \dfrac{J_1(ua)}{ua} & \dfrac{K_1(\alpha a)}{\alpha a} \end{vmatrix} \begin{vmatrix} A \\ F \end{vmatrix} = 0. \tag{4.86c}$$

In Eq. (4.86c) the coefficient $2i/\pi$ has been absorbed in the constant F.

A nonzero solution to Eq. (4.86c) is possible if the determinant of the system is set equal to zero. Enforcing this condition and using Eq. (4.86b) gives

$$\begin{cases} J_0(ua) \dfrac{K_1(\alpha a)}{\alpha a} + \varepsilon_r K_0(\alpha a) \dfrac{J_1(ua)}{ua} = 0, \\ u^2 a^2 + \alpha^2 a^2 = (\varepsilon_r - 1) k_0^2 a^2, \end{cases}$$

which is the dispersion equation for these modes, TM with respect to z.

Azimuthally independent ($n = 0$) TE modes can be treated similarly.

In the general case $n \neq 0$, separation of modes in TE and TM modes is not possible and propagation is hybrid with respect to z.

Additional properties of guided modes are given in Section 7.4.8.

4.6. RADIATION FROM PRESCRIBED SOURCES

Let us consider a homogeneous isotropic medium with a prescribed (phasor) source distribution $\mathbf{J}_0(\mathbf{r})$. The associated (phasor) fields can be

---

21. William Thomson Kelvin: Belfast (Ireland), 1824–Largs, 1907.

computed by solving Maxwell equations (4.46) together with boundary conditions, which are the radiation conditions for an unbounded medium.

We proceed as in Section 3.2.9 and obtain

$$\nabla^2 \mathbf{E} + k^2 \mathbf{E} = \mathbf{M} \cdot \mathbf{J}_0$$

with

$$\mathbf{M} = i\omega\mu \left[\mathbf{I} + \frac{\nabla\nabla}{k^2} \right], \qquad k^2 = \omega^2 \varepsilon \mu,$$

which is the same as Eq. (3.44) evaluated at the carrier angular frequency of the narrowband signal. If the results of Sections 3.2.9 and 3.2.10, i.e., Eqs. (3.46), (3.48), and (3.52), are used again, then

$$\begin{aligned}
\mathbf{E}(r) &= -\iiint_V d\mathbf{V}' \mathbf{M} \cdot \left[\frac{\exp(-ik|\mathbf{r} - \mathbf{r}'|)}{4\pi|\mathbf{r} - \mathbf{r}'|} \mathbf{J}_0(\mathbf{r}') \right] \\
&= -i\omega\mu \left[\mathbf{I} + \frac{\nabla\nabla}{k^2} \right] \cdot \iiint_V d\mathbf{V}' \frac{\exp(-ik|\mathbf{r} - \mathbf{r}'|)}{4\pi|\mathbf{r} - \mathbf{r}'|} \mathbf{J}_0(\mathbf{r}'),
\end{aligned} \qquad (4.87)$$

where $\mathbf{M}$ has been taken outside the integral because it operates on the variable $\mathbf{r}$.

Computation of the field at large distances from the sources is of interest. In Fig. 4.35,

$$\begin{aligned}
|\mathbf{r} - \mathbf{r}'| &= \sqrt{(\mathbf{r} - \mathbf{r}') \cdot (\mathbf{r} - \mathbf{r}')} = \sqrt{r^2 + r'^2 - 2\mathbf{r} \cdot \mathbf{r}'} \\
&\approx r - \hat{\mathbf{r}} \cdot \mathbf{r}' + \frac{1}{2r} [r'^2 - (\hat{\mathbf{r}} \cdot \mathbf{r}')^2] \\
&= r - \hat{\mathbf{r}} \cdot \mathbf{r}' + \frac{1}{2r} |\hat{\mathbf{r}} \times \mathbf{r}'|^2,
\end{aligned} \qquad (4.88)$$

because $\hat{\mathbf{r}} \cdot \mathbf{r}' = |OS|$ and $r'^2 - (\hat{\mathbf{r}} \cdot \mathbf{r}')^2 = |QS|^2 = |\hat{\mathbf{r}} \times \mathbf{r}'|^2$. We take $|\mathbf{r} - \mathbf{r}'| \approx r$ in the denominator of the integral in (4.87), but save all three terms of the series expansion (4.88) in the phase exponent, due to the presence of the factor k. The last terms in Eq. (4.88) is negligible for any direction $\hat{\mathbf{r}}$ if

$$\frac{k}{2r} |\hat{\mathbf{r}} \times \mathbf{r}'|^2 \leqslant \frac{k(d/2)^2}{2r} \ll 1,$$

where the origin O lies at the center of the source region and d is its maximum transverse dimension. For an acceptable maximum phase change of $\pm \pi/8$ we

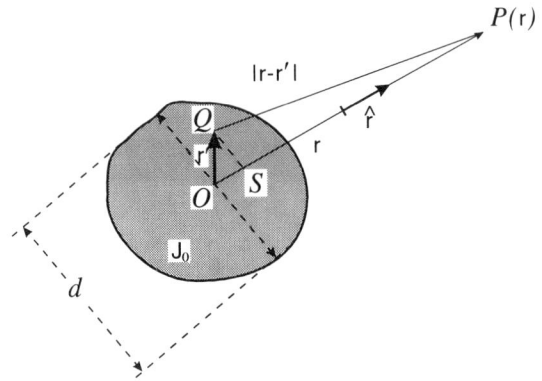

FIGURE 4.35. Radiation from sources $\mathbf{J}_0$.

obtain

$$r > 2d^2/\lambda, \qquad (4.89)$$

which defines the *Fraunhofer*[22] *radiation region*, and Eq. (4.87) becomes

$$\mathbf{E}(\mathbf{r}) = -i\omega\mu \left[\mathbf{I} + \frac{\nabla\nabla}{k^2} \right] \cdot \frac{\exp(-ikr)}{4\pi r} \iiint_V d\mathbf{V}' \mathbf{J}_0(\mathbf{r}') \exp(ik\hat{\mathbf{r}} \cdot \mathbf{r}')$$

$$\approx -i\omega\mu \frac{\exp(-ikr)}{4\pi r} [\mathbf{I} - \hat{\mathbf{r}}\hat{\mathbf{r}}] \cdot \iiint_V d\mathbf{V}' \mathbf{J}_0(\mathbf{r}') \exp(ik\hat{\mathbf{r}} \cdot \mathbf{r}'), \qquad (4.90)$$

because $\nabla \sim -ik\hat{\mathbf{r}}$ for $kr \gg 1$. Equation (4.90) shows that in the radiation region the field is orthogonal to the direction of propagation.

Condition (4.89) may not be verified, but higher-order terms in the series expansion (4.88) can still be neglected: this means that all three terms in Eq. (4.88) must be retained and we are in the *Fresnel region*. Moving even closer to the sources we enter the *near-field region*, where higher order terms in Eq. (4.88) must be added.

4.6.1. The Elementary Electric Dipole

In Fig. 4.36, we consider an elementary electric dipole whose moment, $\mathbf{P} = Q\Delta z\hat{\mathbf{z}}$, is related to the current by the equation

$$-I + i\omega Q = 0, \qquad I\Delta z = i\omega P, \qquad (4.91)$$

---

22. Joseph von Fraunhofer: Straubing (Germany), 1787–Münich, 1826.

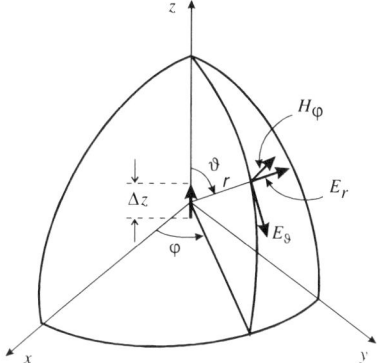

FIGURE 4.36. The field associated with an elementary electric dipole.

obtained by integrating the current density continuity equation

$$\nabla \cdot \mathbf{J} + i\omega\rho = 0$$

over the volume of Fig. 4.37. Accordingly, we can model the source as a current density

$$\mathbf{J} = I\delta(x)\delta(y)\,\text{rect}\left(\frac{z}{\Delta z}\right)\hat{\mathbf{z}}.$$

Hence Eq. (4.87) with $\Delta z \to 0$ (elementary dipole) and Eq. (4.46a) yield

$$\begin{cases} \mathbf{E} = -i\omega\mu I \Delta z \left[\mathbf{I} + \dfrac{\nabla\nabla}{k^2}\right] \cdot \left[\dfrac{\exp(-ikr)}{4\pi r}\,\hat{\mathbf{z}}\right], \\ \mathbf{H} = -\dfrac{1}{i\omega\mu}\,\nabla \times \mathbf{E} = I\Delta z\,\nabla \times \left[\dfrac{\exp(-ikr)}{4\pi r}\,\hat{\mathbf{z}}\right]. \end{cases}$$

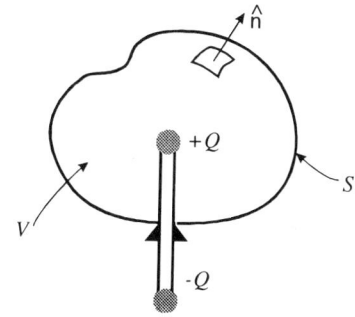

FIGURE 4.37. Integration volume to derive the connection between a dipole moment and the associated current.

In the spherical coordinate system of Fig. 4.36,

$$E_r = \zeta \frac{I\Delta z}{2\pi} \left[\frac{1}{r^2} + \frac{1}{ikr^3} \right] \cos\theta \exp(-ikr), \quad (4.92a)$$

$$E_\theta = \zeta \frac{I\Delta z}{4\pi} \left[\frac{ik}{r} + \frac{1}{r^2} + \frac{1}{ikr^3} \right] \sin\theta \exp(-ikr), \quad (4.92b)$$

$$H_\phi = \frac{I\Delta z}{4\pi} \left[\frac{ik}{r} + \frac{1}{r^2} \right] \sin\theta \exp(-ikr), \quad (4.92c)$$

which are the phasor counterparts of Eqs. (2.41).

The radial components of the Poynting vector in a lossless medium ($k = \beta$),

$$\mathbf{S}\cdot\hat{\mathbf{r}} = \frac{1}{2} E_\theta H_\phi^* = \frac{1}{2} \zeta \frac{|I|^2(\Delta z)^2}{(4\pi)^2} \sin^2\theta \left[\frac{\beta^2}{r^2} - \frac{i}{\beta r^5} \right]$$

$$= S_1 + iS_2,$$

exhibits a real and an imaginary part. The former is responsible for the radiated power, which is obtained by integration over a closed sphere around the dipole:

$$\oiint_A S_1 dA = \frac{1}{2} \zeta \frac{|I|^2(\beta\Delta z)^2}{6\pi},$$

and is independent of the distance r. The latter is responsible for the reactive power,

$$\oiint_A S_2 dA = -\frac{1}{2} \zeta \frac{|I|^2(\beta\Delta z)^2}{6\pi} \frac{1}{(\beta r)^3},$$

and decreases with distance. This power is negative, so that there is an excess of stored electric energy in the neighbor of the electric dipole.

NARROWBAND SIGNALS AND PHASOR FIELDS 215

The real power is formed by the *radiative fields*,

$$\begin{cases} E_\theta = i\zeta \dfrac{I\Delta z}{2\lambda r} \sin\theta \exp(-i\beta r), & (4.93a) \\ H_\phi = \dfrac{E_\theta}{\zeta}, & (4.93b) \end{cases}$$

which decay as $1/r$. The reactive power is formed by the other components, the *reactive fields*, which decay faster than $1/r$.

The radiative fields predominate over the reactive ones if

$$\frac{\beta}{r} > \frac{1}{r^2} \quad \text{and} \quad \frac{\beta}{r} > \frac{1}{\beta r^3},$$

i.e.,

$$r > \frac{1}{\beta} = \frac{\lambda}{2\pi}, \tag{4.94}$$

and are also referred to as *far fields*, unlike the others, the *near fields*. Note that condition (4.94) depends on the carrier frequency.

In order to derive the correct transition of dipole fields (4.92) to the low-frequency limit, we substitute $I\Delta z$ by $i\omega P$ and then set $\omega \to 0$:

$$\begin{cases} E_r = \dfrac{1}{2\pi\varepsilon r^3} \cos\theta, \\ E_\theta = \dfrac{1}{4\pi\varepsilon r^3} \sin\theta, \end{cases}$$

which are the correct expressions for the static field of an electric dipole. For these reasons the radiative, far fields are also referred to as *dynamic*, while the reactive, near fields are *quasi-static*.

4.6.2. The Elementary Magnetic Dipole

Application of the duality relationships (1.74) and (1.75a),

$$\mathbf{E} \to \zeta \mathbf{H}, \quad \mathbf{H} \to -\frac{1}{\zeta}\mathbf{E}, \quad I \to \frac{1}{\zeta}I_m,$$

to Eqs. (4.92) provides the formal expression for the field associated with an elementary magnetic dipole:

$$\begin{cases} H_r = \dfrac{1}{\zeta}\dfrac{I_m \Delta z}{2\pi}\left[\phantom{\dfrac{ik}{r}+}\dfrac{1}{r^2}+\dfrac{1}{ikr^3}\right]\cos\theta\exp(-ikr), & (4.95a) \\[2mm] H_\theta = \dfrac{1}{\zeta}\dfrac{I_m \Delta z}{4\pi}\left[\dfrac{ik}{r}+\dfrac{1}{r^2}+\dfrac{1}{ikr^3}\right]\sin\theta\exp(-ikr), & (4.95b) \\[2mm] E_\phi = -\dfrac{I_m \Delta z}{4\pi}\left[\dfrac{ik}{r}+\dfrac{1}{r^2}\phantom{+\dfrac{1}{ikr^3}}\right]\sin\theta\exp(-ikr). & (4.95c) \end{cases}$$

As shown is Section 2.3.2, an elementary current loop generates fields equivalent to those of a magnetic dipole; see Eq. (2.47). We again derive this result here.

With the aid of Fig. 4.38, we compute the electric field associated with the current loop by means of the integral (4.87). Hence

$$\mathbf{J}_0 = I\delta(z)\delta(r-a)\hat{\boldsymbol{\phi}}'$$

and

$$\mathbf{E}(\mathbf{r}) = -i\omega\mu a I\left[\mathbf{I}+\dfrac{\nabla\nabla}{k^2}\right]\cdot\int_0^{2\pi} d\phi'\,\dfrac{\exp(-ik|\mathbf{r}-\mathbf{r}'|)}{4\pi|\mathbf{r}-\mathbf{r}'|}\,\hat{\boldsymbol{\phi}}', \qquad (4.96)$$

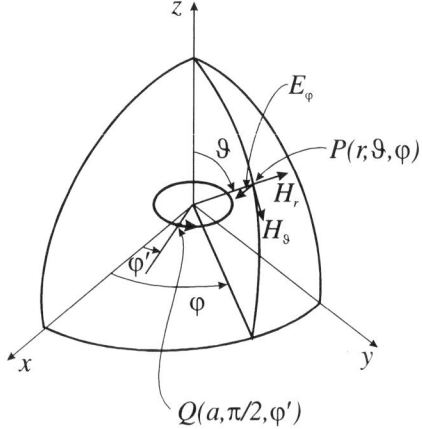

FIGURE 4.38. Computation of the field associated with an elementary current loop.

where

$$|\mathbf{r} - \mathbf{r}'| = |PQ|$$
$$= \sqrt{(r\sin\theta\cos\phi - a\cos\phi')^2 + (r\sin\theta\sin\phi - a\sin\phi')^2 + r^2\cos^2\theta}$$
$$\approx r - a\sin\theta(\cos\phi\cos\phi' + \sin\phi\sin\phi') = r - a\sin\theta f(\phi,\phi').$$

Note that the far-field condition $r \to \infty$ has not been enforced; the series expansion has been justified only by the elementary loop assumption $a \to 0$.

We expand the Green's function in Eq. (4.96) up to terms of order a:

$$\frac{\exp(-ik|\mathbf{r}-\mathbf{r}'|)}{4\pi|\mathbf{r}-\mathbf{r}'|} \approx \exp(-ikr)\left[1 + \frac{ikr+1}{4\pi r^2}\right] a\sin\theta f(\phi,\phi').$$

On substituting

$$\hat{\boldsymbol{\phi}}' = -\sin\phi'\hat{\mathbf{x}} + \cos\phi'\hat{\mathbf{y}}$$

in Eq. (4.96), and performing the integration we obtain

$$\mathbf{E}(\mathbf{r}) = -i\omega\mu a I\left[\mathbf{I} + \frac{\nabla\nabla}{k^2}\right] \cdot \left[\frac{ikr+1}{4\pi r^2}\exp(-ikr)\pi a\sin\theta\hat{\boldsymbol{\phi}}\right]$$
$$= -i\omega\mu\frac{\pi a^2 I}{4\pi}\left[\frac{ik}{r} + \frac{1}{r^2}\right]\sin\theta\exp(-ikr)\hat{\boldsymbol{\phi}}.$$

Comparison with Eq. (4.95c) shows that

$$I_m\Delta z = i\omega U = i\omega\mu\pi a^2 I,$$

where U is the magnetic dipole moment, a result which coincides with that of Section 2.3.2; see Eq. (2.47).

4.6.3. The Elementary Huygens Source

Let us consider two elementary dipoles, one electric and the other magnetic, mutually orthogonal, as depicted in Fig. 4.39A. This configuration is interesting from the viewpoint of the equivalence theorem of Section 4.3.7: any elementary portion $\Delta S = \Delta x \times \Delta y$ of the surface surrounding the sources may be considered as equivalent to the cross configuration of an electric dipole of moment

$$\mathbf{J}_S \Delta y \Delta x = i\omega \mathbf{P} = \hat{\mathbf{z}} \times \mathbf{H}_0 \Delta S$$

and a magnetic dipole of moment

$$\mathbf{K}_S \Delta x \Delta y = i\omega \mathbf{U} = -\hat{\mathbf{z}} \times \mathbf{E}_0 \Delta S,$$

where E_0 and H_0 are the tangential components of the field over the surface:

$$\mathbf{E}_0 = E_o \hat{\mathbf{x}} \quad \text{and} \quad \mathbf{H}_0 = H_0 \hat{\mathbf{y}}.$$

If $E_0 = \zeta H_0$, i.e., if the fields on ΔS are those of a plane wave progressing along the surface normal $\hat{\mathbf{z}}$ (see Fig. 4.39B), then the elementary radiating area ΔS is referred to as a *Huygens source* and

$$i\omega \mathbf{P} = -H_0 \Delta S \hat{\mathbf{x}} \quad \text{and} \quad i\omega \mathbf{U} = -E_0 \Delta S \hat{\mathbf{y}}.$$

Consider the spherical coordinate system of Fig. 4.40. We wish to compute the electric far field associated with the electric dipole $\mathbf{P} = P\hat{\mathbf{x}}$. To this end Eq. (4.93a) must be expressed in a form independent of the choice of the z-axis:

$$\mathbf{E} = \zeta \frac{\omega \hat{\mathbf{r}} \times \mathbf{P} \times \hat{\mathbf{r}}}{2\lambda r} \exp(-ikr).$$

For $\mathbf{P} = P\hat{\mathbf{x}}$ we readily obtain

$$\mathbf{E}_e = \zeta \frac{\omega P}{2\lambda r} \exp(-ikr) \hat{\mathbf{r}} \times \hat{\mathbf{x}} \times \hat{\mathbf{r}}$$

$$= \zeta \frac{\omega P}{2\lambda r} \exp(-ikr)(\cos\theta \cos\phi \hat{\boldsymbol{\theta}} - \sin\phi \hat{\boldsymbol{\phi}}), \quad (4.97a)$$

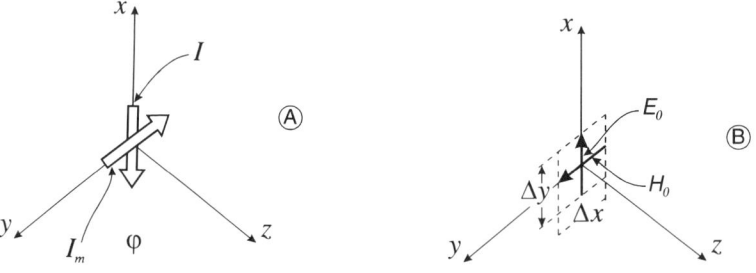

FIGURE 4.39. A Huygens source and application of the equivalence theorem for computing radiation from an elementary area.

NARROWBAND SIGNALS AND PHASOR FIELDS 219

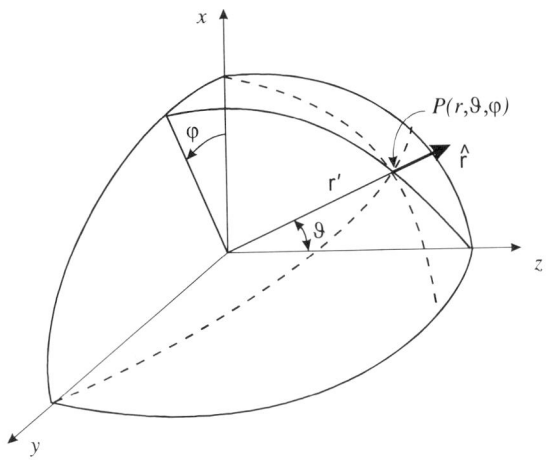

FIGURE 4.40. Computation of the far field associated with an elementary Huygens source.

as follows by expressing the unit vectors $\hat{\mathbf{x}}$ and $\hat{\mathbf{y}}$ in terms of spherical coordinates via Eqs. (A.45). Similarly, the electric far field associated with the magnetic dipole $\mathbf{U} = U\hat{\mathbf{y}}$ is given by

$$\mathbf{E}_m = \frac{\omega U}{2\lambda r} \exp(-ikr)\hat{\mathbf{y}} \times \hat{\mathbf{r}}$$

$$= \frac{\omega U}{2\lambda r} \exp(-ikr)(\cos\phi\hat{\boldsymbol{\theta}} - \cos\theta \sin\phi\hat{\boldsymbol{\phi}}). \quad (4.97b)$$

When $\zeta P = U = -E_0 \Delta S/i\omega$, the total electric field associated with the Huygens elementary area is

$$\mathbf{E}_p = \mathbf{E}_e + \mathbf{E}_m = i\frac{E_0 \Delta S}{2\lambda r} \exp(-ikr)(1 + \cos\theta)(\cos\phi\hat{\boldsymbol{\theta}} - \sin\phi\hat{\boldsymbol{\phi}}). \quad (4.97c)$$

If we change the orientation of the fields over ΔS, so that the electric field $\mathbf{E}_0$ is directed along $\hat{\mathbf{y}}$ and $\mathbf{H}_0$ along $-\hat{\mathbf{x}}$, the electric far field of the Huygens source becomes

$$\mathbf{E}_q = i\frac{E_0 \Delta S}{2\lambda r} \exp(-ikr)(1 + \cos\theta)(\sin\phi\hat{\boldsymbol{\theta}} + \cos\phi\hat{\boldsymbol{\phi}}).$$

The fields $\mathbf{E}_p$ and $\mathbf{E}_q$ are orthogonal, i.e.,

$$\mathbf{E}_p \cdot \mathbf{E}_q^* = 0,$$

and can be used as polarization bases (see Section 4.2.2).

Note that Huygens sources do not radiate in the backward direction ($\theta° = 180°$).

4.6.4. Radiation from Planar Sources

Let us consider a planar current distribution $\mathbf{J}_S(x, y)$ [ampere/m] as depicted in Fig. 4.41. The far field associated with this source can be computed via Eq. (4.90):

$$\mathbf{E} = -i\omega\mu \frac{\exp(-i\beta r)}{4\pi r} [\boldsymbol{I} - \hat{\mathbf{r}}\hat{\mathbf{r}}]$$
$$\cdot \iint dx\,dy\,\mathbf{J}_S(x, y) \exp(i\beta x \sin\theta \cos\phi + i\beta y \sin\theta \sin\phi),$$

because

$$\hat{\mathbf{r}} = \sin\theta \cos\phi\hat{\mathbf{x}} + \sin\theta \sin\phi\hat{\mathbf{y}} + \hat{\mathbf{z}}\cos\theta,$$

see Eqs. (A.44) and Eq. (4.90). If we set

$$\beta \sin\theta \cos\phi = u \quad \text{and} \quad \beta \sin\theta \sin\phi = v,$$

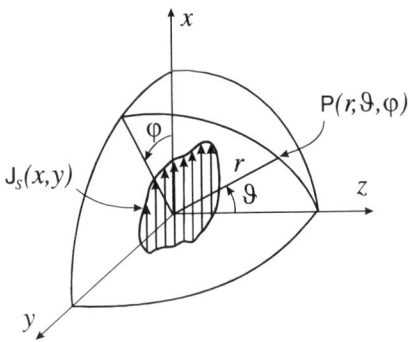

FIGURE 4.41. Radiation from planar currents.

the integral

$$\hat{\mathbf{E}}(u, v) = \iint dx\, dy \exp(iux + ivy) \mathbf{J}_S(x, y)$$

is seen to be a double FT in the conjugate variables (x, y) and (u, v).

If the source distribution is limited between $-a \leq x \leq a$ and $-b \leq y \leq b$, the vector field $\hat{\mathbf{E}}(u, v)$ is a bandlimited function and sampling theorem (4.17) applies:

$$\mathbf{E}(u, v) = \Sigma_n \Sigma_m \mathbf{E}_{nm} \operatorname{sinc}(ua - n\pi) \operatorname{sinc}(vb - m\pi), \quad (4.98a)$$

where

$$\mathbf{E}_{nm} = -i\omega\mu \frac{\exp(-\beta r)}{4\pi r} [\mathbf{I} - \hat{\mathbf{r}}\hat{\mathbf{r}}] \cdot \iint dx\, dy \exp\left(in\pi \frac{x}{a} + im\pi \frac{y}{b}\right) \mathbf{J}_S, \quad (4.98b)$$

and are the field samples along the sampling direction

$$u_n = \frac{n\pi}{a} = \beta \sin\theta_{nm} \cos\phi_{nm}, \qquad v_m = \frac{m\pi}{b} = \beta \sin\theta_{nm} \sin\phi_{nm},$$

i.e.,

$$\beta a \sin\theta_{nm} = \sqrt{(n\pi)^2 + (m\pi)^2}, \qquad \cos\phi_{nm} = \frac{n\pi}{\sqrt{(n\pi)^2 + (m\pi)^2}}. \quad (4.98c)$$

Examination of Eq. (4.98a) shows that the far field is reconstructed starting from its values along the sampling direction (4.98c), the interpolating functions being the sampling ones. Oversampling can be used to improve the convergence of the series expansion (4.98a); see Section 4.1.6 and Eq. (4.18).

Additional material about sampling expansion of radiated fields can be found in Section 8.6.1.

4.6.5. Radiation from Linear Arrays

In its simplest form a linear array is an equispaced sequence of radiators (see Fig. 4.42). If $\mathbf{E}_n$ is the field radiated by a single radiator in the plane of the figure, then the total field is given by

$$\mathbf{E} = \sum_{0}^{N-1} \mathbf{E}_n \exp(-i\beta nd \sin\phi). \quad (4.99)$$

The equation is the one-dimensional discrete counterpart of Eq. (4.90): the sources are localized at $\mathbf{r}' = -nd\hat{\mathbf{x}}$, the operator $-i\omega\mu[\mathbf{I} - \hat{\mathbf{r}}\hat{\mathbf{r}}]$ is included in the expression for $\mathbf{E}_n$, and $\hat{\mathbf{x}} \cdot \hat{\mathbf{r}} = \sin\phi$.

If all the radiators are identical and are driven with a progressive phase,

$$\mathbf{E}_n = \mathbf{E}_0 \exp[in\psi],$$

then Eq. (4.99) becomes

$$\mathbf{E} = \mathbf{E}_0 \sum_n^{N-1} \exp(-in\gamma), \quad \text{with } \gamma = \beta d \sin\phi - \psi,$$

where the finite series is seen to be a geometric progression:

$$f(\phi) = \sum_n^{N-1} \exp(-in\gamma) = \frac{1 - \exp(-iN\gamma)}{1 - \exp(-i\gamma)}$$

$$= \exp[-i(N-1)\gamma/2] \frac{\sin(N\gamma/2)}{\sin(\gamma/2)}. \quad (4.100)$$

Skipping the phase term, the amplitude of the function in Eq. (4.100) is depicted in Fig. 4.43. We note that γ spans $-\infty$ to $+\infty$ along the horizontal axis of Fig. 4.43. However, the *visible region*, i.e., the γ interval which corresponds to measurable values of the field, is limited by the condition $|\sin\phi| \leq 1$.

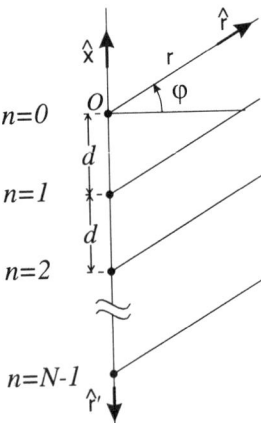

FIGURE 4.42. Computation of the field associated with a linear array.

The maximum value of the absolute value $|f(\phi)|$,

$$f_{\max} = N,$$

is obtained when

$$\frac{\gamma}{2} = \frac{1}{2}(\beta d \sin \phi_m - \psi) = m\pi, \quad m = 0, \pm 1, \ldots,$$

so that the direction of maximum radiation can be controlled by means of proper phasing of the array currents (*electronic steering* of the beam). The structure of the field consists of a *main lobe* (corresponding to the maximum) followed by *secondary lobes* whose amplitude first decreases ($|\gamma| \leq \pi$) and then increases again. The latter is periodic and characterized by equal amplitude maxima (*grating lobes*) at angles given by

$$\frac{d}{\lambda} \sin \phi_m - \frac{\psi}{2\pi} = m.$$

If we require the amplitude of secondary lobes to decrease monotonically in the visible region, $|\sin \phi| \leq 1$, then

$$\frac{|\gamma|}{2} = \frac{1}{2}|\beta d \sin \phi - \psi| \leq \frac{\pi}{2}, \quad |\sin \phi| \leq 1,$$

i.e.,

$$\frac{d}{\lambda} + \frac{|\psi|}{2\pi} \leq \frac{1}{2}, \quad -\pi \leq \psi \leq \pi,$$

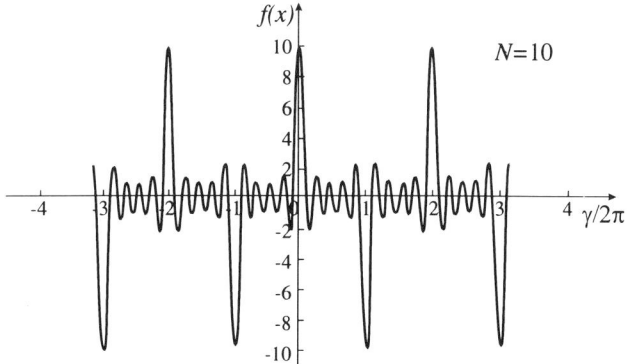

FIGURE 4.43. Plot of the amplitude of the far field associated with a linear array as a function of $\gamma/2\pi$. Number of array elements $N = 10$; $f(x) = f(\gamma/2\pi)$.

which sets a constraint on the array-element spacing. This condition has been obtained by setting $\sin\phi = -1$ for $\psi > 0$ and $\sin\phi = 1$ for $\psi < 0$.

If $\psi = 0$, the first maximum ($m = 0$) is obtained when $\phi_0 = 0, \pi$ (*broadside* configuration). The main-lobe angular width $\Delta\phi$ is given by

$$\frac{N}{2}\beta d \sin\Delta\phi \approx \frac{N}{2}\beta d\Delta\phi = \pi,$$

when $N\beta d \gg 1$. Again, the wider the array width Nd, the smaller the angle within which most of the radiation is concentrated.

If $\psi = \beta d$, the direction of maximum radiation becomes $\phi_0 = 90°$ (*endfire* configuration). Note that radiation is not symmetric with respect to $\phi = 0, \pi$, $\phi_0 = -90°$ does not correspond to a maximum.

When $\mathbf{E}_n$ exhibits a more involved dependence on n, a closed-form expression for the finite summation (4.99) cannot be obtained in general. It is noteworthy that $\mathbf{E}(\beta d \sin\phi)$ is the FT of a finite sequence and a sampling representation similar to Eq. (4.17) is possible.

We proceed as in Section 4.1.6. Assume, for simplicity, that the number of radiators is $2N + 1$, in which case Eq. (4.99) becomes

$$\mathbf{E}(u) = \sum_{-N}^{N} \mathbf{E}_n \exp(inu), \quad u = \beta d \sin\phi,$$

by setting the origin of coordinates at the center of the array. Starting from the ($|N|$-limited) sequence $\mathbf{E}_n$, we generate the periodic sequence $\mathbf{E}'_n$, where $\mathbf{E}'_{n+2N+1} = \mathbf{E}'_n$. We define a window sequence $I_n = 1$, $|n| < N$, so that $\mathbf{E}'_n = \mathbf{E}_n I_n$; $\mathbf{E}(u)$ is obtained as the convolution of the FTs of $\mathbf{E}'_n$ and I_n.

The periodic sequence $\mathbf{E}'_n$ is first expanded in a Fourier series,

$$\mathbf{E}'_n = \sum_{-\infty}^{+\infty} \mathbf{S}_m \exp\left[-i\pi\frac{m}{N}n\right],$$

and then Fourier-transformed to yield

$$\sum_{-\infty}^{+\infty} \mathbf{E}'_n \exp(inu) = 2\pi \sum_{-\infty}^{+\infty} \mathbf{S}_m \delta\left(u - \pi\frac{m}{N}\right).$$

The Fourier transform of the sequence I_n is given by

$$\sum_{-N}^{N} \exp(inu) = \frac{\sin[(2N+1)u/2]}{\sin[u/2]}.$$

Convolving these two FTs leads to

$$\mathbf{E}(u) = 2\pi \sum_{m=-\infty}^{+\infty} \mathbf{S}_m \frac{\sin[(2N+1)(u - m\pi/N)/2]}{\sin[(u - 2m\pi/N)/2]}$$

$$= \sum_{m=-\infty}^{+\infty} \mathbf{E}(u_m) \frac{\sin[(2N+1)(u - u_m)/2]}{(2N+1)\sin[(u - u_m)/2]}, \quad (4.101a)$$

$$u_m = \beta d \sin \phi_m = \frac{2m\pi}{N}. \quad (4.101b)$$

Equation (4.101a) applies to arrays and is analogous to Eq. (4.98a), which applies to continuous apertures of finite extent. As in Section 4.6.4, the far field is reconstructed starting from its values along the sampling directions (4.101b), but the interpolation is performed using *Dirichlet*[23] *functions* instead of the classical sampling functions (4.17). Also, in this case, oversampling can be implemented.

Additional material on arrays is given in Section 8.7.

4.7. SUMMARY AND SELECTED REFERENCES

In this chapter the important case of narrowband signals is treated [4.1–4.7]. The difference between the (almost) phasor fields considered here and the spectral fields of Chapter 3 is explained. Notation and techniques may be similar, but the two domains reside on a different footing and involve different application. Spectral fields are an intermediate step to the computation of space–time transients, unlike phasor fields which are appropriate to sinusoidal or almost monochromatic signals and allow a simplified analysis of related problems. This distinction is not always fully appreciated.

Narrowband signals (Section 4.1), their properties (Section 4.1.2–4.1.5), and their representations (Sections 4.1.6 and 4.1.7) are a prerequisite to complex vectors, which are studied extensively and presented in Section 4.2. Field polarization [1.4] and its coherence [1.4, 1.8] are considered in Sections 4.2.3 and 4.2.4. The discussion is based on Stokes parameters, the original paper in which they appeared being listed under [4.9]. In addition, the polarization bases [4.8], important for the definition of copolar and cross-polar components of the field, are discussed in Section 4.2.2.

Maxwell's equations in phasor form are presented in Section 4.3, which is the monochromatic counterpart of the time-domain Sections 1.1, 1.3, and 1.5.

Plane-wave propagation is dealt with in Section 4.4, it is the phasor-field analogy of Sections 2.1 and 3.2.5–3.2.8. This section includes narrowband-

---

23. Johann Peter Gustav Lejeune Dirichlet: Düren (Germany), 1805–Göttingen, 1859.

signal transmission and scrambling over a dispersive channel (Sections 4.4.1 and 4.4.2), as well as propagation in a magnetized plasma (Section 4.4.3). Guided propagation is examined in Section 4.5, including unbounded structures as the half-space (Section 4.5.1), the grounded dielectric slab (Section 4.5.2), and bounded structures, such as the circular pipe and the dielectric fiber (Section 4.5.4). Emphasis is placed on the construction of the field solution and the imposition of boundary conditions; the material is presented later in Sections 7.3–7.5 in a considerably expanded manner and from a different perspective.

Section 4.6 is devoted to radiation, and includes elementary antennas (Sections 4.6.1, 4.6.2, and 4.6.3), apertures as an application of sampling expansion (Section 4.6.4), and linear arrays (Section 4.6.5). Also in this case, the presentation is focused on the properties rather than on the use of the solutions of the problems under consideration. The latter material is extended and re-examined in Chapter 8, with emphasis on the engineering viewpoint.

References

[4.1] A. T. Adams, *Electromagnetism for Engineers*, Renold Press, New York (1971).
[4.2] C. A. Balanis, *Advances in Engineering Electromagnetics*, Wiley, New York (1989).
[4.3] D. K. Cheng, *Field and Wave Electromagnetics*, Addison-Wesley, Mass (1983).
[4.4] R. F. Harrington, *Time-Harmonic Electromagnetic Fields*, McGraw-Hill, New York (1961).
[4.5] M. A. Phonus, *Applied Electromagnetics*, McGraw-Hill, New York (1978).
[4.6] N. N. Rao, *Basic Electromagnetics and Applications*, Prentice Hall, NJ (1972).
[4.7] L. C. Shen and J. A. Kong, *Applied Electromagnetism*, Brooks Cole, CA (1983).
[4.8] A. C. Ludwig, "The definition of cross-polarization," *IEEE Trans. Antennas Propagat.* **21**, 116–119 (1973).
[4.9] G. G. Stokes, "On the composition and resolution of streams of polarized light from different sources," in *Mathematical and Physical Papers*, Cambridge University Press, London (1883).

5
High-Frequency Fields

5.1. ASYMPTOTIC FORM OF MAXWELL EQUATIONS

We consider a narrowband signal (see Chapter 4) in a lossless time-invariant isotropic medium of parameters $\varepsilon_0\varepsilon$ and $\mu_0\mu$, whose relative permittivity ε and permeability μ may be space-dependent. We define the free-space propagation constant $k_0 = \sqrt{\varepsilon_0\mu_0}$, the free-space intrinsic impedance $\zeta_0 = \sqrt{\mu_0/\varepsilon_0}$, and the relative refractive index $n = \sqrt{\varepsilon\mu}$, space-dependent in general.

We assume that the functions governing the fields and inductions are of the type $\mathbf{E}(\mathbf{r})\exp[-ik_0 L(\mathbf{r})]$, ..., and substitute into the source-free ($\mathbf{J}_0 = 0$) Maxwell equations (4.46) to obtain

$$\begin{cases} \nabla L \times \mathbf{E} - \mu\zeta_0\mathbf{H} = \dfrac{1}{ik_0}\nabla \times \mathbf{E}, & (5.1\text{a}) \\[6pt] \zeta_0\mathbf{H} \times \nabla L - \varepsilon\mathbf{E} = -\dfrac{1}{ik_0}\nabla \times \zeta_0\mathbf{H}, & (5.1\text{b}) \\[6pt] \mathbf{E}\cdot\nabla L = \dfrac{1}{ik_0}\dfrac{\nabla\cdot\varepsilon\mathbf{E}}{\varepsilon}, & (5.1\text{c}) \\[6pt] \zeta_0\mathbf{H}\cdot\nabla L = \dfrac{1}{ik_0}\dfrac{\nabla\cdot\mu\zeta_0\mathbf{H}}{\mu}, & (5.1\text{d}) \end{cases}$$

where use has been made of Eqs. (A.12) and (A.10) to compute the curl and divergence, respectively.

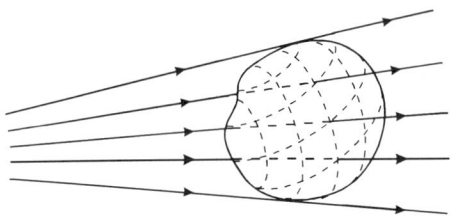

FIGURE 5.1. A rayfront and its associated ray congruence.

In the high-frequency limit $\omega \to \infty$ and $k_0 \to \infty$. We set $\mathbf{E} \to \mathbf{E} + O[1/k_0]$ and $\mathbf{H} \to \mathbf{H} + O[1/k_0]$, and to the lowest (zero) order in $1/k_0$ Eqs. (5.1) become

$$\begin{cases} \nabla L \times \mathbf{E} - \mu \zeta_0 \mathbf{H} = 0, & \text{(5.2a)} \\ \zeta_0 \mathbf{H} \times \nabla L - \varepsilon \mathbf{E} = 0, & \text{(5.2b)} \\ \mathbf{E} \cdot \nabla L = 0, & \text{(5.2c)} \\ \zeta_0 \mathbf{H} \cdot \nabla L = 0, & \text{(5.2d)} \end{cases}$$

which are the *high-frequency limit* of Maxwell equations.

In a lossless medium we can take $\mathbf{E}$, $\mathbf{H}$, and L real, and Eqs. (5.2c–d) show that $(\mathbf{E}, \zeta_0 \mathbf{H})$ are orthogonal to ∇L. Cross-multiplication of Eq. (5.2a) by ∇L and substitution from Eq. (5.2b) lead to

$$\nabla L \times \mathbf{E} \times \nabla L = (\nabla L \cdot \nabla L)\mathbf{E} = |\nabla L|^2 \mathbf{E} = \varepsilon \mu \mathbf{E} = n^2 \mathbf{E}.$$

The nonzero solution for $\mathbf{E}$ requires that

$$|\nabla L|^2 = n^2 \qquad (5.3)$$

This is the *eikonal equation*, L being the eikonal.[1]

A system of curves that fills a portion of space in such a way that, in general, a single curve passes through each point is called a *congruence*. If

$$\nabla L = |\nabla L|\hat{\mathbf{l}},$$

is the *ray vector*, then Eq. (5.3) yields

$$\nabla L = n\hat{\mathbf{l}}, \qquad (5.4)$$

Electromagnetic propagation in the high-frequency limit is described by the *ray congruence* (5.4), orthogonal to the *rayfronts* $L(\mathbf{r}) = $ const; see Fig. 5.1. Note that Eq. (5.4) is a first-order differential equation and must be solved with prescribed boundary conditions $L(\mathbf{r}_0)$ on the surface $\mathbf{r} = \mathbf{r}_0$.

---

1. The term *eikonal* is derived from the Greek word εικών, which means *image*.

HIGH-FREQUENCY FIELDS

Substitution of Eq. (5.4) into Eq. (5.2a) leads to

$$n\hat{\mathbf{l}} \times \mathbf{E} - \mu\zeta_0\mathbf{H} = 0, \qquad (5.5a)$$

which shows that the three vectors $(\hat{\mathbf{l}}, \mathbf{E}, \mathbf{H})$ are mutually orthogonal (see Fig. 5.2) and

$$\sqrt{\varepsilon}|\mathbf{E}| = \sqrt{\mu\zeta_0}|\mathbf{H}|, \qquad (5.5b)$$

so that the field is locally a plane wave along the ray path.

In general, the amplitude and polarization of the field change along the ray. However, Eqs. (5.2) are unable to provide any information about these variations, so we must resort again to Eqs. (5.1).

Cross-multiplication of Eq. (5.1a) by ∇L and substitution from Eq. (5.1b) leads to

$$\nabla L \times \mathbf{E} \times \nabla L - n^2 \mathbf{E} + \mu \frac{\nabla \times \zeta_0 \mathbf{H}}{ik_0} = \frac{(\nabla \times \mathbf{E}) \times \nabla L}{ik_0}.$$

On the other hand,

$$\nabla L \times \mathbf{E} \times \nabla L = |\nabla L|^2 \left(\mathbf{E} - \frac{\mathbf{E} \cdot \nabla L}{|\nabla L|^2} \nabla L \right) = |\nabla L|^2 \mathbf{E} - \frac{1}{ik_0} \frac{\nabla \cdot \varepsilon \mathbf{E}}{\varepsilon} \nabla L,$$

where use has been made of Eq. (5.1c). Accordingly

$$[|\nabla L|^2 - n^2]\mathbf{E} + \frac{1}{ik_0}\left[-\frac{\nabla \cdot \varepsilon\mathbf{E}}{\varepsilon}\nabla L + \mu\nabla \times \zeta_0\mathbf{H} - (\nabla \times E) \times \nabla L \right] = 0. \qquad (5.6)$$

To proceed further we should expand the field in inverse powers of k_0 and substitute into Eq. (5.6). Then, we should equate to zero the coefficients of successive powers of $1/k_0$, leading to a hierarchy of equations for the successive terms of the expansion.

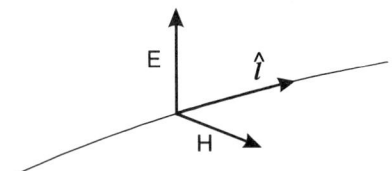

FIGURE 5.2. Geometrical relationships between **E**, **H**, and $\hat{\mathbf{l}}$.

The first term generates the same eikonal equation (5.3). Examination of Eq. (5.6) shows that higher-order terms in the expansion of the electric field multiplying the eikonal equation do not play any further role.

The second term is obtained by setting equal to zero the bracketed term in Eq. (5.6), namely,

$$-\frac{\nabla \cdot \varepsilon \mathbf{E}}{\varepsilon} \nabla L + \mu \nabla \times \zeta_0 \mathbf{H} + \nabla L \times \nabla \times \mathbf{E} = 0, \quad (5.7a)$$

where the zero-order field must be considered.

It is convenient to eliminate the magnetic field $\mathbf{H}$ from Eq. (5.7a). From Eq (5.1a) we have $\nabla L \times \mathbf{E} \approx \mu \zeta_0 \mathbf{H}$, an expression that can be used in Eq. (5.7a) without changing its order. Hence

$$\mu \nabla \times \zeta_0 \mathbf{H} = \mu \nabla \times \left(\frac{1}{\mu} \nabla L \times \mathbf{E}\right) = \nabla \times (\nabla L \times \mathbf{E}) - \nabla \ln \mu \times (\nabla L \times \nabla \times \mathbf{E}),$$

where use has been made of Eq. (A.12). Therefore Eq. (5.7a) assumes the form

$$-\frac{\nabla \cdot \varepsilon \mathbf{E}}{\varepsilon} \nabla L + \nabla \times (\nabla L \times \mathbf{E}) - \nabla \ln \mu \times (\nabla L \times \mathbf{E}) + (\nabla L \times \nabla \times \mathbf{E}) = 0. \quad (5.7b)$$

We further modify Eq. (5.7b) by applying Eq. (A.10) to the first term, Eq. (A.13) to the second, Eq. (A.2) to the third, and Eq. (A.9) with $\nabla L \cdot \mathbf{E} = 0$ to the last term. So finally we derive

$$[\nabla^2 L - \nabla L \cdot \nabla \ln \mu + (2\nabla L \cdot \nabla)]\mathbf{E} + 2(\mathbf{E} \cdot \nabla \ln n)\nabla L = 0.$$

This equation, divided by $\sqrt{\mu}$, can be expressed in a most convenient form by noting that

$$\nabla \frac{1}{\sqrt{\mu}} \mathbf{E} = -\frac{1}{2} \frac{1}{\sqrt{\mu}} \nabla \ln \mu \mathbf{E} + \frac{1}{\sqrt{\mu}} \nabla \mathbf{E}$$

[see Eq. (B.34)] and

$$2\nabla L \cdot \left(\nabla \frac{1}{\sqrt{\mu}} \mathbf{E}\right) = \left(-\nabla L \cdot \nabla \ln \mu + 2 \frac{1}{\sqrt{\mu}} \nabla L \cdot \nabla\right) \mathbf{E},$$

so that

$$(\nabla^2 L + 2\nabla L \cdot \nabla)\frac{\mathbf{E}}{\sqrt{\mu}} + 2\left(\frac{\mathbf{E}}{\sqrt{\mu}} \cdot \nabla \ln n\right)\nabla L = 0. \quad (5.8)$$

It is convenient to introduce the *ray intensity*,

$$I = \frac{1}{2}\frac{\mathbf{E} \cdot \mathbf{E}^*}{\mu} = \frac{1}{2}\zeta_0^2 \frac{\mathbf{H} \cdot \mathbf{H}^*}{\varepsilon} \quad (5.9a)$$

[see Eq. (5.5b)], and the *ray amplitude*

$$A = \sqrt{2I} \quad (5.9b)$$

and *polarization*

$$\mathbf{E} = \sqrt{\mu}A\hat{\mathbf{e}} \quad \text{and} \quad \zeta_0\mathbf{H} = \sqrt{\varepsilon}A\hat{\mathbf{l}} \times \hat{\mathbf{e}}$$

[see Eq. (5.5a)], with

$$\hat{\mathbf{e}} \cdot \hat{\mathbf{e}}^* = \hat{\mathbf{h}} \cdot \hat{\mathbf{h}}^* = 1$$

(see Section 4.2.2). Note that $\hat{\mathbf{e}}$ and $\hat{\mathbf{h}}$ may be complex vectors and describe the field polarization. Then Eq. (5.8) becomes

$$[A\nabla^2 L + 2\nabla L \cdot \nabla A]\hat{\mathbf{e}} + 2A[(\nabla L \cdot \nabla)\hat{\mathbf{e}} + (\hat{\mathbf{e}} \cdot \nabla \ln n)\nabla L] = 0. \quad (5.10)$$

We dot-multiply Eq. (5.10) by $\hat{\mathbf{e}}^*$ and add its complex conjugate, thus getting

$$A\nabla^2 L + 2\nabla A \cdot \nabla L = 0, \quad (5.11a)$$

because

$$\nabla L \cdot \hat{\mathbf{e}}^* = \nabla L \cdot \hat{\mathbf{e}} = 0$$

and

$$\hat{\mathbf{e}}^* \cdot (\nabla L \cdot \nabla)\hat{\mathbf{e}} + \hat{\mathbf{e}} \cdot (\nabla L \cdot \nabla)\hat{\mathbf{e}}^* = n\left[\hat{\mathbf{e}}^* \cdot \frac{d\hat{\mathbf{e}}}{dl} + \hat{\mathbf{e}} \cdot \frac{d\hat{\mathbf{e}}^*}{dl}\right]$$

$$= n\frac{d(\hat{\mathbf{e}} \cdot \hat{\mathbf{e}}^*)}{dl} = 0$$

Equation (5.11a), usually in the form

$$\nabla^2 L + 2\nabla L \cdot \nabla \ln A = 0, \tag{5.11b}$$

is referred to as the *(amplitude) transport equation*; see Section 5.1.1.

In view of Eq. (5.11a) Eq. (5.10) has the form

$$\frac{d\hat{e}}{dl} + (\hat{e} \cdot \nabla \ln n)\hat{l} = 0, \tag{5.12}$$

and is the transport equation for the field polarization. The latter remains constant along a ray for a homogeneous medium ($\nabla \ln n = 0$).

5.1.1. The Transport Equation

From Eq. (5.11a) and (A.10) we obtain

$$\nabla \cdot \left(\frac{1}{2} A^2 \nabla L\right) = \nabla \cdot (In\hat{l}) = 0, \tag{5.13}$$

which is the equivalent of Poynting's theorem (see Section 4.3.1) for ray fields. In fact

$$nI\hat{l} = \tfrac{1}{2} \mathbf{E} \times \mathbf{H}^* = \mathbf{S}$$

[see Eq. (5.9a)] and the electric and magnetic energy densities are equal [see Eq. (5.5b)]. Equation (4.48) becomes $\nabla \cdot \mathbf{S} = 0$ and coincides with Eq. (5.13).

Integration of Eq. (5.13) over a volume bounded by a narrow ray bundle (see Fig. 5.3) leads to

$$I(0)n(0)\, dS(0) = I(l)n(l)\, dS(l), \tag{5.14}$$

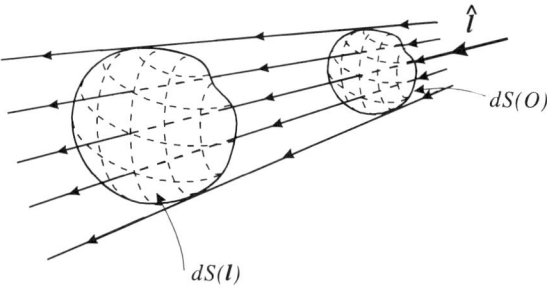

FIGURE 5.3. Geometry of a ray bundle.

so

$$I(l) = I(0) \frac{n(0) \, dS(0)}{n(l) \, dS(l)},$$

which provides the law governing the ray intensity along a generic ray path, once the latter has been determined.

When dS shrinks to a point, or a line, the field intensity diverges, thus implying that approximations involved in the ray descriptions of the field are no longer valid there. From Eq. (5.11b) we have

$$2n \frac{d}{dl} \ln A + \nabla^2 L = 0,$$

hence

$$A(l) = A(0) \exp\left[-\frac{1}{2} \int_0^l dl \, \frac{\nabla^2 L}{n} \right], \tag{5.15}$$

so divergence of A at some l implies $\nabla^2 L \to -\infty$ there. This is the condition that determines the *caustic regions* (points, lines, surfaces) where the ray bundle converges and the ray description becomes invalid.

Crossing a caustic would imply a change in sign of the ray intensity (see Sections 5.1.2 and 5.3.1). However, we require the amplitude of the ray to be a real positive quantity. This is accomplished by introducing a phase jump, equal to $\pi/2$ when line caustics are crossed and π when point caustics are crossed (see Section 5.2.3).

5.1.2. Rays in a Homogeneous Medium

A spatially homogeneous medium is characterized by $n = 1$. If

$$L(x, y, z) = L_x(x) + L_y(y) + L_z(z),$$

then the eikonal equation (5.3) yields

$$\left(\frac{dL_x}{dx} \right)^2 + \left(\frac{dL_y}{dy} \right)^2 + \left(\frac{dL_z}{dz} \right)^2 = 1.$$

Each factor on the left-hand side of this equation is decoupled from the others with respect to the coordinate dependence. Accordingly

$$\left(\frac{dL_x}{dx} \right)^2 = a^2, \quad \left(\frac{dL_y}{dy} \right)^2 = b^2, \quad \text{and} \quad \left(\frac{dL_z}{dz} \right)^2 = c^2,$$

with a, b, and c constants satisfying

$$a^2 + b^2 + c^2 = 1.$$

Integration leads to

$$L_x = ax + a_1, \quad L_y = by + b_1, \quad \text{and} \quad L_z = cz + c_1,$$

in which case

$$L(x, y, z) = L(0) + ax + by + cz$$

and

$$\hat{\mathbf{l}} = \nabla L = a\hat{\mathbf{x}} + b\hat{\mathbf{y}} + c\hat{\mathbf{z}}.$$

We note that $\hat{\mathbf{l}}$ is a constant vector with directional cosines a, b, and c, and that rays are straight lines.

Suppose that a (sufficiently smooth) surface S_0 is prescribed on which the eikonal value is taken to be constant. The ray field solution is constructed *locally* by taking the surface normal $\hat{\mathbf{l}}$, thus determining the coefficients a, b, and c which coincide with the *local* ray directional cosines and fix the *local* ray direction. In terms of the usual coordinate system along the ray (see Section 5.3), rectilinear in this case, with local origin on the surface S_0, we have

$$L(l) = l. \tag{5.16a}$$

Let R_1 and R_2 be the *local* principal radii of curvature (see Section 5.3) of S_0 (see Fig. 5.4). We have

$$dS(0) = R_1 R_2 d\theta_1 d\theta_2 \quad \text{and} \quad dS(l) = (R_1 + l)(R_2 + l)d\theta_1 d\theta_2,$$

and from Eq. (5.14)

$$I(0)R_1 R_2 = I(l)(R_1 + l)(R_2 + l), \tag{5.16b}$$

which provides the change of intensity along the ray. For $l = -R_1$ and $l = -R_2$ we have two caustic points. As we move to other points over the rayfront, the caustic points change their position in general, thus describing two caustic lines. As anticipated in Section 5.1.1, the sign change for $l < -R_1$ (and similarly for $l < -R_2$) is accounted for by introducing a phase jump of $\pi/2$ at the caustic crossing. When $R_1 = R_2$ we have a *focal* point, and the phase jump equals π.

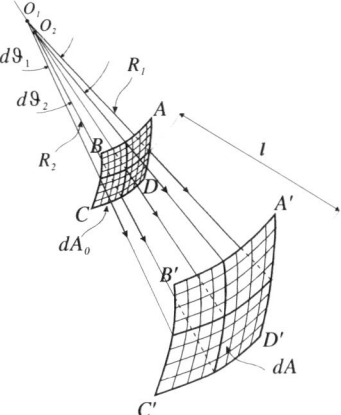

FIGURE 5.4. Ray bundle in a homogeneous medium.

Additional properties of rays in a homogeneous environment are presented in Section 5.3.

5.1.3. Ray Propagation in a Layered Medium

Consider a *layered medium*, i.e., a medium whose refractive index depends only on a single coordinate, say y in cartesian coordinates, as depicted in Fig. 5.5.

For z-independent rays let

$$L(x, y) = L_x(x) + L_y(y),$$

and substitute into Eq. (5.3) to obtain

$$\left(\frac{dL_x}{dx}\right)^2 = b^2 \quad \text{and} \quad \left(\frac{dL_y}{dy}\right)^2 = n^2(y) - b^2,$$

b^2 being a constant. If

$$\hat{\mathbf{l}}(x, y) = \frac{dx}{dl}\hat{\mathbf{x}} + \frac{dy}{dl}\hat{\mathbf{y}},$$

where (x, y) are the ray coordinates, then Eq. (5.4) yields

$$\frac{\partial L}{\partial x} = \frac{dL_x}{dx} = b = n\frac{dx}{dl}$$

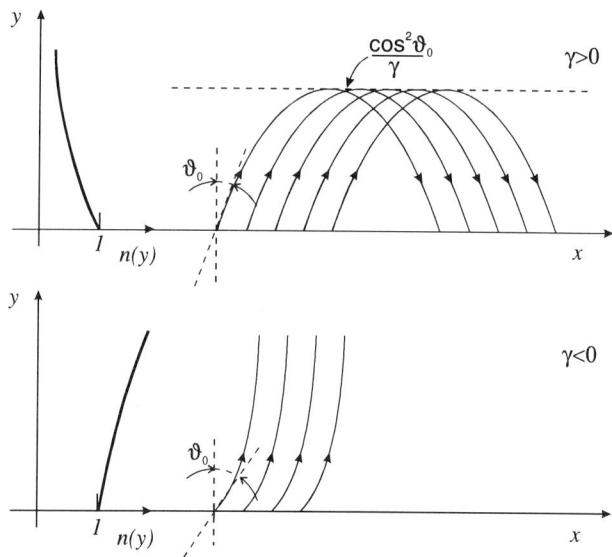

FIGURE 5.5. A layered medium along y, and the associated z-independent ray congruences for decreasing ($\gamma > 0$) and increasing ($\gamma < 0$) values of the refractive index.

and

$$\frac{\partial L}{\partial y} = \frac{dL_y}{dy} = \sqrt{n^2 - b^2} = n\frac{dy}{dl},$$

i.e.,

$$\frac{dx}{dy} = \frac{b}{\sqrt{n^2 - b^2}}, \qquad (5.17a)$$

hence

$$x = \int \frac{b}{\sqrt{n^2 - b^2}}\, dy + x_0, \qquad (5.17b)$$

which is the equation of the ray parametrized in terms of the constant x_0.

As an example, we set

$$n^2(y) = 1 - \gamma y, \qquad y \geqslant 0.$$

HIGH-FREQUENCY FIELDS

Integration of (5.17b) leads to

$$x = -\frac{2b}{\gamma}\sqrt{1 - b^2 - \gamma y} + x_0,$$

namely,

$$y = \frac{1 - b^2}{\gamma} - \frac{\gamma}{4b^2}(x - x_0)^2, \qquad (5.17c)$$

which is the equation of a parabola (see Fig. 5.5).

Boundary conditions can be enforced by specifying the ray angle θ_0 on the plane $y = 0$. Equation (5.17a) evaluated at $y = 0$ gives

$$\left.\frac{dx}{dy}\right|_0 = \frac{b}{\sqrt{1 - b^2}} = \tan\theta_0,$$

or

$$b = \sin\theta_0.$$

For $\gamma > 0$, real values of x require that

$$y \leqslant \frac{1 - b^2}{\gamma} = \frac{\cos^2\theta_0}{\gamma}$$

[see Eq. (5.17c)], the equality defining a caustic (see Fig. 5.5), where the refractive index takes the value

$$n(\cos^2\theta_0/\gamma) = \sqrt{1 - \cos^2\theta_0} = \sin\theta_0.$$

For the considered layered medium, this is the ray-optical counterpart of the limit angle (see Section 4.5.1). Negative values of γ do not pose any restriction on y. Ray paths are depicted in Fig. 5.5.

Similarly, for

$$n^2(y) = 1 - \gamma^2 y^2,$$

we have (see Fig. 5.6)

$$y = \frac{\cos\theta_0}{\gamma}\sin\left[\frac{\gamma(x - x_0)}{\sin\theta_0}\right],$$

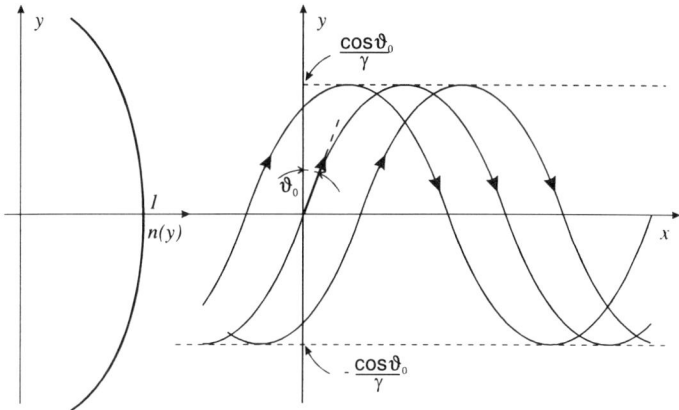

FIGURE 5.6. Duct propagation.

and the rays are *trapped* inside a slab of width $2\cos\theta_0/\gamma$ with

$$n(\cos\theta_0/\gamma) = \sin\theta_0.$$

This phenomenon is referred to as *duct propagation*.

5.1.4. Polarization Change along a Ray

Let us define along a ray three orthogonal unit vectors $\hat{\mathbf{l}}$ (*tangent*), $\hat{\mathbf{n}}$ (*normal*), and $\hat{\mathbf{b}}$ (*binormal*) such that

$$\hat{\mathbf{l}} \times \hat{\mathbf{n}} = \hat{\mathbf{b}},$$

as depicted in Fig. 5.7. The normal $\hat{\mathbf{n}}$ is oriented toward the (local) center of curvature. Changes in these vectors along the curve are described by the Frénet[2] equations:

$$\frac{d\hat{\mathbf{l}}}{dl} = \frac{\hat{\mathbf{n}}}{R}, \qquad \frac{d\hat{\mathbf{n}}}{dl} = -\frac{\hat{\mathbf{l}}}{R} + \frac{\hat{\mathbf{b}}}{\tau}, \qquad \text{and} \qquad \frac{d\hat{\mathbf{b}}}{dl} = -\frac{\hat{\mathbf{n}}}{\tau}, \qquad (5.18)$$

where R is the radius of curvature (the inverse of the curvature) and τ is a measure of the torsion of the ray, i.e., the deviation of the ray path from a planar surface.

---

2. Jean Frederic Frénet: Perigueux (France), 1816–Perigueux, 1900.

HIGH-FREQUENCY FIELDS 239

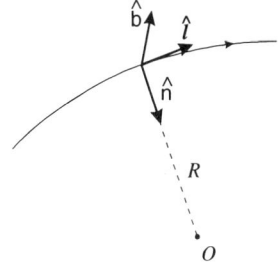

FIGURE 5.7. A ray path and the associated reference coordinate system.

From Eq. (5.12) we obtain

$$\frac{d\hat{\mathbf{e}}\cdot\hat{\mathbf{n}}}{dl} = \frac{de_n}{dl} = \hat{\mathbf{e}}\cdot\frac{d\hat{\mathbf{n}}}{dl} + \hat{\mathbf{n}}\cdot\frac{d\hat{\mathbf{e}}}{dl} = \hat{\mathbf{e}}\cdot\frac{d\hat{\mathbf{n}}}{dl}$$

and

$$\frac{d\hat{\mathbf{e}}\cdot\hat{\mathbf{b}}}{dl} = \frac{de_b}{dl} = \hat{\mathbf{e}}\cdot\frac{d\hat{\mathbf{b}}}{dl} + \hat{\mathbf{b}}\cdot\frac{d\hat{\mathbf{e}}}{dl} = \hat{\mathbf{e}}\cdot\frac{d\hat{\mathbf{b}}}{dl},$$

because dot multiplication of Eq. (5.12) by $\hat{\mathbf{n}}$ and by $\hat{\mathbf{b}}$ leads to

$$\hat{\mathbf{n}}\cdot\frac{d\hat{\mathbf{e}}}{dl} = \hat{\mathbf{b}}\cdot\frac{d\hat{\mathbf{e}}}{dl} = 0.$$

Using the Frénet relations we have

$$\frac{de_n}{dl} = \frac{e_b}{\tau} \quad \text{and} \quad \frac{de_b}{dl} = -\frac{e_n}{\tau},$$

and the polarization of the field does not change *in the* (l, n, b) *coordinate system* if torsion is absent ($\tau \to \infty$).

5.2. RAY PROPERTIES

It was stated in Section 1.1 that a system of curves that fills a portion of space such that, in general, a single curve passes through each point of the space is called a congruence. This system is a *ray congruence* if

$$\nabla \times (n\hat{\mathbf{l}}) = 0, \quad (5.19a)$$

in view of Eq. (5.4). In a homogeneous medium

$$\nabla \times \hat{\mathbf{l}} = 0; \tag{5.19b}$$

in the general case, use of Eq. (A.12) leads to

$$\nabla \ln n \times \hat{\mathbf{l}} \cdot \nabla \times \hat{\mathbf{l}} = 0. \tag{5.19c}$$

The congruence is said to be *normal* if there exists a family of surfaces that cut each of the curves orthogonally. In this case the rays form a *normal congruence* and the surfaces are the rayfronts $L = $ const.

In a homogeneous medium the rays are straight lines (see Section 5.1.2) and are represented by a *linear normal congruence*.

Consider a ray congruence and the closed path C depicted in Fig. 5.8 with the two points (P_1, P_2). By integrating the flux of Eq. (5.19a) over the surface limited by C and using the Stokes theorem we obtain

$$\oint_C d\mathbf{r} \cdot n\hat{\mathbf{l}} = 0, \tag{5.20a}$$

i.e.,

$$\int_{P_1}^{P_2} d\mathbf{r} \cdot n\hat{\mathbf{l}} = \text{cost}(P_1, P_2). \tag{5.20b}$$

Equation (5.20b) is independent of the integration path joining the two

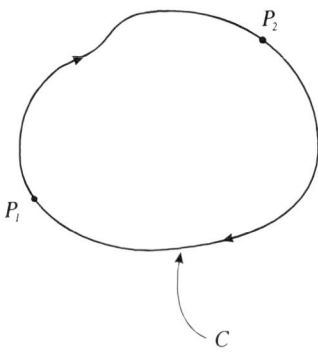

FIGURE 5.8. A closed curve C in the derivation of the Lagrange invariant.

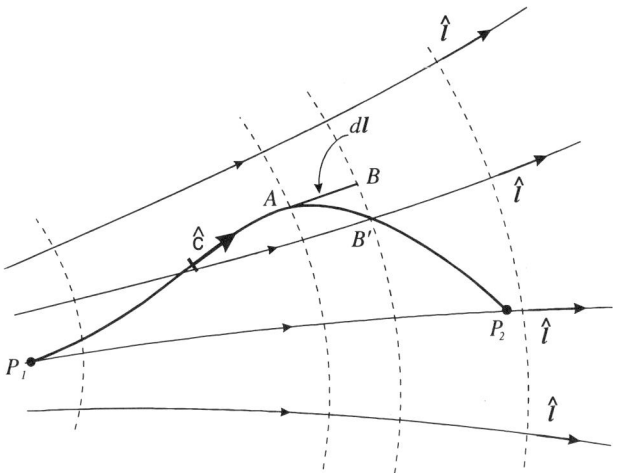

FIGURE 5.9. Derivation of Fermat's principle.

assigned points P_1 and P_2 and is usually referred to as the *Lagrange*[3] *invariant*.[4]

Now let us consider the ray congruence depicted in Fig. 5.9, with the additional curve C joining the two points P_1 and P_2 over the ray path P_1P_2. Equation (5.20b) yields

$$\int_{P_1}^{P_2} n\,dl = \int_C d\mathbf{r}\cdot n\hat{\mathbf{1}}.$$

Similarly, and with reference to the small triangle $AB'B$,

$$dl = \hat{\mathbf{1}} \cdot d\mathbf{r} \leq dc,$$

c being a curvilinear coordinate along C. Accordingly

$$\int_{P_1}^{P_2} n\,dl \leq \int_{P_1}^{P_2} n\,dc, \qquad (5.21)$$

where the two integrals represent the *optical length* between P_1 and P_2 along the actual ray path and a generic path, respectively.

Inequality (5.21) is the mathematical statement of *Fermat's*[5] principle for a single ray congruence: the shortest optical path between two points is the one

---

3. Giuseppe Luigi Lagrange: Torino (Italy), 1736–Paris, 1813.
4. This invariant is the one-dimensional case of a more general integral discussed by J. H. Poincaré.
5. Pierre de Fermat: Beaumont de Lomagne (France), 1601–Castres (France), 1663.

along the actual ray path joining the two points. It is important to stress that this (strong) formulation of the theorem requires that the considered paths lie within a regular neighborhood of the actual path, i.e., within a region where rays do not intersect each other.

Let us now consider the case of two media, of different refractive indexes n_1 and n_2, separated by a smooth surface S (Fig. 5.10). We postulate the existence of an incident and a reflected ray congruence in region 1, and a transmitted ray congruence in region 2.

The incident ray congruence sets up a given eikonal distribution L along the separating surface S, which is the boundary condition for the other two (reflected and transmitted) ray congruences. Let ΔL be the eikonal change from A to B (see Fig. 5.10) over the plane tangent to the surface S. Use of the Lagrange invariant (5.20b) in connection with paths CAB and CB of Fig. 5.10,I

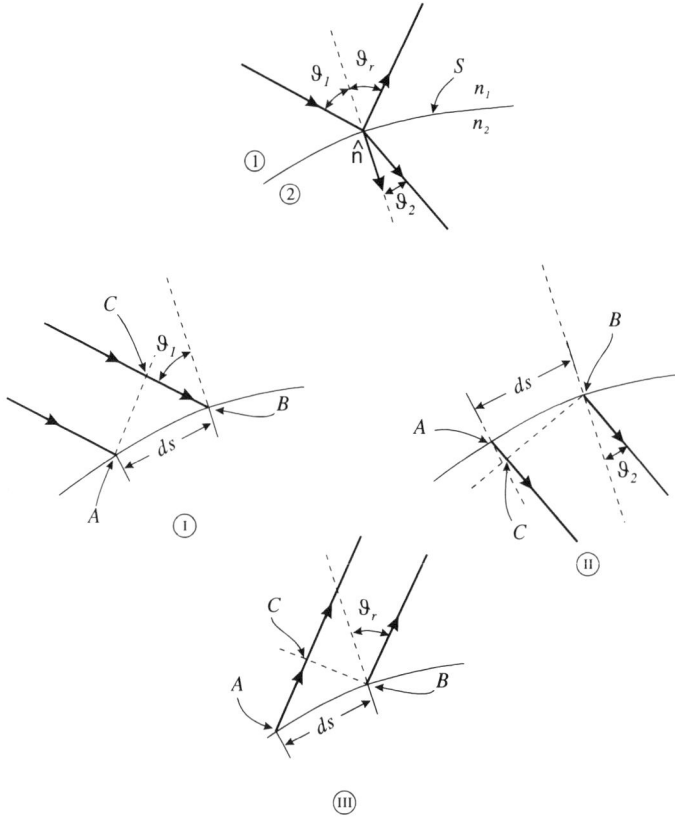

FIGURE 5.10. Derivation of (local) Snell's laws. Incident (I), transmitted (II), and reflected (III) ray congruences.

leads to

$$\Delta L = n_1 \sin \theta_1 \, ds,$$

because there is no change in L along CA, which belongs to the rayfront. Similarly, we obtain (Fig. 5.10,II)

$$\Delta L = n_1 \sin \theta_r \, ds$$

and (Fig. 5.10,III)

$$\Delta L = n_2 \sin \theta_2 \, ds$$

for the reflected and transmitted ray congruences, respectively. Accordingly

$$\theta_r = \theta_1 \qquad (5.22a)$$

$$n_1 \sin \theta_1 = n_2 \sin \theta_2, \qquad (5.22b)$$

which are the *local* generalization of Snell's laws [see Eq. (4.74)]. We conclude that Snell's laws are locally valid in the *local* plane of incidence, defined by the incident ray and the local normal to the surface separating the two media. Obviously, this result is true in the high-frequency regime, i.e., for surface variations smooth compared to the incident wavelength.

It is easy to show that the Lagrange invariant is also valid for the normal congruence composed of incident and reflected, or incident and transmitted rays. For the latter case, and with reference to Fig. 5.11, we have

$$\int_{C_1 + C^+} d\mathbf{r} \cdot n_1 \hat{\mathbf{l}} = 0, \qquad \int_{C_2 + C^-} d\mathbf{r} \cdot n_2 \hat{\mathbf{l}} = 0,$$

$$\oint d\mathbf{r} \cdot n \hat{\mathbf{l}} = \int_{C_1} d\mathbf{r} \cdot n_1 \hat{\mathbf{l}} + \int_{C_2} d\mathbf{r} \cdot n_2 \hat{\mathbf{l}} = 0,$$

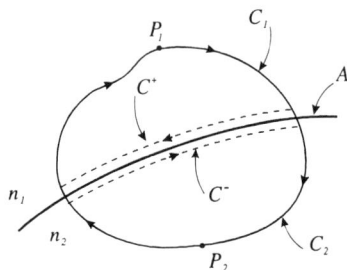

FIGURE 5.11. The Lagrange invariant in the case of refractive index discontinuity.

because the integrals along C^+ and C^- cancel in view of Snell's laws (5.22b). When two congruences are involved, application of the Lagrange invariant requires that the path closes itself on the *same* congruence from which it started; see an example in Section 5.2.1.

Also, Fermat's principle applies for the combination of incident and transmitted, and of incident and reflected ray congruences, although in a weaker form: the optical length along the actual ray path between two given points P_1 and P_2 is *stationary* with respect to neighboring paths joining the same two points. As a check, consider incident and reflected rays in a homogeneous environment as depicted in Fig. 5.12. We choose a local system of coordinates with the (x, z) axes in the incidence plane and the origin of coordinates O in the tangent plane at the specular reflection point according to Snell's laws. If $|P_1 O| = d_1$, $|P_2 O| = d_2$, $Q = Q(x, z)$, and $z = f(x)$ is the equation of the surface in the incidence plane, then

$$s(Q) = |P_1 Q| + |P_2 Q|$$
$$= \sqrt{(d_1 \sin\theta + x)^2 + (d_1 \cos\theta - z)^2} + \sqrt{(d_2 \sin\theta - x)^2 + (d_2 \cos\theta - z)^2}$$

and

$$\left.\frac{ds}{dx}\right|_0 = \frac{d_1 \sin\theta - d_1 \cos\theta f'(0)}{d_1} + \frac{-d_2 \sin\theta - d_2 \cos\theta f'(0)}{d_2} = 0,$$

because $f'(0) = 0$. Accordingly, $s(Q)$ is stationary at the reflection point. Furthermore

$$\left.\frac{d^2 s}{dx^2}\right|_0 = \left[\frac{1}{d_1} + \frac{1}{d_2}\right]\cos^2\theta - 2f''(0)\cos\theta,$$

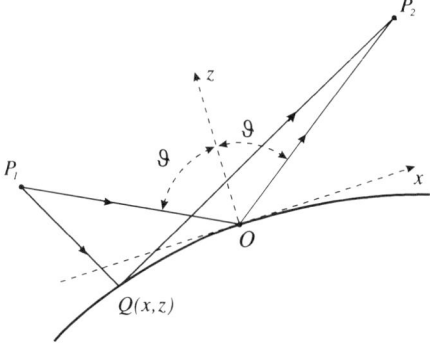

FIGURE 5.12. Fermat's principle in the case of an incident–reflected ray congruence.

and $s(Q)$ is a minimum path for a convex reflecting surface ($f''(0) < 0$), as depicted in Fig. 5.11, but not necessarily so in the concave case ($f''(0) > 0$).

5.2.1. Reflector Antennas

Consider a line source in front of a reflecting cylindrical surface S as depicted in Fig. 5.13. We wish to convert the cylindrical wave radiated by the source onto a plane one. We model the incident field as a cylindrical rayfront emanating from a line source at the origin of coordinates O and request that the rayfront of the reflected ray congruence should be parallel to the plane $z = 0$. Then, the reflected ray congruence should be parallel to the z-axis. The Lagrange invariant (5.20a) is now applied to the incident–reflected ray congruence and to the closed path $OABB'A'O$:

$$|OA| + |AB| - |B'A'| - |A'O| = 0,$$

because the contribution to Eq. (5.20a) along the rayfront $|BB'|$ is zero. Setting $B' = O$ we obtain

$$|OA| + |AB| = |OC| + |CO| = 2f.$$

We observe further that $|OA|, |AB|$ is a direct–reflected ray congruence, so that Snell's law is verified at the reflection point A.

Let $r = r(\theta)$ be the equation of S in cylindrical coordinates. Then

$$|OA| + |AB| = r + r\cos\theta = 2f,$$

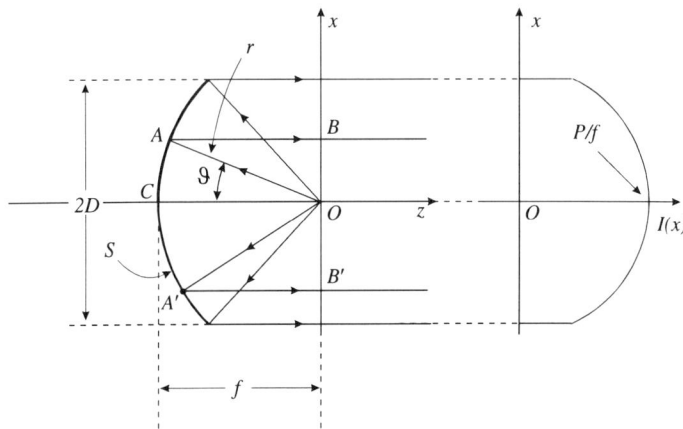

FIGURE 5.13. Transformation of a cylindrical rayfront onto a planar rayfront by means of a metal reflector.

so that

$$r = \frac{2f}{1 + \cos\theta},$$

which is the equation of a parabola.

Let $Pd\theta$ be the power radiated by the line source within an angle $d\theta$. In this case

$$x = r\sin\theta = 2f\,\frac{\sin\theta}{1 + \cos\theta} = 2f\tan\frac{\theta}{2},$$

so

$$dx = \frac{2f}{2\cos^2(\theta/2)}\,d\theta,$$

and the ray intensity I [watt/m] over the plane $z = 0$ is given by

$$I = \frac{Pd\theta}{dx} = \frac{P}{f}\cos^2\frac{\theta}{2} = \frac{P}{f}\,\frac{1}{1 + \tan^2(\theta/2)} = \frac{4Pf}{(2f)^2 + x^2},\quad |x| \leqslant D;$$

hence the ray amplitude distribution is space-limited and tapered (unless the illumination P is properly shaped) over $z = 0$. This distribution does not account for interaction of the rays with the edges of the parabola (see Section 5.4.1) and is usually referred to as the *geometrical optics* (GO) contribution to the total scattered field in the high-frequency regime. To improve the solution, either higher-order ray contributions should be considered (see Section 5.4.2) or the parabolic reflector may be studied as an equivalent aperture (see Section 3.5.3) with the uniform phase and tapered amplitude distribution provided by the PO solution. This computational procedure is referred to as the *equivalent aperture* approach.

Additional material concerning reflector antennas can be found in Section 8.6.

5.2.2. Lens Antennas

Consider a line source in front of a shaped dielectric medium of refractive index $n = \sqrt{\varepsilon}$, as depicted in Fig. 5.14. In this case too we wish to convert the cylindrical wave radiated by the source onto a plane one; the plane $z = 0$ is required to be a rayfront.

As in Section 5.2.1, the optical length along any ray path from the source point O to the emerging planar rayfront is constant and equal to $|OC| + n|CO'|$.

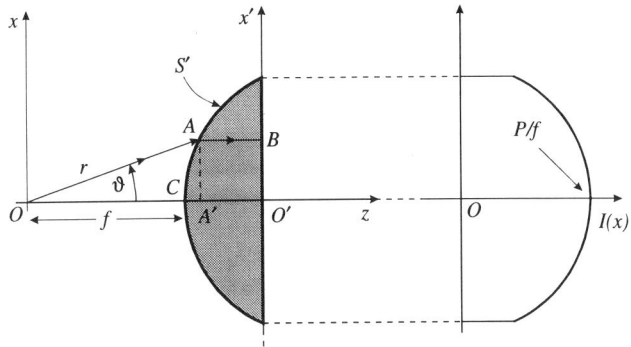

FIGURE 5.14. Transformation of a cylindrical rayfront onto a planar rayfront by means of a dielectric lens.

Let $r = r(\theta)$ be the equation of S in cylindrical coordinates. Then

$$r + n|AB| = f + (r\cos\theta - f)n + n|A'O'|,$$

so that

$$r = f\frac{n-1}{n\cos\theta - 1},$$

which is the equation of a hyperbola. The shaped dielectric cylinder of Fig. 5.14 acts as a dielectric lens converting cylindrical onto planar rayfronts.

Let $d\theta$ be the power radiated by the line source within an angle $d\theta$. In this case

$$x' = r\sin\theta = f\frac{(n-1)\sin\theta}{n\cos\theta - 1},$$

so

$$dx' = (n-1)f\frac{n - \cos\theta}{(n\cos\theta - 1)^2}\,d\theta,$$

and the ray intensity I [watt/m] over the output plane of the lens is given by

$$I = \frac{P\,d\theta}{dx'} = \frac{P}{(n-1)f}\frac{(n\cos\theta - 1)^2}{n - \cos\theta},$$

if refections on the surface S and on the output plane of the lens are neglected.

Again, the transmitted ray intensity turns out to be space-limited and tapered, unless the illumination is properly shaped along θ.

5.2.3. Guided Propagation

Consider electromagnetic structures uniform along the z-axis, such as those depicted in Fig. 5.15. We assume guided propagation along the z-axis,

$$k_0 L(\mathbf{t}, z) = k_0 L_t(\mathbf{t}) + \beta z, \qquad \mathbf{t} = x\hat{\mathbf{x}} + y\hat{\mathbf{y}},$$

and, from Eq. (5.3),

$$k_0^2 |\nabla_t L_t|^2 = k_0^2 n^2 - \beta^2 \tag{5.23}$$

in each of the homogeneous sections of the structure.

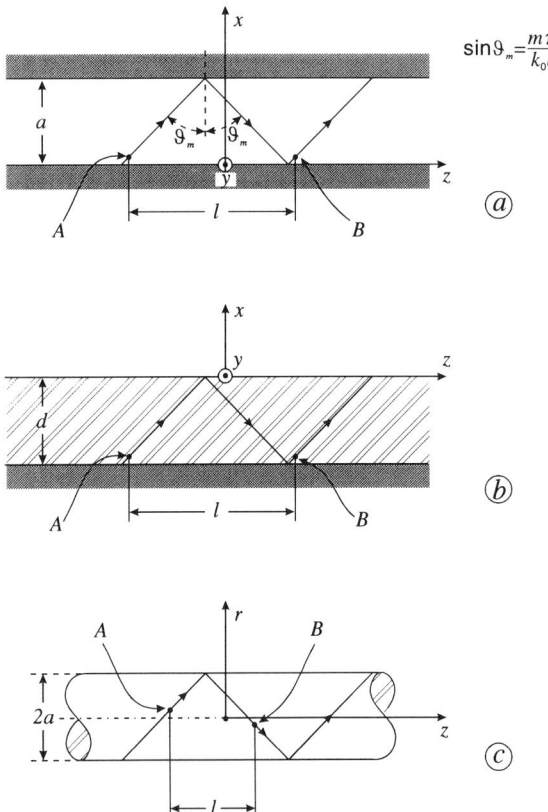

FIGURE 5.15. Guiding structures: (a) parallel-plate waveguide; (b) grounded dielectric slab; (c) circular fiber.

For the parallel-plate rayguide of Fig. 5.15a, with y-independent ray congruences, we have from Eq. (5.23)

$$k_0 \frac{dL_t}{dx} = \sqrt{k_0^2 - \beta^2}, \quad \text{i.e., } k_0 L_t(x) = \pm \sqrt{k_0^2 - \beta^2}\, x = \pm ux,$$

and there are two ray congruences within the guide. These are connected with equal angles θ in order to fulfill Snell's law (5.22a) on the perfectly reflecting boundaries, thus forming the *propagating ray congruence* within the guide.

The guiding property of the structure is now enforced by means of the *consistency condition*. Consider two points on a ray, A and B, with coincident transverse coordinates and spacing l; see Fig. 5.15. The rayfield must be exactly the same, but for the phase shift term $\exp(-i\beta l)$. In this way, the entire transverse distribution of the rayfield remains unchanged while progressing along z with the phase term $\exp(-i\beta z)$. If x is the transverse coordinate of these two corresponding points A and B (see Fig. 5.15a), then the phase change from A to B equals

$$\left[-ua - \beta \frac{l}{2}\right] + \left[-ua - \beta \frac{l}{2}\right]$$

plus the phase change at the reflection boundaries. We take the latter jump equal to twice the phase γ of the Fresnel reflection coefficient at each boundary, in view of the local plane-wave behavior of the ray. The consistency condition requires that the transverse phase change along the ray augmented by the phase jump at the two reflection boundaries should equal an integral number of 2π:

$$-\sqrt{k_0^2 - \beta^2}\, 2a + 2\gamma = 2m\pi, \quad m = 0, 1, \ldots. \qquad (5.24)$$

In the case considered $\gamma = \pi$ (y-polarized E field) and $\gamma = 0$ (y-polarized H field). The consistency condition (5.24) then becomes

$$\beta_m^2 = k_0^2 - \left(\frac{m\pi}{a}\right)^2,$$

with $m = 0, 1, \ldots$ for a y-polarized H field and $m = 1, 2, \ldots$ for a y-polarized E field. We recover the same dispersion equation of Section 4.5.3 with

$$\sin \theta_m = \frac{m\pi}{k_0 a}.$$

For the guiding structure of Fig. 5.15b with y-independent ray congruences, a damped field for $x > 0$, and an electric field polarized along the y-axis (TE ray modes), Eq. (5.23) yields

$$\varepsilon_r k_0^2 L_t(x) = ux, \qquad -d \leqslant x \leqslant 0,$$
$$k_0^2 L_t(x) = -i\alpha x, \qquad 0 \leqslant x,$$

so that the rays make the complex angle

$$\cos\theta = -i\frac{\alpha}{k_0}, \qquad \theta = \frac{\pi}{2} + i\sinh^{-1}\frac{\alpha}{k_0}$$

with respect to the x-axis outside the dielectric slab and the field is exponentially damped there. From Eqs. (4.77a) and (4.76a) we have

$$\Gamma(0) = \frac{u + i\alpha}{u - i\alpha} = \exp[i\gamma(0)], \qquad \gamma(0) = 2\tan^{-1}\frac{\alpha}{u},$$
$$\Gamma(-d) = -1 = \exp[i\gamma(-d)], \qquad \gamma(-d) = \pi,$$

and the consistency condition becomes

$$-u2d + 2\tan^{-1}\frac{\alpha}{u} + \pi = 2m\pi,$$

i.e.,

$$\tan\left(ud - \frac{\pi}{2}\right) = -\cotan ud = \frac{\alpha d}{ud},$$

which leads to the exact dispersion equation (4.81) of Section 4.5.2.

TM ray propagation can be studied similarly.

For the dielectric fiber of Fig. 5.15c, and ϕ-independent ray congruences, Eq. (5.23) yields

$$\varepsilon_r k_0 L_t(r) = ur, \qquad r \leqslant a,$$
$$k_0 L_t(r) = -i\alpha r, \qquad r \geqslant a,$$

when exponential decay of the field outside the fiber is enforced.

For the magnetic field polarized along the azimuthal coordinate ϕ, Eqs. (4.77a) and (4.76b) lead to

$$\Gamma(a) = -\frac{i\alpha\varepsilon_r + u}{-i\alpha\varepsilon_r + u} = \exp[i\gamma(a)], \qquad \text{i.e., } \gamma(a) = \pi + 2\tan^{-1}\frac{\alpha\varepsilon_r}{u}.$$

HIGH-FREQUENCY FIELDS

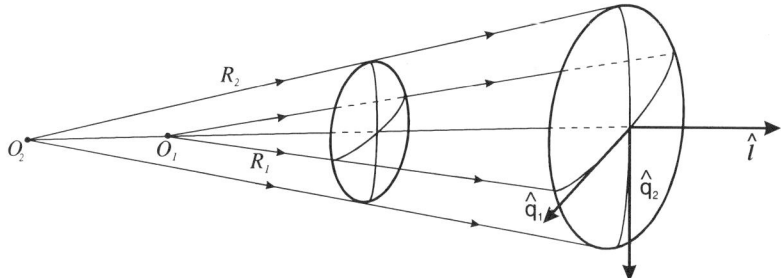

FIGURE 5.16. The ray coordinate system.

The ray paths all lie in the axial plane of the fiber to comply with the ϕ-independence of the ray pattern. The consistency condition is now enforced for any two points A and B (see Fig. 5.15c) which are characterized by the same transverse coordinate. The fiber axis $r = 0$ is a caustic for each of the ray congruences and an additional phase jump of $\pi/2$ should be added (see Sections 5.1.1 and 5.3.1). Accordingly

$$-u2a + 2\tan^{-1}\frac{\alpha\varepsilon_r}{u} + 2\pi - \frac{\pi}{2} = 2m\pi,$$

so

$$(ua)\tan\left(ua + \frac{\pi}{4}\right) = (\alpha a)\varepsilon_r,$$

which is the asymptotic expression for the exact dispersion equation from (4.86c) when $\alpha a \gg 1$ and $ua \gg 1$, as follows by making use of Eqs. (D.20a) and (D.22).

5.3. THE RAY COORDINATE SYSTEM

In this section we consider rays in a homogeneous environment; reflecting boundaries may be present. To keep track of ray propagation it is convenient to introduce a suitable coordinate system. One of these, the (principal) *ray coordinate system* $\hat{\mathbf{q}}_1 \times \hat{\mathbf{q}}_2 = \hat{\mathbf{l}}$, is depicted in Fig. 5.16. The unit vector $\hat{\mathbf{l}}$ is directed along ∇L, while the unit vectors $\hat{\mathbf{q}}_1$ and $\hat{\mathbf{q}}_2$ lie in the principal planes of the local rayfront, characterized by the principal radii of curvature R_1 and R_2, respectively.

The intersection of the rayfront with any plane containing the ray is approximated locally by the (osculating) circle[6]: in the principal planes, the

---

6. Alternatively, a quadratic (parabolic) approximation is often used.

radii corresponding to these circles are R_1 and R_2, respectively. Consider the first principal plane, and the section of the rayfront containing the origin $(0,0,0)$; see Fig. 5.17. If $(q_1, 0, -d_1)$ are the coordinates of a point on the rayfront, then

$$d_1(R_1 - d_1) = q_1^2,$$

i.e.,

$$d_1 \approx \frac{q_1^2}{R_1}, \qquad (5.25a)$$

because we assume that $d_1 \ll R_1$, i.e., that the rayfront is smooth and far away from O_1. In the other principal plane we obtain similarly

$$d_2 = \frac{q_2^2}{R_2}. \qquad (5.25b)$$

It is convenient to define the *curvature matrix* $\boldsymbol{Q}_0$ of the rayfront at the point $(0, 0, 0)$. Its representation is

$$\boldsymbol{Q}_0(0) \to \begin{vmatrix} \dfrac{1}{R_1} & 0 \\ 0 & \dfrac{1}{R_2} \end{vmatrix} \qquad (5.26a)$$

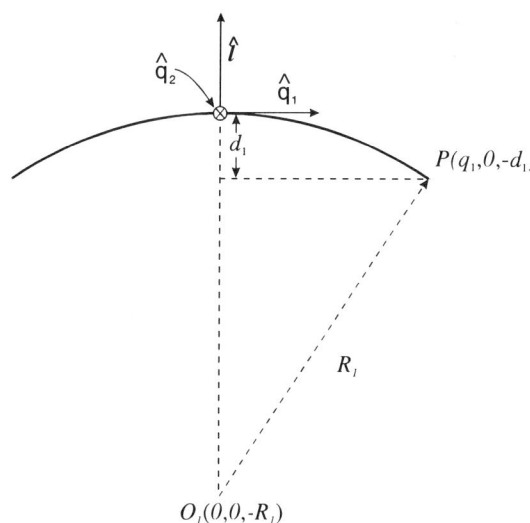

FIGURE 5.17. Description of a rayfront in the ray coordinate system.

when the matrix is referred to the coordinate system introduced. If we move to the point $(0, 0, l)$, we immediately obtain

$$Q_o(l) \to \begin{vmatrix} \dfrac{1}{R_1 + l} & 0 \\ 0 & \dfrac{1}{R_2 + l} \end{vmatrix} = \begin{vmatrix} q_{01} & 0 \\ 0 & q_{02} \end{vmatrix} \qquad (5.26b)$$

and we have

$$Q_o^{-1}(0) \to \begin{vmatrix} R_1 & 0 \\ 0 & R_2 \end{vmatrix}, \quad Q_o^{-1}(l) \to \begin{vmatrix} R_1 + l & 0 \\ 0 & R_2 + l \end{vmatrix} = Q_o^{-1}(0) + Il. \qquad (5.27a)$$

Generalization of Eqs. (5.25) to a generic point $(q_1, q_2, l - d)$ over the rayfront is

$$d = \tilde{\mathbf{q}} \cdot Q_o(l) \cdot \mathbf{q}, \qquad (5.28a)$$

with

$$\mathbf{q} = \begin{vmatrix} q_1 \\ q_2 \end{vmatrix} \quad \text{and} \quad \tilde{\mathbf{q}} = |q_1 q_2|. \qquad (5.28b)$$

This point $(q_1, q_2, l - d)$ and the $\hat{\mathbf{l}}$-axis determine a plane; the corresponding radius of curvature of the rayfront is given by

$$R = \frac{\tilde{\mathbf{q}} \cdot \mathbf{q}}{\tilde{\mathbf{q}} \cdot Q_o(l) \cdot \mathbf{q}}. \qquad (5.29)$$

The ray congruence need not be referred to a coordinate system whose axes $\hat{\mathbf{q}}_1$ and $\hat{\mathbf{q}}_2$ lie in the principal planes; sometimes, different choices are most convenient. Consider new axes which are rotated around $\hat{\mathbf{I}}$ through an angle α with respect to the $\hat{\mathbf{q}}_1$-axis. In this new reference system the curvature matrix takes the form

$$Q \to R \cdot Q_o \cdot \tilde{R} \to \begin{vmatrix} q_{11} & q_{12} \\ q_{21} & q_{22} \end{vmatrix}, \qquad (5.30)$$

with

$$R \to \begin{vmatrix} \cos \alpha & \sin \alpha \\ -\sin \alpha & \cos \alpha \end{vmatrix},$$

and is a rotation matrix. Hence

$$q_{11} = q_{01}\cos^2\alpha + q_{02}\sin^2\alpha, \quad q_{22} = q_{01}\sin^2\alpha + q_{02}\cos^2\alpha,$$

$$q_{12} = q_{21} = (q_{02} - q_{01})\sin\alpha\cos\alpha.$$

This new system is the most general ray coordinate system.

Conversely, when the matrix Q is given in the form (5.30) in a prescribed coordinate system, we can seek the appropriate rotation that renders the matrix diagonal:

$$R \cdot Q \cdot \tilde{R} \to \begin{vmatrix} q_{01} & 0 \\ 0 & q_{02} \end{vmatrix} = Q_0,$$

and determine the principal planes. By enforcing the off-diagonal terms of the matrix $R \cdot Q \cdot \tilde{R}$ to be zero, we derive

$$\tan 2\alpha = \frac{2q_{12}}{q_{11} - q_{22}}. \tag{5.31a}$$

Then

$$\begin{cases} q_{01} = q_{11}\cos^2\alpha + q_{22}\sin^2\alpha + 2q_{12}\sin\alpha\cos\alpha, \\ q_{02} = q_{11}\sin^2\alpha + q_{22}\cos^2\alpha - 2q_{12}\sin\alpha\cos\alpha. \end{cases} \tag{5.31b}$$

Equations (5.28a) and (5.29) are clearly valid in the (general) ray coordinate system: we only need to make the substitution $Q_0 \to Q$. Also

$$\det(Q) = \det(Q_0) = \frac{1}{(R_1 + l)(R_2 + l)}, \tag{5.32}$$

because $\det(R) = \det(\tilde{R}) = 1$. And, surprisingly enough, relation (5.27a) remains valid, too. In fact

$$Q^{-1}(0) = [R \cdot Q_0(0) \cdot \tilde{R}]^{-1} = \tilde{R}^{-1} \cdot Q_0^{-1} \cdot R^{-1} = R \cdot Q_0^{-1}(0) \cdot \tilde{R}$$

$$= Q_0^{-1}(0) + (R_2 - R_1)A.$$

$$A \to \sin\alpha \begin{vmatrix} \sin\alpha & \cos\alpha \\ \cos\alpha & -\sin\alpha \end{vmatrix}.$$

The corresponding relation for $Q^{-1}(l)$ is obtained by the change $R_1 \to R_1 + l$, $R_2 \to R_2 + l$:

$$Q^{-1}(l) = Q_0^{-1}(l) + (R_2 - R_1)A.$$

Accordingly

$$Q^{-1}(l) - Q^{-1}(0) = Q_0^{-1}(l) - Q_0^{-1}(0) = l\mathbf{I} \tag{5.27b}$$

in view of Eq. (5.27a).

5.3.1. High-Frequency Propagation in a Homogeneous Environment

Ray propagation in a homogeneous environment was considered in Section 5.1.2. The results can be cast in a simple form when reference is made to the ray coordinate system.

Equations (5.16b), (5.9b), and (5.32) yield

$$\begin{cases} \mathbf{E}(l) = \sqrt{\dfrac{\det[Q^{-1}(0)]}{\det[Q^{-1}(l)]}}\ \mathbf{E}(0)\exp(-ikl), \\ \zeta_0 \mathbf{H}(l) = \hat{\mathbf{l}} \times \mathbf{E}(l). \end{cases} \tag{5.33}$$

Also, the field does not change its polarization along the ray in a homogeneous environment; see Section 5.1.4.

If $R_1 = R_2 = R$, $R + l = r$, and $L(0) = R$, we have the situation of Fig. 5.18A:

$$\mathbf{E}(\mathbf{r}) = \frac{R}{r}\ \mathbf{E}(R)\exp[-ik_0(r - R)]. \tag{5.34a}$$

Equation (3.34a) may represent a spherical ray congruence *diverging* from the *focal point* $r = 0$. But we can also consider rays that first converge to ($\hat{\mathbf{r}} < 0$), and then diverge from ($\hat{\mathbf{r}} > 0$), the point $r = 0$ (see Fig. 5.18A). Note that the ray coordinate r should first take negative and then positive values if Eq. (3.34a) is used. Accordingly, the field amplitude would also attain negative values. This is avoided by ignoring the amplitude change of sign and introducing a phase jump equal to π when the ray passes through the focal point $r = 0$, thus accounting for the sign change. Similarly, if $R_2 \to \infty$, $R_1 = R$, $R_1 + l = r$, and $L(0) = R$, then

$$\mathbf{E}(\mathbf{r}) = \sqrt{\frac{R}{r}}\ \mathbf{E}_0(R)\exp[-ik_0(r - R)], \tag{5.34b}$$

and we obtain a cylindrical ray congruence with a focal line at $r = 0$ (see Fig. 5.16B). Again, the absolute value $|r|$ is used in the square root and the ray passage across the focal line is accounted for by a phase jump of $\pi/2$.

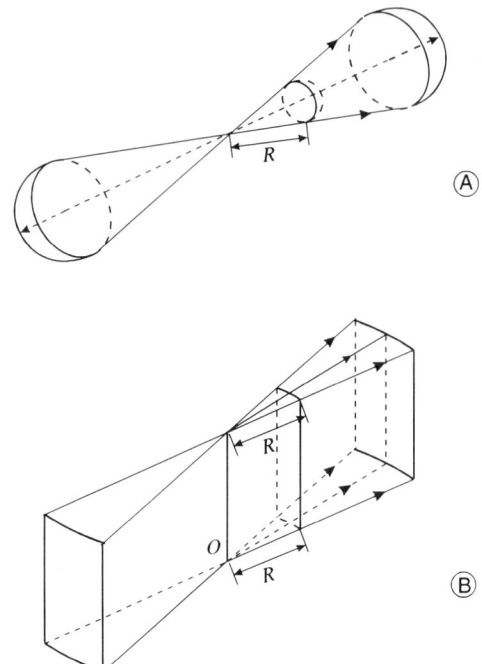

FIGURE 5.18. Spherical (A) and cylindrical (B) ray congruences.

5.3.2. Ray Amplitude at Reflection Boundaries: The Two-Dimensional Case

Let us consider a cylindrical rayfront incident on a cylindrical surface S separating two homogeneous media of different refractive index, as depicted in Fig. 5.19. The incident ray is characterized by (local) curvature $1/R$ and incidence angle θ, and the surface by (local) curvature $1/R_S$.

For an incident ray bundle of angular width $\Delta\beta$, the corresponding width of the reflected ray bundle is $\Delta\beta + 2\Delta\alpha$ (see Fig. 5.19). The common segment intercepted on S by these two ray bundles is given by $R_S \Delta\alpha$, where

$$R_S \Delta\alpha = \frac{R\Delta\beta}{\cos\theta}, \quad \text{i.e.,} \quad R_S \Delta\alpha = \frac{R'(\Delta\beta + 2\Delta\alpha)}{\cos(\theta + \Delta\theta + \Delta\alpha)} \approx \frac{R'(\Delta\beta + 2\Delta\alpha)}{\cos\theta}.$$

The solution for $1/R'$ is

$$\frac{1}{R'} = \frac{1}{R} + \frac{2}{R_S \cos\theta}, \qquad (5.35)$$

which provides the curvature $1/R'$ of the reflected ray bundle.

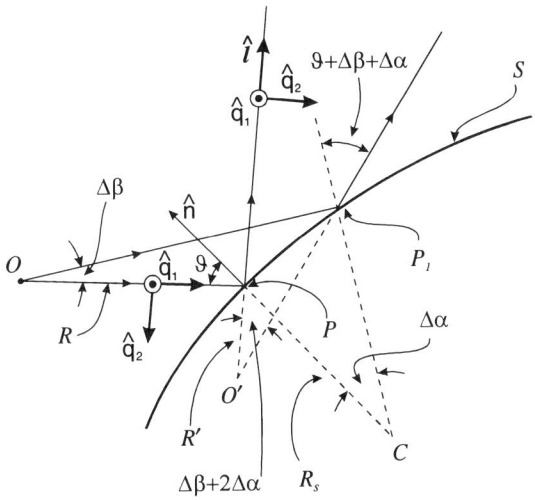

FIGURE 5.19. Ray reflection from a cylindrical boundary.

Let $\hat{\mathbf{q}}_1 \times \hat{\mathbf{q}}_2 = \hat{\mathbf{l}}$ be the ray coordinate system, with $\hat{\mathbf{q}}_1$ orthogonal to the plane of incidence, and let

$$\Gamma \to \begin{vmatrix} \Gamma_\perp & 0 \\ 0 & \Gamma_\| \end{vmatrix} \qquad (5.36)$$

be the local (electric field) *reflection coefficient matrix*, where $\Gamma_\perp$ and $\Gamma_\|$ are given by Eqs. (4.78) in view of the plane-wave local character of the ray field. Then, the reflected field based on this GO ray representation is given by

$$\begin{cases} \mathbf{E}_r = \sqrt{\dfrac{R'}{R' + l}} \; \Gamma \cdot \mathbf{E}_i \exp(-ik_0 l) \\ \zeta \mathbf{H}_r = \hat{\mathbf{l}} \times \mathbf{E}_r, \end{cases} \qquad (5.37)$$

where $\mathbf{E}_i$ is the value of the incident field on the surface. Note that $\Gamma_\perp = -1$ and $\Gamma_\| = +1$ when the reflecting surface is perfectly conducting.

The transmitted ray congruence can be treated similarly.

Figure 5.20 shows the case of a plane ray congruence ($R \to \infty$) incident upon a perfectly conducting circular cylinder of radius $R_S = a$. The reflected ray intensity along the direction θ is given by

$$I(\theta) = \frac{R'}{R' + l} \frac{|\mathbf{E}_i|^2}{2\zeta} = \frac{a \cos(\theta/2)}{a \cos(\theta/2) + 2l} \frac{|\mathbf{E}_i|^2}{2\zeta},$$

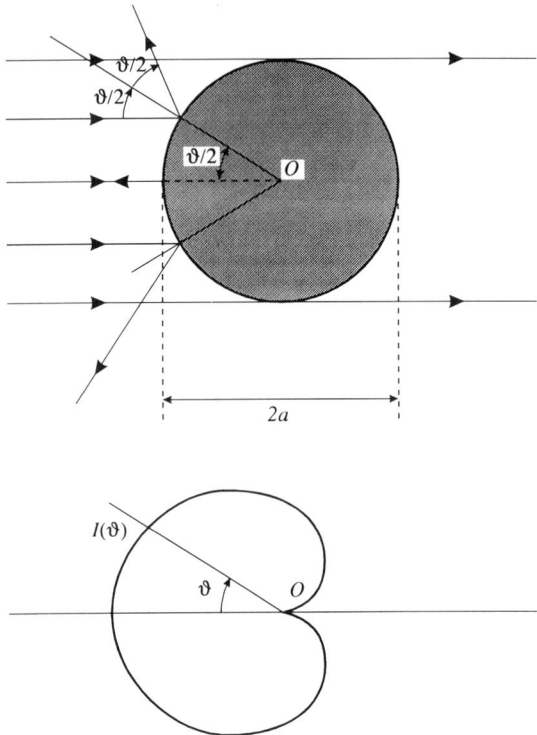

FIGURE 5.20. Reflection of a plane ray congruence by a perfectly conducting circular cylinder.

where l is measured from the cylinder surface; its angular distribution at large distances from the cylinder is depicted in Fig. 5.20. Note the lit-shadow ($\theta = \pm 180°$) discontinuity in the rear region of the cylinder, where GO is no longer a valid approximation (see Section 5.4).

The total scattered power (per unit length along the cylinder axis) is given by

$$P_S = \int_{-\pi}^{\pi} d\theta [R'(\theta) + l] I(\theta) = \int_{-\pi}^{\pi} d\theta \, \frac{a}{2} \cos \frac{\theta}{2} = \frac{|\mathbf{E}_i|^2}{2\zeta} 2a,$$

and depends only on the cross-section dimension of the cylinder.

5.3.3. Ray Amplitude at Reflection Boundaries: The Three-Dimensional Case

Reflection of a ray congruence at a reflection boundary in the general three-dimensional case can be studied by extending the results of Section 5.3.2. A detailed derivation is not given; only the final expressions are quoted.

In Fig. 5.21, incident and reflected rays are referred to their coordinate systems $(\hat{\mathbf{q}}_1^i, \hat{\mathbf{q}}_2^i, \hat{\mathbf{l}}_i)$ and $(\hat{\mathbf{q}}_1^r, \hat{\mathbf{q}}_2^r, \hat{\mathbf{l}}_r)$, respectively. The reflecting surface is described by the curvature matrix C, referred to the local coordinate system $(\hat{\mathbf{c}}_1, \hat{\mathbf{c}}_2, \hat{\mathbf{n}})$. The common origin of the three coordinate systems is the reflection point O. Note that $\hat{\mathbf{q}}_1^i = \hat{\mathbf{c}}_1 = \hat{\mathbf{q}}_1^r$.

With the chosen coordinate systems, the generalization of Eqs. (5.37) to the case at hand is

$$\begin{cases} \mathbf{E}_r(l_r) = \sqrt{\dfrac{\det[\mathbf{Q}_r^{-1}(0)]}{\det[\mathbf{Q}_r^{-1}(l_r)]}}\ \Gamma \cdot \mathbf{E}_i \exp(-ikl_r), \\ \zeta \mathbf{H}_r(l_r) = \hat{\mathbf{l}}_r \times \mathbf{E}_r, \end{cases} \quad (5.38)$$

where $\mathbf{E}_i$ is the value of the incident electric field over the boundary, Γ is the reflection matrix (5.36),

$$\mathbf{Q}_r^{-1}(l_r) = \mathbf{Q}_r^{-1}(0) + Il_r$$

[see Eq. (5.27b)], and

$$\mathbf{Q}_r(0) = \begin{vmatrix} -1 & 0 \\ 0 & 1 \end{vmatrix} \cdot \mathbf{Q}_i(0) \cdot \begin{vmatrix} -1 & 0 \\ 0 & 1 \end{vmatrix} + \begin{vmatrix} \cos\theta & 0 \\ 0 & 1 \end{vmatrix} \cdot C \cdot \begin{vmatrix} 1 & 0 \\ 0 & \dfrac{1}{\cos\theta} \end{vmatrix}$$

$$= \begin{vmatrix} q_{11} + c_{11}\cos\theta & -q_{12} + c_{12} \\ -q_{12} + c_{12} & q_{22} + \dfrac{2c_{22}}{\cos\theta} \end{vmatrix} \quad (5.39)$$

is the reflected ray curvature matrix, q_{pq} being the entries of $\mathbf{Q}_i(0)$.

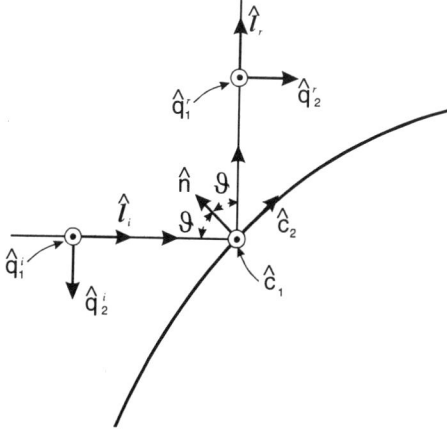

FIGURE 5.21. Reflection of a ray congruence by a surface discontinuity: the three-dimensional case.

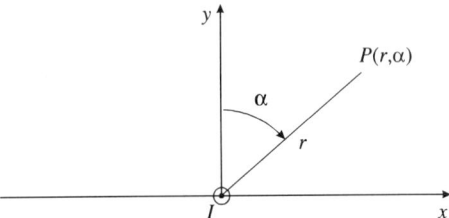

FIGURE 5.22. Asymptotic evaluation of the field associated with a line source.

Transition to the two-dimensional case is readily accomplished by setting $c_{22} = 1/R_S$, $q_{22} = 1/R$, and $c_{11} \to \infty$, thus recovering Eqs. (5.37).
The transmitted ray congruence can be dealt with similarly.

5.4. ASYMPTOTIC FORM OF FIELD REPRESENTATIONS

Rays were introduced in Section 5.1 as the asymptotic solution of (differential) Maxwell equations. In the limit $k_0 \to \infty$ the dominant term of the solution is referred to as the *geometrical optics* contribution, as mentioned earlier in Section 5.2.1. As an alternative procedure we can start from an integral representation of the field and again evaluate its asymptotic value as $k \to \infty$.[7] For instance, the field can be expressed in terms of its plane-wave spectrum (see Section 3.5). Examination of the asymptotic form of the pertinent integral reveals that it coincides with the field associated with a ray congruence: for localized sources, this ray congruence is spherical [see Eq. (3.85)].

Two examples are now treated. We first consider the line source

$$\mathbf{J}_0(\mathbf{r}) = I\delta(x)\delta(y)\hat{\mathbf{z}}$$

depicted Fig. 5.22. The corresponding spectrum, $\hat{\mathbf{J}}_0(\mathbf{k})$, is easily computed (see Section 3.4):

$$\hat{\mathbf{J}}_0(\mathbf{k}) = \int_{-\infty}^{+\infty} d\mathbf{r} \exp(-i\mathbf{k}\cdot\mathbf{r})\mathbf{J}_0(\mathbf{r}) = 2\pi I\delta(w)\hat{\mathbf{z}}.$$

We can neglect the w-dependence in the field evaluation, which amounts to omitting the factor $2\pi\delta(w)$ and setting $w = 0$ in the field spectrum. Equation (3.73a) yields

$$\hat{\mathbf{E}}(\mathbf{k}) = \frac{-i\omega\mu I}{u^2 + v^2 - \omega^2\varepsilon\mu}\hat{\mathbf{z}} = \hat{E}(\mathbf{k})\hat{\mathbf{z}}.$$

---

7. We assume a homogeneous (isotropic) medium and use k instead of k_0 to simplify the notation.

HIGH-FREQUENCY FIELDS

Hence the phasor of the electric field is finally given by

$$E(x, y) = \frac{-i\omega\mu}{(2\pi)^2} I \int_{-\infty}^{+\infty} du \int_{-\infty}^{+\infty} dv \, \frac{\exp(iux + ivy)}{u^2 + v^2 - k^2},$$

with $\omega^2 \varepsilon \mu = k^2$.

Contour integration in the complex $v = v' + iv''$ plane is now in order, as depicted in Fig. 5.23. As in Section 3.4.2, the proper position of the poles with respect to the integration contour is obtained by introducing small conductive losses. For $y \geq 0$ the integration contour can be closed in the upper half-plane, and the integral evaluated in terms of its residue at $v = -\sqrt{k^2 - u^2}$. Thus

$$E(x, y) = -\frac{\omega\mu}{4\pi} I \int_{-\infty}^{+\infty} du \, \frac{\exp(iux - i\sqrt{k^2 - u^2}\, y)}{\sqrt{k^2 - u^2}}. \quad (5.40a)$$

For subsequent analysis it is convenient first to change the integration variable, $u \to -u$, in Eq. (5.40a) and set

$$x = r \sin \alpha, \quad y = r \cos \alpha, \quad u = k \sin \xi.$$

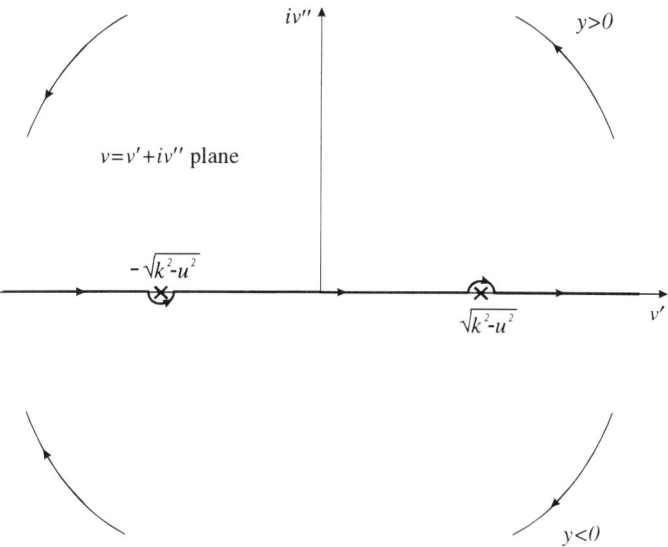

FIGURE 5.23. Integration contour in the complex v-plane for the evaluation of the field associated with the line source of Fig. 5.22.

Then, Eq. (5.40a) can be written as

$$E(r, \alpha) = -\frac{\omega\mu}{4\pi} I \int_C d\xi \exp[-ikr\cos(\xi - \alpha)], \qquad (5.40b)$$

where the *Sommerfeld*[8] *integration contour* C is obtained from the relation

$$\sin(\xi' + i\xi'') = \sin\xi' \cosh\xi'' + i\cos\xi' \sinh\xi'' = \frac{u}{k}$$

as u runs from $-\infty$ to $+\infty$, and is depicted in Fig. 5.24. Note that the vertical branches have been chosen in such a way that along them the factor

$$-iky\cos\left(\pm\frac{\pi}{2} + i\xi''\right) = \mp ky\sinh\xi''$$

ensures convergence of the integral (5.40b) for $y \geqslant 0$.

When $kr \to \infty$, asymptotic evaluation of the integral (5.40b) is possible following the approach of Section 3.5. The argument of the exponential is stationary for $\xi = \alpha$, and the integration contour can be deformed along the *steepest descent path* (SDP), i.e., a contour which ensures the maximum decay of the exponential term around the stationary point (see Section D.5 in Appendix D). If

$$\xi - \alpha = s,$$

then around $s = 0$

$$-i\cos(\xi - \alpha) \approx -i + i\frac{s^2}{2},$$

and

$$E(r) \sim -\frac{\omega\mu}{4\pi} I \exp(-ikr) \int_{-\infty}^{+\infty} ds \exp\left(i\frac{kr}{2}s^2\right)$$

$$= -\frac{\omega\mu}{4} I \sqrt{\frac{2}{\pi kr}} \exp(-ikr)\exp\left(i\frac{\pi}{4}\right), \qquad (5.41)$$

where use has been made of Eq. (C.5). Identical results are obtained for $y \leqslant 0$.

8. Arnold Sommerfeld: Königsberg (Prussia), 1868–Münich, 1951.

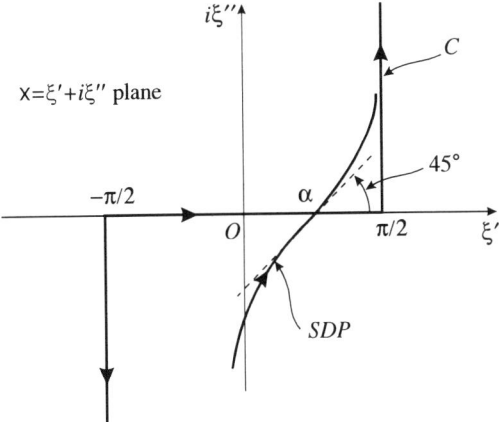

FIGURE 5.24. Sommerfeld integration contour in the complex ζ-plane.

Equation (5.41) represents a diverging cylindrical ray congruence with focal line coincident with the line source [see Eq. (5.34b)]. We conclude that the contribution of the *saddle point* $\zeta = \alpha$ (see Section D.5) to the spectral integral generates a *geometrical optics* ray species. Clearly, deformation of the Sommerfeld integration contour of Fig. 5.24 onto the steepest descent path has been possible due to the lack of singularities of the integral around the saddle point $\zeta = \alpha$. Should this not be the case, additional contributions arise in the evaluation of the spectral integral, thus generating further ray species in addition to the geometrical optics ones. As example is given in Section 5.4.5.

As a second example we consider a line source located at point T above a perfectly conducting infinite plane, as depicted in Fig. 5.25. The source illuminates the plane and surface currents are induced therein. The field scattered to the receiving point Q can be computed as a superposition of the elementary contributions of the current linear density induced over the surface.

We set the origin of the coordinate system at the specular reflection point and employ Eq. (5.41) to compute the incident field at point P over the conducting surface:

$$\mathbf{E}(x, 0) = -\frac{\omega\mu I}{4} \exp\left(i\frac{\pi}{4}\right) \sqrt{\frac{2}{\pi k R}} \exp(-ikR)\hat{\mathbf{z}},$$

$$R = \sqrt{(x + S)^2 + D^2}.$$

In the high-frequency limit the incident field is locally plane, and reflection laws can be applied. The tangent component of the *total* (incident plus

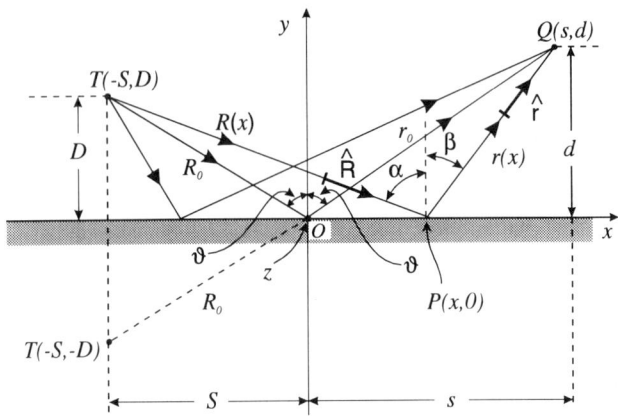

FIGURE 5.25. Reflection of a cylindrical wave from a perfectly conducting plane.

reflected) magnetic field is twice as large as the incident field. This assumption is normally referred to as the *physical optics* (PO) approximation. Accordingly, the induced surface current is given by

$$\mathbf{J} = \frac{2E}{\zeta} \hat{\mathbf{y}} \times (\hat{\mathbf{R}} \times \hat{\mathbf{z}})$$

$$= -\frac{kI}{2} \exp\left(i\frac{\pi}{4}\right) \cos\alpha \sqrt{\frac{2}{\pi k R}} \exp(-ikR)\hat{\mathbf{z}},$$

$$r = \sqrt{(x-s)^2 + d^2}.$$

The electric field scattered to Q is obtained by superposition:

$$\mathbf{E}(Q) = \frac{\omega\mu I}{4} \frac{i}{\pi} \hat{\mathbf{z}} \int_{-\infty}^{+\infty} dx \frac{\cos\alpha}{\sqrt{Rr}} \exp[-ik(R+r)], \qquad (5.42)$$

where use has been made of Eq. (5.41).

Again, a stationary phase evaluation of the integral is possible in the high-frequency limit $k \to \infty$. The phase is stationary when

$$\frac{d}{dx}(R+r) = \frac{x+S}{R} + \frac{x-s}{r} = 0,$$

i.e., when $x = 0$, because

$$\frac{s}{r_0} = \frac{S}{R_0} = \sin\theta, \qquad r_0 = r(0), \qquad \text{and} \qquad R_0 = R(0);$$

see Fig. 5.25. In the neighborhood of $x = 0$

$$R \approx R_0 + x\sin\theta + \frac{1}{2}\frac{x^2}{R_0}\cos^2\theta,$$

$$r \approx r_0 - x\sin\theta + \frac{1}{2}\frac{x^2}{r_0}\cos^2\theta,$$

and

$$\mathbf{E}(Q) \sim \frac{\omega\mu I}{4}\frac{i}{\pi}\frac{\cos\theta}{\sqrt{r_0 R_0}}\hat{\mathbf{z}}\exp[-ik(r_0 + R_0)]$$

$$\times \int_{-\infty}^{+\infty} dx \exp\left(-i\frac{1}{2}\frac{r_0 + R_0}{r_0 R_0}kx^2\cos^2\theta\right)$$

$$= \frac{\omega\mu I}{4}\exp\left(i\frac{\pi}{4}\right)\sqrt{\frac{2}{\pi k(r_0 + R_0)}}\exp[-ik(r_0 + R_0)]\hat{\mathbf{z}},$$

where we have employed (the complex conjugate of) Eq. (C.5).

The scattered field is represented by a cylindrical ray congruence; again, the saddle-point contribution has generated a geometrical optics ray species. This ray species is consistent with image theory over the perfectly conducting plane (see Fig. 5.25). Furthermore, the stationary phase evaluation is also consistent with Fermat's principle (see Section 5.2), because the ray path TOQ is minimum among all possible ray paths joining points T and Q via the *scattering surface*.

The above results can still be applied to gently modulated surfaces by first searching for the stationary points, and then asymptotically evaluating for each of them the scattering integral with a local quadratic approximation for the surface, making use of results of Section 5.3.3. However, if the scattering surface is finite and integration limits of Eq. (5.42) are finite, or the surface curvature varies rapidly as well as the induced current linear density, and PO cannot be applied, then additional contributions appear in the asymptotic evaluation of the scattering integral (5.42) and new ray species are generated. An example is given below in Section 5.4.1.

5.4.1. Scattering by a Conducting Half-Plane

Consider the problem depicted in Fig. 5.26: scattering from a perfectly conducting half-plane excited by an incident plane wave with parallel polarization.

We first examine the half-space $y \geq 0$. The total field is the sum of the incident field and of that scattered by the current $\mathbf{J}(x)$ induced on the half-plane. For the scattered electric field

$$\mathbf{E}_S(Q) = -\frac{\omega\mu}{4}\exp\left(i\frac{\pi}{4}\right)\int_{-\infty}^{0} \mathbf{J}(x)\sqrt{\frac{2}{\pi k r'}}\exp(-ikr') = E_S\hat{\mathbf{z}},$$

$$r' = r'(x) = \sqrt{(r\cos\phi - x)^2 + r^2\sin\phi}.$$

The induced currents are expected to approach the value

$$\mathbf{J}_\infty(x) = 2\hat{\mathbf{y}} \times \mathbf{H}_i(x) = 2\sin\phi_0\frac{E_0}{\zeta}\exp(ikx\cos\phi_0)\hat{\mathbf{z}} = J_\infty\hat{\mathbf{z}},$$

$$\mathbf{E}_0 = \mathbf{E}_i(0)$$

(see Fig. 5.26) away from the half-plane edge (the PO approximation), but this is certainly not the case close to it. However, it is instructive to continue to use the PO approximation, and then to compare the obtained solution with the exact one, whose derivation is cumbersome but available. Accordingly

$$E_S = -\frac{kE_0}{2}\sin\phi_0\exp\left(i\frac{\pi}{4}\right)\int_0^\infty dx\sqrt{\frac{2}{\pi k r'}}\exp(ikx\cos\phi_0 - ikr'). \quad (5.43)$$

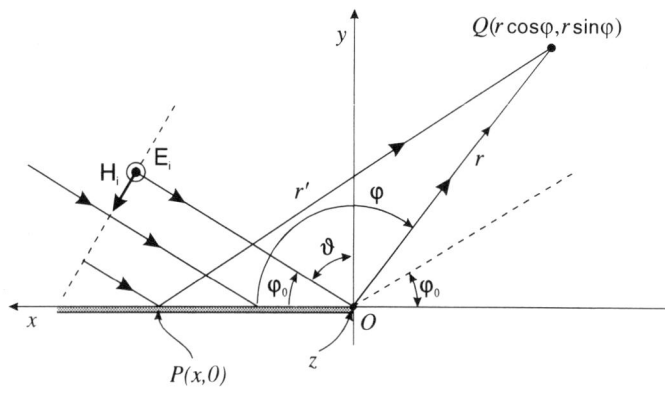

FIGURE 5.26. Scattering from a conducting half-plane.

HIGH-FREQUENCY FIELDS

We consider the two region $0 \leq \phi \leq \pi - \phi_0$ and $\pi - \phi_0 \leq \phi \leq \pi$. In the first region (see Fig. 5.27) the phase term exhibits a stationary point x_s when

$$\frac{d}{dx}[x \cos \phi_0 - r'] = \cos \phi_0 + \frac{r \cos \phi - x}{r'} = 0,$$

i.e.,

$$x_s = r \cos \phi + r'_s \cos \phi_0 = |Q'O| + |OS|, \qquad r'_s = r'(x_s),$$

which corresponds to the (specular) reflection point (see Fig. 5.27).

We note that the stationary point lies within the integration limits of Eq. (5.43) and use of stationary phase asymptotic techniques for the integral evaluation may be foreseen. We have

$$E_S = -\frac{kE_0}{2} \sin \phi_0 \exp\left(i\frac{\pi}{4}\right) \left\{ \int_{-\infty}^{+\infty} dx \sqrt{\frac{2}{\pi k r'}} \exp(ikx \cos \phi_0 - ikr') \right.$$
$$\left. - \int_{-\infty}^{0} dx \sqrt{\frac{2}{\pi k r'}} \exp(ikx \cos \phi_0 - ikr') \right\} = E_1 + E_2,$$

and only the first integral exhibits the stationary point within the integration contour.

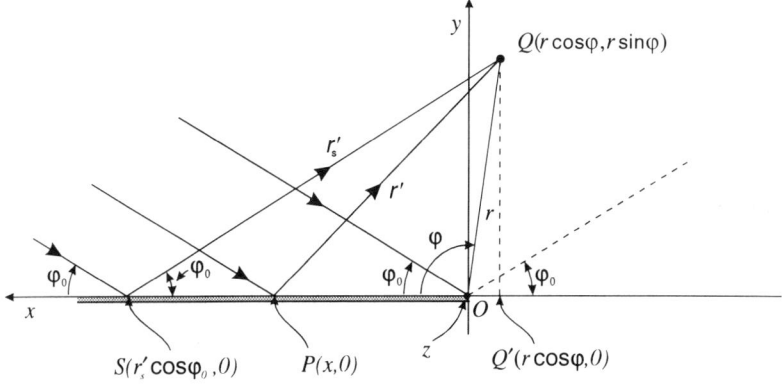

FIGURE 5.27. Scattering from a conducting half-plane (continued): the case $0 \leq \phi \leq \pi - \phi_0$.

The first integral E_1 can be evaluated asymptotically by expanding the phase term around the stationary point:

$$x \cos \phi_0 - r' \approx x_s \cos \phi_0 - r'_s - \frac{1}{2} \frac{r \sin^2 \phi}{r_s'^3} (x - x_s)^2,$$

$$E_1 \sim - \frac{kE_0 \exp(ikx_s \cos \phi_0)}{2} \sin \phi_0 \exp\left(i \frac{\pi}{4}\right)$$

$$\cdot \exp(-ikr'_s) \sqrt{\frac{2}{\pi k r'_s}} \int_{-\infty}^{+\infty} dx \exp\left(-i \frac{kr^2 \sin^2 \phi}{2r_s'^3} x^2\right)$$

$$= -E_0 \exp(ikx_s \cos \phi_0) \exp(-ikr'_s) = -E_i(S) \exp(-ikr'_s) = E_r, \quad (5.44)$$

where use has been made of Eq. (C.5) and of the geometrical relation

$$r'_s \sin \phi_0 = r \sin \phi.$$

Accordingly, Eq. (5.44) represents the GO reflected ray contribution, which exists throughout the region $0 \leq \phi \leq \pi - \phi_0$ up to the *reflection boundary* (RB)

$$\phi = \pi - \phi_0; \quad (5.45)$$

see Fig. 5.28.

For the second integral E_2, whose phase term does not exhibit any stationary point within the integration contour, expansion around $x = 0$ is appropriate:

$$x \cos \phi_0 - r' \approx x \cos \phi_0 - r + x \cos \phi - \frac{x^2}{2r} \sin^2 \phi.$$

The change of variables $x \to -x$ yields

$$E_2 \sim \frac{kE_0}{2} \sin \phi_0 \exp\left(i \frac{\pi}{4}\right) \sqrt{\frac{2}{\pi k r}} \exp(-ikr)$$

$$\cdot \int_0^\infty dx \exp\left[-ik(\cos \phi_0 + \cos \phi)x - i \frac{k \sin^2 \phi}{2r} x^2\right]. \quad (5.46a)$$

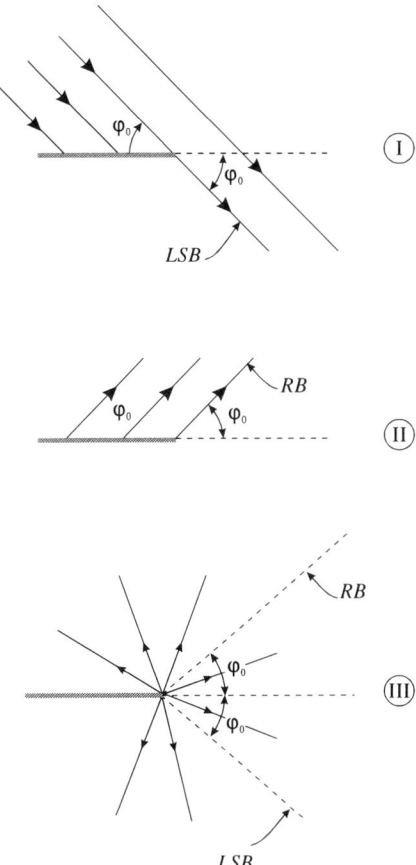

FIGURE 5.28. Scattering from a conducting half-plane (continued): (I) incident ray congruence; (II) reflected ray congruence; (III) diffracted ray congruence.

Integration by parts leads to

$$E_2 \sim E_0 \sqrt{\frac{1}{2\pi kr}} \exp\left(-i\frac{\pi}{4}\right) \exp(-ikr) \frac{\sin\phi_0}{\cos\phi_0 + \cos\phi} + 0\left(\frac{1}{k^{3/2}}\right),$$

where the higher-order terms are negligible in the high-frequency limit $k \to \infty$.

Equation (5.46a) represents a cylindrical ray congruence with a caustic line at the origin (see Fig. 5.28); this ray species is *not* present in the GO solution. The ray congruence exhibits the azimuthal pattern

$$D(\phi, \phi_0) = \frac{\sin\phi_0}{\cos\phi_0 + \cos\phi}, \qquad (5.47)$$

which becomes infinite for $\phi = \pi - \phi_0$, i.e., along the RB, where integration by parts is not allowed, and a more refined evaluation of the integral is needed; see Section 5.4.3.

In the second region $\pi - \phi_0 \leq \phi \leq \pi$ we proceed similarly. The stationary point lies outside the integration limits of Eq. (5.43), and the scattered field can be computed by following the same procedure used for Eq. (5.46a):

$$E_S = E'_2 \sim -\frac{kE_0}{2} \sin\phi_0 \exp\left(i\frac{\pi}{4}\right)\sqrt{\frac{2}{\pi kr}} \exp(-ikr)$$

$$\cdot \int_0^\infty dx \exp\left[+ikx(\cos\phi_0 + \cos\phi) - i\frac{k\sin^2\phi}{2r}x^2\right]$$

$$\sim E_0 \sqrt{\frac{1}{2\pi kr}} \exp\left(-i\frac{\pi}{4}\right)\exp(-ikr) \frac{\sin\phi_0}{\cos\phi_0 + \cos\phi} = E_2. \quad (5.46b)$$

We get the *same* cylindrical ray congruence (5.46a); see Fig. 5.28; again the evaluation is invalid near the RB.

Although the asymptotic form of the diffraction integrals of Eqs. (5.46) becomes invalid for $\phi \approx \pi - \phi_0$, they can be evaluated exactly at the RB. For $\phi = \pi - \phi_0$ we have $\cos\phi + \cos\phi_0 = 0$, and

$$E_2 = \tfrac{1}{2}E_i = -\tfrac{1}{2}E_r; \qquad E'_2 = -\tfrac{1}{2}E_i = \tfrac{1}{2}E_r. \quad (5.48)$$

Scattered field computation in the half-space $y \leq 0$ can be dealt with similarly. Again, the total field is the sum of the incident field (which exists everywhere in the space) and of that scattered by the currents induced on the half-plane. Also, in this case it is convenient to consider the two regions $\pi \leq \phi \leq \pi + \phi_0$ and $\pi + \phi_0 \leq \phi \leq 2\pi$, which are separated by the *lit-shadow boundary* (LSB),

$$\phi = \pi + \phi_0; \quad (5.49)$$

see Fig. 5.28.

In the first region $\pi \leq \phi \leq \pi + \phi_0$ no stationary point is present in the integral (5.43) and we obtain the *same* expression E_2 of Eq. (5.46b).

In the second region $\pi + \phi_0 \leq \phi \leq 2\pi$ the stationary point is again present in the integral (5.43). Its asymptotic evaluation leads to the sum of two terms: the first term is equal and opposite to the incident field, which therefore cancels out in this region; the second term is equal to E_2.

If $E_i(\phi)$ is the *incident* field, $E_r(\phi)$ the *reflected* field given by Eq. (5.44), and $E_d(\phi)$ the *diffracted* field given by Eq. (5.46a), then the *total* field representation is summarized in Table 5.1 and in Fig. 5.29, where the amplitude

HIGH-FREQUENCY FIELDS

TABLE 5.1. Field and Ray Species in the Half-Plane Problem

Regions	$0 \leqslant \phi < \pi - \phi_0$	$\pi - \phi_0 \leqslant \phi \leqslant \pi + \phi_0$	$\pi + \phi_0 \leqslant \phi \leqslant 2\pi$
Fields	$E_i + E_r + E_d$	$E_i + E_d$	E_d

of the incident (I), reflected (II), and diffracted (III) ray congruences are shown at a prescribed (though arbitrary) distance from the half-plane edge.

The direct rays are discontinuous at the LSB, as well as the reflected rays at the RB. However, the asymptotic evaluation of the scattered field generates a new ray species, to be identified (at least away from reflection and lit-shadow boundaries) as a *cylindrical edge ray*. Although their asymptotic evaluation diverges at the two boundaries [see Eqs. (5.46)], the edge rays can be evaluated

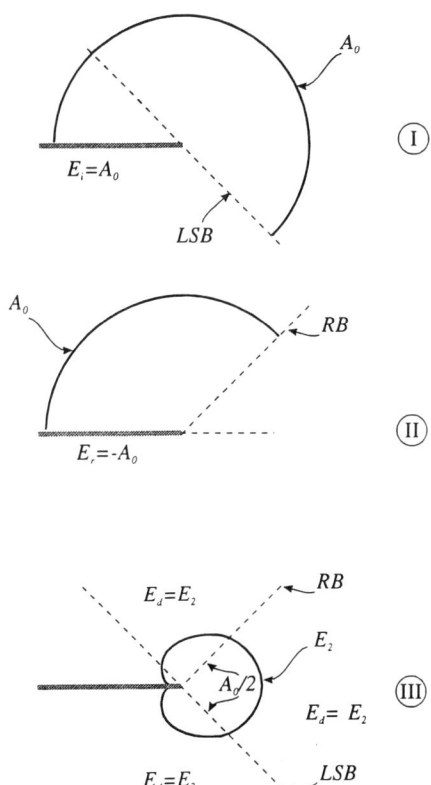

FIGURE 5.29. Scattering from a conducting half-plane (continued). Amplitude of the (I) incident E_i, (II) reflected E_r, and (III) diffracted E_d field at a prescribed distance from the edge. Solid lines represent the amplitude only and should not be confused with the rayfronts; see Fig. 5.28.

exactly there [see Eq. (5.48)]. They are discontinuous in such a way as to compensate for direct and reflected ray discontinuities, thus rendering the *total* field continuous and equal to one-half of the incident one. We recall that the asymptotic results of this section are obtained by using the PO approximation and are usually referred to as the *asymptotic physical optics* (APO) solution to the problem.

5.4.2. The Edge Ray

It was shown in Section 5.4.1 that the asymptotic evaluation of the diffraction integral in the case of a scattering object with a sharp edge (the half-plane) leads to a new scattered ray species, the edge ray, which is linearly proportional to the (complex) amplitude of the incident ray on the edge via an *edge diffraction coefficient*. These rays are consistent with a generalized form of Fermat's principle: the ray path is the shortest path connecting the source with the field point via, but not crossing, the half-plane. This is only one possible situation when additional ray species, in addition to the GO rays, can be predicted (see Fig. 5.30). The relevant rays are obtained by referring to

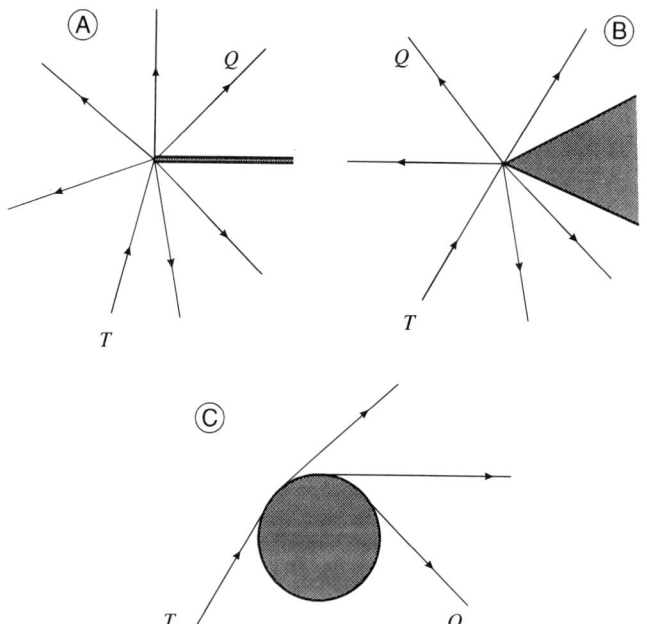

FIGURE 5.30. Rays of the geometrical theory of diffraction and the corresponding canonical problems: (A) half-plane, (B) wedge, (C) circular cylinder.

canonical cases, i.e., geometrical objects that coincide *locally* with the actual scattering bodies. The resulting diffraction integral must be evaluated asymptotically in the high-frequency limit $k \to \infty$ and the result interpreted in terms of new ray congruences, with the exclusion of possible transition regions where the ray description becomes invalid and smoothing transition functions must be used (see Section 5.4.3). This extension of GO has been named the *Geometrical Theory of Diffraction* (GTD) by Keller,[9] one of its founders. For the half-plane of Section 5.4.1, and an incident electric field parallel to the edge, the exact solution of the problem leads to the following expression for the scattered field:

$$\begin{cases} \mathbf{E}_d = \sqrt{\dfrac{2}{\pi k r}} \exp(-ikr) E_0 D_\parallel(\phi, \phi_0) \hat{\mathbf{z}}, \\ \zeta \mathbf{H}_d = \hat{\mathbf{l}} \times \mathbf{E}_d, \end{cases} \quad (5.50a)$$

with

$$D_\parallel(\phi, \phi_0) = -\frac{\exp(-i\pi/4)}{4} \left[\frac{1}{\cos[(\phi - \phi_0)/2]} - \frac{1}{\cos[(\phi + \phi_0)/2]} \right], \quad (5.51a)$$

which is the parallel polarization (or *soft*) edge diffraction coefficient. In Eqs. (5.50a), $\hat{\mathbf{l}}$ is the usual unit vector along the ray and E_0 the amplitude of the incident field at the edge. Equation (5.51a) can easily be expressed in the form

$$D_\parallel(\phi, \phi_0) = \exp\left(-i\frac{\pi}{4}\right) \frac{\sin(\phi/2)\sin(\phi_0/2)}{\cos\phi + \cos\phi_0}, \quad (5.51b)$$

which should be compared[10] to Eq. (5.47) obtained by using the APO approximation.

When the magnetic incident field is parallel to the edge, we obtain similarly

$$\begin{cases} \mathbf{E}_d = \sqrt{\dfrac{2}{\pi k r}} \exp(-ikr) E_0 D_\perp(\phi, \phi_0) \hat{\boldsymbol{\phi}}, \\ \zeta \mathbf{H}_d = \hat{\mathbf{l}} \times \mathbf{E}_d, \end{cases} \quad (5.50b)$$

---

9. Joseph Bernard Keller: Paterson (New Jersey, USA), 1923.
10. It is noteworthy that the two diffraction coefficients are normalized differently. When an identical normalization technique is used, Eq. (5.47) is $\cos(\phi_0/2)/\sin(\phi/2)$ times Eq. (5.51b).

with

$$D_\perp(\phi_1, \phi_0) = -\frac{\exp(-i\pi/4)}{4}\left[\frac{1}{\cos[(\phi+\phi_0)/2]} + \frac{1}{\cos[(\phi-\phi_0)/2]}\right]$$
$$= -\exp\left(-i\frac{\pi}{4}\right)\frac{\cos(\phi/2)\cos(\phi_0/2)}{\cos\phi + \cos\phi_0}, \qquad (5.51c)$$

which is the perpendicular polarization (or *hard*) edge diffraction coefficient.

In the case of oblique incidence, we apply GO rules of reflection: the edge rays are confined over a conical surface (*Keller's cone*) such that the angles of incidence and reflection are equal (see Fig. 5.31). In the ray coordinate system with $\hat{\mathbf{q}}_2$ in the plane of incidence or diffraction (i.e., $\hat{\mathbf{q}}_1$ tangent to the diffraction cone) we have

$$\begin{cases} \mathbf{E}_d = \sqrt{\dfrac{2}{\pi k r}}\exp(-ikr)\boldsymbol{D}(\phi_1\phi_0)\cdot\mathbf{E}_0, \\ \zeta\mathbf{H}_d = \hat{\mathbf{l}}\times\mathbf{E}_d, \end{cases} \qquad (5.52)$$

where

$$\boldsymbol{D} \to \frac{1}{\sin\beta}\begin{vmatrix} D_\perp & 0 \\ 0 & D_\parallel \end{vmatrix}, \qquad (5.53a)$$

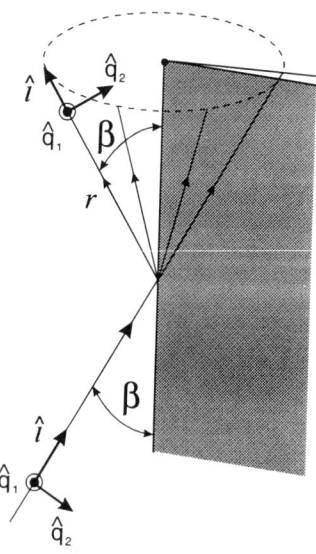

FIGURE 5.31. Diffraction of a ray obliquely incident over a straight edge.

which is the *edge diffraction matrix*. Equation (5.52) are also valid for the wedge case, provided that the appropriate values of $D_\perp$ and $D_\parallel$ are entered in Eq. (5.53a).

In the case of an edge with local curvature $1/R$ (see Fig. 5.32), we should include the appropriate divergence factor in the edge diffraction matrix:

$$\boldsymbol{D} \to \sqrt{\frac{R}{R+r}}\, \boldsymbol{D}. \tag{5.53b}$$

5.4.3. Transition Functions

It has already been noted that the edge ray description becomes invalid near the RB and LSB; *transition functions* are required to account for a smooth passage from cylindrical (away from the boundaries) to plane (close to and on the boundaries) wave species. There is a simple mathematical reason why the ray description breaks down: near the two boundaries the phase linear term

$$kx(\cos\phi_0 + \cos\phi) \approx 0$$

in the integrals (5.46), and integration by parts is no longer allowed.

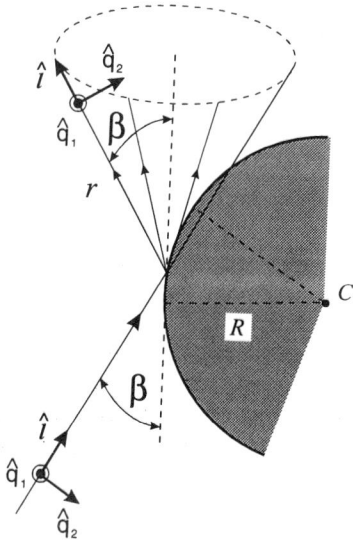

FIGURE 5.32. Diffraction of a ray obliquely incident over a curved edge.

Let us consider again the integral (5.46a). Its phase term can be written as follows:

$$-k(\cos\phi + \cos\phi_0)x - \frac{k\sin^2\phi}{2r}x^2$$

$$= -k\left(\frac{x\sin\phi}{\sqrt{2r}} + \frac{\cos\phi + \cos\phi_0}{\sin\phi}\sqrt{\frac{r}{2}}\right)^2 + k\left(\frac{\cos\phi + \cos\phi_0}{\sin\phi}\right)^2 \frac{r}{2}.$$

If

$$\eta = \sqrt{\frac{kr}{2}}\frac{\cos\phi + \cos\phi_0}{\sin\phi} \qquad (5.54a)$$

is substituted into Eq. (5.46a), then

$$E_2 \sim \frac{kE_0}{2}\frac{\sin\phi_0}{\sin\phi}\exp\left(i\frac{\pi}{4}\right)\sqrt{\frac{2}{\pi k r}}\exp(-ikr)$$

$$\cdot \sqrt{\frac{2r}{k}}\exp(i\eta^2)\int_0^\infty dt\,\exp[-i(t-\eta)^2]$$

$$= E_0 \frac{\sin\phi_0}{\cos\phi_0 + \cos\phi}\exp\left(-i\frac{\pi}{4}\right)\sqrt{\frac{1}{2\pi kr}}\exp(-ikr)\cdot T(\eta),$$

where the change of variable

$$\sqrt{\frac{k}{2r}}\,x\sin\phi \to t$$

has been introduced and

$$T(\eta) = \exp\left(i\frac{\pi}{2}\right)2\eta\exp(i\eta^2)\int_\eta^\infty dt\,\exp(-it^2). \qquad (5.54b)$$

The scattered field is given by the product of two factors. The first is the cylindrical ray congruence provided by the asymptotic evaluation of Eq. (5.46a), which diverges at the RB ($\eta = 0$). The second is a function that is zero at the RB, smoothing out the scattered field there, and is referred to as the *transition function*.

In fact, when $\eta \gg 1$ integration by parts of Eq. (5.54b) leads to

$$T(\eta \gg 1) \sim 1,$$

and the cylindrical ray congruence is fully recovered. On the other hand, when $\phi \to \pi - \phi_0$ and $\eta \ll 1$,

$$T(\eta \to 0) \approx 2\eta \exp\left(i\frac{\pi}{2}\right) \frac{\sqrt{\pi}}{2} \exp\left(-i\frac{\pi}{4}\right)$$

using Eq. (C.5) to compute the integral,

$$E_2 \sim \frac{E_0}{2} \exp(-ikr) = -\frac{1}{2} E_r(r, \phi_0),$$

and a plane-wave ray congruence is obtained.

Similar results are obtained for $\phi > \pi - \phi_0$.

The angular transition region $\pm \Delta\phi$ across $\pi - \phi_0$ can be estimated by setting $\eta = 1$:

$$\sqrt{\frac{kr}{2}} \frac{\cos(\pi - \phi_0 - \Delta\phi) + \cos\phi_0}{\sin(\pi - \phi_0 - \Delta\phi)} \approx \sqrt{\frac{kr}{2}} \Delta\phi = 1,$$

hence

$$\Delta\phi \approx \sqrt{\frac{2}{kr}}.$$

5.4.4. The Slope Diffraction Coefficient

Consider the case of a straight edge illuminated by a spatially varying field, such that its value on the edge is zero (see Fig. 5.33). In this case the edge diffracted field would be zero, in spite of the currents induced on the half-plane, which are not identically zero and provide a scattered field. A more refined analysis of the problem is necessary.

Let $E_i(x)$ be the incident electric field distribution over the plane $y = 0$. We have

$$E_i(x) = \frac{1}{2\pi} \int_{-\infty}^{+\infty} du' \exp(iu'x) E(u') \tag{5.55}$$

with $u' = k \cos\phi'$ and $u_0 = k \cos\phi_0$, where $E(u')$ is the corresponding plane-wave spectrum. Equation (5.55) represents the incident-wave illumination in terms of a superposition of plane waves; their (spectral) amplitude at $y = 0$ is given by $E(u')/2\pi$.

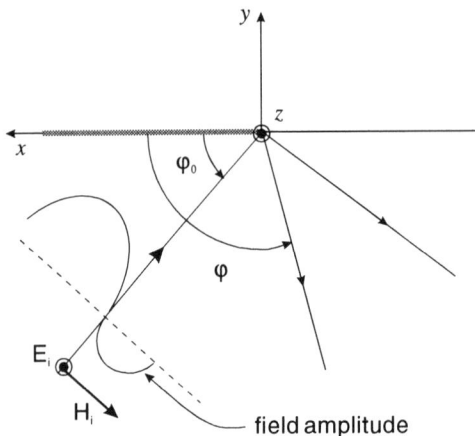

FIGURE 5.33. Scattering from a half-plane with nonuniform illumination.

For each plane-wave constituent the corresponding edge ray is obtained by using the appropriate (soft) edge diffraction coefficient $D(u, u')$ in Eqs. (5.51). Accordingly

$$E_d(x) = \sqrt{\frac{2}{\pi k r}} \exp(-ikr) \cdot \frac{1}{2\pi} \int_{-\infty}^{+\infty} du' E(u') D(u, u'), \quad \text{with } u = k \cos \phi, \quad (5.56)$$

where ϕ is the observation direction.

Assume now that the incident wave is spatially slow varying around the angular direction ϕ_0, so that it is close to a plane wave in the neighborhood of $u' = u_0$. Its spectrum is therefore confined near $u' = u_0$, and a series expansion of $D(u', u)$ around u_0 is appropriate for computing the integral in Eq. (5.56). In this case

$$\frac{1}{2\pi} \int_{-\infty}^{+\infty} du' E(u') D(u, u') \approx \frac{1}{2\pi} \int_{-\infty}^{+\infty} du' E(u') \cdot \left[D(u, u_0) + \frac{\partial D(u, u_0)}{\partial \mu_0} (u' - u_0) \right]$$

$$= D(u, u_0) \frac{1}{2\pi} \int_{-\infty}^{+\infty} du' E(u')$$

$$+ \frac{\partial D(u, u_0)}{\partial u_0} \frac{1}{2\pi} \int_{-\infty}^{+\infty} du' (u' - u_0) E(u'). \quad (5.57)$$

For the first term

$$D(u, u_0) \frac{1}{2\pi} \int_{-\infty}^{+\infty} du' E(u') = D(u, u_0) E_i(0)$$

[see Eq. (5.55)], which is the standard edge diffraction contribution, equal to zero if $E_i(0) = 0$. The additional term is expressed in its appropriate form by noting that

$$\frac{\partial}{\partial x} E_i(x) \exp(-iu_0 x) = \frac{\partial e_i(x)}{\partial x} = \frac{\partial}{\partial x} \frac{1}{2\pi} \int_{-\infty}^{+\infty} du' \exp[i(u' - u_0)x] E(u')$$

$$= \frac{i}{2\pi} \int_{-\infty}^{+\infty} du' \exp[i(u' - u_0)x](u' - u_0) E(u').$$

If we set $x = 0$ and substitute into Eq. (5.57), then the final result is derived in the form

$$E_d(\phi) = \sqrt{\frac{2}{\pi k r}} \exp(-ikr) \left[D(\phi, \phi_0) E_i(0) + \frac{1}{ik} \frac{\partial D(\phi, \phi_0)}{\partial \cos \phi_0} \frac{\partial e_i}{\partial x}\bigg|_0 \right]. \quad (5.58a)$$

The quantity

$$D_s(\phi, \phi_0) = \frac{1}{ik} \frac{\partial D(\phi, \phi_0)}{\partial \cos \phi_0} \quad (5.59)$$

is the *slope diffraction coefficient*. Its value times the derivative of the incident field with the *linear phase term removed*,

$$\frac{\partial e_i}{\partial x} = \frac{\partial}{\partial x} E_i(x) \exp(-ikx \cos \phi_0), \quad (5.58b)$$

and evaluated at the edge, accounts for the (spatial) modulation of the illumination. Note that D_s is of higher order than D, due to the presence of the factor $1/k$.

5.4.5. The Lateral Ray

The current line source at point T over a dielectric half-plane is depicted in Fig. 5.34. Geometrical optics predict a direct and a reflected ray, the latter weighted by the appropriate Fresnel reflection coefficient (4.78a), at the receiving point P. We wish to explore the possibility of (higher-order) ray species and a more refined analysis of the problem is in order.

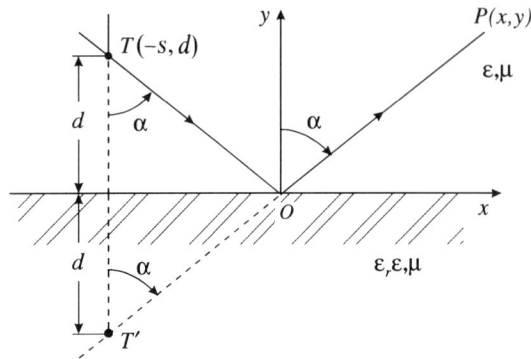

FIGURE 5.34. A current line source over a dielectric half-space.

We start from the plane-wave expansion (5.40a) and evaluate the plane-wave spectrum of the incident field at the interface plane $y = 0$:

$$E_i(u) = -\frac{\omega\mu I}{2} \frac{\exp(ius - i\sqrt{k^2 - u^2}d)}{\sqrt{k^2 - u^2}}, \qquad u = k\sin\alpha.$$

The corresponding plane-wave spectrum of the reflected field is equal to $E_i(u)$ times the reflection coefficient (4.78a):

$$\Gamma(u) = \frac{\sqrt{k^2 - u^2} - \sqrt{k^2\varepsilon_r - u^2}}{\sqrt{k^2 - u^2} + \sqrt{k^2\varepsilon_r - u^2}}, \qquad k^2 = \omega^2\varepsilon\mu.$$

Proceeding as in Section 5.4, the field scattered at the point P is

$$E(r, \alpha) = -\frac{\omega\mu I}{4\pi} \int_C d\xi \, \frac{\cos\xi - \sqrt{\varepsilon_r - \sin^2\xi}}{\cos\xi + \sqrt{\varepsilon_r - \sin^2\xi}} \exp[-ikr\cos(\xi - \alpha)], \quad (5.60)$$

where $r = |T'P| = |T0| + |0P|$ is the distance from the source image and C is the Sommerfield integration contour.

The presence of the square root in Eq. (5.60) requires branch cuts in the complex plane $\xi = \xi' + i\xi''$, as depicted in Fig. 5.35. The appropriate location of the branch points $\xi = \pm\alpha_b$, where

$$\alpha_b = \sin^{-1}\sqrt{\varepsilon_r}, \qquad (5.61)$$

is obtained by letting ε_r possess a vanishingly small imaginary part, $\varepsilon_r \to \varepsilon_r - i\varepsilon'_r$, with $\varepsilon'_r \to 0$.

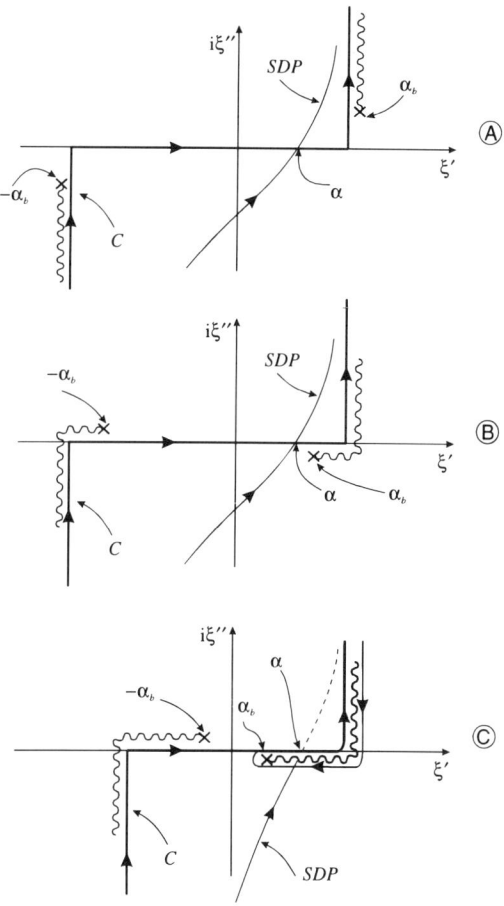

FIGURE 5.35. Integration contour in the complex ζ-plane: (A) $\varepsilon_r > 1$, (B) $\varepsilon_r < 1$, $|\alpha| < |\alpha_b|$, (C) $\varepsilon_r < 1$, $|\alpha| > |\alpha_b|$.

The integral (5.60) exhibits a stationary phase point at $\zeta = \alpha$. For any α if $\varepsilon_r > 1$, and for $|\alpha| < \alpha_b$ if $\varepsilon_r < 1$, the integration contour C can be fully deformed onto the SDP (see Fig. 5.35A–B). This leads to the expected reflected (cylindrical) ray congruence, obtained by setting $\zeta = \alpha$ in the reflection coefficient and asymptotically evaluating the resulting integral in Eq. (5.60). If $\varepsilon_r < 1$ and $|\alpha| < \alpha_b$ (see Fig. 5.35C), deformation of C onto the SDP is still possible, provided that this proceeds onto the lower Riemann sheet (dotted line). The integration contour then returns to the first sheet by crossing the branch cut at infinity, and runs around the latter to reach the point $\pi/2 + i\infty$ on the appropriate side of the branch. A new ray species is generated, represented analytically by the branch-cut contribution. Note that in this case

α_b coincides with the limit angle of Section 4.5.1.
We note that

$$\sqrt{\sin^2\alpha_b - \sin^2\xi} = \pm i\sqrt{\sin^2\xi - \sin^2\alpha_b},$$

so

$$\Gamma(\xi) = \frac{\cos^2\xi - \sin^2\xi + \sin^2\alpha_b \mp 2i\cos\xi\sqrt{\sin^2\xi - \sin^2\alpha_b}}{\cos^2\alpha_b}$$

in the upper and lower part of the cut, respectively. The branch-cut contribution to the integral (5.60) is given by

$$E_b = \frac{i\omega\mu I}{\pi\cos^2\alpha_b}\int_{\alpha_b}^{\pi/2+i\infty} d\xi\cos\xi\sqrt{\sin^2\xi - \sin^2\alpha_b}\exp[-ikr\cos(\xi-\alpha)]. \quad (5.62a)$$

A series expansion near the branch point $\xi = \alpha_b$ is appropriate in the high-frequency limit $kr \to \infty$. We have

$$\eta = \sin\xi - \sin\alpha_b, \qquad \sqrt{\sin^2\xi - \sin^2\alpha_b} \approx \sqrt{2\sin\alpha_b}\cdot\sqrt{\eta},$$

and

$$\cos(\xi - \alpha) = \cos[\sin^{-1}(\eta + \sin\alpha_b) - \alpha]$$
$$\approx \cos(\alpha - \alpha_b) + \eta\,\frac{\sin(\alpha - \alpha_b)}{\cos\alpha_b} - \frac{\eta^2\cos\alpha}{2\cos^3\alpha_b}.$$

With the aid of Fig. 5.36 we have further

$$r\cos\alpha = r_1\cos\alpha_b + r_3\cos\alpha_b \qquad \text{and} \qquad r\sin\alpha = r_1\sin\alpha_b + r_2 + r_3\sin\alpha_b,$$

so

$$r\cos(\alpha - \alpha_b) = r\cos\alpha\cos\alpha_b + r\sin\alpha\sin\alpha_b = r_1 + r_2\sqrt{\varepsilon_r} + r_3,$$
$$r\sin(\alpha - \alpha_b) = r\sin\alpha\cos\alpha_b - r\cos\alpha\sin\alpha_b = r_2\cos\alpha_b,$$

and the integral becomes

$$E_b \approx \frac{i\omega\mu I}{\pi\cos^2\alpha_b}\sqrt{2\sin\alpha_b}\exp[-ik(r_1 + r_2\sqrt{\varepsilon_r} + r_3)]$$
$$\cdot\int_0^\infty d\eta\sqrt{\eta}\exp\left[-ik\left(r_2\eta + \frac{1}{2}\frac{r_1+r_3}{\cos^2\alpha_b}\eta^2\right)\right]. \quad (5.62b)$$

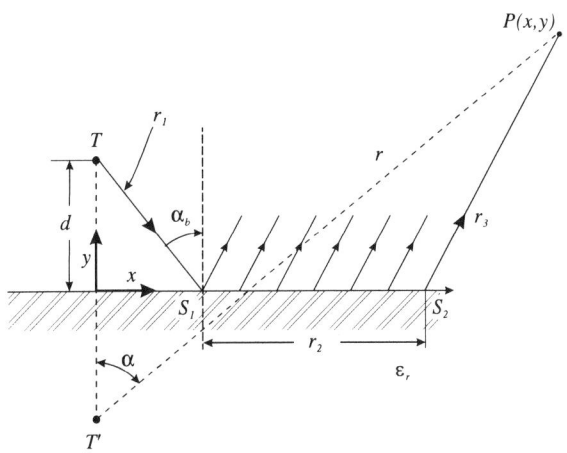

FIGURE 5.36. Ray path of the lateral ray.

The integral in Eq. (5.62b) can be evaluated by means of the change of variable $\sqrt{\ } \to t$ and two successive integrations by parts. Hence

$$2 \int_0^\infty dt\, t^2 \exp\left[-ikr_2 t^2 + ik\frac{r_1+r_3}{2\cos^2\alpha_b} t^4\right]$$
$$= -2 \int_0^\infty dt\, \frac{t \exp[-ikr_2 t^2 + ik(r_1+r_3)t^4/2\cos^2\alpha_b]}{-ikr_2 t + 2i(r_1+r_3)/2\cos^2\alpha_b}$$
$$\approx \frac{2}{ikr_2} \int_0^\infty dt \exp\left[-ikr_2 t^2 + ik\frac{r_1+r_3}{2\cos^2\alpha_b} t^4\right].$$

If $(r_1+r_3)/2\cos^2\alpha_b \ll r_1$, i.e., the source and the observation point are close to the interface compared to their horizontal distance, then we can neglect the second term in the exponential and finally have

$$E_b \sim \frac{2\omega\mu I}{\sqrt{\pi}} \exp\left(-i\frac{\pi}{4}\right) \frac{\varepsilon_r}{1-\varepsilon_r} \frac{\exp[-ik(r_1 + r_2\sqrt{\varepsilon_r} + r_3)]}{(k\sqrt{\varepsilon_r} r_2)^{3/2}}, \qquad (5.63)$$

where use has been made of Eq. (C.5) in evaluating the integral.

The field associated with the branch-cut contribution can be modeled as a ray moving along the path r_1 to the dielectric interface, and then progressing along the surface in the second medium, being continually reflected at the critical angle α_b in the electrically denser medium; see Fig. 5.36.

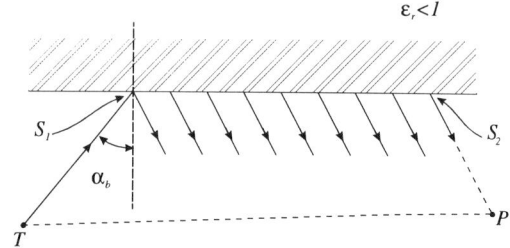

FIGURE 5.37. Up-over-and-down mechanism of propagation.

This new *lateral ray* species is important in subsurface propagation; the dominant radio link is mainly due to this *up-over-and-down* mechanism of propagation (see Fig. 5.37), because the denser medium (water, terrain) is usually lossy. The direct signal is exponentially damped along the path $|TP|$; on the contrary, the up-over-and down signal is exponentially damped along the (usually much shorter) path $|TS_1| + |S_2P|$, and only geometrically attenuated along the path $|S_1S_2|$; see Eq. (5.63).

5.4.6. The Creeping Ray

Consider a perfectly conducting cylinder excited by a plane wave, as depicted in Fig. 5.38. Geometrical optics predict a cylindrical reflected ray, in addition to the incident one [see Eq. (5.37) and Fig. 5.20]. The reflected ray is zero at the LSB (see Fig. 5.20), but the incident ray is discontinuous there. To smooth out this discontinuity and provide some illumination also in the shadow region, a more refined analysis of the problem is necessary. In this case, too, an additional contribution to the GO field is obtained.

The first step is to expand the incident plane wave in cylindrical coordinates:

$$\mathbf{E}_i = E_0 \exp(ikx)\hat{\mathbf{z}} = E_0 \exp(ikr\cos\theta)\hat{\mathbf{z}}$$
$$= \sum_{n=-\infty}^{+\infty} i^n J_n(kr) \exp(in\theta)\hat{\mathbf{z}} = E_i\hat{\mathbf{z}}, \qquad (5.64)$$

with the aid of Eq. (C.16).

Examination of Eq. (5.64) suggests the following representation for the scattered field:

$$\mathbf{E}_s = \sum_{n=-\infty}^{+\infty} i^n a_n H_n^{(2)}(kr) \exp(in\theta)\hat{\mathbf{z}} = E_s\hat{\mathbf{z}},$$

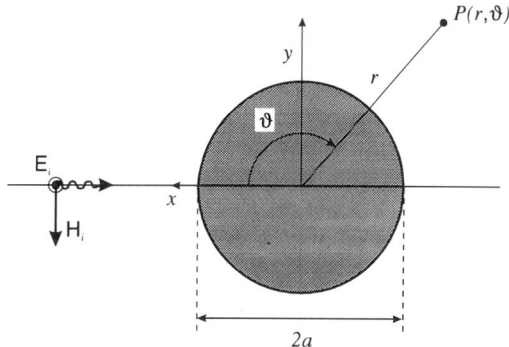

FIGURE 5.38. A plane wave incident normally upon a perfectly conducting circular cylinder.

whose terms individually satisfy the radiation condition at infinity [see Eq. (D.21)]. The expansion coefficients a_n are determined by enforcing the boundary condition at $r = a$,

$$E(a, \theta) = E_i(a, \theta) + E_s(a, \theta) = 0,$$

i.e.,

$$a_n = - \frac{J_n(ka)}{H_n^{(2)}(ka)} E_0.$$

The total field is given by

$$E(r, \theta) = E_0 \sum_{-\infty}^{+\infty} i^n \left[J_n(kr) - \frac{J_n(ka)}{H_n^{(2)}(ka)} H_n^{(2)}(kr) \right] \exp(in\theta). \qquad (5.65a)$$

We first represent the Bessel functions $J_n(ka)$ in Eq. (5.65a) in terms of the two Hankel functions [see Eq. (D.2)]. Then, by setting

$$\begin{aligned} B_n &= i^n \frac{H_n^{(1)}(kr)H_n^{(2)}(ka) - H_n^{(1)}(ka)H_n^{(2)}(kr)}{2H_n^{(2)}(ka)} \\ &= \frac{i^n}{2} \left[H_n^{(1)}(kr) - \frac{H_n^{(1)}(ka)}{H_n^{(2)}(ka)} H_n^{(2)}(kr) \right], \end{aligned} \qquad (5.66)$$

Eq. (5.65a) transforms as follows:

$$E(r, \theta) = E_0 \sum_{-\infty}^{+\infty} B_n(kr) \exp(in\theta). \qquad (5.65b)$$

In order to apply asymptotic techniques used throughout Section 5.4, it is convenient to express the series (5.65b) in integral form. This is accomplished by generating a function in the complex v-plane (see Fig. 5.39) which exhibits simple poles at $v = v_n = n\pi$, and then integrating along a closed contour around these singularities. This procedure yields

$$\oint_C dv f(v) \frac{\exp(iv\theta)\cos v\pi}{\sin v\pi} = 2\pi i \sum_n \frac{f(n)\exp(in\theta)\cos n\pi}{\pi \cos n\pi},$$

where the function $f(v)$ is singularity-free near the real v'-axis. Comparison with Eq. (5.65b) with $B_n = f(n)$ gives

$$E(r, \theta) = \frac{E_0}{2i} \oint_C dv\, B_v \frac{\cos v\pi \exp(iv\theta)}{\sin v\pi}.$$

We now split the integration contour C onto its lower part, from $-\infty$ to $+\infty$, and its upper part, from $+\infty$ to $-\infty$. Along the latter we introduce the change of variables $v \to -v$ and note that

$$B_{-v} = B_v$$

on employing Eq. (D.15). Summation of the two contributions leads to

$$E(r, \theta) = \frac{E_0}{i} \int_{-\infty}^{+\infty} dv\, B_v \frac{\cos v\pi \cos v\theta}{\sin v\pi}, \tag{5.67}$$

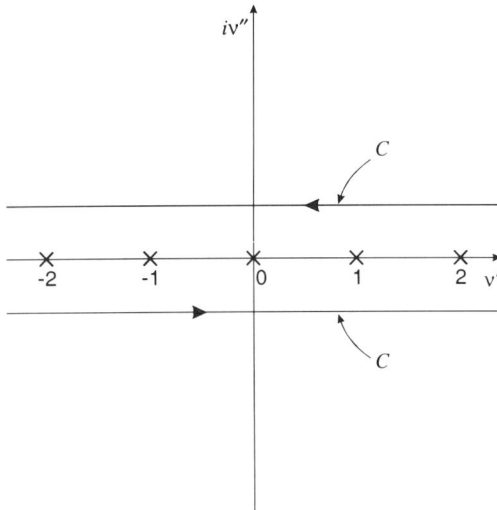

FIGURE 5.39. Integration contour for transforming an infinite series to an integral.

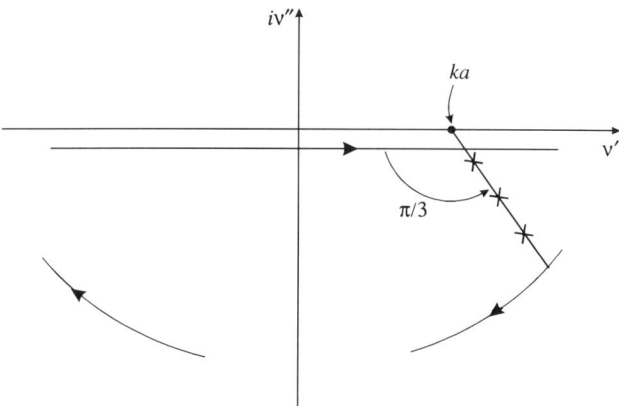

FIGURE 5.40. Integration contour for generation of creeping rays.

where the integration contour runs just under the real v'-axis, depicted in Fig. 5.40. Equation (5.67) is an integral representation of *all* the field, thus containing both the incident and scattered contributions.

A preliminary manipulation of integral (5.67) is convenient before performing its asymptotic evaluation. We have from Eq. (5.66)

$$B_v = i^v \frac{H_v^{(1)}(kr) + H_v^{(2)}(kr)}{2} - i^v \left[\frac{H_v^{(1)}(ka)}{H_v^{(2)}(ka)} + 1\right] \frac{H_v^{(2)}(kr)}{2}$$
$$= i^v J_v(kr) - i^v B_v'. \tag{5.68}$$

The first factor of Eq. (5.68) is substituted into Eq. (5.67) to recover the integral counterpart of the series (5.64) for the incident field. Accordingly, the integral

$$E_s = -\frac{E_0}{i} \int_{-\infty}^{+\infty} dv\, i^v B_v' \frac{\cos v\pi \cos v\theta}{\sin v\pi} \tag{5.69a}$$

represents the scattered field. This can be further simplified before proceeding to its asymptotic evaluation, because

$$\frac{\cos v\pi \cos v\theta}{\sin v\pi} = \frac{\cos v\theta}{\tan v\pi} \sim \frac{\cos v\theta}{i}, \tag{5.70}$$

when v attains even a small imaginary part,[11] a condition that is met in the

---

11. We have $\tan(x + iy) \approx i - i(1 - \tanh y) = 0.9i$ for $y = 0.46\pi$.

following analysis. Accordingly,

$$E_S = E_S^+ + E_S^-, \qquad (5.71a)$$

if

$$E_S^\pm = \frac{E_0}{2} \int_{-\infty}^{+\infty} dv \exp\left(iv\frac{\pi}{2}\right) B_v' \exp(\pm iv\theta). \qquad (5.71b)$$

The scattered field (5.71a) should contain the reflected ray contribution (5.37), but we show that an additional ray species is generated when we move close to the cylinder, particularly in the shadow region.

Let us first recover the reflected ray (5.37) as the saddle-point contribution to the integrals (5.71b) when $kr \to \infty$. We evaluate the v range where stationary phase points may be present. For $|v| \gg ka \gg 1$, the second row of Table 5.2 shows that

$$\frac{H_v^{(1)}(ka)}{H_v^{(2)}(ka)} \approx -1, \qquad B_v' \approx 0.$$

We conclude that the search of stationary phase points for the integrand in Eqs. (5.71b) can be limited to the range $|v| \lesssim ka$, where the third row of Table 5.2 is appropriate. Hence

$$\frac{H_v^{(1)}(ka)}{H_v^{(2)}(ka)} \approx \exp\left[2i\left(\sqrt{(ka)^2 - v^2} - v\cos^{-1}\frac{v}{ka} - \frac{\pi}{4}\right)\right].$$

When $kr \to \infty$, we use the first row of Table 5.2 to represent $H_v^{(2)}(kr)$, and Eq. (5.71b) becomes

$$E_S^\pm(r, \theta) \sim \frac{E_0}{4} \sqrt{\frac{2}{\pi kr}} \exp(-ikr) \exp\left(-i\frac{\pi}{4}\right)$$

$$\cdot \int_{-\infty}^{+\infty} dv \exp\left[2i\left(\sqrt{(ka)^2 - v^2} - v\cos^{-1}\frac{v}{ka}\right) + iv(\pi \pm \theta)\right], \qquad (5.72)$$

which exhibits a stationary phase point when

$$-2\cos^{-1}\frac{v}{ka} + (\pi \pm \theta) = 0,$$

TABLE 5.2. Asymptotic Expansion of Hankel Functions

v, x	$H_v^{(2)}(x)$	$H_v^{(1)}(x)$
1 $\quad |x| \gg |v|$ $\quad |x| \to \infty$	$\sqrt{\dfrac{2}{\pi x}} \exp\left[-i\left(x - v\dfrac{\pi}{2} - \dfrac{\pi}{4}\right)\right]$	$\sqrt{\dfrac{2}{\pi x}} \exp\left[i\left(x - v\dfrac{\pi}{2} - \dfrac{\pi}{4}\right)\right]$
2a $\quad |v| \gg |x|$ $\quad |v| \to \infty, \ |\arg v| \leqslant \pi/2$	$\sqrt{\dfrac{2}{\pi v}} \, i \left[\dfrac{2v}{ex}\right]^v$	$-\sqrt{\dfrac{2}{\pi v}} \, i \left[\dfrac{2v}{ex}\right]^v$
2b $\quad |v| \gg |x|$ $\quad |v| \to \infty, \ \pi/2 \leqslant \arg v \leqslant 3\pi/2$	$-\sqrt{\dfrac{2}{\pi v}} \left[\dfrac{2v}{ex}\right]^v$	$\sqrt{\dfrac{2}{\pi v}} \left[\dfrac{2v}{ex}\right]^v$
3 $\quad x > v$ $\quad |x| \to \infty$ $\quad |v| \to \infty$	$\sqrt{\dfrac{2}{\pi(x^2 - v^2)^{1/2}}} \exp\left[-i\left(\sqrt{x^2 - v^2} - v\cos^{-1}\dfrac{v}{x} - \dfrac{\pi}{4}\right)\right]$	$\sqrt{\dfrac{2}{\pi(x^2 - v^2)^{1/2}}} \exp\left[i\left(\sqrt{x^2 - v^2} - v\cos^{-1}\dfrac{v}{x} - \dfrac{\pi}{4}\right)\right]$

TABLE 5.3. Successive Zeros of the Airy Function $Ai(-x)$

p	1	2	3	4	5
x_p	2.338	4.088	5.521	6.787	7.944

namely,

$$v = \mp ka \sin\frac{\theta}{2}.$$

Deformation of the integration contour along the path of steepest descent is applicable: v attains an imaginary part, thus validating the approximation of Eq. (5.70). Asymptotic evaluation of E_S^- and E_S^+ leads to the reflected ray (5.37).

When the hypothesis $kr \to \infty$ is relaxed, i.e., we move closer to the cylinder boundary, a different evaluation of the integral (5.71b) is most convenient. For the scattered field contribution E_S^- we close the integration contour in the lower half part of the complex v-plane (see Fig. 5.40) due to the presence of the factor $\exp(-iv\theta + iv\pi/2)$, $\theta > \pi/2$. The integral is evaluated in terms of its residues at the singularities of the integrand:

$$H_v^{(2)}(ka) = 0, \tag{5.73}$$

in the lower half plane, which are known to be poles with the following asymptotic expression:

$$v_p \sim ka + \left(\frac{ka}{2}\right)^{1/3} \exp\left(-i\frac{\pi}{3}\right) x_p \tag{5.74}$$

for $ka \gg 1$, and are presented in Fig. 5.40. In Eq. (5.74), the factors x_p are the successive zeros of the Airy[12] function $Ai(-x)$ and are given in Table 5.3. Hence

$$E_S^-(r, \theta) = -\frac{\pi i}{2} E_0 \Sigma_p C_p(ka) \exp\left(iv_p\frac{\pi}{2} - iv_p\theta\right) H_{v_p}^{(2)}(kr), \tag{5.75}$$

where $C_p(ka)$ are the residues of $H_v^{(1)}(ka)/H_v^{(2)}(ka)$. Again, the presence of an imaginary part in the pole values validates approximation (5.70).

Each term of the series (5.75) exhibits a fast varying (complex) phase term ψ_p whose total value is obtained by using the third row of Table 5.2 to expand

---

12. George Biddel Airy: Alniwiek, Northumberland (UK), 1801–Greenwich, 1892.

the function $H_n^{(2)}(kr)$:

$$\psi_p = -\sqrt{(kr)^2 - v_p^2} + v_p \cos^{-1}\frac{v_p}{kr} - v_p\theta + v_p\frac{\pi}{2}. \quad (5.76)$$

Due to the presence of the imaginary part in expression (5.74) for the poles, the residue coresponding to the pole closest to the real v'-axis dominates.

Assuming $v_1 \approx ka$, we have from Eq. (5.76)

$$\psi_1 = -k\left[\sqrt{r^2 - a^2} - a\cos^{-1}\frac{a}{r} - a\frac{\pi}{2} + a\theta\right]$$
$$= -k|BP| - ka\theta_1, \quad (5.77)$$

which suggests the propagation model depicted in Fig. 5.41: a ray is generated at the tangent point A, propagates along the cylinder, and detaches itself along the tangent to reach the field point. This is a *creeping ray* that moves around the cylinder and continuously sheds light around, in particular in the shadow region. To improve the model we write

$$\psi_1(v_1) \approx \psi_1(ka) + \psi_1'(ka)(ka/2)^{1/3}\exp(-i\pi/3)x_p,$$

which enables Eq. (5.77) to be expressed in the alternative form

$$\psi_1 = -k|BP| - k_1 a\theta,$$
$$k_1 = k + (0.8k/a^2)^{1/3} - i(4.14k/a^2)^{1/3}. \quad (5.78)$$

The ray is exponentially damped as it progresses along the cylinder with a propagation constant slightly larger than that of the surrounding medium.

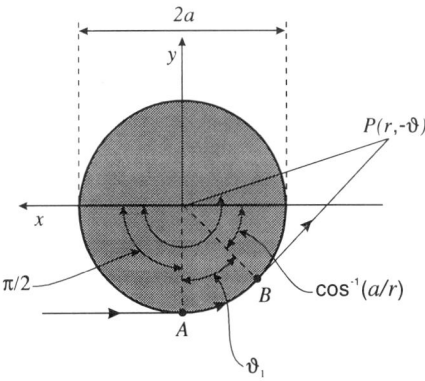

FIGURE 5.41. Creeping ray paths.

Identical results are obtained for the integral E^+, leading to creeping rays propagating in the opposite sense: clockwise and counterclockwise creeping rays are excited along the cylindrical structure.

This transformation of an integral, Eq. (5.71b), in a rapidly converging series, Eq. (5.75) and its E^+ counterpart, via contour integration and residue evaluation, is referred to as a Watson[13] transform.

5.5. SUMMARY AND SELECTED REFERENCES

This chapter provides solutions to Maxwell's equations when $k \to \infty$, i.e., when the medium parameters change slowly, the objects are large, and the observation points are remote in terms of the field wavelength. These assumptions lead to simple (asymptotic) solutions of Maxwell's equations in many cases and shed considerable light on the physics of interaction between the field and its environment. This insight is invaluable for any more accurate, usually numerical (see Chapter 6) study of the applied electromagnetics problem. These asymptotic solutions can be obtained along two lines: starting from either a differential or an integral formulation of the problem.

The first approach is considered in Section 5.1 and the ray properties are exploited in Sections 5.2 and 5.3. It is a classical approach which accounts for the transition from a field to a ray description of propagation and leads to the eikonal and transport equations. An exhaustive textbook on the subject is listed under [5.1]. Clearly, part of the material may appear in some of the textbooks cited in Chapter 1. Illustrative examples include propagation along a layered medium (Section 5.1.3), in particular duct propagation; polarization change along a ray (Section 5.1.4); geometrical optics treatment of reflector antennas and lenses (Sections 5.2.1 and 5.2.2); and guided propagation (Section 5.2.3), where ray techniques are compared to modal techniques exploited in Section 4.5.4. The ray coordinate system is presented and studied in detail in Section 5.3.

The second approach is examined in Section 5.4, where new ray species are systematically discovered by asymptotic evaluation of the field spectral integral [3.7]. The important canonical case of scattering by a conductive half-plane is described in Sections 5.4.1 and 5.4.2. The original paper of Keller is listed under [5.2]; in [5.3] more information can be found as far as applications are concerned.

The geometrical theory of diffraction has been subsequently highly developed [5.4], covering a large variety of problems, and has been modified along many different lines [5.5]. Uniform solutions via the use of transition functions (Section 5.4.3), higher-order ray contributions due to spatial transients of the

---

13. George Neville Watson: Wesward, Devon, (UK), 1886–Leamington, 1965.

incident field (Section 5.4.4), etc., are now available [5.4]. Much effort has been devoted to dominate the field behavior in caustic regions, by mean of *ad hoc* procedures [5.4, 5.5]. Alternative solutions to those given in Sections 5.4.3 and 5.4.4 are available [5.5].

Although all these procedures are interesting, they add complexity to the pure ray description, which somehow loses its simple, intuitive physical appeal. Perhaps it may be desirable to use the unsophisticated ray description to understand the propagation or scattering model, and then to make use of numerical procedures for the final design. Alternative approaches to GTD have also been suggested [5.6, 5.7].

As examples of the generation of additional ray species, the lateral ray and the creeping ray are presented in Sections 5.4.5 and 5.4.6, respectively. More information may be found in [5.4] and [5.8].

References

[5.1] M. Born and E. Wolf, *Principles of Optics*, Pergamon Press, New York (1970).

[5.2] J. B. Keller, "A geometrical theory of diffraction," in *Calculus of Variations and its Applications*, Proc. Symp. Appl. Math. **8**, pp. 27–52, McGraw-Hill, New York (1952).

[5.3] R. G. Kouyoumjian, "The geometrical theory of diffraction and its applications," in *Numerical and Asymptotic Techniques in Electromagnetics* (R. Mittra, ed.), Springer-Verlag, Berlin (1975).

[5.4] G. L. James, *Geometrical Theory of Diffraction for Electromagnetic Waves*, Peter Peregrinus, England (1976).

[5.5] P. H. Pathak, "Techniques for high frequency problems," in *Antenna Handbook: Theory, Application and Design* (Y. T. Wo and S. W. Lee, eds.), Van Nostrand Reinhold, New York (1988).

[5.6] P. Y. Ufimtsev, "Methods of edge waves in the physical theory of diffraction," translation prepared by the US Air Force Foreign Technology Division, Wright Patterson AFB, Ohio (1971).

[5.7] P. Y. Ufimtsev, "Theory of acoustical edge waves," *J. Acoust. Soc. Amer.* **86**, 463–474 (1989).

[5.8] S. Solimeno, B. Cosignani, and P. Di Porto, *Guiding, Diffraction and Confinement of Optical Radiation*, Academic Press, Orlando (1986).

6
The Numerical Domain

6.1. GENERAL CONSIDERATIONS

Solutions of Maxwell equations such as those presented in Chapters 2 through 5 are available in a very limited number of canonical cases that may or may not conveniently match analysis, synthesis, and design problems encountered in engineering applications. There is no doubt that these solutions are very important because they shed light on the physics of the problem and are the necessary basis to understand more complicated situations. Concepts like propagation, dispersion, reflection and scattering, finite signal velocity, and so on, can be learned and appreciated only by examining field analytical solutions.

However, the design of electromagnetic components and systems often requires accurate solutions of field problems that cannot be treated in terms of canonical cases. It is necessary to resort to computational techniques that provide the solution in a numerical form. And several *solvers* are available, i.e., numerical codes that are intended to model the problem at hand. This is accomplished by discretization of Maxwell equations, or their equivalent form, to reach the desired solution.

A first possibility is to start from an integral form of Maxwell's equations. Consider a known electromagnetic field $(\mathbf{E}_i, \mathbf{H}_i)$ in a homogeneous isotropic medium impinging on a metal surface S, as depicted in Fig. 6.1. The surface current $\mathbf{J}_S(\mathbf{r}')$ is induced over S and generates the scattered field $(\mathbf{E}_S, \mathbf{H}_S)$:

$$\mathbf{E}_S(\mathbf{r}) = -i\omega\mu \left[\mathbf{I} + \frac{\nabla\nabla}{k^2} \right] \cdot \iint_S dS' G(\mathbf{r} - \mathbf{r}') \mathbf{J}_S(\mathbf{r}'), \quad (6.1a)$$

$$\mathbf{H}_S(\mathbf{r}) = \nabla \times \iint_S dS' G(\mathbf{r} - \mathbf{r}') \mathbf{J}_S(\mathbf{r}') \quad (6.1b)$$

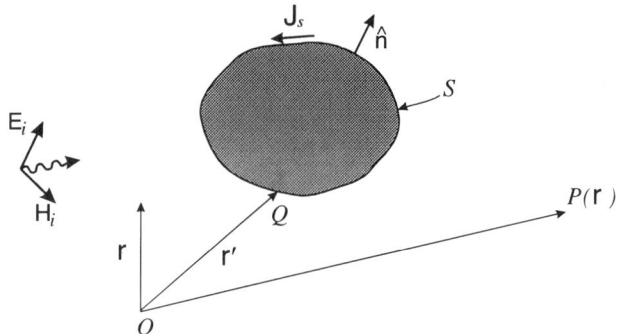

FIGURE 6.1. Deriving the integral form of Maxwell's equations.

[see Eqs. (4.87)], where

$$G(\mathbf{r} - \mathbf{r}') = \frac{\exp(-ik|\mathbf{r} - \mathbf{r}'|)}{4\pi|\mathbf{r} - \mathbf{r}'|} \quad (6.2)$$

is the Green's function appropriate to the problem (note that either phasors or spectral fields are considered). Boundary conditions on the surface S involve the *total* field ($\mathbf{E} = \mathbf{E}_i + \mathbf{E}_S, \mathbf{H} = \mathbf{H}_i + \mathbf{H}_S$): either $\hat{\mathbf{n}} \times \mathbf{E} = 0$ is required to be zero on S, or $\hat{\mathbf{n}} \times \mathbf{H}$ should be equal to the induced surface current density. These conditions are enforced by moving the point $P(\mathbf{r})$ over the surface (see Fig. 6.1) and making use of either Eq. (6.1a) or Eq. (6.1b), thus generating an integral equation in the unknown induced surface current density $\mathbf{J}_S$ (see Section 6.2). This, in turn, is computed by setting a convenient grid over S and transforming the integral into a finite summation. A linear system in the unknown discrete values of $\mathbf{J}_S$ at the grid points is obtained.

The time-domain solution can be determined either by Fourier transformation of its frequency-domain counterpart, or by starting directly from the Fourier transform of the original integral equations (6.1).

The above technique (and its most convenient modifications) is referred to as the *method of moments* (MoM) and is treated in Section 6.2.

A second possible technique can be employed. We consider the *functional* of the fields:

$$\mathscr{F}(\mathbf{E}) = \frac{1}{2}\iiint_V dV \left[\nabla \times \frac{1}{\mu_r}\nabla \times \mathbf{E} - k^2\varepsilon_r\mathbf{E}\right] \cdot \mathbf{E} - \iiint_V dV \mathbf{J} \cdot \mathbf{E}, \quad (6.3)$$

where $k^2 = \omega^2\varepsilon\mu$, while ε_r and μ_r are the relative permittivity and permeability (which may be space-dependent) of the medium, $\mathbf{J}$ is the source current density, and V the volume of interest. We show in Section 6.3 that this functional (or its conveniently modified form, see Section 6.3) is *stationary*, i.e.,

$$\delta \mathscr{F} = 0, \qquad (6.4)$$

to first order in $\delta \mathbf{E}$, if the electric field is a solution of the first two Maxwell's equations (4.46a–b) with appropriate boundary conditions on the surface S bounding V. In order to obtain the field solution, the volume V is subdivided into nonoverlapping patches, or *elements*, for which convenient vector functions $\mathbf{Q}_i$ are defined and used to represent the field therein. Imposition of the stationary condition, Eq. (6.4), leads again to a linear system of equations in the unknown coefficients of the field expansion. This technique, referred to as the *finite element method* (FEM), is examined in Section 6.3. As in the case of MoM, the time-domain solution can be obtained by Fourier transformation of its steady-state counterpart.

A third numerical technique is based on direct discretization of Maxwell's equations, where the derivatives are approximated by their numerical counterpart, as discussed in Section 6.4. In the frequency domain, a convenient spatial grid is set over the volume of interest, the derivatives are evaluated numerically,

$$\left. \frac{dE_x}{dy} \right|_i = \frac{E_x(y_{i+1}) - E_x(y_{i-1})}{2\Delta y},$$

$$\dots \quad \dots \quad \dots$$

and again a system of linear equations in the unknown values of the fields at the grid points is obtained. This technique is referred to as the *finite difference method* (FDM) and is probably the most natural and intuitive approach. We consider it after the others, because its direct implementation in the time domain (see Section 6.4) does not require the usual matrix inversion. Time updating of the field is obtained via a local, rather than a global, procedure.

In order to be attractive, the above-mentioned procedures (together with their extensions, modifications, etc.) should be general, computationally efficient, easy to use, and, last but not least, make use of limited computational facilities. Accordingly, the algorithm is an important, but absolutely not exhaustive, part of the solver; man–machine interfaces, computer graphics, display features, etc., are equally if not more important parts of the solver.

In spite of these considerations, knowledge of the algorithm is fundamental to judge the performance of the solver and to avoid the acritical acceptance of numerical results. In the following, the basic philosophy of the most popular algorithms is discussed.

6.1.1. Matrix Equations

The aforementioned numerical techniques aimed at solving electromagnetic problems lead very often (if not always) to the solution of linear systems of equations of either inhomogeneous,

$$M \cdot \mathbf{x} = \mathbf{a}, \tag{6.5a}$$

or homogeneous,

$$M \cdot \mathbf{x}_i = \xi_i \mathbf{x}_i, \tag{6.5b}$$

type. In Eqs. (6.5) the (usually square $N \times N$) matrix M of entries m_{ij} is the discretized form of the original differential or integral operator. In Eq. (6.5a), $\mathbf{x}$ and $\mathbf{a}$ are $N \times 1$ vectors representing either the field and source values at the grid points, or their expansion coefficients in a convenient basis. In Eq. (6.5b), ξ_i and $\mathbf{x}_i$ are respectively the *eigenvalues* and *eigenvectors* of the homogeneous system corresponding to the possible source-free solutions of the field compatible with the boundary conditions.

Equations (6.5) are of standard type. Equation (6.5a) requires the *inversion* of the matrix M, so that the unknown field is given by

$$\mathbf{x} = M^{-1} \cdot \mathbf{a}$$

while Eq. (6.5b) requires the solution of an algebraic equation,

$$(M - \xi_i I) \cdot \mathbf{x}_i = 0,$$

so that the eigenvalues can be obtained. In both cases the difficulty is of *numerical* type; it resides in the *size* and *structure* of M, which may pose severe problems to do with the computer memory and processing time. The relevant requirements may be alleviated by the form of the matrix, which may be *symmetric*, *sparse* (low number of entries), or *banded* (entries grouped along the diagonal).

Several numerical codes are available for both inverting M and computing the eigenvalues ξ. All of them involve the *efficient* implementation of direct or iterative methods for solving matrix or algebraic equations.

6.1.2. Matrix Inversion

Let us set

$$M = L \cdot U, \tag{6.6}$$

where L and U are lower and upper triangular matrixes, respectively. Substitution into Eq. (6.5a) leads to

$$L \cdot U \cdot x = L \cdot y = a.$$

The solution for x amounts to successively inverting the two matrix equations

$$L \cdot y = a \qquad (6.7a)$$

and

$$U \cdot x = y. \qquad (6.7b)$$

Their triangular nature enables the solution of Eqs. (6.7) to be obtained easily by a forward,

$$l_{11} y_1 = a_1, \qquad l_{22} y_2 = -l_{21} y_1 + a_2, \ldots,$$

and a backward,

$$u_{NN} x_N = y_N, \qquad u_{N-1,N-1} x_{N-1} = -u_{N-1,N} x_N + y_{N-1}, \ldots,$$

propagation procedure.

The decomposition (6.6) can be easily obtained by first setting

$$u_{ii} = 1,$$

and then performing the matrix product $L \cdot U$ row by column:

$$\sum_{1}^{N} l_{pi} u_{iq} = m_{pq}, \qquad (6.8)$$
$$p = 1, \quad q = 1, 2, \ldots, N; \quad p = 2, \quad q = 1, 2, \ldots, N, \quad \ldots .$$

In Eq. (6.8), each successive relation contains only a single new unknown compared to the previous ones, so that the solution can be very easily attained.

The above procedure of direct type is known as the *L–U decomposition method* and leads to efficiently implemented numerical codes. Alternative procedures of iterative type are also available.

6.1.3. Eigenvalue Computation

An iterative procedure for computing the eigenvalues of the eigenvalue equation,

$$\det(M - \xi I) = 0, \qquad (6.9)$$

is the Jacobi[1] method.

Consider a diagonal matrix $\boldsymbol{D}$. It follows from the equation

$$\det(\boldsymbol{D} - \xi_i \boldsymbol{I}) = 0$$

that its eigenvalues coincide with the diagonal entries. Furthermore, each of its eigenvectors, $\mathbf{x}_i$, has only a single entry different from zero, and is taken equal to unity when the normalization

$$\tilde{\mathbf{x}}_i^* \cdot \mathbf{x}_i = 1$$

is enforced. We can form a matrix with the eigenvectors to yield

$$[\mathbf{x}_1, \mathbf{x}_2, \ldots, \mathbf{x}_N] = \boldsymbol{I}. \tag{6.10}$$

Consider now a unitary matrix $\boldsymbol{R}$,

$$\tilde{\boldsymbol{R}}^* \cdot \boldsymbol{R} = \boldsymbol{I},$$

and multiply Eq. (6.5b) by $\boldsymbol{R}$, thus obtaining

$$\boldsymbol{R} \cdot (\boldsymbol{M} \cdot \mathbf{x}) = \boldsymbol{R} \cdot (\boldsymbol{M} \cdot \tilde{\boldsymbol{R}}^* \cdot \boldsymbol{R}) \cdot \mathbf{x} = (\boldsymbol{R} \cdot \boldsymbol{M} \cdot \tilde{\boldsymbol{R}}^*) \cdot (\boldsymbol{R} \cdot \mathbf{x}) = \xi \boldsymbol{R} \cdot \mathbf{x}.$$

The eigenvalues of the new matrix,

$$\boldsymbol{M}' = \boldsymbol{R} \cdot \boldsymbol{M} \cdot \tilde{\boldsymbol{R}}^*, \tag{6.11a}$$

coincide with those of $\boldsymbol{M}$, and its eigenvectors $\mathbf{x}'$ are simply obtained as

$$\mathbf{x}' = \boldsymbol{R} \cdot \mathbf{x}. \tag{6.11b}$$

If we can diagonalize $\boldsymbol{M}'$ by repeated application of the *similarity transformation*, Eq. (6.11a), then the diagonal terms of $\boldsymbol{M}'$ provide (within machine precision) the required eigenvalues.

Let m_{pq} be the largest off-diagonal entry of $\boldsymbol{M}$, and $\boldsymbol{R}$ a unit matrix with the exception of the following terms

$$r_{pp} = r_{qq} = \cos \phi \quad \text{and} \quad r_{pq} = -r_{qp} = \sin \phi,$$

where ϕ is to be determined. It is easy to check by inspection that

$$\tilde{\boldsymbol{R}}^* \cdot \boldsymbol{R} = \boldsymbol{I}$$

1. Karl Gustav Jacobi: Potsdam (Germany), 1804–Berlin, 1851.

and

$$m'_{pq} = (m_{qq} - m_{pp}) \sin\phi \cos\phi + m_{pq}(\cos^2\phi - \sin^2\phi)$$
$$= \frac{m_{qq} - m_{pp}}{2} \sin 2\phi + m_{pq} \cos 2\phi.$$

If

$$\tan 2\phi = \frac{2m_{pq}}{m_{pp} - m_{qq}},$$

then $m'_{pq} = 0$.

Repeated application of this procedure tends to progressively eliminate the off-diagonal terms of the original matrix M. The empty position may be filled again during the procedure, but these elements tend to be smaller, and the matrix tends to become diagonal to machine precision. The diagonal terms provide the eigenvalues, and the corresponding eigenvectors are obtained via Eqs. (6.10) and (6.11b):

$$[\mathbf{x}_1 \mathbf{x}_2 \ldots \mathbf{x}_N] = \tilde{R}_K^* \cdot \ldots \cdot \tilde{R}_2^* \cdot \tilde{R}_1^*,$$

where K is the number of successive transformations.

The Jacobi method may require a large number of iterations to converge. Other more efficient methods, especially for symmetric matrixes, are available.

6.1.4. Matrix Condition

We shall now examine the stability of the matrix inversion procedure. The (quadratic) *norm* of the matrix is defined as follows. Let

$$\widetilde{(M \cdot \mathbf{x})^* \cdot (M \cdot \mathbf{x})} = \tilde{\mathbf{x}}^* \cdot \tilde{M}^* \cdot M \cdot \mathbf{x} = M_x^2 \tilde{\mathbf{x}}^* \cdot \mathbf{x}, \quad (6.12)$$

where the scalar M_x^2 depends on $\mathbf{x}$. The matrix norm M is the maximum value attainable by M_x for all possible values of $\mathbf{x}$:

$$M = \sup(M_x).$$

The matrix $\tilde{M}^* \cdot M$ is hermitian because

$$\widetilde{(\tilde{M}^* \cdot M)^*} = (\tilde{M} \cdot M^*)^* = \tilde{M}^* \cdot M,$$

and exhibits real nonnegative eigenvalues ξ_i^2:

$$\tilde{M}^* \cdot M \cdot \mathbf{x}_i = \xi_i^2 \cdot \mathbf{x}_i. \tag{6.13}$$

If Eq. (6.13) is substituted into Eq. (6.12), then we obtain

$$\tilde{\mathbf{x}}^* \cdot \tilde{M}^* \cdot M \cdot \mathbf{x}_i = \xi_i^2 \cdot \tilde{\mathbf{x}}_i^* \cdot \tilde{\mathbf{x}}_i,$$

i.e.,

$$M_{x_i}^2 = \xi_i^2.$$

The set $\mathbf{x}_i$ forms a basis for the domain of $\tilde{M}^* \cdot M$, and any other vector $\mathbf{x}$ can be represented by a linear combination[2] of the eigenvectors $\mathbf{x}_i$. It follows that $\sup(M_x^2) = \xi_M^2$, where ξ_M^2 is the maximum of the eigenvalues ξ_i^2, and

$$M = \sqrt{\xi_M^2} = \xi_M. \tag{6.14a}$$

The norm m of M^{-1} is also of interest. We multiply Eq. (6.13) by $(\tilde{M}^* \cdot M)^{-1}$ to obtain

$$\mathbf{x}_i = \xi_i^2 (\tilde{M}^* \cdot M)^{-1} \cdot \mathbf{x}_i = \xi_i^2 M^{-1} \cdot (\tilde{M}^*)^{-1} \cdot \mathbf{x}_i = \xi_i^2 M^{-1} \cdot (M^{-1})^* \cdot \mathbf{x}_i,$$

where use has been made of Eqs. (B.41) and (B.40). It follows that the eigenvalues of $M^{-1} \cdot (M^{-1})^*$ equal $1/\xi_i^2$, and the norm m of M^{-1} coincides with the square root of the minimum eigenvalue, ξ_m^2:

$$m = \sqrt{\xi_m^2} = \xi_m. \tag{6.14b}$$

The matrix can be inverted provided that $\xi_m \neq 0$.

The stability of the matrix M as far as its inversion is concerned is described by the *condition number*,

$$\eta = \frac{\xi_M}{\xi_m} \geqslant 1, \tag{6.15}$$

and inversion stability increases as η is reduced. To clarify this point, consider the linear system (6.5a) subject to a perturbation of order ε in both the matrix and the known term entries:

$$M \to M + \varepsilon M', \quad \mathbf{a} \to \mathbf{a} + \varepsilon \mathbf{a}',$$

$$(M + \varepsilon M') \cdot \mathbf{x}(\varepsilon) = \mathbf{a} + \varepsilon \mathbf{a}'. \tag{6.16}$$

---

2. Should two or more eigenvectors coincide, the Gram–Schmidt orthogonalization procedure can be applied; see Section 7.4.4.

Subtracting Eq. (6.5a) from Eq. (6.16) yields

$$M \cdot [\mathbf{x}(\varepsilon) - \mathbf{x}] + \varepsilon M' \cdot \mathbf{x}(\varepsilon) = \varepsilon \mathbf{a}',$$
$$\mathbf{x}(\varepsilon) - \mathbf{x} = \varepsilon M^{-1} \cdot (-M' \cdot \mathbf{x} + \mathbf{a}'),$$

where the left-hand side represents the error in the solution generated by the matrix and known term perturbations.

An overall estimate of relative error to first order in ε is provided by the norm:

$$\frac{\|\mathbf{x}(\varepsilon) - \mathbf{x}\|}{\|\mathbf{x}\|} \leqslant \varepsilon \|M^{-1}\| \cdot \frac{(\|M'\| \|\mathbf{x}\| + \|\mathbf{a}'\|)}{\|\mathbf{x}\|}$$
$$\approx \varepsilon \|M^{-1}\| \|M\| \cdot \left(\frac{\|M'\|}{\|M\|} + \frac{\|\mathbf{a}'\|}{\|\mathbf{a}\|}\right)$$
$$\leqslant \varepsilon \eta \left(\frac{\|M'\|}{\|M\|} + \frac{\|\mathbf{a}'\|}{\|\mathbf{a}\|}\right), \qquad (6.17)$$

where we employed the Schwartz inequality (D.34) extended to matrixes as well as Eqs. (6.14) and (6.15).

Equation (6.17) states that the relative error in the norm of $\mathbf{x}$ may be magnified by the condition number η, given by Eq. (6.15), when compared to the relative error in the norms of $\mathbf{a}$ and M. In this sense η quantifies the sensitivity of the inversion of Eq. (6.5a).

6.2. THE METHOD OF MOMENTS

Consider the *integral equations*

$$s(x) = \alpha(x) f(x) + \int dx' g(x, x') f(x'), \qquad (6.18)$$

where $s(x)$, $\alpha(x)$, and $g(x, x')$ are known functions and $f(x)$ an unknown function. The function $g(x, x')$ often depends only on the difference $x - x'$, $g(x, x') \to g(x - x')$, and is referred to as the *kernel* of the integral. This can be evaluated between fixed or x-dependent limits: in the former case Eq. (6.18) is of Fredholm[3] type, and in the latter of Volterra[4] type. The equation is of first, second, and third type if $\alpha = 0$, $\alpha = \text{const}$, and $\alpha = \alpha(x)$, respectively. In the following we consider Fredholm equations and values of x and x' bounded by (a, b).

---

3. Ivar Fredholm: Stockholm (Sweden), 1866 – Morby (Stockholm), 1927.
4. Vito Volterra: Ancona (Italy), 1860 – Roma, 1940.

A numerical solution to Eq. (6.18) is obtained by first expanding the unknown function $f(x)$ in terms of a convenient set of functions,

$$f(x) = \sum_n f_n q_n(x), \tag{6.19}$$

where the *expansion functions* $q_n(x)$ are known while the expansion coefficients are unknown, and then substituting Eq. (6.19) into Eq. (6.18) to yield

$$s(x) = \sum_n f_n \alpha(x) q_n(x) + \sum_n f_n g_n(x), \tag{6.20}$$

where

$$g_n(x) = \int_a^b dx' g(x, x') q_n(x').$$

Then, we choose another convenient set of functions, the *weighting* (or *test*) *functions* $\psi_n(x)$, which are used to multiply Eq. (6.19) as a preliminary step for further integration over (a, b):

$$s_m = \sum_n f_n r_{mn} + \sum_n f_n g_{mn}, \tag{6.21}$$

where

$$s_m = \int_a^b dx \, s(x) \psi_m(x),$$

and

$$r_{mn} = \int_a^b dx \, \alpha(x) q_n(x) \psi_m(x), \qquad g_{mn} = \int_a^b dx \, \psi_m(x) \int_a^b dx' g(x, x') q_n(x').$$

Equation (6.21) represents a linear system in the unknowns f_n with coefficients $r_{mn} + g_{mn}$ and known terms s_m, when the two indexes n and m span their allowed values (O, N) and (O, M), respectively. If $N = M$, the system is squared and techniques presented in Section 6.1 can be used for its inversion.[5]

A particularly simple choice for the weighting functions is the set of Dirac pulses,

$$\psi_n(x) = \delta(x - x_n),$$

---

5. If $M > N$ the system is overdetermined, because the number of equations is larger than the number of unknowns, and the matrix is rectangular. Also, in this case, a generalized solution can be obtained, e.g., in the mean-square sense.

THE NUMERICAL DOMAIN

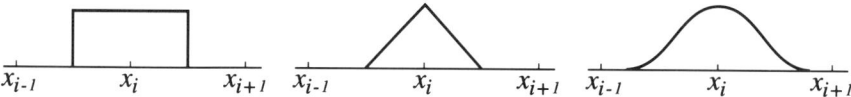

FIGURE 6.2. A few expansion and weighting functions.

while the coefficients in Eq. (6.21) are simply equal to

$$s_m = s(x_m), \qquad r_{mn} = \alpha(x_m)q_n(x_m), \qquad g_{mn} = g_n(x_m).$$

In this case Eq. (6.21) corresponds to forcing Eq. (6.18) to satisfy the set of discrete points x_m, and the procedure is referred to as *point matching*, or the *collocation method*. Other convenient expansion and weighting function sets are the rectangular, triangular, and sinusoidal pulses; see Fig. 6.2. Also, very often expansion and weighting functions are the complex conjugate of each other, such that

$$q_n(x) = \psi_n^*(x).$$

Should q_n and ψ_n belong to the same domain of the unknown field, the procedure is referred to as the *Garlekin*[6] *method*.

The above techniques, with all their modifications, are referred to as the *method of moments* (MoM).

6.2.1. The Electromagnetic Field Integral Equations

Let us consider again Fig. 6.1, where a metal body is excited by an impinging field $(\mathbf{E}_i, \mathbf{H}_i)$. This field is the one that would be present if the scatterer were removed (the *incident field*) and is either directly prescribed or given in terms of its sources (radiating in free space). As presented in Section 6.1, the surface current density $\mathbf{J}_S(\mathbf{r}')$ is induced over the body surface S, while the scattered field, namely,

$$\mathbf{E}_S(\mathbf{r}) = -i\omega\mu \left[\mathbf{I} + \frac{\nabla\nabla}{k^2} \right] \cdot \oiint_S dS' G(\mathbf{r} - \mathbf{r}')\mathbf{J}_S(\mathbf{r}'), \tag{6.21a}$$

$$\mathbf{H}_S(\mathbf{r}) = \nabla \times \oiint_S dS' G(\mathbf{r} - \mathbf{r}')\mathbf{J}_S(\mathbf{r}'), \tag{6.21b}$$

is generated; see Eqs. (6.1) and (6.2). Equation (6.21a) is the surface equivalent of Eq. (4.87), while Eq. (6.21b) is obtained by computing the curl of Eq. (6.21a).

---

6. Boris Grigorievich Galerkin: Polotsk (Russia), 1871–Moscow, 1945.

Before enforcing the boundary conditions on the surface S, a convenient modification of the integrals (6.21) is in order. We have

$$\nabla' \cdot (G\mathbf{J}_S) = \nabla'G \cdot \mathbf{J}_S + G\nabla' \cdot \mathbf{J}_S = -\nabla G \cdot \mathbf{J}_S + G\nabla' \cdot \mathbf{J}_S$$
$$= -\nabla \cdot (G\mathbf{J}_S) + G\nabla' \cdot \mathbf{J}_S,$$

where a prime implies the derivative with respect to the variable $\mathbf{r}'$. Eq. (6.21a) transforms as follows

$$\nabla \cdot \oiint_S dS' G\mathbf{J}_S = \oiint_S dS' \nabla \cdot (G\mathbf{J}_S)$$
$$= -\oiint_S dS' \nabla' \cdot (G\mathbf{J}_S) + \oiint_S dS' G\nabla' \cdot \mathbf{J}_S = \oiint_S dS' G\nabla' \cdot \mathbf{J}_S$$
$$= -i\omega \oiint_S dS' G\rho_S, \qquad (6.22a)$$

because the vector $G\mathbf{J}_S$ lies on the surface S, and the integral of its divergence over the closed surface S is zero. In deriving Eq. (6.22a) we have also used the surface current continuity equation

$$\nabla \cdot \mathbf{J}_S + i\omega\rho_S = 0.$$

Similarly, we use Eq. (A.12) and obtain from Eq. (6.21b)

$$\nabla \times \oiint_S dS' G\mathbf{J}_S = \oiint_S dS' \nabla \times (G\mathbf{J}_S) = -\oiint_S dS' \mathbf{J}_S \times \nabla G$$
$$= \oiint_S dS' \mathbf{J}_S \times \nabla' G. \qquad (6.22b)$$

The imposition of boundary conditions requires that the observation point $\mathbf{r}$ approaches the surface $S: \mathbf{r} \to \mathbf{r}'$. Due to the singularities appearing in the integrands of Eqs. (6.21), we exclude the point $\mathbf{r}' = \mathbf{r}_S$ by means of a small hemisphere* of radius $\Delta \to 0$ as depicted in Fig. 6.3, so Eqs. (6.21) become

$$\mathbf{E}_S(\mathbf{r}_S) = -\iint_{S'} dS' \left[i\omega\mu G\mathbf{J}_S - \frac{\rho_S}{\varepsilon} \nabla' G \right] - \iint_{S_\Delta} dS' \left[i\omega\mu G\mathbf{J}_S - \frac{\rho_S}{\varepsilon} \nabla' G \right], \qquad (6.23a)$$

7. This is the case if the surface S is smooth at $\mathbf{r}' = \mathbf{r}$, otherwise the appropriate portion of the spherical surface must be used.

THE NUMERICAL DOMAIN

$$\mathbf{H}_S(\mathbf{r}_S) = -\iint_{S'} dS' \nabla' G \times \mathbf{J}_S - \iint_{S_\Delta} dS' \nabla' G \times \mathbf{J}_S, \qquad (6.23b)$$

where the first integrals are the principal values with $\mathbf{r}' \neq \mathbf{r}_S$ and use has been made of Eqs. (6.22).

Evaluation of the integral over the hemisphere S_Δ is in order. In the limit $\Delta \to 0$ we have

$$\iint_{S_\Delta} dS' G \mathbf{J}_S \to 0 \qquad (6.24a)$$

$$\nabla' G \to -\frac{1}{4\pi\Delta^2}(-\hat{\mathbf{n}}),$$

$$\iint_{S_\Delta} dS' \rho_S \nabla' G \to \frac{1}{2}\rho_S(\mathbf{r}_S)\hat{\mathbf{n}}, \qquad (6.24b)$$

$$\iint_{S_\Delta} dS' \nabla' G \times \mathbf{J}_S \to \frac{1}{2}\hat{\mathbf{n}} \times \mathbf{J}_S(\mathbf{r}_s). \qquad (6.24c)$$

As discussed in Section 6.1, boundary conditions require either that

$$\hat{\mathbf{n}} \times [\mathbf{H}_i(\mathbf{r}_S) + \mathbf{H}_S(\mathbf{r}_S)] = \mathbf{J}_S(\mathbf{r}_S) \qquad (6.25a)$$

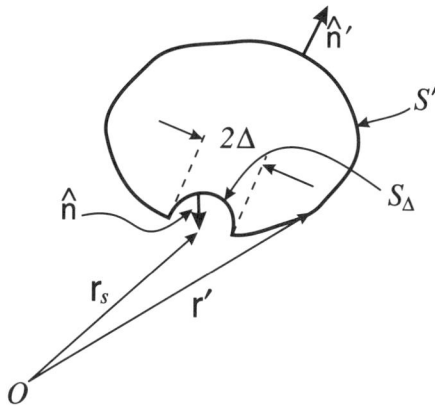

FIGURE 6.3. Derivation of the integral equations for the scattered field.

[see Eq. (1.36)], or

$$\hat{n} \times [\mathbf{E}_i(\mathbf{r}_S) + \mathbf{E}_S(\mathbf{r}_S)] = 0 \qquad (6.25b)$$

[see Eq. (1.37)].
In the former case, substitution of Eqs. (6.23b) and (6.24) into Eq. (6.25a) leads to

$$\mathbf{J}_S(\mathbf{r}_S) = 2\hat{n} \times \mathbf{H}_i(\mathbf{r}_S) + 2\hat{n} \times \iint_{S'} dS' \mathbf{J}_S \times \nabla' G; \qquad (6.26a)$$

in the latter case

$$\hat{n} \times \mathbf{E}_i(\mathbf{r}_S) = \hat{n} \times \iint_{S'} dS' \left[i\omega\mu G \mathbf{J}_S - \frac{\rho_S}{\varepsilon} \nabla' G \right]. \qquad (6.26b)$$

Equation (6.26a) can be used to evaluate the unknown surface current $\mathbf{J}_S$, and is usually referred to as the *magnetic field integral equation* (MFIE) because $\mathbf{J}_S = \hat{n} \times \mathbf{H}$. Once $\mathbf{J}_S$ is computed, the scattered field is obtained via Eqs. (6.21) or (6.24).

Equation (6.26b) is usually referred to as the *electric field integral equation* (EFIE), and can also be used to evaluate the unknown surface current $\mathbf{J}_S$, hence the scattered electromagnetic field.

The numerical solution of either Eq. (6.26a) or Eq. (6.26b) can be obtained by means of the technique presented in Section 6.2 for the scalar case: expansion and weighting functions are chosen for each of the scalar components of $\mathbf{J}_S$ and the MoM procedure implemented. The MFIE is usually preferred in the case of scattering from extended surfaces because it leads to a Fredholm integral equation of the second kind, which exhibits a better conditional matrix. On the contrary, the EFIE may be useful in the case of a wire structure excited by voltages applied at small gaps, which are used to model the incident field.

6.2.2. Scattering by a Metal Strip

Consider a plane wave impinging normally on a metal strip of width $2a$, as depicted in Fig. 6.4. The strip is of infinite extent in the y direction. In this case

$$\mathbf{E}_i(z) = E_0 \hat{y} \exp(ikz) \quad \text{and} \quad \mathbf{H}_i(z) = \zeta E_0 \hat{x} \exp(ikz);$$

THE NUMERICAL DOMAIN

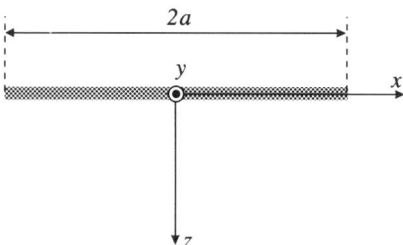

FIGURE 6.4. Scattering by a metal strip.

$\mathbf{J}_S = J_S(x)\hat{\mathbf{y}}$ due to the type of excitation, $\nabla_S \cdot \mathbf{J}_S(x) = 0$, and

$$\iint_{S'} dS' J_S(x') G(x-x', y-y')$$

$$= \int_{-a}^{a} dx' J_S(x') \int_{-\infty}^{+\infty} dy' \frac{\exp(-ik\sqrt{(x-x')^2+(y-y')^2})}{4\pi\sqrt{(x-x')^2+(y-y')^2}}$$

$$= \frac{1}{4i} \int_{-a}^{a} dx' J_S(x') H_0^{(2)}(k|x-x'|),$$

after applying Eq. (C.27). Accordingly, the EFIE (6.26b) simplifies to

$$\frac{k\zeta}{4} \int_{-a}^{a} dx' J_S(x') H_0^{(2)}(k|x-x'|) = E_0. \tag{6.27}$$

Application of the numerical techniques presented in Section 6.2 is now possible. It is anticipated that the results obtained (convergence of the surface current representation, accuracy of the solution, etc.) depend on the choice of expansion and weighting functions, the simplest probably being the set of rectangle and Dirac pulses,

$$q_n(x) = \text{rect.}\left(\frac{x-x_n}{\Delta x}\right) \quad \text{and} \quad \psi_m(x) = \delta(x-x_m),$$

respectively. With this choice Eq. (6.27) can be easily discretized. However, further comment on this point is worthwhile.

The above procedure does not account for any information concerning the surface current distribution, which could be inferred from the presence of the two edges of the strip. In fact, the edge condition (see Section 1.4.3) poses constraints on the field components in the plane orthogonal to the edge and close to it [see Eq. (1.64)]. The x-component of the magnetic field should decay as the inverse of the square root of its distance from the edge; similar behavior is exhibited by the surface current. Accordingly, a term of the type

$$J_e q_e(x) = \frac{J_e}{\sqrt{1-(x/a)^2}}$$

could be anticipated to appear in an appropriate representation for the surface current,

$$J_s(x) \to J_e q_e(x) + J_s(x),$$

the remaining part being regular within $(-a, a)$.

With reference to the new term we have

$$\int_{-a}^{a} dx' \frac{H_0^{(2)}(k|x-x'|)}{\sqrt{1-(x'/a)^2}} = \pi a H_0^{(2)}(k|x-a|)$$

$$- ka\,\mathrm{sgn}(x-x') \int_{-a}^{a} dx' \sin^{-1}\left(\frac{x'}{a}\right) H_1^{(2)}(k|x-x'|), \quad (6.28)$$

after integration by parts and use of Eq. (D.7).

Equation (6.28) is in a form convenient for numerical evaluation. Its inclusion in the integral equation (6.27) is beneficial for both convergence of the representation and conditioning of the resulting system of equations.

6.3. THE FINITE ELEMENT METHOD

Consider a volume V bounded by the surface S as depicted in Fig. 6.5. The medium is isotropic and characterized by (possibly complex) permittivity $\varepsilon_r \varepsilon$ and permeability $\mu_r \mu$, where ε_r and μ_r may be space-dependent. A prescribed source distribution $\mathbf{J}$ is present in V.

We compute the curl of Eq. (4.46a) and employ Eq. (4.46b) to yield

$$\mathbf{M} \cdot \mathbf{E} = -i\omega\mu\mathbf{J}, \qquad (6.29)$$

$$\mathbf{M} \to \nabla \times \frac{1}{\mu_r} \nabla \times - k^2 \varepsilon_r \mathbf{I}, \qquad k^2 = \omega^2 \varepsilon\mu,$$

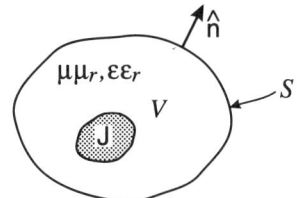

FIGURE 6.5. The finite element method.

and construct the functional[8]

$$\mathscr{F}(\mathbf{E}) = \frac{1}{2} \iiint_V dV (\mathbf{M} \cdot \mathbf{E}) \cdot \mathbf{E} + i\omega\mu \iiint_V dV \mathbf{J} \cdot \mathbf{E}. \qquad (6.30)$$

We compute the first-order variation $\delta\mathscr{F}$ of the functional with respect to the change $\delta\mathbf{E}$ of the vector $\mathbf{E}$,

$$\delta\mathscr{F}(\mathbf{E}) = \frac{1}{2} \iiint_V dV [(\mathbf{M} \cdot \mathbf{E}) \cdot \delta\mathbf{E} + (\mathbf{M} \cdot \delta\mathbf{E}) \cdot \mathbf{E}] + i\omega\mu \iiint_V dV \mathbf{J} \cdot \delta\mathbf{E},$$

and make use of the generalized Eq. (A.28) to obtain

$$\delta\mathscr{F}(\mathbf{E}) = \iiint_V dV [\mathbf{M} \cdot \mathbf{E} + i\omega\mu\mathbf{J}] \cdot \delta\mathbf{E}$$
$$+ \frac{1}{2} \oiint_S dS \frac{1}{\mu_r} [\delta\mathbf{E} \times \nabla \times \mathbf{E} - \mathbf{E} \times \nabla \times \delta\mathbf{E}] \cdot \hat{\mathbf{n}}. \qquad (6.31)$$

We conclude that the functional (6.30) is *stationary* with respect to $\mathbf{E}$,

$$\delta\mathscr{F}(\mathbf{E}) = 0, \qquad (6.32)$$

if the field is a solution of Eq. (6.29) with boundary conditions on S that render the surface integral in Eq. (6.31) equal to zero. Since

$$[\delta\mathbf{E} \times \nabla \times \mathbf{E} - \mathbf{E} \times \nabla \times \delta\mathbf{E}] \cdot \hat{\mathbf{n}} = \delta(\hat{\mathbf{n}} \times \mathbf{E}) \cdot \nabla \times \mathbf{E} - \hat{\mathbf{n}} \times \mathbf{E} \cdot \nabla \times \delta\mathbf{E}$$
$$= -i\omega\mu\mu_r [\delta\mathbf{E} \cdot \mathbf{H} \times \hat{\mathbf{n}} - \mathbf{E} \cdot \delta(\mathbf{H} \times \hat{\mathbf{n}})],$$

---

8. We use the symmetric scalar product. In some cases the complex scalar product may be useful; see Section 6.3.2.

the above condition is satisfied either when

$$\hat{\mathbf{n}} \times \mathbf{E} = 0 \quad \text{on } S, \quad (6.33a)$$

i.e., when the surface is metalized with a perfect conductor, or when

$$\mathbf{H} \times \hat{\mathbf{n}} = 0 \quad \text{on } S, \quad (6.33b)$$

i.e., when the surface S is a perfect magnetic conductor. Most generally, the condition is satisfied when

$$\nabla \times \mathbf{E} \times \hat{\mathbf{n}} = \alpha \hat{\mathbf{n}} \times \hat{\mathbf{E}} \times \hat{\mathbf{n}} \quad \text{on } S. \quad (6.33c)$$

Note that Eq. (6.33c) is consistent with the radiation condition (1.63) at infinity, as follows by taking $\alpha = -i\omega\mu/\zeta$ and $\mu_r = 1$.

The above discussion is important in as much as it shows that the electromagnetic field generated by prescribed sources and fulfilling boundary conditions (6.33) renders the functional (6.30) stationary. However, the latter is not in a convenient form for subsequent analysis, due to the presence of second-order spatial derivatives in the double curl operator. Further elaboration is desirable, and this is accomplished by making use of Eq. (A.27) which transforms Eq. (6.30) to the following equation:

$$\mathscr{F}(\mathbf{E}) = \frac{1}{2} \iiint_V dV \left[\frac{1}{\mu_r} \nabla \times \mathbf{E} \cdot \nabla \times \mathbf{E} - k^2 \varepsilon_r \mathbf{E} \cdot \mathbf{E} \right] + i\omega\mu \iiint dV \mathbf{J} \cdot \mathbf{E}$$

$$- \frac{1}{2} \oiint dS \frac{1}{\mu_r} \hat{\mathbf{n}} \cdot (\mathbf{E} \times \nabla \times \mathbf{E}). \quad (6.34)$$

The surface integral contribution disappears for perfect electric or magnetic boundaries [see Eqs. (6.33a) and (6.33b)], but this is not the case when radiation condition (6.33c) must be applied; see Section 6.3.3.

In order to implement FEM, we divide the volume V into nonoverlapping patches (see Section 6.3.1), or *elements*, within which convenient functions $\mathbf{Q}_p^e$ (see Section 6.3.1) are defined. The *local index* p numbers the basis vectors within each element, and the *element index* e numbers the elements within V. We represent the unknown field $\mathbf{E}$ in each element with the aid of

$$\mathbf{E}^e = \sum_p E_p^e \mathbf{Q}_p^e, \quad (6.35a)$$

where the summation is over the number of elements of the basis. Then, over

THE NUMERICAL DOMAIN 313

the whole volume V,

$$\mathbf{E} = \sum_e \sum_p E_p^e \mathbf{Q}_p^e, \qquad (6.35b)$$

where the new summation is over the number of elements. The two summations that appear in Eq. (6.35b) can be combined into a single one, by assigning a *global index* n to each basis vector:

$$\mathbf{E} = \sum_n E_n \mathbf{Q}_n. \qquad (6.35c)$$

Note, however, that each basis vector $\mathbf{Q}_n$ differs from zero only inside (and on the border of) its pertinent element.

We substitute Eq. (6.35c) into the functional (6.34) and enforce the stationary condition (6.32) by setting

$$\frac{\partial \mathcal{F}}{\partial E_s} = 0. \qquad (6.36)$$

Let us assume, for simplicity, the boundary condition $\hat{\mathbf{n}} \times \mathbf{E} = 0$ on S.[9] Substitution of expansion (6.35c) into Eq. (6.34) subject to condition (6.36) yields

$$\sum_m s_{nm} E_m = i\omega \mu j_n, \qquad (6.37a)$$

where s_{nm} and j_n are known scalars given by

$$s_{nm} = \iiint_V dV \left[\frac{1}{\mu_r} \nabla \times \mathbf{Q}_n \cdot \nabla \times \mathbf{Q}_m - k^2 \varepsilon_r \mathbf{Q}_n \cdot \mathbf{Q}_m \right], \qquad (6.38a)$$

$$j_n = -\iiint_V dV \mathbf{J} \cdot \mathbf{Q}_n. \qquad (6.38b)$$

Note that the former must be computed only once, when the geometry and element shape are prescribed.

If

$$\mathbf{S} \to \{s_{nm}\}, \quad \mathbf{J} \to \{j_n\}, \quad \text{and} \quad \mathbf{E} \to \{E_n\}$$

are a matrix and two column vectors, respectively, then Eq. (6.37a) can be

---

9. This condition can be easily relaxed; see Section 6.3.3.

expressed in the compact form

$$S \cdot E = i\omega\mu J, \qquad (6.37b)$$

to be supplemented (in general) by discretized boundary conditions on S.

We note that the elements of S, as well as those of J, are known quantities that can be computed analytically in most cases when the basis vectors Q_n have been properly chosen.

It is important to recognize that the matrix S is banded when local and element indexes have been properly assembled in the global index numbering system. In fact, when the two indexes n and m are sufficiently far apart (depending on the number of local indexes), the two functions Q_n and Q_m do not overlap and the element $S_{nm} = 0$. The banded property is important in that it is the prerequisite in the use of efficient procedures to invert the linear system (6.37b).

In conclusion inversion of the system (6.37) provides the discretized field E and its reconstruction via Eqs. (6.35).

6.3.1. Elements and Element Bases

The first step in the implementation of FEM is discretization of the field domain. This is a very important task because of its relevance to computer storage requirements, computational time, and also to the accuracy of numerical results. The domain is subdivided into small subdomains, or elements; some possible choices are referred to in Fig. 6.6 for one-, two-, and three-dimensional domains. Element choices increase with the dimensions of the domain. Most

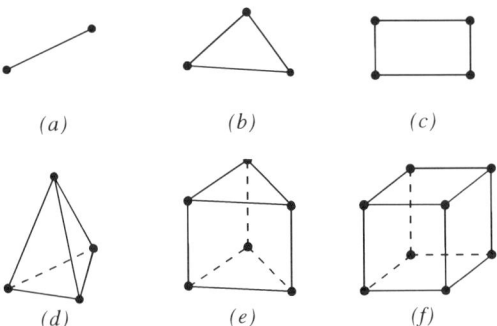

FIGURE 6.6. Finite elements appropriate to one- (a), two- (b, c), and three-dimensional (d, e, f) domains: (a) linear, (b) triangular, (c) rectangular, (d) tetrahedral element, (e) triangular prism, (f) rectangular brick elements.

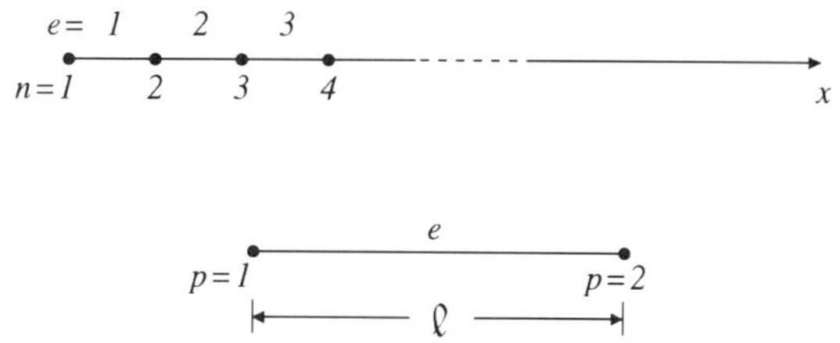

FIGURE 6.7. One-dimensional interpolation. Note the local (p), element (e), and global (n) addresses.

popular are the triangular (for two dimensions) and tetrahedral (for three dimensions) elements, because they can accommodate (at least approximately) curved boundaries. In all cases the elements are not required to maintain the same area: their dimension can be adjusted for a better fit of the boundaries, or better precision in regions of rapid field change.

The second step is the choice of interpolation functions, which should represent the field over the element. For scalar functions, such as field components, it is convenient to make use of *node-based* (scalar) interpolation, and the unknowns in the FEM are the values of the (scalar) field at the nodes of the elements. For vector functions it is more convenient to use *edge-based* (vector) interpolation, and the unknowns in the FEM are the components of the field along the edges of the element. In both cases the interpolation is selected to be of polynomial type (linear, quadratic, or of higher order). In the following we refer only to linear interpolation, which is simple but requires a larger number of elements for comparable numerical precision.

Let us first consider the scalar case.

In the simple one-dimensional case (see Fig. 6.7) we have on each element (whose address e is omitted)

$$E(x) = ax + b; \quad E(0) = E_1 = ax_1 + b, \quad E(l) = E_2 = ax_2 + b.$$

On replacing a and b by expressions in terms of E_1 and E_2, we obtain

$$E(x) = \frac{E_2 - E_1}{l} x - \frac{E_2 x_1 - E_1 x_2}{l} = E_1 \frac{x_2 - x}{l} + E_2 \frac{x - x_1}{l}$$

and (momentarily restoring the element address)

$$E^e(x) = \sum_{p=1}^{2} E_p^e Q_p^e(x), \tag{6.39}$$

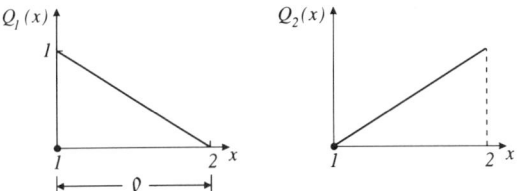

FIGURE 6.8. Basis functions for linear interpolation over a line element.

where the element bases are given by

$$Q_1(x) = \frac{x_2 - x}{l} \quad \text{and} \quad Q_2(x) = \frac{x - x_1}{l} \tag{6.40}$$

and presented in Fig. 6.8.

A similar procedure is followed for the two-dimensional case (see Fig. 6.9).[10] In this case we set

$$E(x) = ax + by + c,$$

and express the constants a, b, and c as functions of the (scalar) field values $E_1 = E(x_1)$, $E_2 = E(x_2)$, and $E_3 = E(x_3)$. On rearranging the final expressions we have

$$E^e(x) = \sum_{p=1}^{3} E_p^e Q_p^e(x), \tag{6.41}$$

where (omitting the element address)

$$Q_1(x) = \frac{1}{2\Delta}[\alpha_1 + \beta_1 x + \gamma_1 y] \tag{6.42}$$

with

$$\alpha_1 = x_2 y_3 - y_2 x_3, \quad \beta_1 = y_2 - y_3, \quad \gamma_1 = x_3 - x_2. \tag{6.43}$$

The element area is indicated by Δ and the other basis elements are obtained by the index permutation $1 \to 2 \to 3 \to 1$, all of them presented graphically in Fig. 6.10.

Scalar bases for the three-dimensional case are obtained similarly.

---

10. We consider triangular elements and interpolation functions with the three (unknown) values of the field at the three nodes. Should rectangular elements be used, bilinear interpolation functions may be used to include the four values of the field at the four nodes.

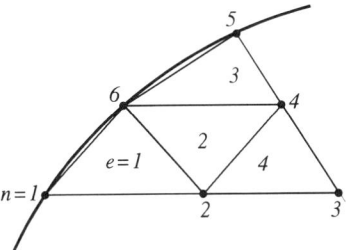

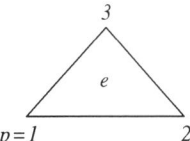

FIGURE 6.9. Two-dimensional interpolation over tetrahedral elements. Note again the local (p), element (e), and global (n) addresses.

We turn now to the vector case, where edge-based vector interpolation is most convenient, because their appropriate choice allows imposition of the condition $\nabla \cdot \varepsilon_r \mathbf{E} = 0$ in source-free regions. This is already enforced in the variational approach, as it follows by taking the divergence of Eq. (6.29), but it is *not* so in the FEM implementation, because the trial function is expanded in bases whose first derivatives are not necessarily continuous on the element boundaries. For this reason the solution of Eq. (6.32) is referred to as the *weak solution* and spurious nonphysical fields may be added to the physical ones.

The situation may be corrected if edge-based divergence-free vectors are used to represent the trial field. In the two-dimensional case with triangular elements (see Fig. 6.11), let us consider the vector (again omitting the global index)

$$\mathbf{Q}_{12} = Q_1 \nabla Q_2 - Q_2 \nabla Q_1,$$

where Q_1 and Q_2 are the (node-based) interpolation functions [Eq. (6.42)]. Hence

$$\nabla \cdot \mathbf{Q}_{12} = 0,$$

as follows by using Eq. (A.10) and noting that $\nabla^2 Q_1 = \nabla^2 Q_2 = 0$ [see Eq. (6.42)].

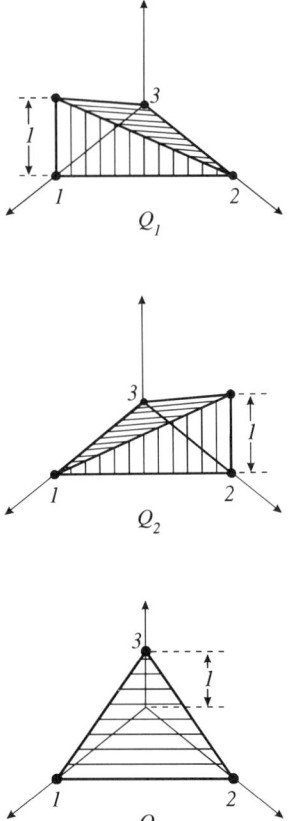

FIGURE 6.10. Plots of the three basis functions for tetrahedrl element.

The component of $\mathbf{Q}_{12}$ along the edge $p = 1$ (see Fig. 6.11) is now computed. Examination of Fig. 6.10 shows that Q_1 and Q_2 change linearly from 0 to 1, and from 1 to 0, respectively, on moving from node 2 to node 1:

$$\hat{\mathbf{t}}_1 \cdot \nabla Q_1 = -\frac{1}{l_1}, \quad \hat{\mathbf{t}}_2 \cdot \nabla Q_2 = \frac{1}{l_1},$$

with

$$Q_1 + Q_2 = 1,$$

on the edge $p = 1$. Accordingly

$$\mathbf{Q}_{12} \cdot \hat{\mathbf{t}}_1 = \frac{1}{l_1}$$

THE NUMERICAL DOMAIN

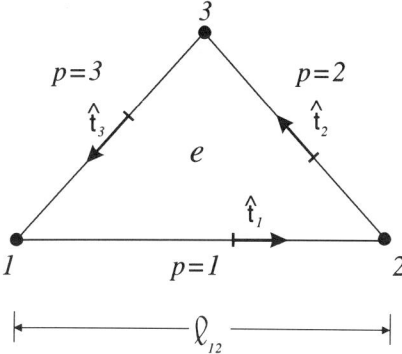

FIGURE 6.11. Edge-based vector bases. Note that the local address (p) now refers to the edge rather than to the node.

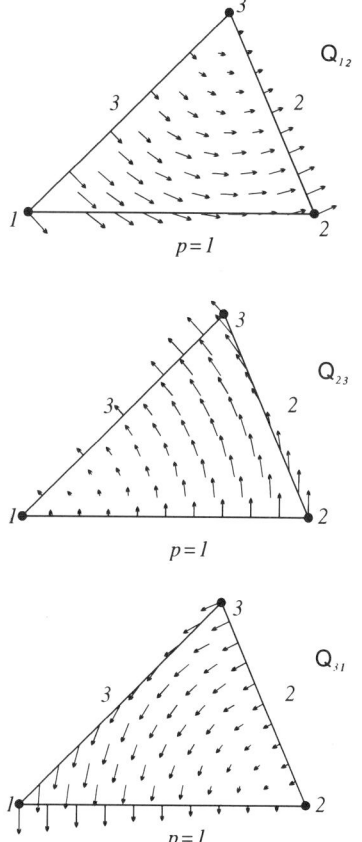

FIGURE 6.12. Vector basis functions for the triangular element.

on the edge $p = 1$ and the vector $\mathbf{Q}_{12}$ exhibits a constant component there. In addition

$$\mathbf{Q}_{12} \cdot \hat{\mathbf{t}}_2 = 0 = \mathbf{Q}_{12} \cdot \hat{\mathbf{t}}_3,$$

on the edges $p = 2$ and $p = 3$, respectively, because Q_1 and Q_2 are zero there (see Fig. 6.11 again). We conclude that the (adimensional) vectors

$$\mathbf{Q}_1^e = l_1 \mathbf{Q}_{12}^e, \qquad \mathbf{Q}_2^e = l_2 \mathbf{Q}_{23}, \qquad \mathbf{Q}_3^e = l_3 \mathbf{Q}_{31}$$

form convenient bases for representing a divergence-free trial field on the element. This basis is represented pictorially in Fig. 6.12.

Divergence-free vector basis functions for the three-dimensional case are obtained similarly.

6.3.2. Guided-Wave Propagation

Consider the uniform isotropic waveguide of arbitrary cross section depicted in Fig. 6.13. We assume the $\exp(-i\beta z)$ z-dependence for the fields (see Section 4.6):

$$\mathbf{E}(x, y, z) \to \mathbf{E}(x, y) \exp(-i\beta z),$$

with boundary condition

$$\hat{\mathbf{n}} \times \mathbf{E} = 0$$

on C.

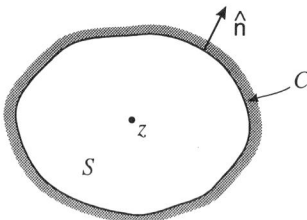

FIGURE 6.13. Uniform waveguide with arbitrary cross section and space-varying relative permittivity $\varepsilon_r(x, y)$.

THE NUMERICAL DOMAIN

In this case the functional (6.34) can be limited to the cross section of the guide and Eq. (6.30) becomes

$$\mathscr{F}(\mathbf{E}) = \iint_S dS[\nabla \times \mathbf{E} \cdot \nabla \times \mathbf{E}^* - k^2 \varepsilon_r \mathbf{E} \cdot \mathbf{E}^*], \quad (6.44)$$

where we have assumed $\mu_r = 1$ and the complex conjugate scalar product.

Equation (6.44) shows that the functional $\mathscr{F}(\mathbf{E})$ is a real quantity (if ε_r is real) with the chosen (complex conjugate) scalar product, equal (but for an inessential factor) to the electromagnetic energy per unit length of the guide.[11]

It is convenient to decompose the electric field in its transverse, $\mathbf{E}_t$, and longitudinal, $E_z \hat{\mathbf{z}}$, components:

$$\mathbf{E} = \mathbf{E}_t + E_z \hat{\mathbf{z}},$$
$$\nabla \times \mathbf{E} = \nabla_t \times \mathbf{E}_t + \nabla_t E_z \times \hat{\mathbf{z}} - i\beta \mathbf{E}_t \times \hat{\mathbf{z}},$$

in which case Eq. (6.44) transforms to

$$\mathscr{F}(\mathbf{E}) = \iint_S dS[\nabla_t \times \mathbf{E}_t \cdot \nabla_t \times \mathbf{E}_t^*$$
$$+ (\nabla_t E_z - i\beta \mathbf{E}_t) \cdot (\nabla_t E_z - i\beta \mathbf{E}_t)^* - k^2 \varepsilon_r \mathbf{E} \cdot \mathbf{E}^*].$$

If we refer to normalized lengths, $x \to \beta x, \ldots$, and let $E_z \to -iE_z$, then Eq. (6.44) (apart from the inessential multiplicative factor β^2) becomes

$$\mathscr{F}(\mathbf{E}) = \iint_S dS\{\beta^2 [\nabla_t \times \mathbf{E}_t \cdot \nabla_t \times \mathbf{E}_t^* - (\nabla_t E_z + \mathbf{E}_t) \cdot (\nabla_t E_z + \mathbf{E}_t)^*] - k^2 \varepsilon_r \mathbf{E} \cdot \mathbf{E}^*\}. \quad (6.45)$$

Equation (6.45) is in the appropriate form for FEM implementation. Vector and scalar fields can be expanded in the vector and scalar bases of Section 6.3.1 and a matrix equation in the eigenvalue β^2 is obtained.

6.3.3. Absorbing Boundary Conditions

Consider the scattering problem depicted in Fig. 6.14. The incident field $(\mathbf{E}_i, \mathbf{H}_i)$ is impinging on a perfectly conducting metal body which scatters the

---

11. In this case, stationarity of the functional (6.44) can be identified with a maximum or minimum. However, we can always change the sign of $\mathscr{F}(\mathbf{E})$, thus ensuring that the condition $\delta \mathscr{F} = 0$ is a minimum of $\mathscr{F}(\mathbf{E})$.

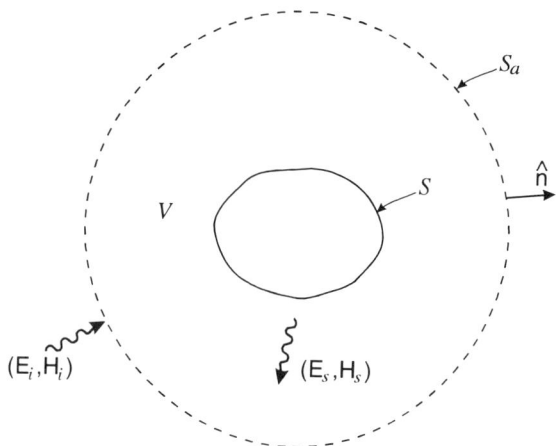

FIGURE 6.14. A scattering problem to be studied via FEM.

field ($\mathbf{E}_S, \mathbf{H}_S$). The surrounding medium can be lossy and inhomogeneous, and is unbounded.

Although there is no restriction on the size of the volume V in the functional of Eq. (6.34), practical implementation of FEM requires the volume to be finite. The simplest possibility is to limit the volume by a spherical surface[12] S_a sufficiently remote from the scatterer so that the radiation condition (see Section 1.4.2) for the scattered field can be applied therein.

The requirement about the size of the sphere may be too onerous, leading to an exceedingly large dimension of the final matrix FEM implementation. More elaborate radiation conditions, referred to as *absorbing boundary conditions*, can be used for smaller distances from the scatterer at the expense of a more elaborate formulation of the problem. In the following we consider the simplest case of a conventional radiation condition at infinity.

We wish to derive a stationary functional representation for the scattered field. We start from the formulation of Eq. (6.30) specified for the field $\mathbf{E}_S$ ($\mathbf{J} = 0$),

$$\mathscr{F}(\mathbf{E}_S) = \frac{1}{2} \iiint\limits_V dV \left[\left(\nabla \times \frac{1}{\mu_r} \nabla \times \mathbf{E}_S \right) \cdot \mathbf{E}_S - k^2 \varepsilon_r \mathbf{E}_S \cdot \mathbf{E}_S \right],$$

but we should check if the surface integral in Eq. (6.31) is equal to zero also in the case under consideration. In fact, $\mathbf{E}_S$ is *not* the total electric field.

---

12. In many cases a spherical surface is not the most convenient choice, and appropriate modifications of the radiation condition must be introduced.

THE NUMERICAL DOMAIN

In Fig. 6.14, we see that the surface bounding the volume V is the sum of the surfaces of the metal scatterer, S, and of the large sphere, S_a. The radiation conditions apply on the latter surface, and the surface integral in Eq. (6.31) is equal to zero; see Eq. (6.33c) and the related discussion. This is at variance with the surface S, where the tangential component of the scattered field should compensate for the same component of the incident field:

$$\hat{\mathbf{n}} \times \mathbf{E}_S = -\hat{\mathbf{n}} \times \mathbf{E}_i.$$

Accordingly, $\hat{\mathbf{n}} \times \mathbf{E}_S$ is prescribed over S, with the requirement that $\delta \mathbf{E}_S$ is zero there. We conclude that the functional $\mathscr{F}(\mathbf{E}_S)$ is stationary, like that of Eq. (6.30), and we make use of Eq. (A.27) to obtain

$$\mathscr{F}(\mathbf{E}_S) = \frac{1}{2} \iiint_V dV \left(\frac{1}{\mu_z} \nabla \times \mathbf{E}_S \cdot \nabla \times \mathbf{E}_S - k^2 \varepsilon_r \mathbf{E}_S \cdot \mathbf{E}_S \right)$$
$$- \frac{1}{2} \iint_{S+S_a} dS \hat{\mathbf{n}} \cdot (\mathbf{E}_S \times \nabla \times \mathbf{E}_S) \frac{1}{\mu_r}. \quad (6.46)$$

Let us examine the surface integral in Eq. (6.46). We assume $\mu_r = 1$ at large distances from the scatterers and enforce the boundary conditions

$$\hat{\mathbf{n}} \times \nabla \times \mathbf{E}_S = \frac{i\omega\mu}{\zeta} \hat{\mathbf{n}} \times \mathbf{E}_S \times \hat{\mathbf{n}} \quad \text{on } S_a.$$

As far as the second surface integral is concerned, we consider again the total field,

$$\mathbf{E} = \mathbf{E}_S + \mathbf{E}_i,$$

and manipulate the integrand as follows:

$$\hat{\mathbf{n}} \cdot (\mathbf{E}_S \times \nabla \times \mathbf{E}_S) = \hat{\mathbf{n}} \cdot [(\mathbf{E} - \mathbf{E}_i) \times \nabla \times (\mathbf{E} - \mathbf{E}_i)]$$
$$= (\hat{\mathbf{n}} \times \mathbf{E}) \cdot \nabla \times (\mathbf{E} - \mathbf{E}_i) - (\hat{\mathbf{n}} \times \mathbf{E}_i) \cdot \nabla \times \mathbf{E} + \hat{\mathbf{n}} \cdot (\mathbf{E}_i \times \nabla \times \mathbf{E}_i). \quad (6.47)$$

Boundary conditions on the metal scatterer make the first and second terms in Eq. (6.47) equal to zero, because $\hat{\mathbf{n}} \times \mathbf{E} = 0$ on S and $-\hat{\mathbf{n}} \times \mathbf{E}_i \cdot \nabla \times \mathbf{E} = \hat{\mathbf{n}} \times \mathbf{E} \cdot \nabla \times \mathbf{E}_i$, see Eq. (A.9) with $\mathbf{E} = E\hat{\mathbf{n}}$. The last term is fixed and does not play any role in the enforcement of the functional stationary condition. Accordingly, and omitting inessential factors, FEM can be applied to the functional

$$\mathcal{F}(E_S) = \iiint_V dV \left[\frac{1}{\mu_r} \nabla \times \mathbf{E}_S \cdot \nabla \times \mathbf{E}_S - k^2 \varepsilon_r \mathbf{E}_S \cdot \mathbf{E}_S \right]$$
$$+ ik \oiint_{S_a} dS(\hat{\mathbf{n}} \times \mathbf{E}_S) \cdot (\hat{\mathbf{n}} \times \mathbf{E}_S) \quad (6.48a)$$

together with the boundary conditions

$$\hat{\mathbf{n}} \times \mathbf{E}_S = -\hat{\mathbf{n}} \times \mathbf{E}_i \quad \text{on } S. \quad (6.48b)$$

6.4. THE FINITE DIFFERENCE METHOD

As stated in Section 6.1, the FDM transforms Maxwell's equations from their differential form to their *finite difference* form. This is accomplished by approximating the differential operator by means of its incremental ratio. For any differentiable function $f(\xi)$ we have

$$f(\xi \pm \Delta\xi) = f(\xi) \pm f'(\xi)\Delta\xi + f''(\xi)\frac{(\Delta\xi)^2}{2}$$

and

$$f(\xi + \Delta\xi) - f(\xi - \Delta\xi) = f'(\xi)2\Delta\xi + 0[(\Delta\xi)^3].$$

Accordingly

$$f'(\xi) = \frac{f(\xi + \Delta\xi) - f(\xi - \Delta\xi)}{2\Delta\xi} + 0[(\Delta\xi)^2]. \quad (6.49)$$

Equation (6.49) is applied to both time and space derivatives. In this case the method is also referred to as the finite difference time domain (FDTM).

Consider plane-wave propagation [Eq. (2.1b)] along the z-axis. Let

$$t_n = n\Delta t \quad \text{and} \quad z_m = m\Delta z$$

be the discretized values of the time and space coordinates. It is convenient to link the electric and magnetic fields by computing the former at times $(n+1/2)\Delta t$ and spaces $m\Delta z$, and the latter at times $(n\Delta t)$ and spaces $(m+1/2)\Delta z$

THE NUMERICAL DOMAIN

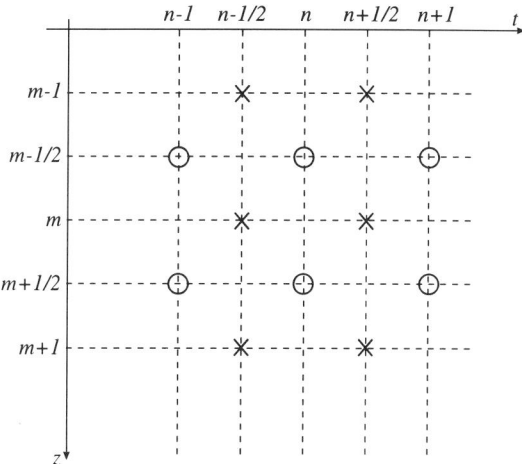

FIGURE 6.15. Leapfrog scheme for FDM. Electric fields are represented by crosses and magnetic fields by circles.

(see Fig. 6.15). Set

$$e[(n + 1/2)\Delta t, m\Delta z] = e_m^{n+1/2} \quad \text{and} \quad h[n\Delta t, (m + 1/2)\Delta z] = h_{m+1/2}^n.$$

Then Eqs. (2.1b) are discretized as follows:

$$\begin{cases} e_m^{n+1/2} = e_m^{n-1/2} - \dfrac{\Delta t}{\varepsilon_m \Delta z}(h_{m+1/2}^n - h_{m-1/2}^n), \\ h_{m+1/2}^n = h_{m+1/2}^{n-1} - \dfrac{\Delta t}{\mu_{m+1/2}\Delta z}(e_{m+1}^{n-1/2} - e_m^{n-1/2}), \end{cases} \quad (6.50)$$

where ε and μ may be space-dependent, while

$$\varepsilon_m = \varepsilon[m\Delta z] \quad \text{and} \quad \mu_{m+1/2} = \mu[(m + 1/2)\Delta z].$$

Equations (6.50) are usually referred to as the *leapfrog scheme* and, with their three-dimensional generalization, allow space–time updating of the fields. Note that the time-updated value of one field, say $h_{m+1/2}^n$, is computed by using the values of the other field at the preceding time, $(n - 1/2)\Delta t$, but displaced spatially to the right, $(m + 1)\Delta z$, and to the left, $m\Delta z$. Note again that the fields are computed over two interlaced grids.

Spatial interlacing of the fields in the three-dimensional case is depicted later in Fig. 6.17 for phasor fields (see Section 6.4.2).

Some comments about the choice of the elementary cell size, Δz, and Δt are in order. Clearly, this choice depends on the space–time shape of the source field and on the spatial change of the medium where the field propagates. In general, the *numerical* (space or time) *frequency*, i.e., $1/\Delta z$ or $1/\Delta t$, should be small compared to the maximum (space or time) frequency, $1/\lambda$ or $1/T$, respectively, of the electromagnetic signal; a ratio of between 1/10 and 1/20 is usually sufficient. However, this is a statement which is not always easy to apply, because to guess the spectral characteristic of the desired solution is not an easy matter. A cut-and-try procedure is normally applied, by comparing results obtained with different discretization, until convergence is reached.

The choice of cell size has an impact on the stability and numerical dispersion of the solution, as discussed in the next section.

6.4.1. Stability and Numerical Dispersion

Let us consider a homogeneous medium and a z-dependence for the fields of $\exp(-ikz)$ type,

$$e_p^q \to e^q \exp(-ikp\Delta z), \qquad h_s^r \to h^r \exp(-iks\Delta z).$$

Substitution into Eqs. (6.50) yields

$$\begin{cases} e^{n+1/2} = e^{n-1/2} + \dfrac{\Delta t}{\varepsilon \Delta z} 2i \sin(k\Delta z/2) h^n, \\ h^n = h^{n-1} + \dfrac{\Delta t}{\mu \Delta z} 2i \sin(k\Delta z/2) e^{n-1/2}, \end{cases}$$

which can be expressed in the following form:

$$\begin{vmatrix} e^{n+1/2} \\ h^n \end{vmatrix} = \boldsymbol{T} \cdot \begin{vmatrix} e^{n-1/2} \\ h^{n-1} \end{vmatrix},$$

with

$$\boldsymbol{T} \to \begin{vmatrix} 1 - 4\left(\dfrac{c\Delta t}{\Delta z}\right)^2 \sin^2 \dfrac{k\Delta z}{2} & 2i \sin \dfrac{k\Delta z}{2} \\ 2i \sin \dfrac{k\Delta z}{2} & 1 \end{vmatrix} \qquad (6.51)$$

We require the time-updated field values to be a phase-shifted replica of the corresponding previous values:

$$e^{n+1/2} = \xi e^{n-1/2}, \qquad h^n = \xi h^{n-1},$$

$$\xi = \exp(i\omega \Delta t). \qquad (6.52)$$

Accordingly

$$\det[\mathbf{T} - \xi \mathbf{I}] = 0,$$

$$\xi = \left(1 - 2\eta^2 \sin^2 \frac{k\Delta z}{2}\right) \pm i\sqrt{1 - \left(1 - 2\eta^2 \sin^2 \frac{k\Delta z}{2}\right)^2},$$

$$\eta = \frac{c\Delta t}{\Delta z}. \tag{6.53}$$

When condition (6.52) is applied, we obtain

$$\cos \omega \Delta t = 1 - 2\eta^2 \sin^2 \frac{k\Delta z}{2} \tag{6.54a}$$

together with

$$\eta^2 \leq 1. \tag{6.54b}$$

Equation (6.54b) corresponds to the *stability condition*; only if $c\Delta t \leq \Delta z$ is the time-updating process described by Eqs. (6.50) stable. Should this condition not be met ω may acquire an imaginary part, in which case the updating process diverges.

Equation (6.54a), which may be written in its equivalent form

$$\frac{1 - \cos \omega \Delta t}{2} = \sin^2 \frac{\omega \Delta t}{2} = \eta^2 \sin^2 \frac{kc\Delta t}{2\eta}, \tag{6.55}$$

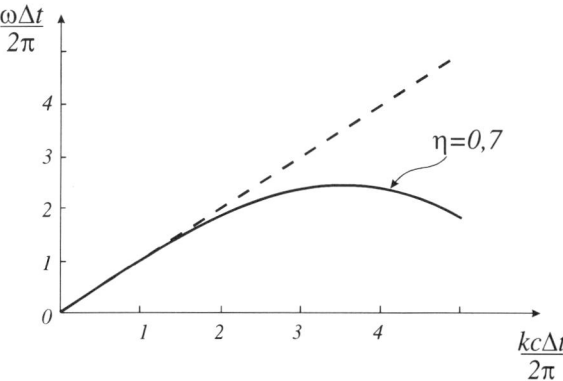

FIGURE 6.16. Dispersion equation for free-space propagation in the discretized space. The dashed line refers to the continuous case.

is the *dispersion equation* of the discretized wave propagation. When $\eta \to 0$ we recover the (continuum) dispersion equation,

$$\omega^2 = (kc)^2,$$

as well as when $\eta = 1$ (theoretical case $c\Delta t = \Delta z$). It is noted that propagation becomes dispersive as η increases (see Fig. 6.16), although the medium is local (see Section 1.2) and therefore nondispersive. Dispersion is an artifact induced by the discretization, and can be controlled by an appropriate choice of $c\Delta t$ compared to Δz.

6.4.2. FDM in the Frequency Domain

Although FDM is particularly suited to time-domain implementation, its use in the transformed frequency domain is certainly possible. The time-domain solution can be recovered, if desired, by FT inversion.

We start from Maxwell equations (3.15) and define a cartesian grid

$$x_n = n\Delta x, \qquad y_m = m\Delta y, \qquad z_s = s\Delta z; \tag{6.56}$$

see Fig. 6.17. If we set

$$E_x(n\Delta x, m\Delta y, s\Delta z) = E_x(n, m, s),$$

and similarly for the other fields and source components, then

$$[E_z(n, m+1, s+1/2) - E_z(n, m, s+1/2)] - [E_y(n, m+1/2, s+1) - E_y(n, m+1/2, s)]$$
$$= -i\omega\mu(n, m+1/2, s+1/2)H_x(n, m+1/2, s+1/2),$$
$$[H_z(n+1/2, m+1/2, s) - H_z(n+1/2, m-1/2, s)] - [H_y(n+1/2, m, s+1/2)$$
$$- H_y(n+1/2, m, s-1/2)]$$
$$= i\omega\varepsilon(n+1/2, m, s)E_x(n+1/2, m, s) + J_{0x}(n+1/2, m, s), \tag{6.57}$$
$$\cdots \cdots \cdots \cdots$$

which, together with the appropriate boundary conditions, provide the linear system to be inverted for field sample retrieval.

In Eqs. (6.57) the left-hand sides represent the numerical evaluation of the curl in component form. Note that the field components are interlaced, as shown in Fig. 6.17.

THE NUMERICAL DOMAIN

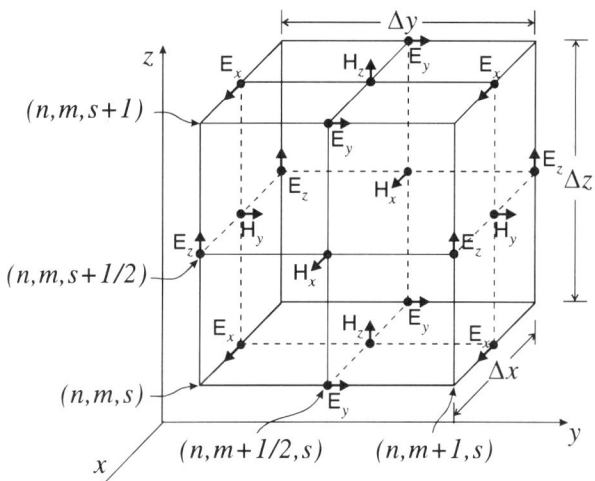

FIGURE 6.17. The interlaced parallelepiped grid for the implementation of FDM in rectangular coordinates.

6.5. SUMMARY AND SELECTED REFERENCES

This chapter is devoted to the presentation of some of the techniques used for numerical solution of Maxwell's equations: the method of moments (MoM, Section 6.2), the finite element method (FEM, Section 6.3), and the finite difference method (FDM, Section 6.4). Each of these methods, and of their many extensions and modifications, have been implemented in the form of solvers, i.e., appropriate numerical codes, which are not provided in this chapter but are usually available. However, it is important to be aware of the techniques [6.1–6.3] adopted and of their limitations, and this is the scope of the chapter. It is also noted that the last step in the majority of procedures is to invert a linear system of equations. A brief discussion on the subject [6.4] is given in Sections 6.1.2 to 6.1.4.

MoM [6.5, 6.6] was traditionally first used in electromagnetics, and is considered in Section 6.2. Electric (EFIE) and magnetic (MFIE) field integral equations are derived in Section 6.2.1 and the importance of appropriate choices for expansion and test functions is discussed in Section 6.2.2.

Most recently, much attention has been devoted to FEM [6.7, 6.8] which is examined in Section 6.3. Both node-based and edge-based interpolation functions are presented, emphasizing the importance of the latter in constructing a divergence-free field. This is an important tool with which to avoid spurious cavity modes [6.9, 6.10]. For the important case of unbounded structures (Section 6.5.3), only the conventional radiation condition is considered; more elaborate approaches are discussed in [6.7] and in [6.11, 6.12]

FDM [6.13], based on the Yee scheme [6.14], is considered in Section 6.4, together with its stability and inherent artifacts (Section 6.4.1).

The discipline of numerical solution of Maxwell's equation is a rapidly expanding area, and only the very basic concepts have been presented in this chapter. Nevertheless, it is believed that this minimal knowledge is an important prerequisite for serious use of available electromagnetic solvers.

References

[6.1] D. S. Jones, *Methods in Electromagnetic Wave Propagation*, IEEE Press, New York (1994).
[6.2] M. N. O. Sadiku, *Numerical Techniques in Electromagnetics*, CRC Press, Boca Raton (1992).
[6.3] S. Nakamura, *Computational Methods in Engineering and Science*, Wiley, New York (1977).
[6.4] G. H. Golub and C. F. Van Loan, *Matrix Computations*, Johns Hopkins University Press (1989).
[6.5] R. F. Harrington, *Field Computation by the Moment Method*, McMillan, New York (1968).
[6.6] J. J. H. Wang, *Generalized Moment Method in Electromagnetics*, Wiley, New York (1991).
[6.7] J. Jin, *The Finite Element Method in Electromagnetics*, Wiley, New York (1993).
[6.8] G. Pelosi and P. Silvester, *Finite Element Wave Electromagnetics*, IEEE Press, New York (1994).
[6.9] K. Hayata, M. Koshiba, M. Eguchi, and M. Suzuki, "Vectorial finite-element method without any spurious solution for dielectric waveguiding problems using transverse magnetic field component," *IEEE Trans. Microwave Theory Tech.* **MTT-34**, 1120–1124 (1986).
[6.10] A. J. Kobelanski and J. P. Webb, "Eliminating spurious modes in finite-element waveguide problems by using divergence-free fields," *Electron. Lett.* **22**, 569–570 (1986).
[6.11] G. Mur, "Absorbing boundary conditions for the finite difference approximation of the time domain electromagnetic field equations," *IEEE Trans. Elec. Comp.* **EMC-23**, 377–382 (1981).
[6.12] A. Bayliss, M. Gunzburger, and E. Turkel, "Boundary conditions in the numerical solution of electric equations in exterior regions," *SIAM J. Appl. Math.* **42**, 430–451 (1982).
[6.13] K. Kunz and D. J. Loebbers, *The Finite Difference Time-Domain Method for Electromagnetics*, CRC Press, Boca Raton (1993).
[6.14] K. S. Yee, "Numerical solution of initial boundary value problems in solving Maxwell equations in isotropics media," *IEEE Trans. Antennas Propagat.* **AP-14**, 302–307 (1986).

7
Engineering Topics: Propagation

7.1. GENERAL CONSIDERATIONS

In this and the following two chapters, electromagnetic propagation, radiation, and scattering is treated from an engineering point of view. Accordingly, this section should be considered as an introduction to Chapters 8 and 9, too.

The foundations of electromagnetic theory were presented earlier in chapters 1 and 2 and solution techniques in Chapters 3 through 6. This has been supplemented by a large number of case studies, highlighting the physics of the model, the computational procedures, and the properties of the solution. All the above material is a necessary background for any engineer operating in the area of applied electromagnetics. There is no doubt that he/she should understand and be aware of all the physics involved in his/her design; but his/her main duty is to face an overall systems problem whose performance must be judged in terms of simple, possibly scalar, parameters. It is also highly desirable that these parameters should be related to observable quantities—the *integral measure* of the fields—that render the description of the physical process simple and attractive, with emphasis on only those features that play a dominant role in the design problem. As an example, a (partial) list of engineering observables and parameters applicable in applied electromagnetics is given in Table 7.1.

This chapter is devoted to electromagnetic propagation from an engineering standpoint. Here and in the subsequent two chapters most of the presentation is relative to a narrowband field, and phasor formalism is used.

TABLE 7.1. Engineering Observables and Parameters in Applied Electromagnetics

Observable or parameter	Description
Voltage, current	Appropriate integral measure of electric and magnetic field. Used in equivalent circuits.
Impedance, admittance	Ratio between voltage and current phasors, or vice versa. Used in equivalent circuits.
Directivity, gain	Describe the ability of a radiator to focus the electromagnetic energy.
Receiving area	Describes the ability of an antenna to collect incoming electromagnetic radiation.
Scattering cross section	Describes the ability of an object to scatter incoming electromagnetic energy.
Quality factor	Describes the ability of a cavity to sustain its resonant oscillation

7.2. TRANSMISSION LINES

The oldest hardware implementation of a *transmission line* (TL) system is composed of two parallel metal wires driven by either a voltage or a current generator, as depicted in Fig. 7.1. The equations that describe the space–time change of electromagnetic perturbation along the wires are derived here. Their importance resides in their ability to model a large variety of electromagnetic propagation situations.

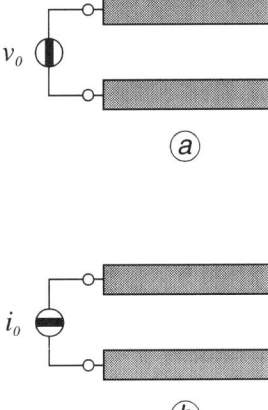

FIGURE 7.1. A two-wire TL driven by (A) a voltage and (B) a current generator.

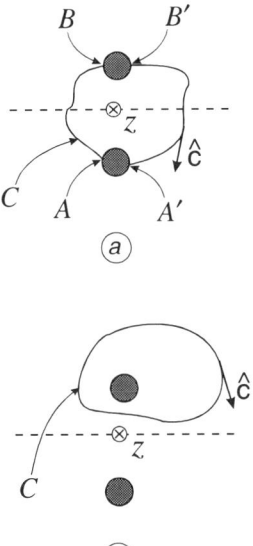

FIGURE 7.2. Cross section of the two-wire TL.

With reference to Fig. 7.2, we first derive simple scalar measures of the electromagnetic field. We assume the two wires to be perfectly conducting and exploit electromagnetic propagation which is TEM with respect to the propagation z-axis. The surrounding medium is isotropic and homogeneous. In this case (Fig. 7.2a)

$$\oint_C d c \mathbf{e} \cdot \hat{\mathbf{c}} = 0 = \int_A^B d c \mathbf{e} \cdot \hat{\mathbf{c}} + \int_{B'}^{A'} d c \mathbf{e} \cdot \hat{\mathbf{c}}$$

[see Eq. (1.5)], because $h_z = b_z = 0$ (TEM wave) and the line integral along BB' and $A'A$ is zero (perfectly conducting wires). Accordingly, the line integral

$$v(z, t) = -\int_A^B d c \mathbf{e} \cdot \hat{\mathbf{c}} \qquad (7.1)$$

is uniquely defined, being independent of the endpoints A and B on the two wires and the integration path. Note that the integration path must lie on the transverse plane $z = $ const. The scalar $v(z, t)$ is referred to as the *TL voltage*, and is an observable quantity.

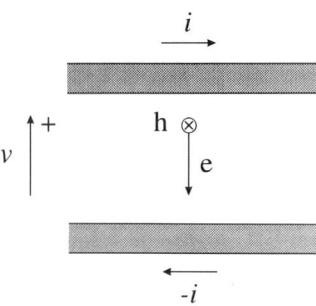

FIGURE 7.3. Definition of positive and negative values of voltages and currents along the TL.

If $i(z, t)$ is the flux along $\hat{z}$ of the surface current over the wire, then we also have (Fig. 7.2b)

$$\oint_C d c \mathbf{h} \cdot \hat{\mathbf{c}} = i(z, t) \tag{7.2}$$

[see Eq. (1.6)], because $e_z = d_z = 0$ (TEM wave). The integral (7.2) does not depend on the integration contour C provided that it does not intersect or include the second wire, thus uniquely defining the scalar $i(z, t)$. This is an observable quantity and is referred to as the *TL current*. As before, the integration path is constrained to lie in the transverse plane. For a source excitation odd[1] with respect to the symmetry plane, the current along the second wire is just equal to $-i$. Positive values of voltages and currents are depicted in Fig. 7.3.

Consider now a longitudinal section of the TL, as depicted in Fig. 7.4. Equation (1.5) applied to the closed path $A'ABB'A'$ yields

$$\oint_C d c \mathbf{e} \cdot \hat{\mathbf{c}} = -v(z, t) + v(z + \Delta z, t) = -\frac{\partial \psi}{\partial t} \Delta z, \tag{7.3a}$$

where ψ [weber/m] is the flux of the magnetic induction within the two wires, per unit TL length. Similarly, Eq. (1.8) applied to the volume ΔV of Fig. 7.4 containing one wire yields

$$-i(z, t) + i(z + \Delta z, t) + \frac{\partial q}{\partial t} \Delta z, \tag{7.3b}$$

---

1. Even excitation is also possible, in principle. In this case the voltage of the two wires is the same, and is measured with respect to its value at infinity. The contour C encircles the two wires and the TL current is the sum of the two currents. Even excitation is common in coaxial cables; see Section 7.4.3.

ENGINEERING TOPICS: PROPAGATION

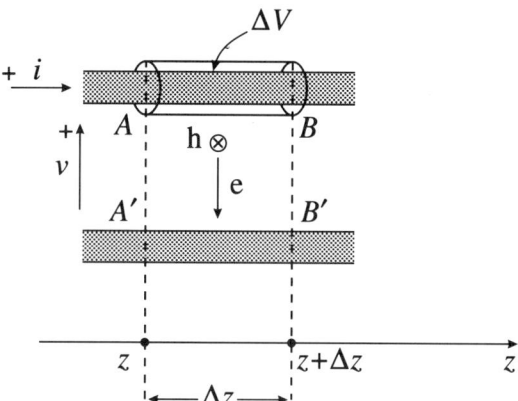

FIGURE 7.4. Longitudinal section of the TL.

where q [coulomb/m] is the charge over the wire, per unit TL length. This charge equals the flux ϕ [coulomb/m] of the electric induction, normal to the wire [see Eq. (1.3)], per unit TL length.

In the limit $\Delta z \to 0$, Eqs. (7.3) reduce to

$$\begin{cases} -\dfrac{\partial v}{\partial z} = \dfrac{\partial \psi}{\partial t}, \\ -\dfrac{\partial i}{\partial z} = \dfrac{\partial \phi}{\partial t}, \end{cases} \qquad (7.4)$$

which are the most general form of the *transmission line equations*.

The next step is to set up a relation between voltages and currents on the one hand, and electric and magnetic induction fluxes on the other, i.e., to set up appropriate constitutive relations (see Section 1.2). If

$$\psi = Li \quad \text{and} \quad \phi = Cv, \qquad (7.5)$$

where L [henry/m] and C [coulomb/m] are the induction and capacitance per unit length of the TL, respectively, then Eqs. (7.4) become

$$\begin{cases} -\dfrac{\partial v}{\partial z} = L\dfrac{\partial i}{\partial t}, \\ -\dfrac{\partial i}{\partial z} = C\dfrac{\partial v}{\partial t}, \end{cases} \qquad (7.6)$$

when the system is time-invariant (L and C do not change with t); or

$$\begin{cases} -\dfrac{\partial \psi}{\partial t} = \dfrac{1}{C}\dfrac{\partial \psi}{\partial z}, \\ -\dfrac{\partial \phi}{\partial t} = \dfrac{1}{L}\dfrac{\partial \phi}{\partial z}, \end{cases} \quad (7.7)$$

when the system is space-invariant (L and C do not change with z).[2]

Examination of Section 2.2 shows that v, i are the corresponding scalars of the fields $\mathbf{e}, \mathbf{h}$ for time-invariant media, while ϕ, ψ are the corresponding scalars of the inductions $\mathbf{d}, \mathbf{b}$ for space-invariant media.

When L and C are constant, Eqs. (6.6) and (6.7) can be easily integrated. The solution to Eqs. (7.6) is

$$\begin{cases} v(z, t) = v^+(ct - z) + v^-(ct + z), \\ R_0 i(z, t) = v^+(ct - z) - v^-(ct + z), \end{cases} \quad (7.8)$$

where

$$c = \dfrac{1}{\sqrt{LC}} \quad \text{and} \quad R_0 = \sqrt{\dfrac{L}{C}}$$

and are the *propagation speed* and *characteristic resistance* of the TL, respectively. Analogously, the solution to Eqs. (7.7) is

$$\begin{cases} \phi(z, t) = \phi^+(z - ct) + \phi^-(z + ct), \\ \dfrac{1}{R_0}\psi(z, t) = \phi^+(z - ct) - \phi^-(z + ct), \end{cases} \quad (7.9)$$

7.2.1. The Telegraphists' Equations

In the frequency domain, or for phasor signals, Eqs. (7.6) transform as follows:

$$\begin{cases} -\dfrac{dV}{dz} = i\omega L I, \\ -\dfrac{dI}{dz} = i\omega C V, \end{cases} \quad (7.10a)$$

---

[2]. Equations (7.6), which are the most popular set, are obviously valid also in the case of a space-invariant medium.

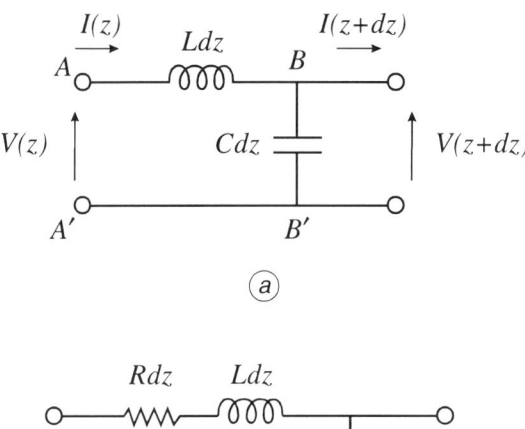

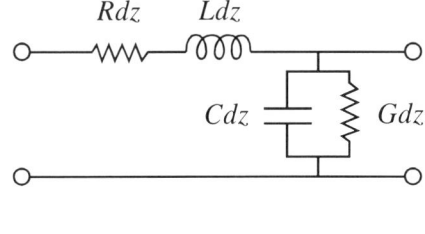

FIGURE 7.5. Circuit model for an elementary cell of the TL: the lossless (A) and lossy (B) cases.

and suggest the circuit model depicted in Fig. 7.5a. The two equations (7.10a) are obtained by applying the second and first Kirchhoff's laws to the circuit $A'ABB'A'$ and to the node B, respectively. This (lossless) circuit suggests an easy extension to the lossy case by including the series resistance R [ohm/m] and shunt conductance G [siemens/m] per unit length of the TL, as depicted in Fig. 7.5b. The generalized equations are

$$\begin{cases} -\dfrac{dV}{dz} = (R + i\omega L)I, \\ -\dfrac{dI}{dz} = (G + i\omega C)V, \end{cases} \qquad (7.10b)$$

which are formally identical to Eqs. (7.10a) if

$$L + \frac{R}{i\omega} \to L \quad \text{and} \quad C + \frac{G}{i\omega} \to C.$$

It is noted that pure TEM modes cannot exist if $R \neq 0$ and/or $G \neq 0$, so that the lossy case should be considered as a small perturbation of the lossless case:

$$\frac{R}{\omega L} \ll 1 \quad \text{and} \quad \frac{G}{\omega c} \ll 1.$$

The linear system (7.10b) is sometimes referred to as the *telegraphists'* (or *telephonists'*) *equations*, and was originally derived from the circuit model of Fig. 7.5 rather than from the field equations. It relies on the *static* parameters L and C, and its similarity with the plane-wave equations of Sections 3.2.5 and 4.4 is evident. The solution is

$$\begin{cases} V(z) = V^+ \exp(-ikz) + V^- \exp(ikz), \\ Z_0 I(z) = V^+ \exp(-ikz) - V^- \exp(ikz), \end{cases} \quad (7.11a)$$

where

$$k = \omega\sqrt{LC} \quad \text{and} \quad Z_0 = \sqrt{\frac{L}{C}} \quad (7.12)$$

and are *propagation constant* and *characteristic impedance* of the TL, respectively. In Eqs. (7.12), L and C include R and G, should the TL be lossy.

In the lossless case both $k = \beta$ and $Z_0 = R_0$ are real; in the small-losses cases we have

$$k = \omega\sqrt{LC} \sqrt{\left(1 + \frac{R}{i\omega L}\right)\left(1 + \frac{G}{i\omega C}\right)}$$

$$\approx \omega\sqrt{LC} - i\frac{1}{2}\left(\frac{R}{R_0} + GR_0\right) = \beta - i\alpha$$

and

$$Z_0 = \sqrt{\frac{L}{C}} \sqrt{\frac{1 + (R/i\omega L)}{1 + (G/i\omega C)}} \approx R_0 - i\frac{R - GR_0^2}{2\beta} = R_0 + iX_0.$$

In all practical cases we can take $Z_0 \approx R_0$, and this is often tacitly assumed in the following. As far as k is concerned, α cannot usually be neglected unless the TL length l is such that

$$\alpha l \ll 1.$$

ENGINEERING TOPICS: PROPAGATION

TABLE 7.2. High-Frequency TL Parameters[a]

	1 (two wires)	2 (coaxial cable)
L	$\dfrac{\mu_0}{\pi}\cosh^{-1}\dfrac{D}{2a}$	$\dfrac{\mu_0}{2\pi}\ln\dfrac{b}{a}$
C	$\dfrac{\pi\varepsilon_0}{\cosh^{-1}(D/2a)}$	$\dfrac{2\pi\varepsilon_0\varepsilon_r}{\ln(b/a)}$
R	$\approx \dfrac{2}{\sigma\pi a^2},\quad \delta>a$ $\approx \dfrac{1}{\sigma\pi a\delta},\quad \delta<a$	$\approx \dfrac{1}{\sigma\pi}\left(\dfrac{1}{a^2}+\dfrac{1}{b^2}\right),\quad \begin{array}{l}\delta>a\\ \delta>b\end{array}$ $\approx \dfrac{1}{2\pi\sigma\delta}\left(\dfrac{1}{a}+\dfrac{1}{b}\right),\quad \begin{array}{l}\delta<a\\ \delta<b\end{array}$
G	0	$G=\omega C\tan\Lambda$

[a]Case 1. Two wires in air, with interaxial spacing D and radius a.
Case 2. Coaxial cable with inner conductor of radius a and outer conductor of radius b and thickness Δ.
Skin depth: $\delta=\sqrt{2/\omega\mu_0\sigma}$. Tangent loss: $\tan\Lambda=\varepsilon_2/\varepsilon_1$.

Note that even in the large-loss case Z_0 is real and equal to R_0, i.e., $Z_0=R_0$, if the *Heaviside condition*

$$\frac{R}{\omega L}=\frac{G}{\omega C}$$

is fulfilled. Values of TL parameters are given in Table 7.2.

Equations (7.11a) are the *traveling-wave* form of voltages and currents along the TL: they highlight *progressive* (direct) and *regressive* (reflected) waves along the TL system. A voltage and current representation alternative to Eqs. (7.11a) is provided by the *standing-wave* solution. In the lossless case

$$\begin{cases} V(z)=V_0\cos\beta z - iR_0 I_0\sin\beta z, \\ I(z)=I_0\cos\beta z - i\dfrac{V_0}{R_0}\sin\beta z, \end{cases} \quad (7.11\text{b})$$

where

$$V_0=V^+ + V^- \quad\text{and}\quad R_0 I_0=V^+ - V^-$$

are the voltage and current at $z=0$, respectively. Plots of $|V(z)|$ and $|I(z)|$ for a short-circuited ($V_0=0$) and open-circuited ($I_0=0$) TL are depicted in Fig. 7.6.

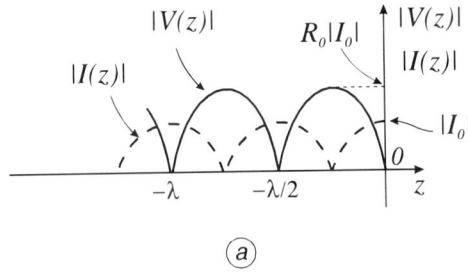

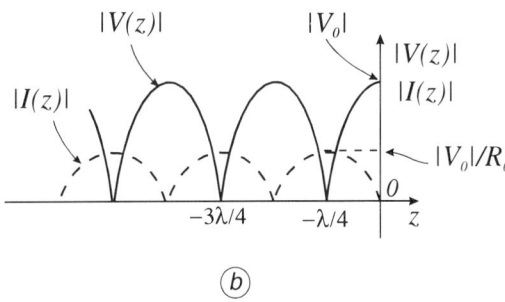

FIGURE 7.6. Plots of $|V(z)|$ and $|I(z)|$ for a short-circuited (A) and open-circuited (B) TL.

Finally, Eqs. (7.12) yield

$$kZ_0 = \omega L \quad \text{and} \quad k/Z_0 = kY_0 = \omega C,$$

which are used to set Eqs. (7.10a) in

$$\begin{cases} -\dfrac{dV}{dz} = ikZ_0 I, \\ -\dfrac{dI}{dz} = ikY_0 V, \end{cases} \quad (7.10c)$$

which are the TL equations in terms of the *dynamic* parameters k and Z_0.

7.2.2. Reflection Coefficient and Impedance

If

$$\Gamma(z) = \frac{V^- \exp(ikz)}{V^+ \exp(-ikz)} = \Gamma(0) \exp(2ikz) \quad (7.13a)$$

ENGINEERING TOPICS: PROPAGATION

is the *reflection coefficient* [see also Eqs. (3.36)], then Eqs. (7.11a) become

$$\begin{cases} V(z) = V^+ \exp(-ikz)[1 + \Gamma(z)], \\ R_0 I(z) = V^+ \exp(-ikz)[1 - \Gamma(z)]. \end{cases} \quad (7.11c)$$

Use of Eqs. (7.11c) leads to the following expression for the power flux along the TL:

$$P(z) = \frac{1}{2} V(z) I^*(z)$$

$$= \frac{1}{2} \frac{|V^+|^2}{R_0} [(1 - |\Gamma(z)|^2) + (\Gamma(z) - \Gamma^*(z))] = P_1 + iP_2. \quad (7.14)$$

We note that the definition of the reflection coefficient is related to a reference direction: in Eq. (7.13a) this is taken coincident with the positive sense of the z-axis and leads to the conventional expression, Eq. (7.13a). Most generally, we can define two reflection coefficients,

$$\vec{\Gamma}(z) = \frac{V^- \exp(ikz)}{V^+ \exp(-ikz)} \quad \text{and} \quad \overleftarrow{\Gamma}(z) = \frac{V^+ \exp(-ikz)}{V^- \exp(ikz)}, \quad (7.15)$$

$\vec{\Gamma}$ in the same sense and $\overleftarrow{\Gamma}$ in the opposite sense of the reference axis. These definitions assure that the module squared of the reflection coefficient does not exceed unity if it is defined toward a load, and vice versa if it is defined toward a source.

Let us compute the real power fluxes

$$\vec{P}_1 = \text{Re}\left[\frac{1}{2} VI^*\right] = \frac{|V^+|^2}{2R_0}(1 - |\vec{\Gamma}|^2)$$

and

$$\overleftarrow{P}_1 = \text{Re}\left[-\frac{1}{2} VI^*\right] = \frac{|V^-|^2}{2R_0}(1 - |\overleftarrow{\Gamma}|^2)$$

along the positive ($\vec{P}_1$) and negative ($\overleftarrow{P}_1$) directions of the z-axis. When the TL is loaded along its positive direction, $\vec{P}_1 \geq 0$, which implies $|\vec{\Gamma}|^2 \leq 0$, while if it is loaded along its negative direction, $\overleftarrow{P}_1 \geq 0$, and $|\overleftarrow{\Gamma}|^2 \leq 0$; and vice versa in the case of a source.

A TL is *matched* if $\Gamma(z) = 0$, which is obtained if $\Gamma(0) = 0$ [see Eq. (7.13a)]. This is usually a desirable condition because the power flux, given by Eq. (7.14), is purely real and constant along the TL; voltage and current amplitudes are constant as well [see Eqs. (7.11c) and Fig. 7.7a].

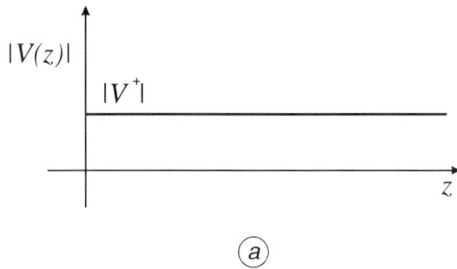

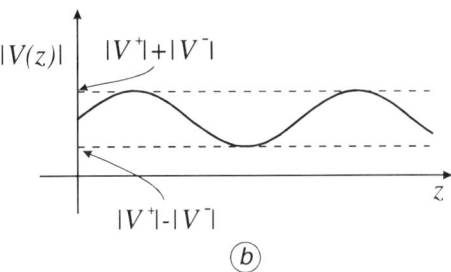

FIGURE 7.7. Voltage amplitude along a matched (A) and mismatched (B) TL.

If the TL is not matched, then voltage and current amplitudes change (with spatial period $\lambda = 2\pi/\beta$) along the line; see Fig. 7.7b. An estimate of the mismatch is provided by the (voltage) *standing-wave* ratio,

$$SWR = \frac{|V^+| + |V^-|}{|V^+| - |V^-|} = \frac{1 + |\Gamma|}{1 - |\Gamma|}, \tag{7.16}$$

a real scalar related simply to the reflection coefficient. In the matched case $SWR = 1$. For small values of $|\Gamma|$, the reflected power equals the incident power times $(SWR - 1)^2/4$.

The reflection coefficient is a parameter appropriate to the traveling-wave solution (7.11a). Similarly, and with reference to the standing-wave solution (7.11b), we define the *input impedance* of the TL:

$$Z(z) = \frac{V(z)}{I(z)} = R_0 \frac{Z(0) - iR_0 \tan kz}{R_0 - iZ(0) \tan kz}, \tag{7.17a}$$

ENGINEERING TOPICS: PROPAGATION

whose expression is obtained by employing Eqs. (7.11b) with

$$Z(0) = \frac{V(0)}{I(0)} = \frac{V_0}{I_0},$$

the input impedance at $z = 0$. Note that the definition of the input impedance is related to a reference direction, as in the case of the reflection coefficient. Usually, it is taken coincident with the z-axis, and definition (7.17a) applies. Should this not be the case, we use different definitions and symbols:

$$\vec{Z}(z) = \frac{V(z)}{I(z)} = \vec{R}(z) + i\vec{X}(z)$$

and

$$\overleftarrow{Z}(z) = -\frac{V(z)}{I(z)} = \overleftarrow{R}(z) + i\overleftarrow{X}(z) \tag{7.18}$$

for the two input impedances, with $\vec{Z}$ in the same sense and $\overleftarrow{Z}$ in the opposite sense of the reference axis. These definitions assure that the input resistances are positive if they are defined toward a load:

$$\vec{P}_1(z) = \frac{1}{2}\vec{R}(z)|I(z)|^2 \quad \text{and} \quad \overleftarrow{P}(z) = \frac{1}{2}\overleftarrow{R}(z)|I(z)|^2,$$

and negative if they are defined toward a source.

Equations (7.13a) for the change of the reflection coefficient and (7.17a) for the change of the input impedance along the TL depend on the abscissa z, which requires one to choose the propagation axis z. Expressions that depend only upon a (positive) distance are possible. Based on Fig. 7.8 we obtain

$$\Gamma(z) = \Gamma(-l) = \Gamma_L \exp(-i2kl) \tag{7.13b}$$

and

$$Z(z) = Z(-l) = R_0 \frac{Z_L + iR_0 \tan kl}{R_0 + iZ_L \tan kl}, \tag{7.17b}$$

where l is the *distance* to the point where the reflection coefficient Γ_L and input impedance Z_L are defined. Equations (7.13b) and (7.17b) are orientation-independent and provide a convenient expression for computing $\vec{\Gamma}, \vec{Z}$ and/or $\overleftarrow{\Gamma}, \overleftarrow{Z}$.

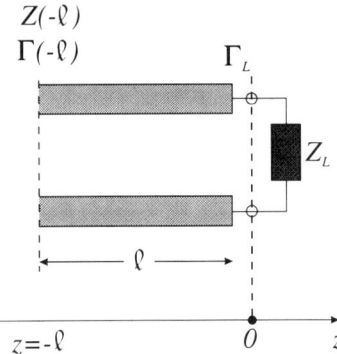

FIGURE 7.8. Definition of input impedance and reflection coefficient.

Equation (7.17a) shows that the TL acts as an impedance transformer. For instance, when $Z(0) = 0$ (short-circuited TL) we have (in the lossless case)

$$Z(-l) = iR_0 \tan \beta l = iX(-l), \qquad (7.19)$$

and the TL is equivalent (at the prescribed frequency) to an inductance $(l < \lambda/4)$ or a capacitance $(\lambda/4 < l < \lambda/2)$.

Similarly, if $l = \lambda/4$ we have

$$Z(-\lambda/4) = \frac{R_0^2}{Z(0)}, \qquad (7.20)$$

and the TL acts as an impedance inverter.

Impedance and reflection coefficients are not unrelated. From Eq. (7.11c) $(R_0 \to Z_0)$

$$\frac{V(z)}{Z_0 I(z)} = \frac{Z(z)}{Z_0} = \frac{1 + \Gamma(z)}{1 - \Gamma(z)}, \qquad (7.21a)$$

$$\Gamma(z) = \frac{Z(z) - Z_0}{Z(z) + Z_0}. \qquad (7.21b)$$

Setting $\Gamma = \Gamma_1 + i\Gamma_2$, $Z_0 = R_0$, and $Z(z)/R_0 = r(z) + ix(z)$ the *normalized* input impedance, we obtain by equating real and imaginary coefficients of Eq.

ENGINEERING TOPICS: PROPAGATION

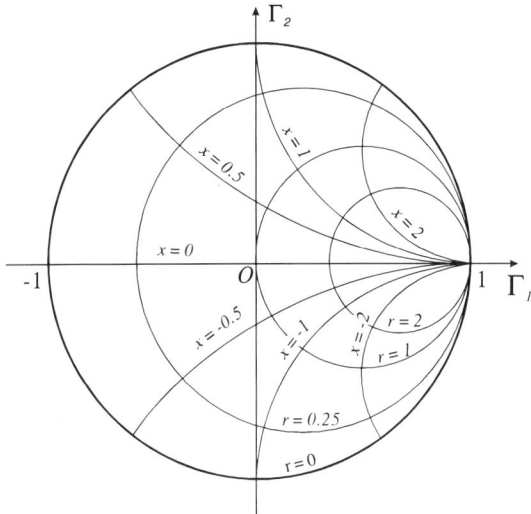

FIGURE 7.9. Smith's chart.

(7.21b), and with some algebra,

$$\Gamma_1^2 + \Gamma_2^2 - \frac{2r}{r+1}\Gamma_1 + \frac{r-1}{r+1} = 0$$

and

$$(\Gamma_1 - 1)^2 + \Gamma_2^2 - \frac{2}{x}\Gamma_2 = 0.$$

For constant values of r and x these equations represent circles in the plane Γ_1, Γ_2. This yields *Smith's chart* of Fig. 7.9, which relates graphically normalized input impedance and reflection coefficient of the TL.

7.2.3. Matching

A TL is *matched* if the reflected wave is absent. If the TL is terminated by its characteristic resistance (or impedance, in general) it is matched as follows from Eqs. (7.21b) and (7.13a): $\Gamma(0) = 0$ and $\Gamma(z) = 0$. An infinitely long TL is matched as well, because the radiation condition (see Section 1.4.2) does not allow the presence of reflected signals.

Techniques are available to match a TL to a mismatched load; see Fig. 7.10. If the load is real, $Z_L = R_L$, a TL section of length $\lambda/4$ (see Fig. 7.10a)

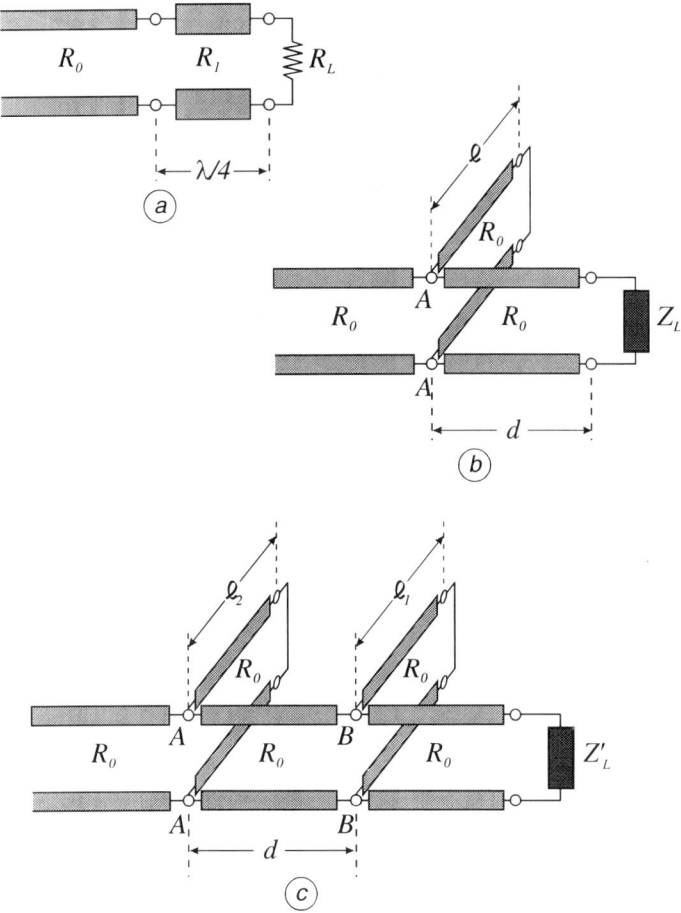

FIGURE 7.10. Matching devices: (a) impedance transformer; (b) single stub adapter; (c) double stub adapter.

with characteristic resistance $R_1 = \sqrt{R_L R_0}$ assures the matching [see Eq. (7.20)]. Should the load be complex, $Z_L = R_L + iX_L$, we can use the *single* or *double stub* matching technique. In the former case (see Fig. 7.10b) the *input admittance* $Y(z) = G(z) + iB(z) = 1/Z(z)$ to the left of the section AA of the TL is set equal to the *characteristic admittance* $G_0 = 1/R_0$ of the TL:

$$G_0 \frac{Y_L + iG_0 \tan \beta d}{G_0 + iY_L \tan \beta d} + \frac{G_0}{i \tan \beta l} = G_0,$$

$$Y_L = G_L + iB_L = 1/Z_L,$$

and the equation is solved for $u = \tan \beta d$ and $v = \tan \beta l$:

$$\begin{cases} [G_L(G_0 - G_L) - B_L^2]u^2 + 2G_0 B_L u - G_0(G_0 - G_L) = 0, \\ v = \dfrac{G_L u}{(G_0 - G_L) - B_L u}. \end{cases}$$

Real solutions for u and v, i.e., for d and l, always exist[3] because the discriminant of the quadratic equation for u is always positive (or zero). When two stubs are used (see Fig. 7.10c), we set equal to G_0 the input admittance to the left of AA:

$$G_0 \frac{[G_L + iB_L - i(G_0/v_1)] + iG_0 u}{G_0 + i[G_L + iB_L - i(G_0/v_1)]u} - i\frac{G_0}{v_2} = G_0,$$

with $v_1 = \tan \beta l_1$, $v_2 = \tan \beta l_2$, $u = \tan \beta d$, and $G_L + iB_L$ the load admittance at BB. For a fixed value of u, matching can be achieved by proper choice of v_1 and v_2. For instance, when $u \to \infty$ ($\beta d = \pi/2$) we obtain

$$(G_0 G_L - G_L^2 - B_L^2)v_1^2 + 2G_0 B_L v_1 - G_0^2 = 0,$$

$$v_2 = \frac{G_L v_1}{G_0 - B_L v_1}.$$

Real solutions for v_1 and v_2, i.e., for l_1 and l_2, do not always exist: it is required that the discriminant of the quadratic equation for v_1 be nonnegative:

$$G_0^2 G_L(G_0 - G_L) \geq 0, \quad \text{i.e., } G_0 \geq G_L.$$

7.2.4. Multisection Transmission Lines

Let us consider the multisection (lossless) TL depicted in Fig. 7.11 and the expanded view of its nth section. The input impedance of the latter, Z_n, is computed via Eq. (7.17b) to be

$$\begin{aligned} Z_n &= R_{0n} \frac{Z_{n+1} + iR_{0n} \tan \beta_n d_n}{R_{0n} + iZ_{n+1} \tan \beta_n d} \\ &= R_{0n} \frac{[Z_{n+1} + R_{0n}] \exp(i\beta_n d_n) + [Z_{n+1} - R_{0n}] \exp(-i\beta_n d_n)}{[Z_{n+1} + R_{0n}] \exp(i\beta_n d_n) - [Z_{n+1} - R_{0n}] \exp(-i\beta_n d_n)} \\ &= R_{0n} \frac{1 + \Gamma_{n+1} \exp(-2i\beta_n d_n)}{1 - \Gamma_{n+1} \exp(-2i\beta_n d_n)}, \end{aligned}$$

---

3. Note that u and v may attain negative values; the corresponding solution for d and l must always be chosen to be positive.

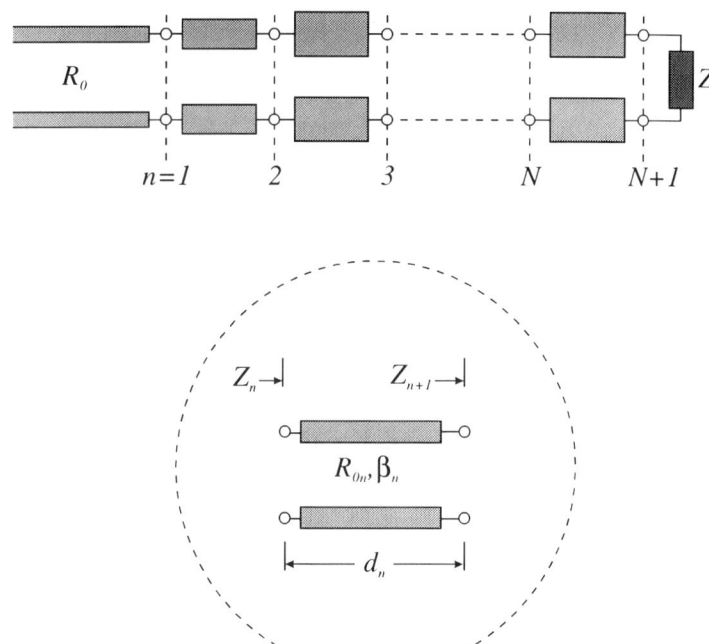

FIGURE 7.11. Multisection TL and expanded view of generic section n.

where

$$\Gamma_{n+1} = \frac{Z_{n+1} - R_{0n}}{Z_{n+1} + R_{0n}}$$

is the reflection coefficient at the output of section n. For the input reflection coefficient of the same section we have with the substitution of Z_n

$$\Gamma_n = \frac{Z_n - R_{0(n-1)}}{Z_n + R_{0(n-1)}} = \frac{\gamma_n + \Gamma_{n+1}\exp(-2i\beta_n d_n)}{1 + \Gamma_{n+1}\gamma_n \exp(-2i\beta_n d_n)},$$

where

$$\gamma_n = \frac{R_{0n} - R_{0(n-1)}}{R_{0n} + R_{0(n-1)}} \qquad (7.22)$$

is the reflection coefficient of two semi-infinite TLs of characteristic resistances R_{0n} and $R_{0(n-1)}$, respectively.

ENGINEERING TOPICS: PROPAGATION

If the multisection TL is aimed to minimize the input reflection coefficient Γ_1, we can reasonably assume that both $|\Gamma_n|$ and $|\gamma_n|$ are small, and so $|\Gamma_n \gamma_n| \ll 1$. Accordingly

$$\Gamma_n \approx \gamma_n + \Gamma_{n+1} \exp(-2i\beta_n d_n),$$

which allows computation of the input reflection coefficient Γ_1 by a simple iteration.

When $\beta_n d_n = \psi = \text{const}$, $\gamma_n = \gamma = \text{const}'$, and also[4]

$$\gamma_N = \frac{Z_L - R_{0(N-1)}}{Z_L + R_{0(N-1)}} = \gamma, \qquad Z_L = R_L,$$

then

$$\Gamma_1 = \gamma \sum_{n=1}^{N} \exp[-2i(n-1)\psi] = \gamma \sum_{n=0}^{N-1} \exp[-2in\psi]$$

$$= \gamma \frac{\exp(-2iN\psi) - 1}{\exp(-2i\psi) - 1},$$

hence

$$|\Gamma_1| = N|\gamma| \left| \frac{\sin N\psi}{N \sin \psi} \right|.$$

On the other hand, Eq. (7.22) yields

$$R_{0n} = \frac{1+\gamma}{1-\gamma} R_{0(n-1)} = \left(\frac{1+\gamma}{1-\gamma}\right)^n R_0,$$

$$R_L = \left(\frac{1+\gamma}{1-\gamma}\right)^N R_0, \qquad N \ln \frac{1+\gamma}{1-\gamma} \approx 2N\gamma = \ln \frac{R_L}{R_0},$$

and

$$|\Gamma_1| \approx \frac{1}{2} \left| \ln \frac{R_L}{R_0} \cdot \frac{\sin(N\psi)}{N\psi} \right|.$$

The multisection TL is effective in reducing the mismatch between the input TL and the load provided that $N\psi$, the *electric length* of the matching device,

---

4. This implies that Z_L is real.

exceeds π and is large compared to the input–output *impedance log jump* $|\ln(R_L/R_0)|$.

7.2.5. Nonuniform Transmission Lines

For a (spatially) nonuniform TL the parameters L and C in Eqs. (7.10a) become z-dependent. We divide the first equation by L, take the derivative with respect to z, and substitute from the second equation. This yields

$$L\frac{d}{dz}\frac{1}{L}\frac{dV}{dz} + k^2 V = 0, \qquad k^2 = \omega^2 L(z) C(z),$$

which is equivalent to

$$\frac{d^2 V}{dz^2} - \frac{d\ln L}{dz}\frac{dV}{dz} + k^2 V = 0.$$

A similar equation is obtained for I.

For TLs with spatially slow-varying parameters the term containing the factor $d\ln L/dz$ can usually be neglected and, in any case, eliminated by the substitution $V = \sqrt{L}\bar{V}$. This leads to

$$\frac{d^2 \bar{V}}{dz^2} + \left[k^2 + \frac{1}{2}\frac{d^2 \ln L}{dz^2} - \frac{1}{4}\left(\frac{d\ln L}{dz}\right)^2 \right] \bar{V} = 0,$$

and shows that the *local* propagation constant of the modified voltage $\bar{V}$ is slightly changed compared to k. However, the additional terms are again negligible for a spatially slow-varying TL, so we consider the equation

$$\frac{d^2 V}{dz^2} + k^2(z) V = 0. \qquad (7.23)$$

Note that this equation is exact when $C = C(z)$ and $L = \text{const}$.

Equation (7.23) can be solved in terms of simple functions in a very limited number of cases, i.e., in terms of propagation constant *profile* $k(z)$. If the inhomogeneous TL is *spatially* slow-varying, we can seek the solution of Eq. (7.23) as a perturbation of the corresponding solution for the homogeneous TL, i.e., by setting

$$V(z) = A(z) \exp[i\phi(z)].$$

Substitution into Eq. (7.23) gives

$$\frac{d^2 A}{dz^2} + i\left[2\frac{dA}{dz}\frac{d\phi}{dz} + A\frac{d^2\phi}{dz^2}\right] + \left[k^2 - \left(\frac{d\phi}{dz}\right)^2\right] A = 0,$$

where the slow-varying assumption can be enforced by neglecting the term $d^2 A/dz^2$. Hence

$$\left(\frac{d\phi}{dz}\right)^2 = k^2(z),$$

which coincides with the eikonal equation (5.3), and

$$\frac{d^2\phi}{dz^2} + 2\frac{d\phi}{dz}\frac{d\ln A}{dz} = 0,$$

which coincides with the transport equation (5.11b) of geometrical optics. Therefore

$$\phi(z) = \pm \int_0^z dz\, k(z), \qquad (7.24a)$$

and

$$\frac{d}{dz}\ln\frac{d\phi}{dz} + \frac{d}{dz}\ln A^2 = 0,$$

$$\sqrt{k(z)}\, A(z) = \sqrt{k(0)}\, A(0), \qquad (7.24b)$$

with $A(0)$ a constant. Equations (7.24) provide the Wentzel[5]–Kramers–Brillouin (WKB) solution to the spatially slow-varying TL.

The general solution for the voltage is obtained by combining the two solutions generated by Eqs. (7.24). The corresponding solution for the current is obtained by using Eqs. (7.10). If we refer to modified voltages and currents, namely,

$$\sqrt{\frac{k(z)}{k(0)}}\, V(z) \to V(z) \quad \text{and} \quad \sqrt{\frac{k(z)}{k(0)}}\, I(z) \to I(z),$$

and we again neglect the term containing the derivative of $\ln k(z)$ with respect

---

5. Gregor Wentzel: 1898–Dusseldorf (Germany), 1898–Ascona (Germany), 1978.

to z, we obtain

$$\begin{cases} V(z) = V^+ \exp[-i\psi(z)] + V^- \exp[i\psi(z)], \\ Z_0(z)I(z) = V^+ \exp[-i\psi(z)] - V^- \exp[i\psi(z)], \end{cases} \quad (7.25a)$$

where

$$\psi(z) = \int_0^z dz\, k(z) \quad \text{and} \quad Z_0(z) = \sqrt{L(z)/C(z)},$$

which are analogous to Eqs. (7.11a). Furthermore,

$$\begin{cases} V(z) = V_0 \cos \psi(z) - iZ_0 I_0 \sin \psi(z), \\ I(z) = I_0 \cos \psi(z) - i\dfrac{V_0}{Z_0} \sin \psi(z), \end{cases} \quad (7.25b)$$

which are analogous to Eqs. (7.11b).

Equation (7.25a) allows the definition of a reflection coefficient, and Eq. (7.25b) of an input impedance, as in the case of a homogeneous TL.

Equations that directly describe the change of input impedance and reflection coefficient along the TL are also obtainable. Let

$$u(z) = \frac{Z(z)}{Z_0(z)}$$

be the *locally* normalized input impedance. Then

$$\frac{du}{dz} = \frac{d}{dz} \frac{V(z)}{Z_0(z)I(z)} = -ik + iku^2 - \frac{dZ_0/dz}{Z_0} u,$$

where use has been made of Eqs. (7.10c). Accordingly

$$\frac{du}{dz} = ik(u^2 - 1) - u \frac{d \ln Z_0}{dz}, \quad (7.26a)$$

which is a Riccati[6] equation whose solution coincides with Eq. (7.17a) in the case of uniform TLs.

The equation for the *local* reflection coefficient,

$$\gamma(z) = \frac{u - 1}{u + 1},$$

---

6. Jacopo Francesco Riccati: Venice (Italy), 1867–Treviso, 1754.

ENGINEERING TOPICS: PROPAGATION

can be obtained similarly. We solve for u and substitute in Eq. (7.26a) to yield

$$\frac{d\gamma}{dz} = 2ik\gamma - \frac{1-\gamma^2}{2}\frac{d\ln Z_0}{dz}, \qquad (7.26b)$$

which is again a Riccati equation whose solution coincides with Eq. (7.13a) in the case of uniform TLs.

For spatially slow-varying TLs, $\gamma^2 \approx 0$, and Eq. (7.26b) linearizes to

$$\frac{d\gamma}{dz} \approx 2ik\gamma - \frac{1}{2}\frac{d\ln Z_0}{dz}.$$

When $k = \beta = $ const (as, for instance, TL in air), we set $\gamma(z) = \gamma_0(z)\exp(2i\beta z)$ and obtain

$$\frac{d\gamma_0}{dz}\exp(2i\beta z) \approx -\frac{1}{2}\frac{d\ln Z_0}{dz}.$$

Hence

$$\gamma(z) = \left\{\gamma(0) - \int_0^z dz\,\frac{1}{2}\frac{d\ln Z_0}{dz}\exp(-2i\beta z)\right\}\exp(2i\beta z).$$

For a nonuniform TL matched at its end, $Z(0) = Z_0(0)$, $\gamma(0) = 0$, and the input reflection coefficient, $\gamma(-l)$, is (but for a phase term) the FT of (one-half of) its characteristic impedance relative rate.

For the (lossless) *exponential line*

$$R_0(z) = R_0(0)\exp(-\alpha z) \quad \text{and} \quad \gamma(-l) = -\frac{\alpha l}{2}\frac{\sin\beta l}{\beta l}\exp(-i\beta l).$$

Again, the tapered TL is effective in reducing the mismatch between the input TL and the output load $R_0(0)$ if βl is close to, or larger than, π, and large compared to the input–output impedance log jump αl (see Section 7.2.4).

7.2.6. Multiconductor Transmission Lines

Equations and results for two-wire TLs can be generalized to multiconductor TLs. Let us consider the case of two wires in a cable configuration, as depicted in Fig. 7.12. Let L_1 and C_1 be the inductance and capacitance per unit length of the first wire with respect to the outer conductor, and L_2 and C_2 the same parameters for the second wire. Also $R_1 = \sqrt{L_1/C_1}$ and $k_1 = \omega\sqrt{L_1 C_1}$

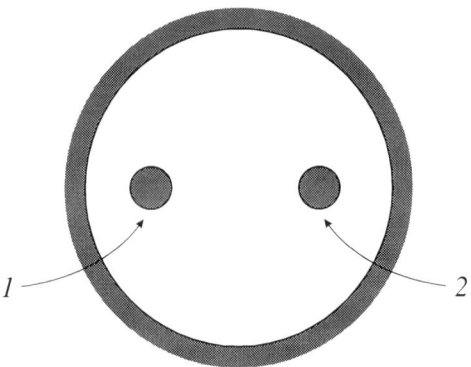

FIGURE 7.12. Multiconductor TL in a cable configuration.

are the corresponding dynamical parameters of the first wire, and R_2 and k_2 those of the second wire. Voltages V_1 and V_2 are measured between each wire and the outer conductor, and are divided by $\sqrt{R_1}$ and $\sqrt{R_2}$, respectively. Currents I_1 and I_2 are multiplied by $\sqrt{R_1}$ and $\sqrt{R_2}$, so that the new quantities

$$\hat{V} = V/\sqrt{R} \quad \text{and} \quad \hat{I} = \sqrt{R}I$$

exhibit the same dimensions [watt$^{1/2}$]. Mutual inductance M and capacitance N (per unit length) between the two wires are normalized to $\sqrt{R_1 R_2}$, where

$$m = \frac{M}{\sqrt{R_1 R_2}} \quad \text{and} \quad n = \frac{N}{\sqrt{R_1 R_2}}.$$

To derive the TL equations appropriate to the problem, we can rely upon the model based on the equivalent circuit of Fig. 7.5 modified to account for the coupling. With $\partial/\partial z \to -i\beta$, we obtain in matrix form

$$\boldsymbol{M} \cdot \mathbf{u} = \beta \mathbf{u},$$

with

$$\mathbf{u} \to \begin{vmatrix} \hat{V}_1 \\ \hat{I}_1 \\ \hat{V}_2 \\ \hat{I}_2 \end{vmatrix} \quad \text{and} \quad \boldsymbol{M} \to \begin{vmatrix} 0 & k_1 & 0 & \omega m \\ k_1 & 0 & \omega n & 0 \\ 0 & \omega m & 0 & k_2 \\ \omega n & 0 & k_2 & 0 \end{vmatrix}.$$

ENGINEERING TOPICS: PROPAGATION

The equation

$$(M - \beta_i I) \cdot \mathbf{u}_i = 0$$

is biquadratic in the unknown propagation constant β_i and provides four eigenvalues,

$$\det(M - \beta_i I) = 0, \qquad \beta_i = \pm \beta_{a,b},$$

and four associated (normalized) eigenvectors corresponding to four independent propagation modes that can be set along the TL: two progressive distinct modes with propagation constants β_a and β_b, respectively, as well as two corresponding regressive modes with propagation constants $-\beta_a$ and $-\beta_b$. Each mode propagates as $\exp(-i\beta_i z)$, $i = a, b$.

When $k_1 = k_2 = k$,[7]

$$\beta_{a,b}^2 = k^2 \pm k\omega(m+n) - \omega^2 mn,$$

and so

$$\beta_{a,b} \approx k \pm \frac{\omega(m+n)}{2} = k \pm \chi,$$

if $\omega m, \omega n \ll k$.

7.2.7. Transmission Line Generators

Consider a TL which is driven by a spatially concentrated current I_G or voltage V_G, as depicted in Fig. 7.13. Equations (7.10a) generalize as follows:

$$\begin{cases} -\dfrac{dV}{dz} = i\omega L I - V_G \delta(z - z'') \\ -\dfrac{dI}{dz} = i\omega C V - I_G \delta(z - z'). \end{cases}$$

Integration of the first equation across z'' and of the second across z' leads to

$$-V(z'' + \Delta z) + V(z'' - \Delta z) = -V_G, \qquad \Delta z \to 0$$

---

7. This is always the case if the cable cross section is homogeneous.

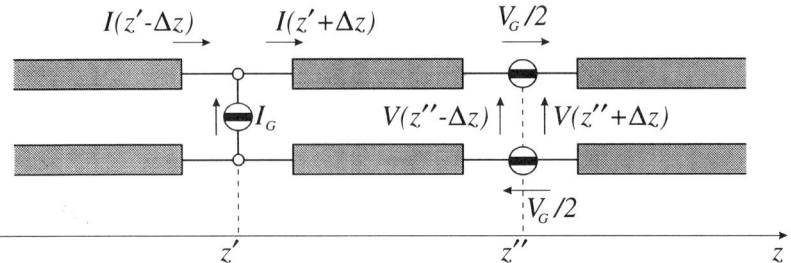

FIGURE 7.13. A TL driven by current and voltage (ideal) generators.

and

$$-I(z' + \Delta z) + I(z' - \Delta z) = -I_G, \quad \Delta z \to 0,$$

and these equations are consistent with the generator signs in Fig. 7.13.[8] Most generally, we consider *distributed sources* $\hat{V}(z)$ [volt/m] and $\hat{I}(z)$ [ampere/m] along the TL, so Eqs. (7.10c) are generalized as follows:

$$\begin{cases} -\dfrac{dV}{dz} = ikZ_0 I - \hat{V}, \\ -\dfrac{dI}{dz} = ikY_0 V - \hat{I}. \end{cases} \quad (7.27)$$

7.3. EQUIVALENT TRANSMISSION LINES: TWO-DIMENSIONAL STRUCTURES

Let us consider a homogeneous, isotropic medium unbounded in the x and y directions (see Fig. 7.14), and plane-wave propagation at angle θ to the z-axis. The x-axis is chosen in such a way that the propagation vector lies in the (x, z) plane. The electromagnetic field is y-independent, so we obtain from Section 4.5

$$\begin{vmatrix} -i\beta & i\omega\mu \\ i\omega\varepsilon & -i\beta \end{vmatrix} \begin{vmatrix} E_x \\ H_y \end{vmatrix} = A \cdot \begin{vmatrix} E_x \\ H_y \end{vmatrix} = \begin{vmatrix} \partial E_z/\partial x \\ 0 \end{vmatrix}, \quad (7.28a)$$

$$\begin{vmatrix} -i\beta & i\omega\mu \\ i\omega\varepsilon & -i\beta \end{vmatrix} \begin{vmatrix} E_y \\ -H_x \end{vmatrix} = A \cdot \begin{vmatrix} E_y \\ -H_x \end{vmatrix} = \begin{vmatrix} 0 \\ \partial H_z/\partial x \end{vmatrix}, \quad (7.28b)$$

---

8. For the sake of graphical simplicity, the full voltage $V_g \delta(z - z'')$ impressed by the generator is usually localized in only one of the two wires.

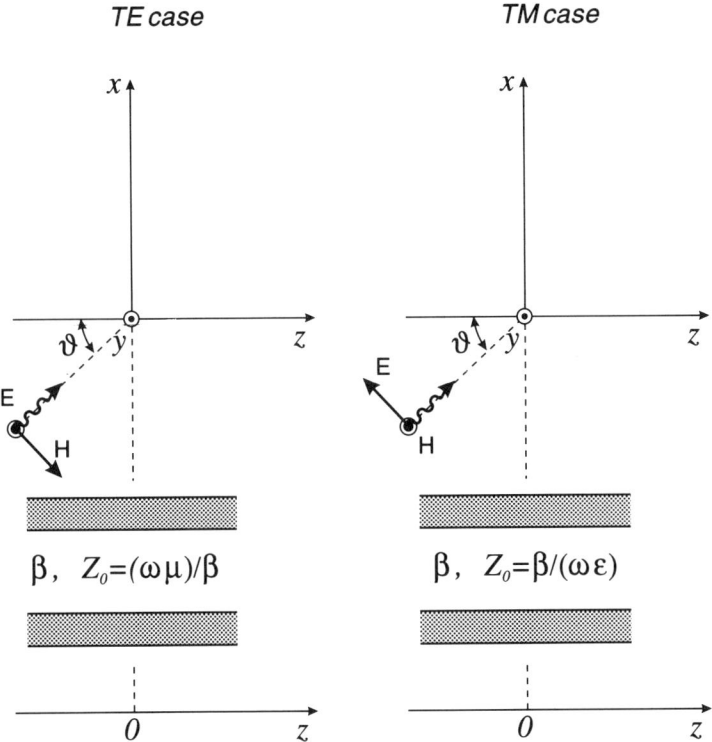

FIGURE 7.14. Equivalent TLs for a homogeneous isotropic medium unbounded in the x and y directions.

which shows that TE [Eqs. (7.28b)] and TM [Eqs. (7.28a)] modes are possible with respect to the propagation direction z.

Equations (4.69) indicate that the dependence of both E_z and H_z on the transverse coordinates is of the type $\exp(-iux)$, $u = k\sin\theta$; this is also the appropriate x-dependence for all field components. Further, Eq. (4.70) shows that

$$\beta = k\cos\theta = \sqrt{k^2 - u^2}.$$

Equations (4.46a–b) projected onto the z-axis yield

$$\frac{\partial E_y}{\partial x} = -iuE_y = -i\omega\mu H_z \quad \text{and} \quad \frac{\partial H_y}{\partial x} = -iuH_y = i\omega\varepsilon E_z. \quad (7.29)$$

Substitution of E_z and H_z into Eqs. (7.28a) and (7.28b), respectively, with

$-i\beta \to \partial/\partial z$, leads to

$$\begin{cases} -\dfrac{\partial E_x}{\partial z} = i\omega\mu\left(1 - \dfrac{u^2}{k^2}\right)H_y, \\ -\dfrac{\partial H_y}{\partial z} = i\omega\varepsilon E_x, \end{cases} \quad (7.30\text{a})$$

and

$$\begin{cases} -\dfrac{\partial E_y}{\partial z} = i\omega\mu(-H_x), \\ -\dfrac{\partial(-H_x)}{\partial z} = i\omega\varepsilon\left(1 - \dfrac{u^2}{k^2}\right)E_y. \end{cases} \quad (7.30\text{b})$$

If

$$Z_0 = \frac{1}{Y_0} = \frac{\beta}{\omega\varepsilon} \quad \text{with } \beta = \sqrt{k^2 - u^2}, \quad (7.31\text{a})$$

then the transformation

$$E_x \to V(z)\exp(-iux), \quad H_y \to I(z)\exp(-iux),$$
$$E_z \to -\frac{u}{\omega\varepsilon}I(z)\exp(-iux) \quad (7.32\text{a})$$

enables Eq. (7.30a) to be expressed in the form

$$\begin{cases} -\dfrac{dV}{dz} = i\beta Z_0 I, \\ -\dfrac{dI}{dz} = i\beta Y_0 V. \end{cases} \quad (7.33)$$

These equations describe TM propagation in a transversely unbounded configuration and are formally identical to TL equations (7.10c), with appropriate values (7.31a) of the propagation constant and characteristic impedance.

For Eqs. (7.30b), we derive the same equations (7.33) with

$$\beta = \sqrt{k^2 - u^2}, \quad Z_0 = \frac{1}{Y_0} = \frac{\omega\mu}{\beta}, \quad (7.31\text{b})$$

and

$$E_y \to V(z) \exp(-iux), \qquad -H_x \to I(z) \exp(-iux),$$
$$H_z \to \frac{u}{\omega\mu} V(z) \exp(-iux), \qquad (7.32b)$$

which describes the propagation of TE waves.

We conclude that the propagation of both TE and TM waves can be studied in terms of an *equivalent transmission line* (ETL).

7.3.1. Multilayer Propagation

Consider plane-wave propagation in the layered structure depicted in Fig. 7.15. The x-dependence for the field is of the type $\exp(-iux)$ everywhere in the layers, in order to fulfil boundary conditions at each layer interface (see Section 4.6).

We model each layer in terms of an ETL of propagation constant

$$\beta_n = \sqrt{k_n^2 - u^2}, \qquad u = k_0 \sin\theta,$$

and characteristic impedance

$$Z_n = \frac{\omega\mu_n}{\beta_n}, \qquad Z_n = \frac{\beta_n}{\omega\varepsilon_n},$$

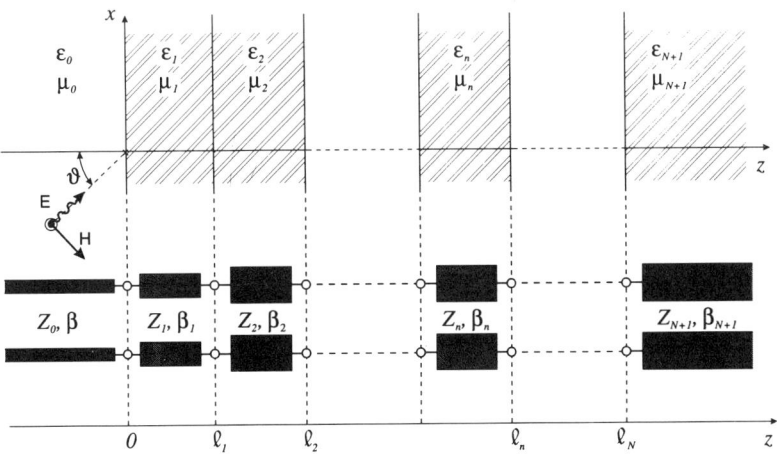

FIGURE 7.15. A layered structure and its equivalent chain of TLs. Incidence of TE type is shown.

for TE and TM incidence, respectively. Accordingly

$$\begin{cases} V_n(z) = \hat{V}_n \cos[\beta_n(z - l_n)] - iZ_n \hat{I}_n \sin[\beta_n(z - l_n)], \\ I_n(z) = \hat{I}_n \cos[\beta_n(z - l_i)] - iY_n \hat{V}_n \sin[\beta_n(z - l_n)], \end{cases} \quad (7.34)$$

$$\hat{V}_n = V_n(l_n), \qquad \hat{I}_n = I_n(l_n),$$

$$l_{n-1} \leqslant z \leqslant l_n, \qquad l_0 = 0, \qquad n = 1, \ldots, N.$$

Continuity of tangential fields at layer interfaces requires continuity of voltages and currents at ETL junctions:

$$\hat{V}_n = V_{n+1}(l_n), \qquad \hat{I}_n = I_{n+1}(l_n),$$

$$n = 1, \ldots, N - 1,$$

which is a linear system of $2(N-1)$ equations in $2N$ unknowns. Two additional equations are needed and are provided by output and input boundary conditions. If Z_L denotes the output impedance,[9] then

$$\hat{V}_N = Z_L \hat{I}_N.$$

If V_i is the *incident* voltage at $z = 0$ and Γ_0 the input reflection coefficient, then the two relations

$$V_1(0) = (1 + \Gamma_0)V_i \quad \text{and} \quad I_1(0) = (1 - \Gamma_0)\frac{V_i}{Z_0}$$

lead to

$$V_1(0) = 2V_i - Z_0 I_1(0),$$

which is the last necessary equation.

Inversion of the complete $2N \times 2N$ linear system with known term V_i determines the coefficients $\hat{V}_n$ and $\hat{I}_n$. These, in turn, provide the solution to multilayer propagation via Eqs. (7.34).

For $N = 1$, we obtain the case of two half-spaces (see Section 4.6.1). This case is modeled in terms of two semi-infinite adjoined ETLs; expressions (4.78) for the Fresnel reflection coefficients coincide with those obtained via the ETL model.

---

9. Should the last layer be indefinite along z, the load impedance Z_L coincides with the characteristic impedance of the last ETL.

7.3.2. Transverse Resonance

Consider the two grounded dielectric slabs depicted in Fig. 7.16 with

$$\mu_i = \mu_2 = \mu, \qquad \varepsilon_1 = \varepsilon, \qquad \varepsilon_2 = \varepsilon_r \varepsilon,$$

and its corresponding ETL model.

The z-dependence of the fields is of type $\exp(-i\beta z)$, this implying that propagation is long this axis. However, it is convenient to consider an ETL along the x-axis, because the medium is homogeneous for $x > 0$ and $x < 0$. Accordingly, we set

$$k^2 = \omega^2 \varepsilon \mu = \beta^2 + u_1^2 \tag{7.35a}$$

and

$$k^2 \varepsilon_r = \omega^2 \varepsilon \mu \varepsilon_r = \beta^2 + u_2^2, \tag{7.35b}$$

u_1 and u_2 being the propagation constants of the two ETLs, respectively. We also have from Eqs. (7.35)

$$u_2^2 = u_1^2 + (\varepsilon_r - 1)k^2 \tag{7.36}$$

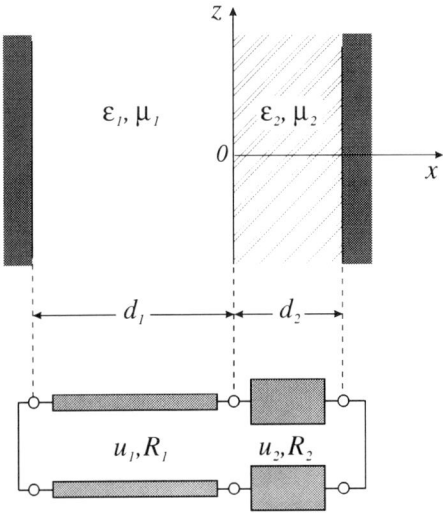

FIGURE 7.16. Propagation along two grounded dielectric slabs.

The two input impedances $\vec{Z}(x)$ and $\overleftarrow{Z}(x)$ in Eq. (7.18) must only differ in sign, due to the continuity of the voltage and of the current at any point along the TL. Accordingly

$$\overleftarrow{Z}(x) + \vec{Z}(x) = 0, \qquad (7.37)$$

which is usually referred to as the *transverse resonance condition*.

Equation (7.37) depends on the transverse wavenumbers u_1 and u_2 but not directly on the field amplitudes, thus conveniently providing, together with Eq. (7.36), the dispersion equation of this two-media guiding structure:

$$R_1 \tan u_1 d_1 + R_2 \tan u_2 d_2 = 0,$$

R_1 and R_2 being the characteristic resistances of the two ETL.

Solutions for equivalent voltages and currents are obtained by enforcing continuity,

$$V_1(0) = V_2(0),$$

and boundary conditions,

$$V_1(-d_1) = V_2(d_2) = 0,$$

in Eqs. (7.11b). Hence

$$\begin{cases} V_1(x) = -iR_1 I_0 \sin\left[u_1\left(\dfrac{d_1}{2} \pm x\right)\right], \\[2mm] I_1(x) = I_0 \cos\left[u_1\left(\dfrac{d_1}{2} \pm x\right)\right]. \end{cases}$$

with I_0 constant. Note that the other continuity equation, $I_1(0) = I_2(0)$, is automatically satisfied if the transverse resonance condition (7.37) is enforced.

The field components are computed by means of Eqs. (7.32).

7.3.3. Propagation along a Grounded Slab

Consider the grounded dielectric slab of relative permittivity ε_r and its associated ETL as depicted in Fig. 7.17. The transverse resonance condition (7.37) leads to

$$iR \tan ud + R_0 = 0,$$

because the second ETL is semi-infinite.

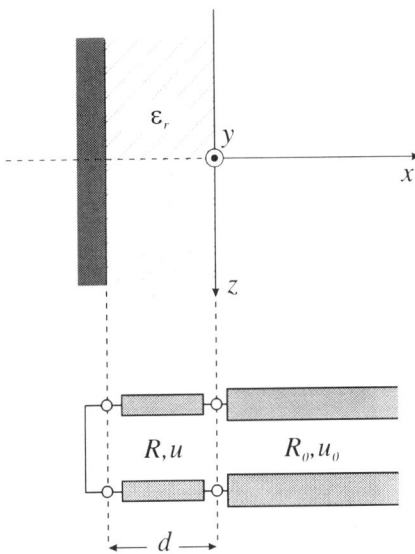

FIGURE 7.17 A grounded dielectric slab and its associated ETL.

As in Section 4.5.2 we seek a field solution which is guided along z and exponentially damped for $x > 0$. We let $u_0 = -i\alpha$,

$$k^2(\varepsilon_r - 1) = u^2 + \alpha^2 \tag{7.38}$$

[see Eq. (7.36)], and we obtain for modes that are TE with respect to x

$$i\frac{\omega\mu}{u}\tan ud + \frac{\omega\mu}{-i\alpha} = 0, \tag{7.39a}$$

which coincides with the first of Eqs. (4.81).

For modes that are TM with respect to x we obtain similarly:

$$i\frac{u}{\omega\varepsilon\varepsilon_r}\tan ud + \frac{-i\alpha}{\omega\varepsilon} = 0. \tag{7.39b}$$

Graphical analysis of the system obtained by considering Eqs. (7.38) and (7.39b) (with the same procedure as in Fig. 4.31) shows that there is a mode with no cut-off. However, in the low-frequency limit $kd \to 0$, Eq. (7.38) shows that $ud \to 0$ and $\alpha d \to 0$. If $\tan ud \approx ud$ in Eq. (7.39b), then substitution into Eq. (7.38) leads to

$$\alpha d \approx \frac{\varepsilon_r - 1}{\varepsilon_r}(kd)^2, \qquad kd \ll 1.$$

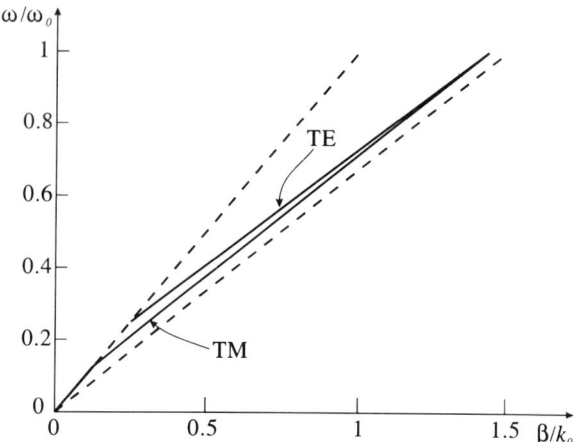

FIGURE 7.18. Brillouin diagram $\omega = \omega(\beta)$ for the fundamental TE and TM modes along a grounded dielectric slab with $\varepsilon_r = 2.5$. The angular frequency and propagation constant are normalized to $\omega_0 = ck_0$ and k_0, respectively, where $k_0 = 2\pi/d$.

We conclude that, in spite of the absence of cut-off, the guiding properties of the structure are negligible for $kd < 1$.

Computation of voltages and currents along the ETLS is straightforward; field components are obtained via Eqs. (7.32). As already shown in Section 4.5.2, fields that are TE with respect to x are also TE with respect to z, and similarly for TM modes.

The numerical solution of Eqs. (7.38), (7.39a) and (7.38), (7.39b) allows computation of the transverse wavenumber for TE and TM modes, respectively. The propagation constant is then given by $\beta^2 = k^2\varepsilon_r - u^2$.

The Brillouin diagram for the fundamental TE and TM modes is given in Fig. 7.18 for the case $\varepsilon_r = 2.5$. The dashed lines represent free-space propagation in the dielectric (lower line) and in vacuo (upper line), respectively. Cut-off is defined as the transition between guided ($\alpha^2 > 0$) and not guided ($\alpha^2 < 0$) propagation. Accordingly, cut-off requires $\alpha = 0$ and the propagation constant β coincides with that of free space; see Eq. (7.35a) with $u_1 = -i\alpha = 0$.

7.4. EQUIVALENT TRANSMISSION LINES: THREE-DIMENSIONAL STRUCTURES

In order to derive the ETL of the most general configurations, it is convenient to refer to the vector equivalent of Eqs. (7.28) extended to y-dependent fields (see Section 4.5.4).

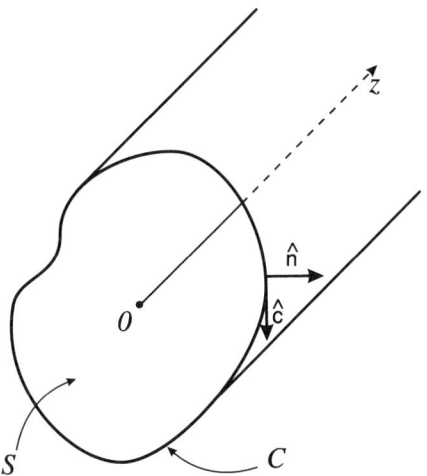

FIGURE 7.19. A uniform waveguide.

We consider electromagnetic fields in a *uniform guide* (see Fig. 7.19): the cross-section geometry does not change with z. The contour C, which is perfectly conducting (a limitation that is relaxed in Section 7.4.6) may partly recede to infinity, thus describing an unbounded guiding configuration as those in Figs. 7.16 and 7.17. Longitudinal conductors may be present in the guide, thus rendering the cross section multiconnected (see Section 7.4.3).

Let

$$\mathbf{E}_t = E_x \hat{\mathbf{x}} + E_y \hat{\mathbf{y}} \quad \text{and} \quad \mathbf{H}_t = H_x \hat{\mathbf{x}} + H_y \hat{\mathbf{y}} \qquad (7.40)$$

be the *transverse field components*. We multiply Eq. (7.28a) and Eq. (7.28b) generalized to y-dependent fields, namely,

$$A \cdot \begin{vmatrix} E_x \\ H_y \end{vmatrix} = \begin{vmatrix} \partial E_z/\partial x \\ \partial H_z/\partial y \end{vmatrix} \quad \text{and} \quad A \cdot \begin{vmatrix} E_y \\ -H_x \end{vmatrix} = \begin{vmatrix} \partial E_z/\partial y \\ -\partial H_z/\partial x \end{vmatrix},$$

by $\hat{\mathbf{x}}$ and $\hat{\mathbf{y}}$, respectively, and add the result to obtain

$$\begin{vmatrix} -i\beta & i\omega\mu \\ i\omega\varepsilon & -i\beta \end{vmatrix} \begin{vmatrix} \mathbf{E}_t \\ \mathbf{H}_t \times \hat{\mathbf{z}} \end{vmatrix} = \begin{vmatrix} \nabla_t E_z \\ \nabla_t H_z \times \hat{\mathbf{z}} \end{vmatrix}. \qquad (7.41)$$

An additional convenience of Eq. (7.41) is that it is independent of the choice of transverse coordinate system.

Equations (4.46a–b) projected onto the z-axis yield

$$\frac{\partial E_y}{\partial x} - \frac{\partial E_x}{\partial y} = -\nabla_t \cdot (\hat{\mathbf{z}} \times \mathbf{E}_t) = -i\omega\mu H_z \tag{7.42a}$$

and

$$\frac{\partial H_y}{\partial x} - \frac{\partial H_x}{\partial y} = \nabla_t \cdot (\mathbf{H}_t \times \hat{\mathbf{z}}) = i\omega\varepsilon E_z, \tag{7.42b}$$

which generalize Eq. (7.29). Substituting E_z and H_z into Eqs. (7.41) with $-i\beta \to \partial/\partial z$ leads to

$$\begin{cases} -\dfrac{\partial \mathbf{E}_t}{\partial z} = i\omega\mu(\mathbf{H}_t \times \hat{\mathbf{z}}) - \dfrac{\nabla_t \nabla_t \cdot (\mathbf{H}_t \times \hat{\mathbf{z}})}{i\omega\varepsilon}, \\ -\dfrac{\partial (\mathbf{H}_t \times \hat{\mathbf{z}})}{\partial z} = i\omega\varepsilon \mathbf{E}_t - \dfrac{[\nabla_t \nabla_t \cdot (\hat{\mathbf{z}} \times \mathbf{E}_t)] \times \hat{\mathbf{z}}}{i\omega\mu}, \end{cases}$$

i.e.,

$$\begin{cases} -\dfrac{\partial \mathbf{E}_t}{\partial z} = i\omega\mu \left[\mathbf{I} + \dfrac{\nabla_t \nabla_t}{k^2} \right] \cdot (\mathbf{H}_t \times \hat{\mathbf{z}}), \\ -\dfrac{\partial \mathbf{H}_t}{\partial z} = i\omega\varepsilon \left[\mathbf{I} + \dfrac{\nabla_t \nabla_t}{k^2} \right] \cdot (\hat{\mathbf{z}} \times \mathbf{E}_t), \end{cases} \tag{7.43}$$

which relate the transverse components of the fields and are again independent of the choice of transverse coordinate system. On the other hand, the *longitudinal* field *components* E_z and H_z are related to the transverse ones via Eq. (7.42).

Solutions for the field in the guide are obtained by setting

$$\mathbf{E}_t(x, y, z) = \mathbf{e}(x, y)V(z) \quad \text{and} \quad \mathbf{H}_t(x, y, z) = \mathbf{h}(x, y)I(z) \tag{7.44}$$

and substituting into Eqs. (7.43). Hence

$$\begin{cases} -\mathbf{e}\dfrac{dV}{dz} = i\omega\mu I \left[\mathbf{I} + \dfrac{\nabla_t \nabla_t}{k^2} \right] \cdot (\mathbf{h} \times \hat{\mathbf{z}}), \\ -\mathbf{h}\dfrac{dI}{dz} = i\omega\varepsilon V \left[\mathbf{I} + \dfrac{\nabla_t \nabla_t}{k^2} \right] \cdot (\hat{\mathbf{z}} \times \mathbf{e}). \end{cases} \tag{7.45}$$

We now study possible TE field solutions ($E_z = 0$), or *modes*. From Eq. (7.42b)

$$\nabla_t \cdot (\mathbf{h} \times \hat{\mathbf{z}}) = \hat{\mathbf{z}} \cdot \nabla_t \times \mathbf{h} = 0,$$

and the first of Eqs. (7.43) shows that $\mathbf{e}$ and $\mathbf{h} \times \hat{\mathbf{z}}$ have the same direction and can be taken proportional[10] of even identical:

$$\mathbf{e} = \mathbf{h} \times \hat{\mathbf{z}}. \tag{7.46}$$

The second of Eqs. (7.43) then shows that $\nabla_t \nabla_t \cdot \mathbf{h}$ should be proportional to $\mathbf{h}$:

$$\nabla_t \nabla_t \cdot \mathbf{h} + k_t^2 \mathbf{h} = 0, \tag{7.47}$$

k_t^2 [1/m^2] being a constant.

Substituting Eqs. (7.46) and (7.47) into Eqs. (7.45) yields

$$\begin{cases} -\dfrac{dV}{dz} = i\beta Z_0 I, \\ -\dfrac{dI}{dz} = i\beta Y_0 V, \end{cases} \tag{7.48}$$

i.e., the same ETL equations (7.33) with

$$\beta = \sqrt{k^2 - k_t^2} \quad \text{and} \quad Z_0 = \frac{1}{Y_0} = \frac{\omega\mu}{\beta}. \tag{7.49a}$$

We conclude that TE propagation can again be modeled in terms of an ETL, whose solution provides the scalar coefficients $V(z)$ and $I(z)$ of the transverse and longitudinal fields, Eqs. (7.44) and (7.42a), respectively. Furthermore, for a guide cross section without inner conductors[11] (*single connected guiding structures*),

$$\oint_{C_1} d c \mathbf{h} \cdot \hat{\mathbf{c}} = 0$$

for any closed path C_1 in the transverse cross section, because longitudinal

---

10. This additional degree of freedom is used in connection with the coaxial cable configuration; see Section 7.4.3.
11. For a multiconnected region the vector function $\mathbf{h}$ should be rendered single-valued by means of an appropriate cut in the guide cross section.

electric currents are absent and $E_z = 0$. Accordingly

$$\mathbf{h} = -\frac{1}{k_t} \nabla_t \psi, \qquad (7.50\text{a})$$

and substituting into Eq. (7.47) gives

$$\nabla_t [\nabla_t^2 \psi + k_t^2 \psi] = 0,$$

i.e.,[12]

$$\nabla_t^2 \psi + k_t^2 \psi = 0. \qquad (7.51\text{a})$$

Boundary conditions require that the electric field tangent to the guide walls is zero on C (see Fig. 7.19):

$$V\mathbf{e} \cdot \hat{\mathbf{c}} = V(\mathbf{h} \times \hat{\mathbf{z}}) \cdot \hat{\mathbf{c}} = -V\mathbf{h} \cdot \hat{\mathbf{n}} = \frac{V}{k_t} \frac{\partial \psi}{\partial n},$$

$$\frac{\partial \psi}{\partial n} = 0 \quad \text{on } C. \qquad (7.52\text{a})$$

We conclude that the transverse field distribution is provided by the solution of the Helmoltz[13] equation (7.51a) with appropriate Neumann boundary conditions (7.52a). The vector fields $\mathbf{h}$ and $\mathbf{e}$ are computed via Eqs. (7.50a) and (7.46) and appear in the transverse field expressions, Eq. (7.44), as well as in the longitudinal field expression, Eq. (7.42a). The representation of TE fields is summarized in Table 7.3.

Modes that are TM can be dealt with similarly. The same ETL equations (7.48) are derived with

$$\beta = \sqrt{k^2 - k_t^2} \quad \text{and} \quad Z_0 = \frac{1}{Y_0} = \frac{\beta}{\omega \varepsilon}, \qquad (7.49\text{b})$$

and the same equation (7.46), $\mathbf{h} = \hat{\mathbf{z}} \times \mathbf{e}$, relates the modal vectors $\mathbf{h}$ and $\mathbf{e}$. We also have

$$\mathbf{e} = -\frac{1}{k_t} \nabla_t \phi, \qquad (7.50\text{b})$$

---

12. The equation should read $\nabla_t^2 \psi + k_t^2 \psi = \text{const}$, whose solution differs from that of Eq. (7.51a) only by an inessential constant.
13. Herman Ludwig Ferdinand von Helmoltz: Potsdam (Germany), 1821–Berlin, 1894.

ENGINEERING TOPICS: PROPAGATION

TABLE 7.3. Summary of TE and TM Modal Expressions

Modes	Vector modal fields	Helmotz equation	ETL parameters	Transverse fields	Longitudinal fields
TE	$\mathbf{h} = -\dfrac{1}{k_t}\nabla_t\psi$	$\nabla_t^2\psi + k_t^2\psi = 0$	$\beta = \sqrt{k^2 - k_t^2}$	$\mathbf{E}_t = e V$	$i\omega\mu H_z = V\nabla_t \cdot \mathbf{h}$ $= V k_t \psi$
	$\mathbf{e} = \mathbf{h} \times \hat{\mathbf{z}}$	$\dfrac{\partial\psi}{\partial n} = 0$ on C	$Z_0 = \dfrac{\omega\mu}{\beta}$	$\mathbf{H}_t = h I$	$E_z = 0$
TM	$\mathbf{e} = -\dfrac{1}{k_t}\nabla_t\phi$	$\nabla_t^2\phi + k_t^2\phi = 0$	$\beta = \sqrt{k^2 - k_t^2}$	$\mathbf{E}_t = e V$	$i\omega\varepsilon E_z = I\nabla_t \cdot \mathbf{e}$ $= I k_t \phi$
	$\mathbf{h} = \hat{\mathbf{z}} \times \mathbf{e}$	$\phi = 0$ on C	$Z_0 = \dfrac{\beta}{\omega\varepsilon}$	$\mathbf{H}_t = h I$	$H_z = 0$
TEM	$\mathbf{e} = -\nabla_t\phi$ $\mathbf{h} = F\hat{\mathbf{z}} \times \mathbf{e}$	$\nabla_t^2\phi = 0$ $\phi = \phi_i$ on C_i	$\beta = k$ $Z_0 = F\zeta^a$	$\mathbf{E}_t = e V$ $\mathbf{H}_t = h I$	$E_z = H_z = 0$

[a] For the choice of the constant F, see Section 7.4.3.

because $H_z = 0$, ϕ again being a solution of the Helmoltz equation

$$\nabla_t^2\phi + k_t^2\phi = 0. \tag{7.51b}$$

Note that the guide cross section may be multiconnected. Boundary conditions for Eq. (7.51b) are again obtained by enforcing the condition that the electric field tangent to the guide walls is zero:

$$V\mathbf{e}\cdot\hat{\mathbf{c}} = -\frac{V}{k_t}\frac{\partial\phi}{\partial c} = 0,$$

and

$$i\omega\varepsilon E_z = I\nabla_t \cdot \mathbf{e} = -\frac{I}{k_t}\nabla_t^2\phi = I k_t \phi = 0$$

on C. The second condition implies the first and Eq. (7.51b) is subject to the Dirichlet condition

$$\phi = 0 \quad \text{on } C. \tag{7.52b}$$

This is no longer true when $E_z = 0$ (in addition to $H_z = 0$) over the entire guide cross section, as discussed subsequently in connection with TEM modes. TM field expressions are listed in Table 7.3.

TEM modes require $E_z = H_z = 0$. If

$$\mathbf{e} = -\nabla_t \phi \qquad (7.50c)$$

because $H_z = 0$ (see TM modes), then

$$\nabla_t^2 \phi = 0, \qquad (7.51c)$$

i.e., the Laplace equation, with boundary condition

$$\frac{\partial \phi}{\partial c} = 0 \quad \text{on } C; \qquad (7.52c)$$

see TM modes again. If the guide cross section is singly connected, then Eq. (7.52c) implies $\phi = \text{const}$ on C, and the only solution to Eq. (7.51c) is $\phi = \text{const}$: no TEM propagation is possible. In the case of multiply connected regions, Eq. (7.52c) implies that ϕ is constant on each of the *separate* boundaries where it can attain different values: TEM modes are possible. We obtain the same ETL equations (7.48) with

$$\beta = k, \qquad (7.49c)$$

because in this case $k_t = 0$. The definition of the ETL characteristic impedance is a question of convenience; see Section 7.4.3. The representation of TEM fields is summarized in Table 7.3.

7.4.1. The Rectangular Waveguide

Let us consider a perfectly conducting metal pipe of rectangular cross section, as depicted in Fig. 7.20. For TE modes, the solution of Eq. (7.51a) can be obtained by the method of separation of variables:

$$\psi(x, y) = (A \cos k_x x + B \sin k_x x) \cdot (C \cos k_y y + D \sin k_y y),$$

with

$$k_x^2 + k_y^2 = k_t^2.$$

Boundary conditions (7.52a) must be satisfied at $(0, y)$ and $(x, 0)$ and require

ENGINEERING TOPICS: PROPAGATION

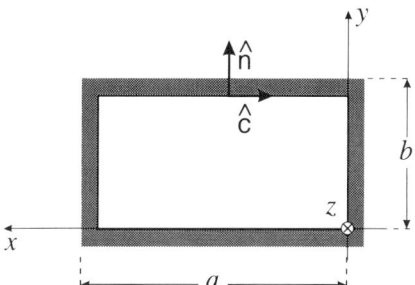

FIGURE 7.20. Cross section of a rectangular waveguide.

$B = D = 0$. When they are satisfied at (a, y) and (x, b), we obtain

$$k_x = \frac{n\pi}{a} \quad \text{and} \quad k_y = \frac{m\pi}{b}, \qquad (7.53a)$$

respectively, with n and m integers. Each choice of these two numbers corresponds to a possible field solution, or mode, in the guide.

The modal functions ψ become

$$\psi_{nm}(x, y) = F_{nm} \cos\left(\frac{n\pi}{a} x\right) \cos\left(\frac{m\pi}{b} y\right), \qquad (7.54a)$$

and the corresponding modal transverse vector functions $\mathbf{e}_{nm}$ and $\mathbf{h}_{nm}$ can be easily computed. The (arbitrary) constant F_{nm} in Eq. (7.54a) can be chosen by enforcing that the total power associated with the mode and crossing the waveguide cross section, namely,

$$\frac{1}{2}\int_0^a dx \int_0^b dy (\mathbf{E}_t \times \mathbf{H}_t^*) \cdot \hat{\mathbf{z}} = \frac{1}{2} V_{nm} I_{nm}^* \int_0^a dx \int_0^b dy (\mathbf{e}_{nm} \times \mathbf{h}_{nm}^*) \cdot \hat{\mathbf{z}},$$

is equal to $\frac{1}{2} V_{nm} I_{nm}^*$, thus requiring that

$$\int_0^a dx \int_0^b dy (\mathbf{e}_{nm} \times \mathbf{h}_{nm}^*) \cdot \hat{\mathbf{z}} = 1. \qquad (7.55)$$

The latter integral, which is extended to the guide cross section S, can be computed directly, or by means of the general procedure in Section 7.4.4. By following the direct approach we first compute the modal vectors

$$\mathbf{h}_{nm} = \frac{F_{nm}}{\sqrt{(n\pi/a)^2 + (m\pi/b)^2}} \left[\left(\frac{n\pi}{a}\right) \sin\left(\frac{n\pi}{a} x\right) \cos\left(\frac{m\pi}{b} y\right) \hat{\mathbf{x}} \right.$$
$$\left. + \left(\frac{m\pi}{b}\right) \cos\left(\frac{n\pi}{a} x\right) \sin\left(\frac{m\pi}{b} y\right) \hat{\mathbf{y}} \right],$$
$$\hat{\mathbf{z}} \cdot (\mathbf{e}_{nm} \times \mathbf{h}^*_{nm}) = (\hat{\mathbf{z}} \times \mathbf{e}_{nm}) \cdot \mathbf{h}^*_{nm} = \mathbf{h}_{nm} \cdot \mathbf{h}^*_{nm},$$

and Eq. (7.55) is computed as follows:

$$\int_0^a dx \int_0^b dy\, \mathbf{h}_{nm} \cdot \mathbf{h}^*_{nm} = \begin{cases} |F_{nm}|^2 \dfrac{ab}{4} & \text{if } n \neq 0, m \neq 0, \\[1em] |F_{nm}|^2 \dfrac{ab}{2} & \text{if } n = 0, m \neq 0 \text{ or } n \neq 0, m = 0. \end{cases}$$

Accordingly

$$F_{nm} = \sqrt{\frac{2}{ab}} \quad \text{for } \text{TE}_{n0} \text{ and } \text{TE}_{0n} \text{ modes,} \qquad (7.56a)$$

$$F_{nm} = \sqrt{\frac{4}{ab}} \quad \text{for } \text{TE}_{nm} \text{ modes with } n \neq 0, m \neq 0. \qquad (7.56b)$$

A double-numerable number of transverse electric modes, TE_{nm}, can propagate along the guide, which operates as a dispersive channel because

$$\omega^2 = c^2 \beta^2 + c^2 k_t^2$$

from Eq. (7.49a). The first propagating mode is TE_{10}, if $a > b$, with a cut-off wavelength at

$$\frac{2\pi}{\lambda_c} = \frac{\pi}{a}, \quad \text{i.e., } \lambda_c = 2a. \qquad (7.57a)$$

Modes that are TM can be dealt with similarly. We have

$$k_x = \frac{n\pi}{a}, \quad k_y = \frac{m\pi}{b}, \qquad (7.53b)$$

ENGINEERING TOPICS: PROPAGATION

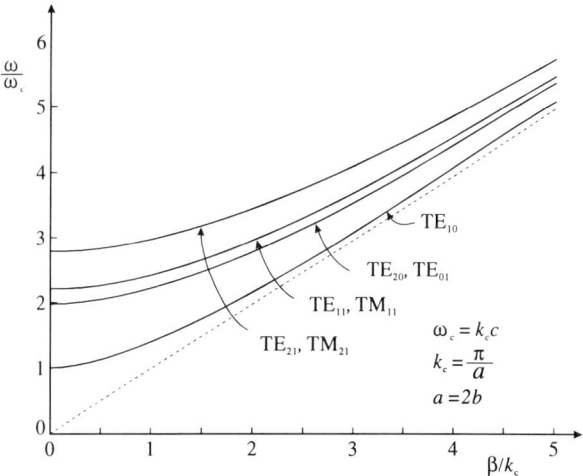

FIGURE 7.21. Brillouin diagrams for the rectangular waveguide. Angular frequency and propagation constant are normalized to cut-off angular frequency ω_c and corresponding free-space propagation constant $k_c = \omega_c\sqrt{\varepsilon\mu}$ of the fundamental TE_{10} mode. The dashed line corresponds to free-space propagation. Waveguide dimension $a = 2b$.

and

$$\phi_{nm} = F_{nm} \sin\left(\frac{n\pi}{a}x\right)\sin\left(\frac{m\pi}{b}y\right), \tag{7.54b}$$

with F_{nm} given by Eq. (7.56b) alone because TM_{0m} and TM_{n0} modes are not present in Eq. (7.54b). TM modes are dispersive as well. The first mode is TM_{11}, whose cut-off wavelength

$$\lambda_c = \frac{2ab}{\sqrt{a^2 + b^2}} \tag{7.57b}$$

is shorter than that of the TE_{10} mode.

Brillouin diagrams corresponding to the rectangular waveguide are given in Fig. 7.21.

7.4.2. The Circular Waveguide

Let us consider a perfectly conducting metal pipe of circular cross section as depicted in Fig. 7.22.

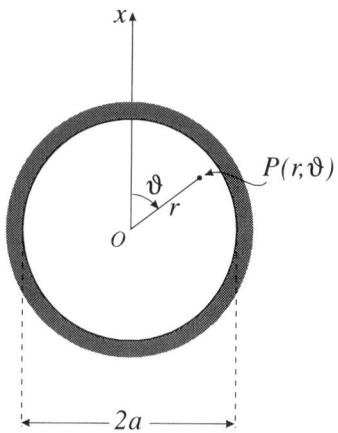

FIGURE 7.22. Cross section of a circular waveguide.

For TE modes, the solution of Eq. (7.51a) can be obtained by the method of separation of variables[14]:

$$\psi(r, \theta) = [AJ_n(k_t r) + BY_n(k_t r)] \exp(\pm in\theta), \qquad n = 0, 1, \ldots,$$

where the θ-periodicity of the field has been satisfied by taking n as an integer; see Section 4.5.4. We also have $B = 0$ because $Y_n(k_t r)$ diverges for $r = 0$. Boundary conditions (7.52a) require

$$J'_n(k_t a) = 0, \qquad k_t a = q_{nm}, \tag{7.58a}$$

q_{nm} being the successive zeros of the derivatives of Bessel functions (see Table 4.4). Accordingly

$$\psi_{nm}(r, \theta) = F_{nm} J_n\left(q_{nm}\frac{r}{a}\right) \exp(\pm in\theta), \tag{7.59a}$$

where the constant F_{nm} is obtained by enforcing the usual normalization (7.55):

$$F_{nm} = \frac{1}{\sqrt{\pi}} \frac{1}{a|J_{n+1}(q_{nm})|} \frac{q_{nm}}{\sqrt{q_{nm}^2 - n^2}} \tag{7.60a}$$

with the aid of Eq. (C.9).

---

14. The symbol θ is used for the angular coordinate to avoid confusion with the subsequently used symbol ϕ; see Eq. (7.59b).

Also in this case a double-numerable number of modes TE_{nm} can propagate along the guide, which operates as a dispersive channel. The first propagating mode is TE_{11}, with a cut-off wavelength given by

$$\frac{2\pi}{\lambda_c} = \frac{q_{11}}{a} = \frac{1.841}{a}, \quad \text{i.e., } \lambda_c = 1.08\pi a.$$

Modes that are TM can be dealt with similarly. Boundary conditions (7.52b) require

$$J_n(k_t a) = 0, \quad k_t a = p_{nm}, \tag{7.58b}$$

p_{nm} being the successive zeros of the Bessel functions (see Table 4.3). Accordingly

$$\phi_{nm}(r, \theta) = F_{nm} J_n\left(p_{nm}\frac{r}{a}\right) \exp(\pm in\theta), \tag{7.59b}$$

where

$$F_{nm} = \frac{1}{\sqrt{\pi}} \frac{1}{a|J_{n+1}(p_{nm})|}, \tag{7.60b}$$

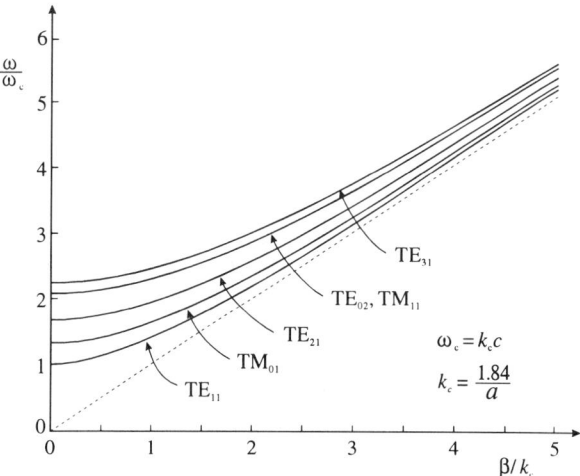

FIGURE 7.23. Brillouin diagrams for the circular waveguide. Angular frequency and propagation constant are normalized to cut-off angular frequency ω_c and corresponding free-space propagation constant $k_c = \omega_c\sqrt{\varepsilon\mu}$ of the fundamental TE_{11} mode. The dashed line corresponds to free-space propagation.

with the aid of Eq. (C.10). TM modes are dispersive as well. The first propagating mode is TM_{01} with

$$\frac{2\pi}{\lambda_c} = \frac{p_{01}}{a} = \frac{2.405}{a} = 0.83\pi a,$$

and its cut-off wavelength is shorter than that of TE_{11}, which is the dominant mode of the guide.

Brillouin diagrams for the circular waveguide are given in Fig. 7.23.

7.4.3. The Coaxial Cable

Let us consider the coaxial cable whose cross section is depicted in Fig. 7.24. This is a doubly connected region and TEM mode propagation is possible (see Section 7.4). The solution of Eq. (7.51c) with $\phi(a) = \phi_a = $ const and $\phi(b) = \phi_b = $ const' is required. Accordingly, ϕ is azimuthally independent and

$$\phi(r) = A \ln r + B.$$

The application of boundary conditions leads to

$$\ln\left(\frac{b}{a}\right)\phi(r) = \phi_a \ln\frac{b}{r} - \phi_b \ln\frac{a}{r}.$$

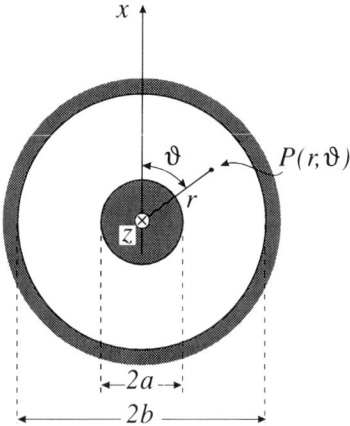

FIGURE 7.24. Cross section of a coaxial cable.

ENGINEERING TOPICS: PROPAGATION

The electromagnetic field is obtained by employing Eqs. (7.44), (7.50c), and (7.46) to yield

$$\mathbf{E}(r, z) = \frac{\phi_a - \phi_b}{r \ln(b/a)} V(z)\hat{\mathbf{r}} = \mathbf{e}(r)V(z)$$

and

$$\mathbf{H}(r, z) = \frac{\phi_a - \phi_b}{r \ln(b/a)} I(z)\hat{\boldsymbol{\theta}} = \mathbf{h}(r)I(z).$$

It is convenient that the function V coincides with the voltage between the inner and outer conductor of the cable,

$$-\int_b^a dr\, \mathbf{E} \cdot \hat{\mathbf{r}} = -V \int_b^a dr\, \mathbf{e} \cdot \hat{\mathbf{r}},$$

and that the function I coincides with the total current along the inner conductor,

$$\int_0^{2\pi} d\theta\, r\mathbf{H} \cdot \hat{\boldsymbol{\theta}} = I \int_0^{2\pi} d\theta\, r\mathbf{h} \cdot \hat{\boldsymbol{\theta}}.$$

To meet these requirements we require the condition

$$-\int_b^a dr\, \mathbf{e} \cdot \hat{\mathbf{r}} = \phi_a - \phi_b = 1,$$

thus obtaining

$$\mathbf{e} = \frac{1}{r}\hat{\mathbf{r}},$$

and we take

$$\mathbf{h} = F\hat{\mathbf{z}} \times \mathbf{e},$$

instead of Eq. (7.46), and enforce the condition

$$\int_0^{2\pi} d\theta\, r\mathbf{h} \cdot \hat{\boldsymbol{\theta}} = \frac{2\pi F}{\ln(b/a)} = 1,$$

so that

$$\mathbf{h} = \frac{\ln(b/a)}{2\pi} \hat{\mathbf{z}} \times \mathbf{e}.$$

Substitution into Eq. (7.43) yields the TL equations (7.48) with

$$\beta = k \quad \text{and} \quad Z_0 = \zeta \frac{\ln(b/a)}{2\pi}, \tag{7.49d}$$

This is in agreement with the results of Table 7.2 for the lossless case because

$$\beta Z_0 = \omega L = \frac{\omega \mu}{2\pi} \ln(b/a) \quad \text{and} \quad \frac{\beta}{Z_0} = \omega C = \frac{2\pi\omega\varepsilon}{\ln(b/a)}.$$

The TL parameters L and C are the (static) values of inductance and capacitance per unit length of the cable, respectively.

7.4.4. Mode Orthogonality and Power Flux

Sections 7.4.1 and 7.4.2 show that an infinite, numerable number of modes can propagate in a waveguide. These modes form the orthogonal basis of a Hilbert space and any (square-integrable) electromagnetic field can be expanded as a superposition of modes.

We first prove that the *eigenvalues* k_t^2 of Eqs. (7.51a) are real. By the use of Eq. (A.25) we have

$$\iint_S dS \psi^* \nabla_t^2 \psi = -k_t^2 \iint_S dS |\psi|^2$$

$$= -\iint_S dS \nabla_t \psi^* \cdot \nabla_t \psi + \oint_C dc \psi^* \frac{\partial \psi}{\partial n},$$

where S is the waveguide cross section and C its boundary. The line integral is zero in view of boundary conditions (7.52a) and

$$k_t^2 = \frac{\iint_S dS |\nabla_t \psi|^2}{\iint_S dS |\psi|^2} \geq 0. \tag{7.61}$$

Similar results apply for TM modes.

For two eigenvalues k_p^2, k_q^2 and corresponding *eigenfunctions* ψ_p, ψ_q we have from Eq. (5.51a)

$$\psi_q^*(\nabla_t^2 \psi_p + k_p^2 \psi_p) = 0 \quad \text{and} \quad \psi_p(\nabla_t^2 \psi_q^* + k_q^2 \psi_q^*) = 0.$$

Subtracting the second equation from the first and using Eq. (A.26), we obtain

$$\iint_S dS [\psi_q^* \nabla_t^2 \psi_p - \psi_p \nabla_t^2 \psi_q^*] = (k_q^2 - k_p^2) \iint_S dS \psi_p \psi_q^*$$

$$= \oint_C dc \left(\psi_q^* \frac{\partial \psi_p}{\partial n} - \psi_p \frac{\partial \psi_q^*}{\partial n} \right) = 0,$$

in view of boundary conditions (7.52a).

If $k_p^2 \neq k_q^2$ for $p \neq q$, then

$$\iint_S dS \psi_p \psi_q^* = 0, \quad p \neq q.$$

The integral can always be normalized to unity for $p = q$ (see Section 7.4.1), thus obtaining

$$\iint_S dS \psi_p \psi_q^* = \delta_{pq}, \tag{7.62}$$

where the modal functions form an *orthonormal set*.

If $k_p^2 = k_q^2$ for $p \neq q$ the modes are *degenerate*, but this degenerate set can be linearly combined in such a way as to become orthogonal (the *Gram*[15]–*Schmidt*[16] *orthogonalization procedure*).

---

15. Jorgen Pedersen Gram: Nustrup (Danemark), 1850– Copenhagen, 1916.
16. Erhard Schmidt: Dorpart (now Tartu, Russia), 1876– Berlin, 1959.

Similar results apply for TM modes. TE and TM eigenfunctions are mutually orthogonal as well.

The eigenfunction orthogonality can be used to show that the vector modes are also orthogonal:

$$\iint_S dS\,\mathbf{h}_p \cdot \mathbf{h}_q^* = \frac{1}{k_p k_q} \iint_S dS\,\nabla_t \psi_p \cdot \nabla_t \psi_q^*$$

$$= \frac{1}{k_p k_q}\left[-\iint_S dS\,\psi_p \nabla_t^2 \psi_q^* + \oint_C dc\,\psi_p \frac{\partial \psi_q^*}{\partial n}\right]$$

$$= \frac{k_q}{k_p} \iint_S dS\,\psi_p \psi_q^* = \delta_{pq}, \qquad (7.63a)$$

where use has been made of Eq. (A.25) and of boundary conditions (7.52a). For TM modes we obtain similarly

$$\iint_S dS\,\mathbf{e}_p \cdot \mathbf{e}_q^* = \delta_{pq}. \qquad (7.63b)$$

Equation (7.63a) is now used to compute the power flux associated with a mode in the guide. Hence

$$P = \frac{1}{2}\iint dS\,\mathbf{E}_t \times \mathbf{H}_t^* \cdot \hat{\mathbf{z}}\,dS = \frac{1}{2} VI^* \iint dS\,\hat{\mathbf{z}} \cdot (\mathbf{e} \times \mathbf{h}^*)$$

$$= \frac{1}{2} VI^* \iint dS\,\mathbf{h} \cdot \mathbf{h}^* = \frac{1}{2} VI^*,$$

and the ETL can be safely used to compute the power propagation along the guide.

Similar results apply for TM and TEM modes.

In a multimode guide the total power flux is the sum of the individual mode powers. If

$$\mathbf{E} = \mathbf{E}_p + \mathbf{E}_q \quad \text{and} \quad \mathbf{H} = \mathbf{H}_p + \mathbf{H}_q,$$

ENGINEERING TOPICS: PROPAGATION

i.e., the superposition of two TE modes, then

$$\frac{1}{2}\iint_S dS\hat{\mathbf{z}} \cdot (\mathbf{E} \times \mathbf{H}^*) = \frac{1}{2}V_p I_p^* + \frac{1}{2}V_q I_q^* + \frac{1}{2}V_p I_q^* \iint_S dS(\hat{\mathbf{z}} \times \mathbf{e}_p) \cdot \mathbf{h}_q^*$$
$$+ \frac{1}{2}V_q I_p^* \iint_S dS(\hat{\mathbf{z}} \times \mathbf{e}_q) \cdot \mathbf{h}_p^*$$

and the mixed terms are zero in view of Eqs. (7.46) and (7.63a).

Similar results are obtained by considering TM modes as well as TE and TM modes.

7.4.5. Waveguide Excitation

Equations (7.63) can be employed conveniently to expand the electromagnetic field in the waveguide modes. As an example, consider the field excited by a longitudinal magnetic source $\mathbf{K} = K\hat{\mathbf{z}}$, $[K] = [\text{volt/m}^2]$, in the guide. The (source-free) equations (7.42) should be substituted by the following ones:

$$-\nabla_t \cdot (\hat{\mathbf{z}} \times \mathbf{E}_t) = -i\omega\mu H_z - K \quad (7.64\text{a})$$

and

$$\nabla_t \cdot (\mathbf{H}_t \times \hat{\mathbf{z}}) = i\omega\varepsilon E_z, \quad (7.64\text{b})$$

while the system (7.43) similarly transforms to

$$\begin{cases} -\dfrac{\partial \mathbf{E}_t}{\partial z} = i\omega\mu \left[I + \dfrac{\nabla_t \nabla_t}{k^2} \right] \cdot (\mathbf{H}_t \times \hat{\mathbf{z}}), \\ -\dfrac{\partial \mathbf{H}_t}{\partial z} = i\omega\varepsilon \left[I + \dfrac{\nabla_t \nabla_t}{k^2} \right] \cdot (\hat{\mathbf{z}} \times \mathbf{E}_t) + \dfrac{\nabla_t K}{i\omega\mu}. \end{cases} \quad (7.65)$$

Inspections of Eq. (7.64b) shows that $E_z = 0$ is an allowed condition and TE modes are excited.

We now expand the source term $\nabla_t K$ in the set of TE modes:

$$\nabla_t K = -i\omega\mu \sum \hat{I}_p \mathbf{h}_p, \quad (7.66)$$

$[\hat{I}] = [\text{ampere/m}]$, where a double summation index is understood.[17] By

---

17. The summation index is usually double. A single index is used symbolically for simplicity, whenever confusion does not arise.

similarly expanding the field we are lead to ETL equations formally identical to those of Section 7.2.7,

$$\begin{cases} -\dfrac{dV_p}{dz} = i\beta Z_0 I_p, \\ -\dfrac{dI_p}{dz} = i\beta Y_0 V_p - \hat{I}_p, \end{cases}$$

whose solution provides the amplitudes V_p and I_p for each mode excited in the guide.

The current modal source, $\hat{I}_p$, is readily obtained on multiplying Eq. (7.66) by $\mathbf{h}_p^*$ and integrating over the guide cross section. Use of Eq. (7.63a) leads to

$$-i\omega\mu\hat{I}_p = \iint_S dS \mathbf{h}_p^* \cdot \nabla_t K = -\frac{1}{k_p}\iint_S dS \nabla_t \psi_p^* \cdot \nabla_t K$$
$$= \frac{1}{k_p}\iint_S dS K \nabla_t^2 \psi_p^* = -k_p \iint_S dS K \psi_p^*,$$

where Eq. (A.25) and boundary conditions (7.52a) have been employed.

For a rectangular waveguide the TE$_{nm}$ modes are all mutually orthogonal, even when the eigenvalues coincide, due to the factorization of the modal functions $\psi_{nm}(x, y) = \psi_n(x)\psi_m(y)$, where ψ_n and ψ_m form independent orthogonal sets. For a small loop excitation (see Fig. 7.25) we have

$$\mathbf{K} = i\omega U \delta\left(y - \frac{b}{2}\right)\delta(x - a)\delta(z)\hat{\mathbf{z}} \approx i\omega\mu I \Delta A \delta\left(y - \frac{b}{2}\right)\delta(x - a)\delta(z)\hat{\mathbf{z}},$$

where ΔA is the small loop area and I its current. The modal source current for the dominant TE$_{10}$ mode is readily obtained:

$$\hat{I}_{10}(z) = \frac{\pi/a}{i\omega\mu}\int_0^a dx \int_0^b dy \cdot K(x, y, z) \sqrt{\frac{2}{ab}} \cos\left(\frac{\pi}{a}x\right)$$
$$= -\frac{\Delta A}{S}\sqrt{\frac{2b}{a}}\pi I \delta(z).$$

Note that the small loop can be modeled as an elementary magnetic dipole as far as the modal functions $\psi_{nm}(x, y)$ of the guide are essentially constant across the loop area. For the computation of higher-order modes a more refined model is needed for the source.

ENGINEERING TOPICS: PROPAGATION

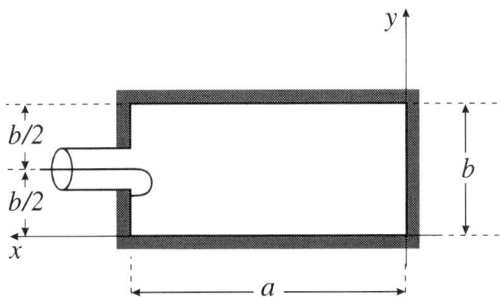

FIGURE 7.25. Rectangular guide excitation.

7.4.6. Waveguide Losses

A waveguide exhibits losses either because the filling material is lossy, or because the guide walls are imperfectly conducting.

Consider the longitudinal section of the guide, as depicted in Fig. 7.26, and a mode propagating along the positive sense of the z-axis. In this case

$$P(z) = \tfrac{1}{2}V(z)I^*(z) = P(0)\exp(-2\alpha z),$$

and the power flux decrease,

$$P(z) - P(z + \Delta z) \approx 2\alpha P(z)\Delta z,$$

is accounted for by the losses that occur in the guide section of length Δz.

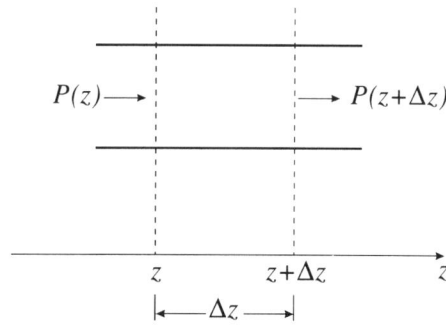

FIGURE 7.26. Evaluation of waveguide losses.

For volume (dielectric) losses, let $\varepsilon = \varepsilon_1 - i\varepsilon_2$. Hence

$$\frac{1}{2}\frac{|V|^2}{R_0}2\alpha\Delta z = \frac{1}{2}\omega\varepsilon_2|V|^2\Delta z \iint_S dS|\mathbf{e}|^2, \qquad (7.67a)$$

which is used to compute the value of the attenuation constant α. A case of interest is the coaxial cable, because a dielectric material is usually necessary to sustain the inner conductor. Here

$$\iint_S dS|\mathbf{e}|^2 = \iint dS \nabla_t\phi \cdot \nabla_t\phi = \oint dc\phi \frac{\partial \phi}{\partial n} = \frac{2\pi}{\ln(b/a)},$$

by using Eq. (A.25) and the results of Section 7.4.3. Accordingly, Eqs. (7.67a) and (7.49d) yield

$$\alpha = \frac{\pi}{\lambda}\frac{\varepsilon_2}{\varepsilon_1}. \qquad (7.67b)$$

We note that this result is valide for small losses, because in Eq. (7.67a) R_0 is assumed real. In fact, Eq. (7.67b) can be most simply obtained via the series expansion

$$k = \omega\sqrt{\mu(\varepsilon_1 - i\varepsilon_2)} \approx \omega\sqrt{\mu\varepsilon_1}\left(1 - i\frac{\varepsilon_2}{2\varepsilon_1}\right).$$

For wall losses we use a perturbative analysis, assuming that on the guide walls the tangential electric field $\mathbf{E}_t$ is proportional to the *unperturbed* tangential magnetic field:

$$\mathbf{E}_t = \zeta_M \mathbf{H} \times \hat{\mathbf{n}} \qquad \text{on } C, \qquad (7.68)$$

where ζ_M is the intrinsic impedance of the metal (the *Leontovic*[18] or *Shchukin*[19]–*Leontovic boundary conditions*). Equation (7.68) is justified intuitively because in a good conductor

$$\left|\varepsilon - i\frac{\sigma}{\omega}\right| \approx |\sigma/\omega| \gg \varepsilon_0,$$

---

18. Mikail Alexandrovich Leontovic: Moscow (Russia), 1903–1981.
19. Aleksander Nikolaevich Shchukin: St. Petersburg (Russia), 1900–1990.

ENGINEERING TOPICS: PROPAGATION

and plane-wave propagation in the metal is always along the normal, irrespective of the incidence angle [see Eq. (4.74)].

If we set

$$\zeta_M = \sqrt{\frac{\mu}{\varepsilon - i\sigma/\omega}} \approx \frac{1}{\sigma\delta}(1 + i) \tag{7.69}$$

[see Eq. (4.79)], then the flux of the real part of the Poynting vector across the guide walls, i.e., the dissipated power, can be computed and equated to the power loss. For TM modes

$$\frac{1}{2}R_0|I|^2 2\alpha\Delta z = \frac{1}{2}\frac{|I|^2}{\sigma\delta}\Delta z \oint_C dc|\mathbf{h}\cdot\hat{\mathbf{c}}|^2, \tag{7.70a}$$

with

$$\oint_C dc|\mathbf{h}\cdot\hat{\mathbf{c}}|^2 = \oint_C dc|\mathbf{e}\times\hat{\mathbf{z}}\cdot\hat{\mathbf{c}}|^2$$

$$= \oint_C dc|\mathbf{e}\cdot\hat{\mathbf{n}}|^2 = \frac{1}{k_t^2}\oint_C dc\left|\frac{\partial\phi}{\partial n}\right|^2,$$

where use has been made of Eq. (7.50b).

For TE modes we must include the additional longitudinal component H_z of the magnetic field, in which case

$$\frac{1}{2}R_0|I|^2 2\alpha\Delta z = \frac{1}{2}\frac{|I|^2}{\sigma\delta}\Delta z\frac{1}{k_t^2}\oint_C dc\left|\frac{\partial\psi}{\partial c}\right|^2 + \frac{1}{2}\frac{k_t^2}{(\omega\mu)^2}\frac{R_0^2|I|^2}{\sigma\delta}\Delta z\oint_C dc|\psi|^2. \tag{7.70b}$$

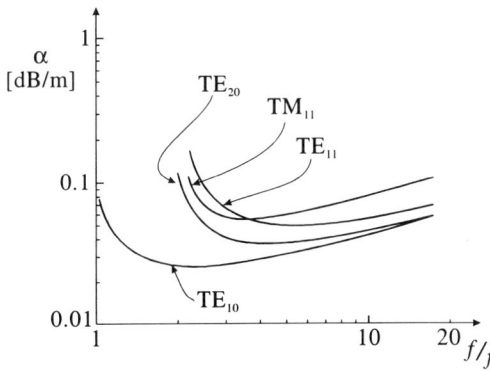

FIGURE 7.27. Plot of attenuation constant vs. operating frequency for several modes in a rectangular copper waveguide: $\sigma = 5.8 \times 10^7$ [siemens/m], $a = 2b = 2$ in. (10.16 cm). The frequency is normalized to the cut-off frequency f_c of the fundamental TE_{10} mode.

Equations (7.70) can be used to evaluate the attenuation constant α. This coefficient increases with frequency for TM modes because $1/\sigma\delta \sim \sqrt{f}$ [see Eq. (4.79)]. This is at variance with TE modes, where the first term in Eq. (7.70b) increases as $(f)^{1/2}$ and the second decreases as $(f)^{-3/2}$. Low-attenuation TE modes require $\partial\psi/\partial c = 0$ on the guide boundary. This constraint is satisfied by azimuthally independent modes in a circular pipe (see Section 5.4.2) which, however, do not comprise the fundamental mode.

Plots of α vs. the operating frequency are given in Fig. 7.27 for the rectangular waveguide.

7.4.7. The Inhomogeneous Rectangular Waveguide

Consider the two-dielectric rectangular waveguide depicted in Fig. 7.28. TE and TM modes with respect to the y-axis are possible. Again, it is convenient to introduce the ETL along y, although true propagation is along the z-axis. The inhomogeneous-guide eigenvalues can be easily computed with the aid of the transverse resonance condition, as in Sections 7.3.2 and 7.3.3.

For TE modes propagating along the positive z-axis we have, but for a constant,

$$\psi(x, z) \sim \cos\left(\frac{n\pi}{a} x\right) \exp(-i\beta z), \tag{7.71a}$$

which is the solution of Eq. (7.51a) subject to the boundary condition

$$\partial\psi/\partial x = 0 \quad \text{at } x = 0, a,$$

and to the radiation condition at infinity. Also,

$$k^2 = \beta^2 + \left(\frac{n\pi}{a}\right)^2 + u_0^2$$

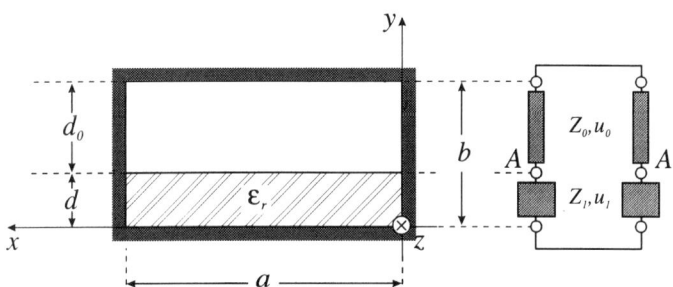

FIGURE 7.28. Inhomogeneously filled rectangular waveguide and its associated ETL.

in the air and

$$k^2\varepsilon_r = \beta^2 + \left(\frac{n\pi}{a}\right)^2 + u^2$$

in the dielectic, which combine to give

$$k^2(\varepsilon_r - 1) = u^2 - u_0^2. \tag{7.72}$$

Now Eq. (7.37) is employed to yield

$$\frac{\tan u_0 d_0}{u_0} + \frac{\tan ud}{u} = 0, \quad d + d_0 = b, \tag{7.73a}$$

by computing the two input impedances of the ETL at section AA.

Equation (7.73a) combined with Eq. (7.72) provide the dispersion equation for the structure and the (infinitely numerable) y-wavenumbers u_{0m} and u_m of the guiding structure.

If $V(x)$ and $I(x)$ represent the modal solution of the ETL in Fig. 7.28, then the corresponding field components are given by

$$H_x\hat{\mathbf{x}} + H_z\hat{\mathbf{z}} = -\frac{1}{k_t}I\nabla_t\psi, \quad i\omega\mu H_y = k_t V\psi,$$

$$\mathbf{E} = \mathbf{H} \times \hat{\mathbf{y}},$$

where

$$k_{t0}^2 = u_{0m}^2 + \left(\frac{n\pi}{a}\right)^2 \quad \text{and} \quad k_t^2 = u_m^2 + \left(\frac{n\pi}{a}\right)^2$$

in the air and dielectric, respectively. We note that the modes are *hybrid* with respect to the propagation direction; E_z and H_z both differ from zero unless $n = 0$, which implies that E_z is equal to zero.

TM modes can be dealt with similarly. In this case

$$\phi(x, z) \sim \sin\left(\frac{n\pi}{a}x\right)\exp(-i\beta z), \tag{7.71b}$$

which is the counterpart of Eq. (7.71a). Equation (7.72) remains unchanged

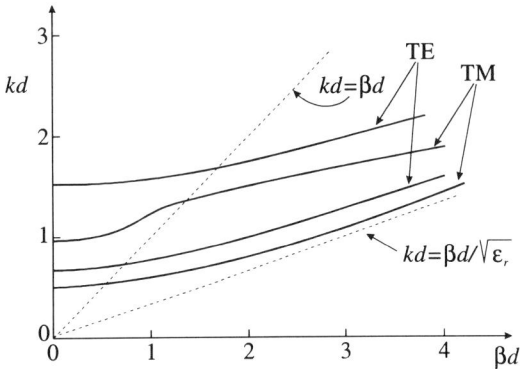

FIGURE 7.29. Normalized Brillouin diagrams kd vs. βd for the first modes of the inhomogeneous waveguide of Fig. 7.28: $a = 2b$, $d = b/2$, $\varepsilon_r = 9$. Dashed curves refer to free-space propagation in vacuo and in a dielectric medium with $\varepsilon_r = 9$. Modes are TE and TM with respect to y.

while Eq. (7.73a) is replaced by

$$u_0 \tan(u_0 d_0) + \frac{u}{\varepsilon_r} \tan(ud) = 0. \tag{7.73b}$$

Also in this case the fields are hybrid with respect to z, unless $n = 0$ ($H_z = 0$).

Brillouin diagrams for the first two TE and TM modes with respect to y are given in Fig. 7.29.

7.4.8. The Fiber

Consider the dielectric fibre guiding structure whose cross section is depicted in Fig. 7.30. As already noted in Section 4.5.4, propagation of pure TE or TM modes with respect to the z-direction is not possible, unless the field is azimuthally independent. The complete field solution is obtained by applying the general procedure of Section 4.5.4. The field is computed in the core $r \leqslant a$ and in the cladding $r \geqslant a$ of the fiber, while continuity of its tangential components must be satisfied at the interface $r = a$.

It is instructive to proceed in a different way, in order to obtain a *radial transmission line* model for the structure, the ETL being along the radial direction $\hat{\mathbf{r}}$.

Let

$$\mathbf{E}(r, \theta, z) \to \exp(-i\beta z)\exp(in\theta)\mathbf{E}(r) \quad \text{and} \quad \mathbf{H}(r, \theta, z) \to \exp(-i\beta z)\exp(in\theta)\mathbf{H}(r).$$

We first consider fields TE with respect to $\hat{\mathbf{z}}$, and compute the z- and

ENGINEERING TOPICS: PROPAGATION

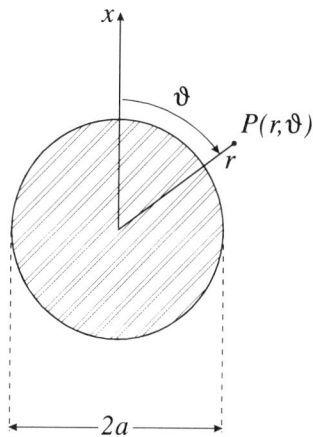

FIGURE 7.30. Cross section of a dielectric fiber.

θ-components of Eq. (4.46a) and the z-component of Eq. (4.46b):

$$\frac{1}{r}\frac{d}{dr}(rE_\theta) - i\frac{n}{r}E_r = -i\omega\mu H_z, \qquad (7.74a)$$

$$i\beta E_r = -i\omega\mu H_\theta, \qquad (7.74b)$$

$$\frac{in}{r}H_z + i\beta H_\theta = i\omega\varepsilon E_r. \qquad (7.74c)$$

These equations may be combined to eliminate E_r and H_θ and yield the first of Eqs. (7.75a) below. We similarly compute the θ-component of Eq. (4.46b) and the r-component of Eq. (4.46a):

$$-i\beta H_r - \frac{dH_z}{dr} = i\omega\varepsilon E_\theta, \qquad (7.74d)$$

$$i\beta E_\theta = -i\omega\mu H_r. \qquad (7.74e)$$

The latter equations enable one to eliminate H_r and lead to the second of Eqs. (7.75a) below. If

$$k_r^2 + \beta^2 = k^2 = \omega^2 \varepsilon \mu,$$

then we have

$$\begin{cases} -\dfrac{1}{r}\dfrac{d}{dr}rE_\theta = i\omega\mu\left(1 - \dfrac{n^2}{k_r^2 r^2}\right)H_z, \\ -\dfrac{d}{dr}H_z = i\omega\varepsilon\dfrac{k_r^2}{k^2}E_\theta, \end{cases} \quad (7.75\text{a})$$

which are the radial ETL equations for the cylindrical guiding structure of Fig. 7.30 when $E_z = 0$. The other field components are readily expressed in terms of E_θ and H_z via Eqs. (7.74b,c,e):

$$\begin{cases} H_\theta = \dfrac{\beta}{k_r}\dfrac{n}{k_r r}H_z, \\ E_r = \dfrac{\omega\mu}{k_r}\dfrac{n}{k_r r}H_z, \\ H_r = -\dfrac{\beta}{\omega\mu}E_\theta. \end{cases} \quad (7.76\text{a})$$

Computation of fields that are TM with respect to z can be dealt with similarly to yield

$$\begin{cases} E_\theta = \dfrac{\beta}{k_r}\dfrac{n}{k_r r}E_z, \\ H_r = -\dfrac{\omega\varepsilon}{k_r}\dfrac{n}{k_r r}E_z, \\ E_r = \dfrac{\beta}{\omega\varepsilon}H_\theta, \end{cases} \quad (7.76\text{b})$$

and

$$\begin{cases} \dfrac{1}{r}\dfrac{d}{dr}rH_\theta = i\omega\varepsilon\left(1 - \dfrac{n^2}{k_r^2 r^2}\right)E_z, \\ \dfrac{dE_z}{dr} = i\omega\mu\dfrac{k_r^2}{k^2}H_\theta, \end{cases} \quad (7.75\text{b})$$

which are the radial ETL equations for the cylindrical structure of Fig. 7.30 when $H_z = 0$.

Equations (7.75) are interesting because they allow computation of the r-dependence of fields for *graded index fibers*, i.e., when the dielectric constant of the core of the fiber becomes r-dependent.

The two ETL equations, (7.75a) and (7.75b), are coupled by boundary conditions at the core–cladding interface $r = a$, and these can be satisfied only when both sets of field components are present.

The solution of Eqs. (7.75a) is obtained by multiplying the second equation by r, taking the derivative with respect to r, and making use of the first equation. This procedure leads to

$$\frac{1}{r}\frac{d}{dr}r\frac{dH_z}{dr} + \left(k_r^2 - \frac{n^2}{r^2}\right)H_z = 0, \tag{7.77}$$

which is a Bessel equation. Accordingly

$$\begin{cases} H_z = PZ_n(k_r r), \\ E_\theta = \frac{i\omega\mu}{k_r}PZ'_n(k_r r), \end{cases} \tag{7.78a}$$

where $Z_n(k_r r)$ is the appropriate solution of Bessel equation (7.77); a prime implies a derivative with respect to the argument, while P is a constant.

For Eqs. (7.75b) we obtain similarly

$$\begin{cases} E_z = QZ_n(k_r r), \\ H_\theta = -\frac{i\omega\varepsilon}{k_r}QZ'_n(k_r r), \end{cases} \tag{7.78b}$$

with Q a constant.

The full field components are obtained via Eqs. (7.78) and (7.76), and their superposition provides the general solution for the n-order modes.

In the core $r \leqslant a$ we take $k_r = u$ with

$$\beta^2 + u^2 = \omega^2\varepsilon\mu\varepsilon_r = k^2\varepsilon_r, \tag{7.79a}$$

$Z_n = J_n(ur)$ which ensures a finite field at $r = 0$, and $\zeta P = A$ and $Q = B$, ζ being the intrinsic impedance of the cladding. The field components transverse to r are readily obtained in the form

$$\begin{cases} E_\theta = A\frac{ik}{u}J'_n(ur) + B\frac{\beta}{u}\frac{n}{ur}J_n(ur), \\ \zeta H_z = AJ_n(ur), \\ -\zeta H_\theta = -A\frac{\beta}{u}\frac{n}{ur}J_n(ur) + B\frac{ik}{u}\varepsilon_r J'_n(ur), \\ E_z = BJ_n(ur). \end{cases} \tag{7.80a}$$

In the cladding we take $k_r = -i\alpha$ with

$$\beta^2 - \alpha^2 = \omega^2 \varepsilon \mu = k^2, \tag{7.79b}$$

$Z_n = H_n^{(2)}(-i\alpha r) \to K_n(\alpha r)$, and $Z_n' \to iK_n'(\alpha r)$ (see Section 4.5.4), which ensures an exponentially decaying field at infinity, and also $\zeta P = C$ and $Q = D$, thus obtaining

$$\begin{cases} E_\theta = -C\dfrac{ik}{\alpha} K_n'(\alpha r) - D\dfrac{\beta}{\alpha}\dfrac{n}{\alpha r} K_n(\alpha r), \\ \zeta H_z = CK_n(\alpha r), \\ -\zeta H_\theta = C\dfrac{\beta}{\alpha}\dfrac{n}{\alpha r} K_n(\alpha r) - D\dfrac{ik}{\alpha} K_n'(\alpha r), \\ E_z = DK_n(\alpha r). \end{cases} \tag{7.80b}$$

Continuity of the tangential components of the field at the core–cladding interface leads to a 4 × 4 homogeneous linear system of equations:

$$\begin{vmatrix} \dfrac{ik}{u} J_n'(ua) & \dfrac{ik}{\alpha} K_n'(\alpha a) & \dfrac{\beta}{u}\dfrac{n}{ua} J_n(ua) & \dfrac{\beta}{\alpha}\dfrac{n}{\alpha a} K_n(\alpha a) \\ J_n(ua) & -K_n(\alpha a) & 0 & 0 \\ -\dfrac{\beta}{u}\dfrac{n}{ua} J_n(ua) & -\dfrac{\beta}{\alpha}\dfrac{n}{\alpha a} K_n(\alpha a) & \dfrac{ik}{u}\varepsilon_r J_n'(ua) & \dfrac{ik}{\alpha} K_n'(\alpha a) \\ 0 & 0 & J_n(ua) & -K_n(\alpha a) \end{vmatrix} \begin{vmatrix} A \\ C \\ B \\ D \end{vmatrix} = 0. \tag{7.81a}$$

The determinant of Eq. (7.81a) set equal to zero, together with

$$k^2(\varepsilon_r - 1) = u^2 + \alpha^2, \tag{7.81b}$$

provides the dispersion equation for the n-order modes propagating along the fiber.

When $n = 0$, the 4 × 4 matrix in Eq. (7.81a) breaks down into two 2 × 2 matrixes yielding the dispersion equation for the azimuthally independent modes, which are either TE or TM with respect to z (see Section 4.5.4). When $n \neq 0$, the above factorization is not possible and a complicated dispersion equation is obtained. Substantial simplification is possible if the dielectric constants of the core and of the cladding are nearly the same, $\varepsilon_r \approx 1$, which is usually the case in most applications.

ENGINEERING TOPICS: PROPAGATION

The recursion relations in Eqs. (D.5) and (D.6) enable Eqs. (7.80a) to be expressed as follows:

$$\begin{cases} E_\theta = \dfrac{ikA + \beta B}{2}\dfrac{J_{n-1}(ur)}{u} - \dfrac{ikA - \beta B}{2}\dfrac{J_{n+1}(ur)}{u}, \\ \zeta H_z = \dfrac{ikA + \beta B}{2ik} J_n(ur) + \dfrac{ikA - \beta B}{2ik} J_n(ur), \\ -\zeta H_\theta = -\dfrac{\beta A - ik\varepsilon_r B}{2}\dfrac{J_{n-1}(ur)}{u} - \dfrac{\beta A + ik\varepsilon_r B}{2}\dfrac{J_{n+1}(ur)}{u}, \\ E_z = \dfrac{ikA + \beta B}{2\beta} J_n(ur) - \dfrac{ikA - \beta B}{2\beta} J_n(ur). \end{cases} \quad (7.82a)$$

Similarly, Eqs. (D.10) and (D.11) enable Eqs. (7.80b) to be expressed as follows:

$$\begin{cases} E_\theta = \dfrac{ikC + \beta D}{2}\dfrac{K_{n-1}(\alpha r)}{\alpha} + \dfrac{ikC - \beta D}{2}\dfrac{K_{n+1}(\alpha r)}{\alpha}, \\ \zeta H_z = \dfrac{ikC + \beta D}{2ik} K_n(\alpha r) + \dfrac{ikC - \beta D}{2ik} K_n(\alpha r), \\ -\zeta H_\theta = -\dfrac{\beta C - ikD}{2}\dfrac{K_{n-1}(\alpha r)}{\alpha} + \dfrac{\beta C + ikD}{2}\dfrac{K_{n+1}(\alpha r)}{\alpha}, \\ E_z = \dfrac{ikC + \beta D}{2\beta} K_n(\alpha r) - \dfrac{ikC - \beta D}{2\beta} K_n(\alpha r). \end{cases} \quad (7.82b)$$

If we set

$$ikA - \beta B = 0 \quad \text{and} \quad ikC - \beta D = 0,$$

then

$$\beta A + ik\varepsilon_r B = -\dfrac{u^2}{ik} B, \qquad \beta C + ikD = \dfrac{\alpha^2}{ik} D,$$

$$\beta A - ik\varepsilon_r B = \dfrac{2\beta^2 + u^2}{ik} B, \qquad \beta C - ikD = \dfrac{2\beta^2 - \alpha^2}{ik} D.$$

If further $\varepsilon_r \approx 1$, then both α^2 and u^2 can be neglected as a suitable approximation [see Eq. (7.81b)], in which case Eqs. (7.82a) reduce to

$$\begin{cases} E_\theta \approx A_1 \dfrac{J_{n-1}(ur)}{u}, \\[4pt] \zeta H_z \approx A_1 \dfrac{J_n(ur)}{ik}, \\[4pt] -\zeta H_\theta \approx -\dfrac{\beta}{ik} A_1 \dfrac{J_{n-1}(ur)}{u}, \\[4pt] E_z \approx A_1 \dfrac{J_n(ur)}{\beta}, \end{cases} \quad \text{for } r \leqslant a,$$

while Eqs. (7.82b) reduce to

$$\begin{cases} E_\theta \approx C_1 \dfrac{K_{n-1}(\alpha r)}{\alpha}, \\[4pt] \zeta H_z \approx C_1 \dfrac{K_n(\alpha r)}{ik}, \\[4pt] -\zeta H_\theta \approx -\dfrac{\beta}{ik} C_1 \dfrac{K_{n-1}(\alpha r)}{\alpha}, \\[4pt] E_z \approx C_1 \dfrac{K_n(\alpha r)}{\beta}, \end{cases} \quad \text{for } r \geqslant a,$$

with constants $A_1 = (ikA + \beta B)/2$ and $C_1 = (ikC - \beta D)/2$.

Now ζH_θ and E_z are proportional to E_θ and ζH_z, respectively, and continuity of the former at the core–cladding interface implies continuity of the latter. Boundary conditions at $r = a$ can be set involving the field components E_θ and ζH_z alone. The dispersion equation becomes

$$\frac{J_{n-1}(ua)}{uaJ_n(ua)} - \frac{K_{n-1}(\alpha a)}{\alpha a K_n(\alpha a)} = 0 \tag{7.83a}$$

and defines, together with Eq. (7.81b), the HE_n modes[20] propagating along the fiber.

---

20. See the footnote corresponding to the EH_n modes.

The other independent solution is given similarly by

$$\begin{cases} E_\theta \approx -B_1 \dfrac{J_{n+1}(ur)}{u}, \\[4pt] \zeta H_z \approx B_1 \dfrac{J_n(ur)}{ik}, \\[4pt] -\zeta H_\theta \approx -\dfrac{\beta}{ik} B_1 \dfrac{J_{n+1}(ur)}{u}, \\[4pt] E_z \approx -B_1 \dfrac{J_n(ur)}{\beta}, \end{cases} \quad \text{for } r \leqslant a,$$

and

$$\begin{cases} E_\theta \approx D_1 \dfrac{K_{n+1}(\alpha r)}{\alpha}, \\[4pt] \zeta H_z \approx D_1 \dfrac{K_n(\alpha r)}{ik}, \\[4pt] -\zeta H_\theta \approx \dfrac{\beta}{ik} D_1 \dfrac{K_{n+1}(\alpha r)}{\alpha}, \\[4pt] E_z \approx -D_1 \dfrac{K_n(\alpha r)}{\beta}, \end{cases} \quad \text{for } r \geqslant a,$$

with

$$\frac{J_{n+1}(ua)}{ua J_n(ua)} + \frac{K_{n+1}(\alpha a)}{\alpha a K_n(\alpha a)} = 0, \tag{7.83b}$$

and defines, together with Eq. (7.81b), the EH_n modes[21] propagating along the fiber.

For a prescribed value of n, Eqs. (7.83) provide an infinite number of roots that can be ordered by increasing values of $u_m a$, $m = 1, 2, \ldots, u_m a$ being the successive roots of Eqs. (7.83). The corresponding field solutions are the EH_{nm} and HE_{nm} modes that propagate along a *weakly* guiding fiber, $\varepsilon_r \approx 1$.

---

21. To justify the EH_n mode notation, we observe that at cut-off $\alpha a \to 0$ and $K_{n+1}(\alpha a)/[\alpha a K_n(\alpha a)] \to 2n/(\alpha a)^2$, upon employing Eq. (D.19b). Also, from Eq. (7.83b) we obtain always at cut-off $0 > J_n(ua) \to 0$, which is the dispersion equation of TM_n modes (also referred to as E_n modes) in a circular metal pipe. The dual HE_n mode notation is used in connection with the other dispersion equation (7.83a).

Cut-off conditions are obtained by letting $(\alpha a)^2 \to 0$, which defines the transition between guided, $(\alpha a)^2 < 0$, and leaky, $(\alpha a)^2 > 0$, modes. The dominant modes with the same dispersion equation[22]

$$\frac{J_0(ua)}{uaJ_1(ua)} - \frac{K_0(\alpha a)}{\alpha a K_1(\alpha a)} = 0, \qquad (ka)^2(\varepsilon_r - 1) = (ua)^2 + (\alpha a)^2,$$

are the HE_{11} and the $EH_{(-1)1}$ modes. For $\alpha a \to 0$, we have from Eq. (D.19)

$$\frac{K_0(\alpha a)}{\alpha a K_1(\alpha a)} \approx -\ln\frac{\alpha a}{2} = \frac{J_0(ua)}{uaJ_1(ua)},$$

and a solution is obtained for $ua \to 0$. This mode does not exhibit cut-off frequency.

The usual representation of the Brillouin diagram $\omega = \omega(\beta)$ is not convenient in the case of fibers with $\varepsilon_r \approx 1$, typical values being $\varepsilon_r \approx 1.02$. As a matter of fact, the dispersion curve for each mode detaches from the line $\omega = \beta c$ (free-space propagation in the cladding) and tends to the line $\omega = \beta c/\sqrt{\varepsilon_r}$ (free-space propagation in the core). These two curves are almost coincident, and examination of the modal dispersion curves becomes difficult. For this reason it is most convenient to use a different normalization as presented in Fig. 7.31, where $ka\sqrt{\varepsilon_r - 1}$ (which is proportional to ω) is depicted vs. β/k. The (normalized) propagation constant ranges from 1 at cut-off to $\sqrt{\varepsilon_r}$ far away from it.

Finally, we note that computation of components E_r and H_r of the field in Eqs. (7.76) and their decomposition into EH and HE modes shows that they exhibit the same (r, θ) dependence of the components E_θ and H_θ when $\varepsilon_r \approx 1$. This allows a compact representation of the transverse fields. The same simple computation yields in the core of the fiber

$$\mathbf{E}_t(r, \theta, z) = -iA_1 \frac{J_{n-1}(ur)}{u} \exp[i(n-1)\theta] \exp[-i\beta z](\hat{\mathbf{x}} + i\hat{\mathbf{y}})$$

and

$$\zeta \mathbf{H}_t(r, \theta, z) = \frac{\beta}{ik} \mathbf{E}_t(r, \theta, z)$$

---

22. These modes are degenerate and can be combined to provide two orthogonal modes; see the end of this section.

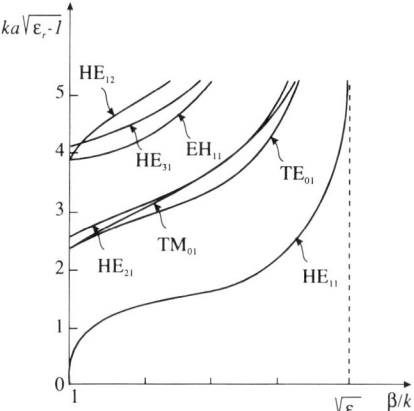

FIGURE 7.31. Normalized Brillouin diagrams $ka\sqrt{\varepsilon_r - 1}$ vs. β/k for the first propagating modes of a dielectric fiber with $\varepsilon_r = 1.02$.

for HE modes, and

$$\mathbf{E}_t(r, \theta, z) = -iB_1 \frac{J_{n+1}(ur)}{u} \exp[i(n+1)\theta] \exp[-i\beta z](\hat{\mathbf{x}} - i\hat{\mathbf{y}})$$

and

$$\zeta \mathbf{H}_t(r, \theta, z) = -\frac{\beta}{ik} \mathbf{E}_t(r, \theta, z)$$

for EH modes, respectively. Similar expressions are obtained in the cladding. We conclude that the transverse fields are circularly polarized with opposite senses in the two sets of modes, and their appropriate combination may be used to generate linearly polarized fields, LP_{pq}, where the first index accounts for the azimuthal variation and the second index for the radial variation.

If we combine HE_{1m} and EH_{-1m} modes which possess the same dispersion relation, we obtain LP_{0m} orthogonal modes that are linearly polarized along either the x- or the y-axis, and include the dominant LP mode. This situation can be approximately met also by combining $HE_{p+1,m}$ and $EH_{p-1,m}$ modes because, asymptotically, they possess the same dispersion equations away from cut-off.

7.5. PLANAR GUIDING CONFIGURATIONS

Some planar guiding configurations are considered in Sections 4.5.2 and 7.3.3. Their ability to propagate low-frequency fields is now investigated. An answer to this question is provided by their cut-off properties.

Consider the grounded slab configuration of Fig. 7.17. As already noted, the dominant TE mode (which is transverse electric with respect to both x and z) exhibits the cut-off wavelength

$$\lambda_c = 4d\sqrt{\varepsilon_r - 1},$$

which renders its possible practical use only in the microwave and optical frequency bands. The dominant TM mode (which is transverse magnetic with respect to x and z) does not exhibit cut-off, but in this case too the attenuation of the field outside the dielectric slab becomes adequate only if $kd > 1$ (see Section 7.3.3). Again, the structure is practically usable only in the high-frequency regime.

On the other hand, planar guiding structures are attractive owing to the advent of printed circuit technology, due to their easy integration. Furthermore, it is desirable to design guiding structures that can be used till the microwave range, but do not suffer low-frequency limitations. This is accomplished by a geometrical modification of two-wire transmission lines, rendering this cut-off free guiding structure compatible with the planarity requirement. Two examples of these planar guiding structures are given in Fig. 7.32.

The *stripline* (Fig. 7.32A) consists of a sandwich of two wide conducting plates bonded to a dielectric slab within which a thin conducting plate is buried. The two external plates are grounded, and the voltage applied to the

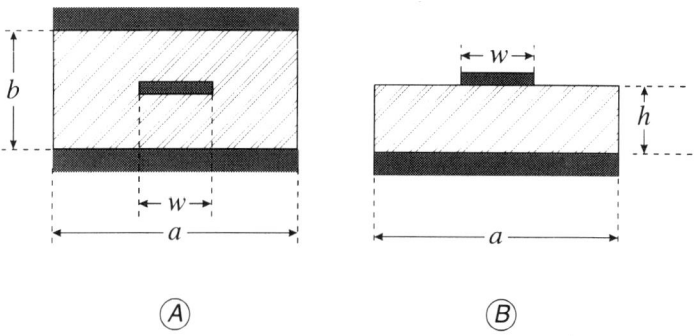

FIGURE 7.32. Planar guiding structures: (A) stripline and (B) microstrip.

ENGINEERING TOPICS: PROPAGATION

inner plate can propagate along the line, as in the coaxial cable configuration, from dc to a usable frequency whose limit is set by the appearance of higher-order modes. The *microstrip* (Fig. 7.32B) is obtained from the stripline by removing the upper plate and one-half of the dielectric material. Again, propagation from dc is possible.

For the above guiding structures, however, no pure TE or TM mode with respect to the propagation direction is possible. A longitudinal component of either the electric and/or the magnetic field is present, and hybrid modes propagate. In the static limit, pure transverse electrostatic and magnetostatic fields are possible, thus implying that the longitudinal components of the field approach zero as $\omega \to 0$. In the low-frequency regime, a quasi-TEM approximation is allowed; propagation is described by the TL equations (7.10), where L and C are the static inductance and capacitance per unit length of the guiding structure. As the frequency increases, the above approximation breaks down and numerical methods must be used to compute the electromagnetic field distribution.

7.5.1. The Effective Dielectric Constant

Let us consider the microstrip depicted in Fig. 7.32B and let C_0 be the static capacitance per unit length of the structure when the dielectric is removed (air-filled microstrip). In this case

$$k_0^2 = \omega^2 \varepsilon_0 \mu_0 = \omega^2/c^2 = \omega^2 L_0 C_0$$

and

$$L_0 = \frac{1}{c^2 C_0}.$$

When the dielectric is present, a reasonable assumption is that the inductance per unit length, L, coincides with L_0, because no magnetic material is involved. This is unlike the capacitance per unit length, C, which attains a different value related to the geometry of the microstrip and to the dielectric constant of the material.

The TL parameters are given by

$$\beta = \omega\sqrt{LC} = k_0\sqrt{\varepsilon_e} \quad \text{and} \quad R_0 = \sqrt{\frac{L}{C}} = \sqrt{\frac{L_0}{C_0}}\frac{1}{\sqrt{\varepsilon_e}},$$

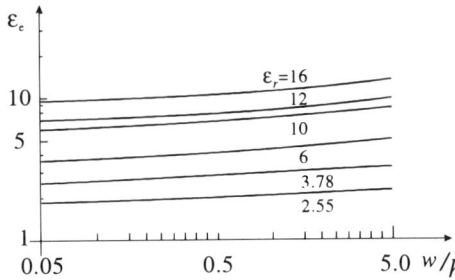

FIGURE 7.33. Effective dielectric constant of a microstrip.

where

$$\varepsilon_e = \frac{C}{C_0}$$

is the *effective dielectric constant* of the material, as far as the TL is concerned.

A plot of ε_e vs. the ratio w/h for the microstrip of Fig. 7.32 is given in Fig. 7.33.

7.6. EQUIVALENT CIRCUITS

Transmission lines, waveguides, and fibers can be loaded by means of lumped or distributed elements, to achieve desirable transfer properties between input and output signals.

Let us consider a bounded volume V (see Fig. 7.34) which is accessible by means of a number of guiding structures. The volume is referred to as a *junction* and prescribed cross sections of the feeding guides define its accessible *ports*. We assume that a single mode is present in each port; this implies that the guiding structures from the junction are sufficiently long so that possible higher-order modes excited within the junction are damped to negligible values. We can define voltages and currents in each port, say V_n and I_n, or incident and reflected field quantities, say a_n and b_n, where

$$\begin{cases} \dfrac{V_n}{\sqrt{R_n}} = a_n + b_n, \\ \sqrt{R_n}\, I_n = a - b_n. \end{cases} \quad (7.84)$$

This set represents the normalized form of Eqs. (7.11a), R_n being the charac-

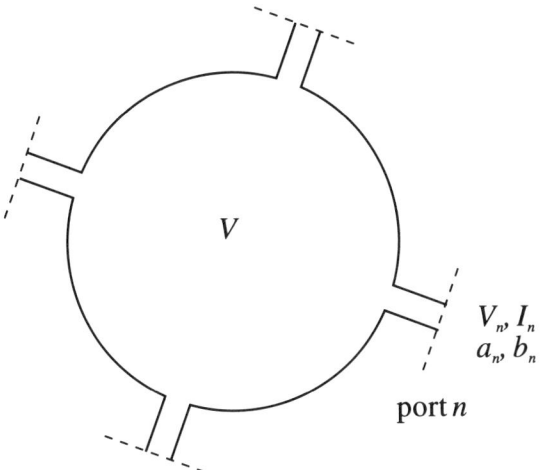

FIGURE 7.34. Geometry of an electromagnetic junction.

teristic resistance of the ETL of port n. Note that (normalized) incident a_n and reflected b_n fields exhibit the same dimension [watt$^{1/2}$].

Voltages and currents, as well as incident and reflected waves at each port, are not unrelated quantities. The former are linearly connected by the port input impedance, and the latter by the port reflection coefficient. However, the impedance and reflection coefficient at each port depends on the load conditions at all the other ports, i.e., on the ports' field values. We can (arbitrarily) select one from the current, voltage, or incident wave as input, and one from the voltage, current, or reflected wave as output, respectively, postulating a linear relationship between input and output quantities in view of the assumed linearity of the system. If

$$\mathbf{V} \to \begin{vmatrix} V_1 \\ \vdots \\ V_N \end{vmatrix}, \quad \mathbf{I} \to \begin{vmatrix} I_1 \\ \vdots \\ I_N \end{vmatrix}, \quad \mathbf{a} \to \begin{vmatrix} a_1 \\ \vdots \\ a_N \end{vmatrix}, \quad \mathbf{b} \to \begin{vmatrix} b_1 \\ \vdots \\ b_N \end{vmatrix}$$

are the voltage, current, incident, and reflected wave column vectors, respectively, then

$$\mathbf{V} = \mathbf{Z} \cdot \mathbf{I}, \tag{7.85a}$$

$$\mathbf{I} = \mathbf{Y} \cdot \mathbf{V}, \tag{7.85b}$$

$$\mathbf{b} = \mathbf{S} \cdot \mathbf{a}, \tag{7.85c}$$

where Z is the *impedance matrix*, Y the *admittance matrix*, and S the *scattering matrix*. These, in turn, are not unrelated. Hence

$$I = Z^{-1} \cdot V,$$

i.e.,

$$Y = Z^{-1}. \tag{7.86a}$$

Furthermore, with the aid of the normalization diagonal matrix,

$$N \to \begin{vmatrix} \sqrt{R_1} & \cdot & 0 \\ \cdot & \cdot & \cdot \\ 0 & \cdot & \sqrt{R_N} \end{vmatrix},$$

we also have

$$V = N \cdot (a + b) = N \cdot (I + S) \cdot a,$$
$$V = Z \cdot I = Z \cdot N^{-1} \cdot (a - b) = Z \cdot N^{-1} \cdot (I - S) \cdot a,$$

i.e.,

$$Z = N \cdot (I + S) \cdot (I - S)^{-1} \cdot N, \tag{7.86b}$$

which generalizes Eq. (7.21a). Also

$$Y = Z^{-1} = N^{-1} \cdot (I - S) \cdot (I + S)^{-1} \cdot N^{-1}, \tag{7.86c}$$

and from Eq. (7.86b)

$$(N^{-1} \cdot Z \cdot N^{-1}) \cdot (I - S) = (I + S),$$

i.e.,

$$S = (Z_N - I) \cdot (Z_N + I)^{-1}, \tag{7.86d}$$

which generalizes Eq. (7.21b) in terms of the *normalized impedance matrix*:

$$Z_N = N^{-1} \cdot Z \cdot N^{-1}.$$

ENGINEERING TOPICS: PROPAGATION

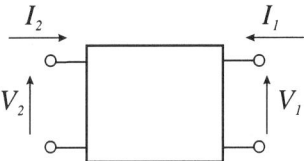

FIGURE 7.35. A two-port junction.

For a two-port system (see Fig. 7.35), in addition to impedance, admittance, and scattering matrices, it is sometimes convenient to define the *transmission* matrix T:

$$\begin{vmatrix} V_2 \\ I_2 \end{vmatrix} = \begin{vmatrix} T_{11} & T_{12} \\ T_{21} & T_{22} \end{vmatrix} \cdot \begin{vmatrix} V_1 \\ -I_1 \end{vmatrix}. \tag{7.85d}$$

Equations (7.85a–b) yield

$$I_1 = Y_{11}V_1 + Y_{12}V_2 \quad \text{and} \quad V_1 = Z_{11}I_1 + Z_{12}I_2,$$

and

$$T \to \begin{vmatrix} -\dfrac{Y_{11}}{Y_{12}} & -\dfrac{1}{Y_{12}} \\ \dfrac{1}{Z_{12}} & \dfrac{Z_{11}}{Z_{12}} \end{vmatrix}. \tag{7.86e}$$

7.6.1. Computation of Matrix Entries

Consider the simple two-port junctions of Fig. 7.36. For the impedance matrix we apply an ideal current generator of prescribed amplitude I at port 1, leaving port 2 open (see Fig. 7.36a). Hence

$$V_1 = Z_{11}I \quad \text{and} \quad V_2 = Z_{21}I,$$

which determine Z_{11} and Z_{21}. The entries Z_{22} and Z_{12} are computed similarly. For the simple TL section of Fig. 7.37

$$Z \to \frac{R_0}{i \sin \beta l} \begin{vmatrix} \cos \beta l & 1 \\ 1 & \cos \beta l \end{vmatrix},$$

following Eq. (7.11b) applied to the case at hand.

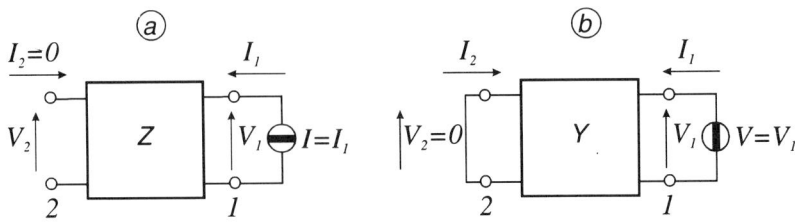

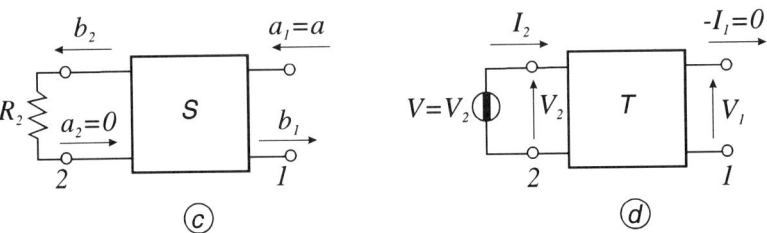

FIGURE 7.36. Computation of junction matrix entries.

For the admittance matrix we refer to Fig. 7.36b:

$$I_1 = Y_{11}V \quad \text{and} \quad I_2 = Y_{21}V,$$

and similarly for the other entries Y_{12} and Y_{22}. For the TL section of Fig. 7.37

$$Y \to \frac{1}{iR_0 \sin \beta l} \begin{vmatrix} \cos \beta l & -1 \\ -1 & \cos \beta l \end{vmatrix}.$$

Note that Eq. (7.86a) is verified.

For the scattering matrix we refer to Fig. 7.36c. Port 2 is matched, therefore

$$b_1 = S_{11}a \quad \text{and} \quad b_2 = S_{21}a,$$

so the entries S_{11} and S_{21} can be determined. The same procedure is applied for entries S_{22} and S_{12}. In the simple case of Fig. 7.37

$$S \to \exp(-i\beta l) \begin{vmatrix} 0 & 1 \\ 1 & 0 \end{vmatrix},$$

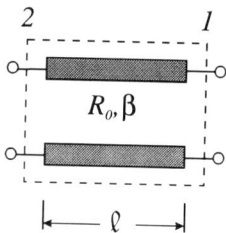

FIGURE 7.37. Transmission line section of length l.

as follows from Eq. (7.11a) applied to the case at hand. It can be easily checked that Eq. (7.86d) is verified.

For the transmission matrix (see Fig. 7.36d) we leave port 1 open, in which case

$$V = T_{11}V_1 \quad \text{and} \quad I_2 = T_{21}V_1,$$

thus evaluating the entries T_{11} and T_{21}. The other two entries are obtained by short-circuiting port 2. In the simple case of Fig. 7.37

$$T \to \begin{vmatrix} \cos \beta l & iR_0 \sin \beta l \\ \dfrac{i \sin \beta l}{R_0} & \cos \beta l \end{vmatrix},$$

which is consistent with Eq. (7.86e).

The above procedures can be applied similarly to the n-ports junction by successively connecting the generator to all ports and properly loading all the other ports.

7.6.2. Junction Matrix Properties

Let us examine the junction of Fig. 7.38, where all ports are open-circuited and ideal current generators I_a and I_b are applied to ports n and m, respectively. Note that each of these ports is open-circuited, too, when the generator is switched off.

We apply the reciprocity theorem of Section 4.3.6 to the junction volume V encompassed by the surface comprising the junction shield S and the port sections S_i. The generators induce fields $\mathbf{E}_a$ and $\mathbf{E}_b$ inside V, but the volume

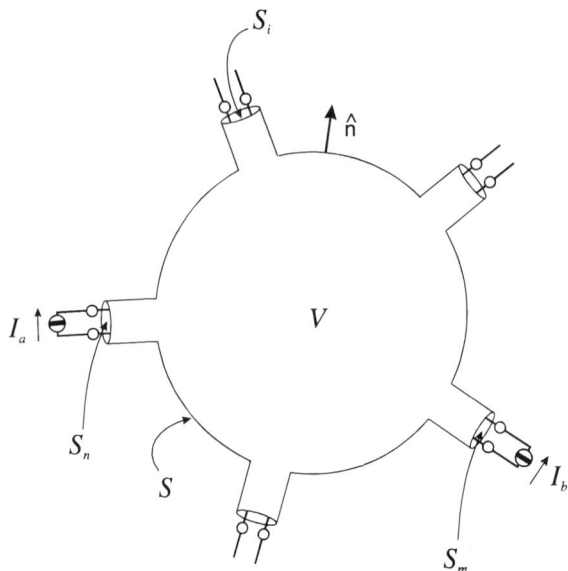

FIGURE 7.38. Application of the reciprocity theorem to a junction.

integral appearing in the reciprocity theorem, Eq. (4.54), namely,

$$\iiint dV [\mathbf{E}_a \cdot \mathbf{J}_b - \mathbf{E}_b \cdot \mathbf{J}_a],$$

is equal to zero because we assume the junction to be passive and there is no generator ($\mathbf{J}_a = \mathbf{J}_b = 0$) inside it. Hence the surface integrals is also equal to zero. The latter can be limited to the port sections n and m only, because we assume the junction to be limited by a perfectly conducting shield ($\hat{\mathbf{n}} \times \mathbf{E} = 0$ on S) and the electric field tangent to all the ports, but n and m, is zero because these are open-circuited ($\hat{\mathbf{n}} \times \mathbf{H} = 0$ on these ports). Accordingly

$$0 = \iint_{S_n + S_m} dS (\mathbf{E}_a \times \mathbf{H}_b - \mathbf{E}_b \times \mathbf{H}_a) \cdot \hat{\mathbf{n}}$$
$$= (V_n^a I_n^b - V_n^b I_a) + (V_m^a I_b - V_m^b I_m^a), \qquad (7.87)$$

where we have used the modal representation for the fields and the normalization[23] of Section 7.4.4. In Eq. (7.87), V_n^a and I_n^a are the voltage and current at port n due to the excitations I_a; similar definitions apply to the other quantities.

We note that $I_n^b = I_m^a = 0$, because the ports n and m are open-circuited and that $V_n^b = Z_{nm} I_b$ and $V_m^a = Z_{mn} I_a$. Substitution into Eq. (7.87) yields

$$Z_{nm} = Z_{mn},$$

provided that the junction does not contain anisotropic materials and the reciprocity theorem can be applied (see Section 4.3.6).

Application of Eqs. (7.86a) and (7.86d) shows that the symmetry property of the impedance matrix is also verified for the admittance and scattering matrixes,

$$Y_{nm} = Y_{mn} \quad \text{and} \quad S_{nm} = S_{mn},$$

because the inverse of a symmetric matrix and the product of symmetric matrices are still symmetric, so that

$$Z = \tilde{Z}, \quad Y = \tilde{Y}, \quad S = \tilde{S}. \quad (7.88)$$

We conclude that *impedance, admittance, and scattering matrixes are symmetric for a reciprocal junction.*

Consider now a reciprocal lossless junction which is also *passive*, implying that it does not contain any generator. We use the Poynting theorem formulation of Eqs. (4.49) where the volume V and surface $A = S + \Sigma S_i$ are depicted in Fig. 7.38. All ports are either excited or loaded, with voltage V_n and current I_n at the generic port n. Integration over the surface A can be limited to the port cross sections ($\hat{\mathbf{n}} \times \mathbf{E} = 0$ on S) to yield

$$-\tfrac{1}{2} \sum_n V_n I_n^* + 2i\omega(W_m - W_e) = 0.$$

Accordingly

$$\sum_n \text{Re}[V_n I_n^*] = 0.$$

First all ports, but port 1, are open-circuited ($I_n = 0$), so we have

$$\text{Re}[V_1 I_1^*] = \text{Re}[Z_{11}] I_1 I_1^* = 0,$$

---

23. The modal vectors **e** and **h** are normalized as in Section 7.4.4, if they are real. Should this not be the case, the alternative formulation of the reciprocity theorem with the complex conjugate product should be used. This is certainly possible in the lossless case.

and Z_{11} is purely imaginary: $Z_{11} \to iX_{11}$. By successively moving the single excited port to $n = 2, 3, \ldots$, we obtain

$$Z_{nn} \to iX_{nn}$$

for any n. Suppose now that all the ports, but n and m, are open-circuited:

$$\mathrm{Re}[V_n I_n^* + V_m I_m^*] = \mathrm{Re}[Z_{nn} I_n I_n^* + Z_{nm} I_m I_n^* + Z_{mm} I_m I_m^* + Z_{mn} I_n I_m^*]$$
$$= \mathrm{Re}[Z_{nm}](I_m I_n^* + I_n I_m^*) = 0, \qquad Z_{nm} = Z_{mn},$$

i.e.,

$$Z_{nm} = Z_{mn} \to iX_{nm},$$

for any n, m. In view of Eq. (7.86a) we cn state that *impedance and admittance matrixes of reciprocal lossless junctions are purely imaginary*:

$$Z \to iX \quad \text{and} \quad Y \to iB. \tag{7.89a}$$

For the scattering matrix S of a lossless junction

$$\tilde{\mathbf{b}} \cdot \mathbf{b}^* = \tilde{\mathbf{a}} \cdot \tilde{S} \cdot S^* \cdot \mathbf{a}^* = \tilde{\mathbf{a}} \cdot \mathbf{a}^*,$$
$$\tilde{S} \cdot S^* = I = \tilde{S}^* \cdot S, \tag{7.89b}$$

i.e.,

$$\sum_k S_{kn}^* S_{km} = \delta_{nm}, \tag{7.89c}$$

which is true also in the absence of reciprocity.

7.6.3. Shift of Port Position

Consider a shift l_n in the reference plane of port n such that at the new position

$$a_n' \to a_n \exp(i\beta l_n) \quad \text{and} \quad b_n' = b_n \exp(-i\beta l_n).$$

If

$$H \to \begin{vmatrix} \exp(-i\beta l_1) & \cdot & \cdot & 0 \\ 0 & \cdot & \cdot & \exp(-i\beta l_N) \end{vmatrix} \tag{7.90}$$

ENGINEERING TOPICS: PROPAGATION

is the (diagonal) *shift matrix*, then

$$\mathbf{a}' = H \cdot \mathbf{a}, \qquad \mathbf{b}' = H^* \cdot \mathbf{b},$$

and

$$\mathbf{a}' = H \cdot \mathbf{a} = H \cdot S \cdot \mathbf{b} = H \cdot S \cdot (H^*)^{-1} \cdot \mathbf{b}'$$
$$= H \cdot S \cdot H \cdot \mathbf{b}',$$

because

$$H \cdot H^* = I.$$

We conclude that a change of reference plane for the junction ports implies transformation of the scattering matrix according to

$$S' = H \cdot S \cdot H,$$
$$S'_{nm} = S_{nm} \exp[i(\psi_n + \psi_m)], \qquad \psi_n = -\beta l_n, \qquad \psi_m = -\beta l_m.$$

The $N \times N$ matrix H contains N independent phases, $\psi_1, \ldots, \psi_N$, that can be used to adjust the phases of some entries of the matrix S'. The maximum number of phase adjustments equals N.

7.6.4. The Three-Port Junction

Consider a three-port *matched* junction, this implying that $b_n = 0$ for any n, if all other ports are matched. Accordingly

$$S_{11} = S_{22} = S_{33} = 0.$$

Enforcement of condition (7.89c) for $n \neq m$ leads to

$$S^*_{31} S_{32} = S^*_{21} S_{23} = S^*_{12} S_{13} = 0,$$

which is fulfilled if

$$S_{32} = S_{21} = S_{13} = 0.$$

Hence Eq. (7.89c) for $n = m$ yields

$$|S_{12}|^2 = |S_{23}|^2 = |S_{31}|^2 = 1.$$

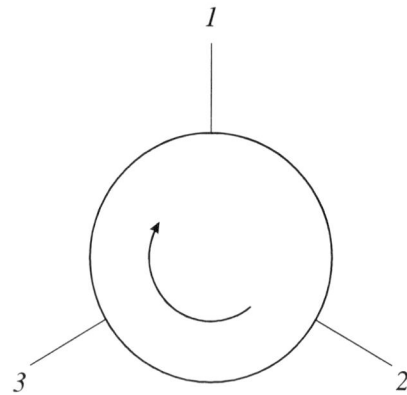

FIGURE 7.39. The circulator.

The scattering matrix of the junction is

$$S \to \begin{vmatrix} 0 & 1 & 0 \\ 0 & 0 & 1 \\ 1 & 0 & 0 \end{vmatrix},$$

because we can render the phases of the three entries equal to zero by means of a shift of the port position (see Section 7.6.3).

A completely matched three-port lossless junction *cannot* be reciprocal and should contain anisotropic material, because its scattering matrix does not fulfil Eq. (7.88). Such a device is called a *circulator* (see Fig. 7.39): the power entering one port is transferred to the neighboring port, according to a prescribed circulation sense. We also conclude that a lossless reciprocal three-point junction cannot be fully matched.

7.6.5. The Four-Port Junction

Let us examine a four-port reciprocal matched junction; its scattering matrix is

$$S \to \begin{vmatrix} 0 & S_{12} & S_{13} & S_{14} \\ S_{12} & 0 & S_{23} & S_{24} \\ S_{13} & S_{23} & 0 & S_{34} \\ S_{14} & S_{24} & S_{34} & 0 \end{vmatrix}.$$

ENGINEERING TOPICS: PROPAGATION

Enforcement of condition (7.89c) for $n = m = 1$ and 4 leads to

$$|S_{12}|^2 + |S_{13}|^2 + |S_{14}|^2 = 1 \quad \text{and} \quad |S_{14}|^2 + |S_{24}|^2 + |S_{34}|^2 = 1, \quad (7.91a)$$

i.e.,

$$|S_{12}|^2 + |S_{13}|^2 = |S_{24}|^2 + |S_{34}|^2. \quad (7.91b)$$

Similarly, with the same condition (7.89c) and $n = m = 2$ and 3,

$$|S_{12}|^2 + |S_{23}|^2 + |S_{24}|^2 = 1 \quad \text{and} \quad |S_{13}|^2 + |S_{23}|^2 + |S_{34}|^2 = 1, \quad (7.91c)$$

i.e.,

$$|S_{12}|^2 + |S_{24}|^2 = |S_{13}|^2 + |S_{34}|^2. \quad (7.91d)$$

Comparison of Eqs. (7.91b) and (7.91c) yields

$$|S_{13}| = |S_{24}| \quad \text{and} \quad |S_{12}| = |S_{34}|. \quad (7.91e)$$

Let α_{nm} be the phase of the entry S_{nm}. By means of the shift matrix (7.90) we can render S_{12} and S_{34} real,

$$\psi_1 + \psi_2 + \alpha_{12} = 0 \quad \text{and} \quad \psi_3 + \psi_4 + \alpha_{34} = 0,$$

i.e.,

$$\psi_2 = -\psi_1 - \alpha_{12}, \quad \psi_4 = -\psi_3 - \alpha_{34},$$

$$S_{12} = S_{34} = s,$$

with s real. Similarly, we can render the phases of S_{13} and S_{24} equal,

$$\psi_1 + \psi_3 + \alpha_{13} = \psi_2 + \psi_4 + \alpha_{24} = \alpha,$$

i.e.,

$$\psi_3 = -\psi_1 - \frac{\alpha_{12} + \alpha_{13} + \alpha_{34} - \alpha_{24}}{2},$$

$$S_{13} = S_{24} = s_1 \exp(i\alpha),$$

with s_1 real. Note that the phase α cannot be modified via the shift matrix $\boldsymbol{H}$ because

$$\alpha = \frac{\alpha_{24} + \alpha_{13} - \alpha_{12} - \alpha_{34}}{2}.$$

Condition (7.89c) is now enforced for $n = 1$ and $m = 4$:

$$ss_1 \exp(i\alpha) + s_1 \exp(-i\alpha)s = 0,$$

$$\alpha = \pm \frac{\pi}{2}; \tag{7.91f}$$

for $n = 1$ and $m = 2$,

$$-iS_{23} + iS_{14}^* = 0;$$

and for $n = 1$ and $m = 3$,

$$S_{23} + S_{14}^* = 0.$$

These last two relations are inconsistent unless

$$S_{23} = S_{14} = 0.$$

In addition, from Eq. (7.91a),

$$s_1 = \sqrt{1 - s^2}. \tag{7.91g}$$

The general structure of the scattering matrix is the following:

$$\boldsymbol{S} \to \begin{vmatrix} 0 & S_{12} & S_{13} & 0 \\ S_{12} & 0 & 0 & S_{13} \\ S_{13} & 0 & 0 & S_{12} \\ 0 & S_{13} & S_{12} & 0 \end{vmatrix}, \tag{7.92a}$$

$$S_{12} = s, \qquad S_{13} = \pm i\sqrt{1 - s^2}. \tag{7.92b}$$

A completely matched reciprocal four-port junction is a *directional coupler* (see Fig. 7.40): each port is connected to only two other ports and decoupled from the remaining one.

ENGINEERING TOPICS: PROPAGATION

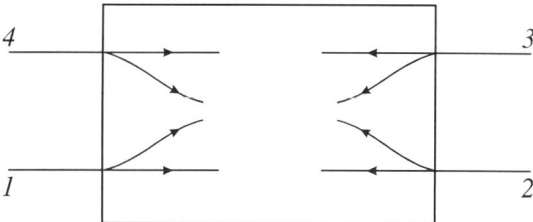

FIGURE 7.40. The Directional Coupler.

7.6.6. The Directional Coupler

In practice, the realization of directional coupler matching and isolation features can only be met approximately, and entries such as S_{11} and S_{14} may differ slightly from zero. A quality measure of the coupler is provided by the *coupling factor*

$$C = 20 \lg |S_{13}| \tag{7.93a}$$

and *directivity factor*

$$D = 20 \lg \frac{|S_{13}|}{|S_{14}|}. \tag{7.93b}$$

The former is a measure of the power transferred from port 1 to the coupled port 3, while the latter is a measure of the power isolation between ports 4 and 3. The two factors (7.93) can be defined for each port and either computed or measured.

The simplest implementation of a directional coupler is obtained by connecting two adjoined waveguides by one or more small holes. The case of two holes in the common narrow wall of the guides is depicted in Fig. 7.41. A simple, yet approximate, model for *small* hole coupling is now presented.

In Fig. 7.41, consider the lower waveguide which is excited by a TE_{10} mode. In the absence of the hole, the magnetic field on the narrow wall is directed along z and given by

$$i\omega\mu H_z = V_t \nabla_t \cdot \mathbf{h} = -\frac{V}{\pi/a} \nabla_t^2 \psi = \frac{\pi}{a} V(z)\psi(0),$$

where $V(z)$ and $\psi(0)$ are the modal functions of the TE_{10} mode (see Section

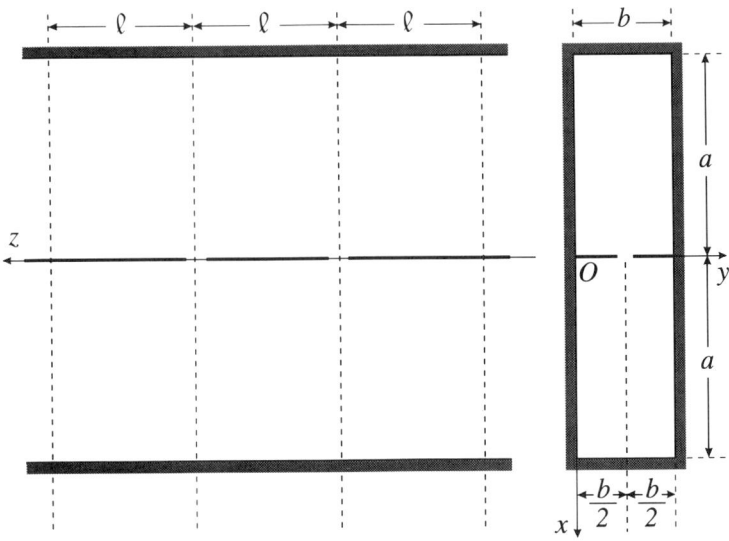

FIGURE 7.41. Coupling through two holes in the common narrow wall of two adjoining rectangular waveguides.

7.4.1). A current linear density,

$$\mathbf{J}_S = \hat{\mathbf{x}} \times \hat{\mathbf{z}} H_z = -\frac{\pi V}{i\omega\mu a}\psi\hat{\mathbf{y}},$$

is present on the wall; see Eq. (3.11) for phasor fields.

The main effect of the hole is to intercept and deviate the wall current lines (see Fig. 7.42). Accordingly, a charge linear density ρ_l appears on the hole rim, evaluated approximately by using the *unperturbed* current density $\mathbf{J}_S$ in the current continuity equation (3.11),

$$\pm \mathbf{J}_S \cdot \hat{\mathbf{y}} \Delta z + i\omega \rho_l \Delta z = 0,$$

$$\rho_l = \mp \frac{\pi}{\omega^2 \mu a} V\psi,$$

where the upper and lower signs refer to the upper and lower rim of the hole, respectively. These charges, in turn, generate an electric field $\mathbf{E}_S$ across the hole, essentially confined in the wall plane and directed along $\hat{\mathbf{y}}$. We can metallize the hole, thus restoring the original waveguide wall, provided that the magnetic surface current density

$$-\mathbf{K}_S = \hat{\mathbf{x}} \times \mathbf{E}_S = \hat{\mathbf{x}} \times \hat{\mathbf{y}} E_S = \hat{\mathbf{z}} E_S$$

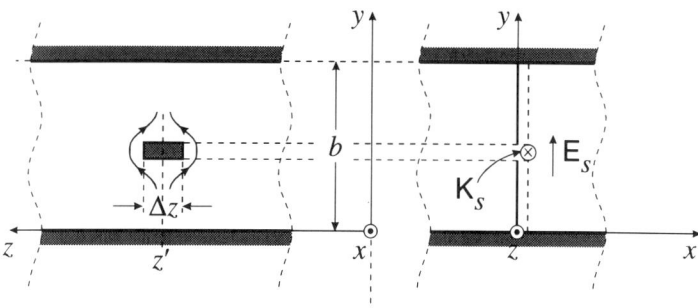

FIGURE 7.42. Expanded view of hole coupling.

is set over the aperture. The tangential electric field is zero over the metallized hole, and attains its value $\mathbf{E}_S$ as the magnetic current $\mathbf{K}_S$ is crossed; see the phasor equivalent of Eq. (1.37) and Fig. 7.42.

The total flux of the magnetic surface current along the z-axis,

$$I_m = \int dy \mathbf{K}_S \cdot \hat{\mathbf{z}} = -\int dy E_S(y),$$

equals the equivalent voltage set up across the hole. This voltage is proportional to the charge (divided by ε) induced on the (upper) rim of the hole via a (nondimensional) constant ζ which is *real*, due to the essentially electrostatic nature of the voltage–charge relation for a hole of small dimension:

$$I_m = \zeta \frac{\rho_l}{\varepsilon} = -\zeta \frac{\pi}{k^2 a} V \psi.$$

We conclude that the presence of a hole centered at $z = z'$ is accounted for by the magnetic current density [weber/m^2],

$$\begin{aligned}\mathbf{K} &= -\zeta \frac{\pi}{k^2 a} V(z) \psi(x) \delta(x) \delta\left(y - \frac{b}{2}\right) \text{rect}\left(\frac{z - z'}{\Delta z}\right) \hat{\mathbf{z}} \\ &\approx -\zeta \frac{\pi \Delta z}{k^2 a} V(z) \psi(x) \delta(x) \delta\left(y - \frac{b}{2}\right) \delta(z - z') \hat{\mathbf{z}} = K \hat{\mathbf{z}},\end{aligned} \quad (7.94)$$

as far as its effect on the TE$_{10}$ mode is concerned. This longitudinal magnetic

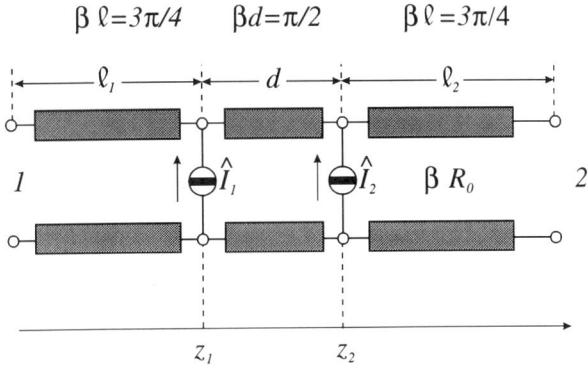

FIGURE 7.43. The ETL of two-hole coupling.

current is modeled on the ETL by the electric current generator

$$\hat{I}\delta(z-z') = \frac{\pi}{i\omega\mu a} \int_0^a dx \int_0^b dy K(x,y)\psi^*(x)$$
$$= -\xi\left(\frac{\pi}{ka}\right)^2 \frac{\Delta z}{i\omega\mu} \frac{2}{ab} V(z)\delta(z-z') = i\frac{q}{R_0} V(z)\delta(z-z')$$

(see Section 7.4.5), with q a real (positive) nondimensional constant.

The ETL for the two-hole case is depicted in Fig. 7.43, where the distance between the two holes is equal to one-fourth of the waveguide wavelength, and the junction ports are those indicated in the figure. The incident voltage at terminal 1 is V^+ and the junction is matched at terminal 2. Each current generator injects voltages $R_0\hat{I}/2$ along the two senses of the ETL. We compute the voltages at $z = z_1, z_2$:

$$\begin{cases} V(z_1) = V^+ \exp\left(-i\frac{3\pi}{4}\right) + i\frac{q}{2}V(z_1) + \frac{q}{2}V(z_2), \\ V(z_2) = V^+ \exp\left(-i\frac{5\pi}{4}\right) + \frac{q}{2}V(z_1) + i\frac{q}{2}V(z_2), \end{cases}$$

i.e.,

$$\frac{V(z_1)}{V^+} = \exp\left(-i\frac{3\pi}{4}\right)\frac{1-iq}{1-iq-q^2/2} \quad \text{and} \quad \frac{V(z_2)}{V^+} = \exp\left(-i\frac{3\pi}{4}\right)\frac{-i}{1-iq-q^2/2}.$$

These values $V(z_1)$ and $V(z_2)$ of the voltages along the ETL allow computation of the equivalent sources $\hat{I}_1$ and $\hat{I}_2$ that model the presence of the two holes.

ENGINEERING TOPICS: PROPAGATION

The entries of the scattering matrix S can now be computed by applying the procedure of Section 7.6.1. Direct and reflected voltages at each port are the superposition of the incident voltage V^+ and of those injected by the currents $\hat{I}_1$ and $\hat{I}_2$ times the appropriate phase shifts. This leads to

$$S_{11} = i\frac{q}{2}\frac{V(z_1)}{V^+}\exp\left(-i\frac{3\pi}{4}\right) + i\frac{q}{2}\frac{V(z_2)}{V^+}\exp\left(-i\frac{5\pi}{4}\right)$$

$$= \frac{iq^2/2}{1 - iq - q^2/2} \approx i\frac{q^2}{2} \tag{7.95a}$$

and

$$S_{12} = 1 + i\frac{q}{2}\frac{V(z_1)}{V^+}\exp\left(-i\frac{5\pi}{4}\right) + \exp\left(-i\frac{5\pi}{4}\right)\frac{q}{2}\frac{V(z_2)}{V^+}\exp\left(-i\frac{3\pi}{4}\right)$$

$$= 1 + iq\frac{1 - iq/2}{1 - iq - q^2/2} \approx 1 + iq - \frac{q^2}{2}. \tag{7.95b}$$

In order to compute the S_{13} and S_{14} entries, we follow the same procedure by considering magnetic sources in the adjoining waveguide. The electric field at the other side of the hole is the same as that already computed and the magnetic current density necessary to sustain it is reversed in sign, because the normal to the guide wall is reversed, too. However, we must change $\delta(x) \to \delta(x - a)$ in Eq. (7.94) if we wish the senses of the reference axes to remain unchanged. This compensates for the change in sign of the magnetic current, because $\psi(a) = -\psi(0)$, and

$$S_{13} = iq - \frac{q^2}{2} \quad \text{and} \quad S_{14} = i\frac{q^2}{2},$$

a result in agreement with that of Eqs. (7.95) with only suppression of the incident-wave contribution.

Matching at the junction requires $S_{11} = 0$, which is obtained in the limit of $q^2/2 \to 0$. Accordingly

$$S_{11} \approx 0, \quad S_{12} \approx 1 + iq, \quad S_{13} \approx -iq, \quad S_{14} \approx 0,$$

and conditions (7.92b) are fulfilled to first order in q. We have also from Eqs. (7.93)

$$C \approx 20\lg\left|iq - \frac{q^2}{2}\right| \approx 20\lg q \,[\text{dB}]$$

and

$$D \approx 20 \lg \frac{|iq - q^2/2|}{|iq^2/2|} \approx 20 \lg \frac{2}{q} \, [\text{dB}].$$

7.6.7. Periodic Structures

If a number of two-port junctions are connected in a serial arrangement, the most convenient way to analyze the resulting structure is to use the transmission matrix (7.85d). If (V_N, I_N) and $(V_1, -I_1)$ are the input and final output voltages and currents, respectively (see Fig. 7.44), then

$$\begin{vmatrix} V_N \\ I_N \end{vmatrix} = T_N \cdot T_{N-1} \cdot \ldots \cdot T_1 \cdot \begin{vmatrix} V_1 \\ -I_1 \end{vmatrix}.$$

If all the junctions are identical, namely, $T_n = T$, and the junction chain is matched at both sides, then we may seek a solution such that the couple (V_n, I_n) is just a scaled version of the couple $(V_{n-1}, -I_{n-1})$:

$$\begin{vmatrix} V_n \\ I_n \end{vmatrix} = T \cdot \begin{vmatrix} V_{n-1} \\ -I_{n-1} \end{vmatrix} = \xi \begin{vmatrix} V_{n-1} \\ -I_{n-1} \end{vmatrix},$$

which requires

$$\det(T - \xi I) = 0 \tag{7.96}$$

for a nonzero field solution.

If ξ reduces to a pure phase shift, i.e.,

$$\xi = \exp(\pm i\gamma l),$$

l being the junction length, then the solution becomes periodic, and an infinite

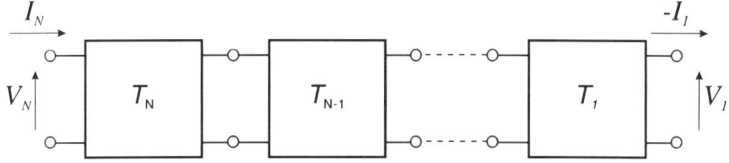

FIGURE 7.44. Serial connection of N two-port junctions.

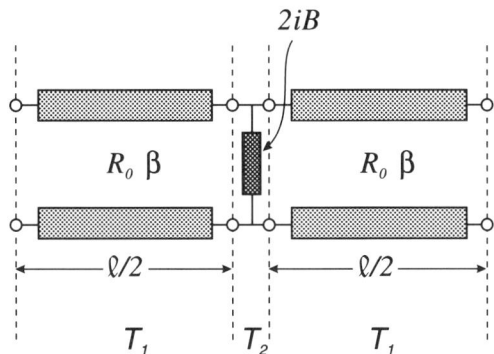

FIGURE 7.45. The elementary cell of a periodic structure.

number of modes are possible (Floquet's[24] theorem), each characterized by the wavenumber

$$\gamma_m \to \gamma + \frac{2m\pi}{l}.$$

Each mode propagates along the periodic structure with propagation constant $\pm \gamma_m$ and they are often referred to as Bloch[25] waves.

Consider the simple junction of Fig. 7.45 with

$$T = T_1 \cdot T_2 \cdot T_1,$$

where

$$T_1 \to \begin{vmatrix} \cos\beta\frac{l}{2} & iR_0 \sin\beta\frac{l}{2} \\ \frac{i}{R_0} \sin\beta\frac{l}{2} & \cos\beta\frac{l}{2} \end{vmatrix} \quad \text{and} \quad T_2 \to \begin{vmatrix} 1 & 0 \\ 2iB & 1 \end{vmatrix}.$$

Condition (7.96) leads to

$$\xi = \cos\beta l - R_0 B \sin\beta l \pm i\sqrt{1 - (\cos\beta l - R_0 B \sin\beta l)^2},$$

---

24. Achille Marie Gaston Floquet: Epinal (France), 1847– Nancy, 1920.
25. Felix Bloch: Zurich (Switzerland), 1905–1983.

with

$$\cos \gamma l = \cos \beta l - R_0 B \sin \beta l.$$

Propagation along the periodic structure is possible if

$$|\cos \beta l - R_0 B \sin \beta l| \leq 1.$$

This condition is frequency-dependent, and the periodic structure exhibits filtering properties. For a coaxial cable loaded periodically by capacitances $2C_l$, we have

$$\beta = k = \omega \sqrt{LC}, \qquad R_0 B = \sqrt{\frac{L}{C}} \omega C_l = k \frac{C_l}{C},$$

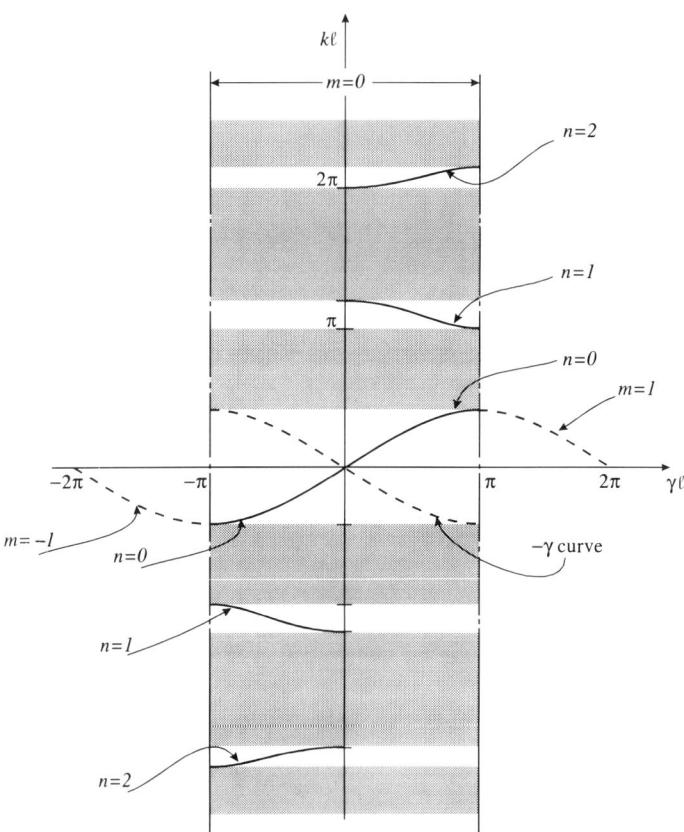

FIGURE 7.46. Normalized Brillouin diagram for a coaxial cable with periodic capacitive loading: $C_l = Cl$, $m = 0$. Stop bands are darkened.

and

$$\cos \gamma l = \cos kl - \frac{C_l}{Cl} kl \sin kl.$$

The corresponding Brillouin diagram is shown in Fig. 7.46 for $C_l = Cl$. The appropriate classification of its branches is now examined.

The region $\pi \leqslant \gamma l \leqslant \pi$ corresponds to the modes with $m = 0$; successive modes in this region can be classified according to another index n. For each mode, the branches are chosen in such a way that $\omega = \omega(\gamma)$ is an odd function (see also Section 3.2.1). Solutions of the dispersion equation corresponding to $+\gamma$ are shown in the figure. For the fundamental mode $n = 0$, the solution corresponding to $-\gamma$ is also shown (dashed line), as well as for $m = 0$ and, in part, for $m \pm 1$, too. This indicates how the different branches are generated for values of m different from zero. Note the stop-band regions and the appearance of modes with opposite phase and group velocities [see Eqs. (4.60) and (4.65)]. This property is employed in some traveling-wave tubes.

7.6.8. Obstacles in Waveguides

Periodic loading of waveguides is obtained by means of obstacles in the waveguide section; they appear as shunt (or series) loads in the ETL. In the absence of losses the loading should be reactive.

As an example, let us consider the (inductive) iris depicted in Fig. 7.47. The incident TE_{10} mode excites along the iris a surface current density $\mathbf{J}_S$, which in turn generates the scattered field. This can be expanded in the waveguide modes: due to the symmetry of the obstacle, only TL_{n0} modes are excited.

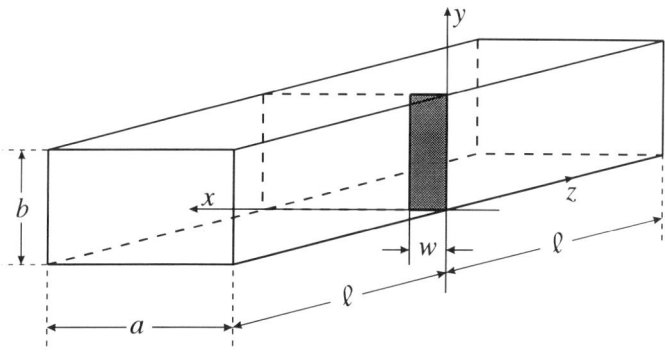

FIGURE 7.47. The inductive iris.

Boundary conditions at $z = 0$ involve the tangential components $\mathbf{E}_t = E\hat{\mathbf{y}}$ and $\mathbf{H}_t = H\hat{\mathbf{x}}$ of the *total* field, the sum of the incident and scattered components. These fields are expanded in terms of the modes as follows:

$$E = V\psi_1(x)\exp(-i\beta_1 z) + \sum_{n}^{1} V_n \psi_n(x)\exp(i\beta_n z), \tag{7.97a}$$

$$H = Y_{01} V\psi_1(x)\exp(-i\beta_1 z) - \sum_{n}^{1} Y_{0n} V_n \psi_n(x)\exp(i\beta_n z), \tag{7.98a}$$

for $z \leq 0$; and

$$E = \sum_{n}^{1} V'_n \psi_n(x)\exp(-i\beta_n z), \tag{7.97b}$$

$$H = \sum_{n}^{1} Y_{0n} V'_n \psi_n(x)\exp(-i\beta_n z), \tag{7.98b}$$

for $z \geq 0$; with

$$\psi_n = \sqrt{\frac{2}{ab}} \sin\left(\frac{n\pi}{a} x\right), \quad \beta_n = \sqrt{k^2 - \left(\frac{n\pi}{a}\right)^2}, \quad Y_{0n} = \frac{\beta_n}{\omega\mu}.$$

Before satisfying the boundary conditions in order to evaluate the expansion coefficients in Eqs. (7.97) and (7.98), let us consider the field distribution far away from the iris. If only the TE_{10} mode can propagate, the other modes being cut-off, and l is sufficiently large, only forward and backward scattered TE_{10} modes are present in the two terminal sections of the junction, besides the incident one, because the modal propagation constants become purely imaginary: $\beta_n \to -i\alpha_n$ for $n \geq 2$. These scattered fields can be accounted for by a shunt admittance Y in the equivalent TL, whose value is determined by requiring that the corresponding reflection coefficient Γ,

$$\Gamma = \frac{Y_0 - (Y + Y_0)}{Y_0 + (Y + Y_0)} = -\frac{Y}{Y + 2Y_0}, \quad \text{where} \quad Y_0 = \frac{\beta_1}{\omega\mu} = Y_{01},$$

must equal the ratio V_1/V, namely,

$$\Gamma = \frac{V_1}{V},$$

and so

$$Y = -Y_0 \frac{2\Gamma}{1 + \Gamma}. \tag{7.99}$$

ENGINEERING TOPICS: PROPAGATION

We conclude that the shunt admittance Y is determined by the modal coefficient V_1, which is however coupled to all the others by the boundary conditions, which are now in order.

The electric field is continuous at $z = 0$, i.e.,

$$E(z^+, x) - E(z^-, x) = 0, \quad (7.100)$$

and the magnetic field is discontinuous due to the presence of the induced surface current:

$$H(z^+, x) - H(z^-, x) = J_S(x), \quad (7.101)$$

the induced current being limited to the iris surface $0 \leqslant x \leqslant w$.

Equations (7.97) and (7.100) yield

$$V + V_1 = V_1' \quad \text{and} \quad V_n = V_n', \quad n \geqslant 2, \quad (7.102)$$

by equating the two modal expansions term by term. From Eqs. (7.98) and (7.101)

$$2 \sum_1^n \frac{\beta_n}{\omega\mu} V_n \psi_n(x) = J_S(x), \quad (7.103)$$

using Eq. (7.102). The coefficients V_n are now readily obtained by multiplying both sides of Eq. (7.103) by $\psi_n(x)$ and integrating over the guide cross section:

$$\frac{2\beta_n}{\omega\mu} V_n = \sqrt{\frac{2b}{a}} \int_0^w dx J_S(x) \sin\left(\frac{n\pi}{a} x\right). \quad (7.104)$$

We now invoke the condition that the tangential electric field must be equal to zero on the iris. Equation (7.104) is employed to express all coefficients $V_n, n \geqslant 2$, as functions of the current integral. Hence from Eq. (7.97a) specified at $z = 0$,

$$\sin\left(\frac{\pi}{a} x\right) + \frac{V_1}{V} \sin\left(\frac{\pi}{a} x\right) + \sum_2^n \frac{\beta_1}{\beta_n} \sin\left(\frac{n\pi}{a} x\right) \int_0^w dx f(x) \sin\frac{n\pi}{a} x = 0, \quad (7.105)$$

for $0 \leqslant x \leqslant w$, where

$$f(x) = \sqrt{\frac{b}{2a}} \frac{\omega\mu}{V\beta_1} J_S(x).$$

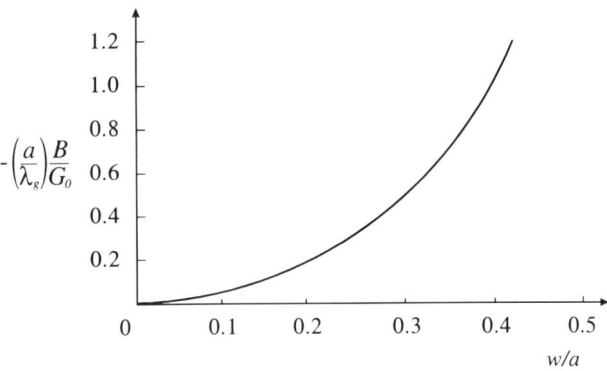

FIGURE 7.48. Equivalent normalized susceptance of an inductive iris in a rectangular waveguide with $a/\lambda = 0.7$.

If

$$F_n = \int_0^w dx f(x) \sin\frac{n\pi}{a} x,$$

then Eq. (7.105) is rewritten as

$$\sin\left(\frac{\pi}{a}x\right) + \frac{V_1}{V}\sin\left(\frac{\pi}{a}x\right) + \sum_{2}^{n} \frac{\beta_1}{\beta_n} F_n \sin\left(\frac{n\pi}{a}x\right) = 0, \qquad (7.106)$$

for $0 \leqslant x \leqslant w$, and can be solved numerically for the unknown coefficients V_1/V and F_n after truncation at a convenient value N of n.[26] An example of such a computation is given in Fig. 7.48.

We may proceed differently, multiplying Eq. (7.106) by $f^*(x)$ and integrating over $0 \leqslant x \leqslant w$, to obtain

$$F_1^*(1 + \Gamma) + i \sum_{2}^{n} \frac{\beta_1}{\alpha_n} F_n F_n^* = 0,$$

$$\alpha_n = \sqrt{\left(\frac{n\pi}{a}\right)^2 - k^2}.$$

---

26. The simplest way to proceed is to enforce condition (7.106) at N points, $0 \leqslant x \leqslant N$; see Chapter 6.

When this is combined with Eq. (7.99) and Eq. (7.104), $n = 1$, we have

$$Y = -2iY_0 \frac{F_1 F_1^*}{\sum_{n \atop 2}(\beta_1/\alpha_n) F_n F_n^*}. \qquad (7.107)$$

This shows that the shunt admittance is purely reactive and of inductive type.

It can be easily shown that Eq. (7.107) is *stationary* with respect to the assumed current distribution: its value does not change to first-order perturbation of the latter. This provides a practical way to compute the shunt susceptance given by Eq. (7.107) without solving Eq. (7.106): *reasonable* values of $f(x)$ can be used and the corresponding values of $Y = -iB$ computed.

It is noteworthy that the formally simple expression (7.107) may require extensive numerical computation, because the series in the denominator may converge slowly. In fact, if the series is truncated at $n\pi w/a < 1$, we have

$$F_n \approx \frac{n\pi}{a} \int_0^w dx f(x) x,$$

and dependence on w would be lost. This remark is relevant when deriving the proper susceptance behavior for $w/a \ll 1$.

7.7. SUMMARY AND SELECTED REFERENCES

This is the first of three chapters devoted to the presentation of problems of electromagnetic transmission, radiation, and scattering from an engineering viewpoint. In particular, this chapter is devoted to propagation along guiding structures and their connections via microwave junctions [7.1–7.6].

The basic guiding structure is the transmission line, in a two-wire or coaxial cable configuration, whose equations are presented in Section 7.1 starting from Maxwell's equations. This is unlike the usual approaches, which prefer a low-frequency derivation [7.7]. In this chapter the equivalent circuit is a result of the obtained equations, and not the starting point from which they are derived. Impedance and reflection coefficient, and matching, nonuniform, and multisection transmission lines are considered in the subsequent Sections 7.2.2 to 7.2.6. In particular, the Riccati equation [7.8] for impedance and reflection coefficient is discussed in Section 7.2.5. Line generators are presented in Section 7.2.7.

The importance of the transmission line equations resides in their ability to model a large variety of guiding structures [7.9, 3.7], offering a powerful unified approach to their study. This model is particularly important for the engineering applications in view of its inherent simplicity, physical intuition, and scalar description of the problem.

The transmission line model is used systematically in Sections 7.3 and 7.4 to describe guided propagation along two-dimensional and three-dimensional structures, respectively. The former include the grounded slab (Section 7.3.3; the latter include rectangular and circular waveguides (Sections 7.4.1 and 7.4.2), the coaxial cable (Section 7.4.3), the inhomogeneous rectangular guide (Section 7.4.7), and the fiber [7.10, 7.11, 5.8] (Section 7.4.8). It is noted that the inhomogeneous structures are also studied via the transmission line formalism, making use of the powerful transverse resonance condition (Section 7.3.2). Waveguide excitation (Section 7.4.5) and waveguide losses (Section 7.4.6) are also examined. Strip lines and microstrips [7.12] are considered in Section 7.5.

Microwave junctions and equivalent microwave circuits are considered in Section 7.6. Properties of impedance, admittance, transmission, and scattering matrixes are exploited in Sections 7.6.1–7.6.3. The circulator [7.13] and directional coupler [7.14] are studied in Sections 7.6.4 and 6.6.5, respectively. As further examples, the equivalent circuit of an iris in a rectangular waveguide is computed in Section 7.6.8, and periodic structures are treated in Section 7.6.7.

References

[7.1] R. E. Collin, *Field Theory of Guided Waves*, McGraw-Hill, New York (1960).
[7.2] R. E. Collin, *Foundations for Microwave Engineering*, McGraw-Hill, New York (1966).
[7.3] R. E. Elliott, *An Introduction to Guided Waves and Microwave Circuits*, Prentice-Hall, Englewood Cliffs, N.J. (1993).
[7.4] A. Ishimaru, *Electromagnetic Wave Propagation, Radiation and Scattering*, Prentice-Hall, Englewood Cliffs, N.J. (1991).
[7.5] N. Marcuvitz, *Waveguide Handbook*, McGraw-Hill, New York (1951).
[7.6] C. G. Montgomery, R. H. Dicke, and E. M. Purcell, *Principles of Microwave Circuits*, McGraw-Hill, New York (1948).
[7.7] S. R. Seshadi, *Fundamentals of Transmission Lines and Electromagnetic Fields*, Addison-Wesley, Reading, Mass. (1971).
[7.8] I. Sugai, "Table of solutions of Riccati's equations," *Proc. IEE* **50**, 2024–2026 (1962).
[7.9] N. Marcuvitz and J. Schwinger, "On the representation of electric and magnetic fields produced by currents and discontinuities in waveguides," *J. Appl. Phys.* **22**, 806–819 (1951).
[7.10] N. S. Kapany and J. J. Burke, *Optical Waveguides*, Academic Press, New York (1972).
[7.11] D. Marcuse, *Light Transmission Optics*, Van Nostrand-Reinhold, New York (1974).
[7.12] S. B. Cohn, "Characteristic impedance of the shielded strip transmission line," *IEEE Trans. Microwave Theory Tech.* **2**, 52–55 (1954).
[7.13] C. E. Fay and R. L. Comstock, "Operation of the ferrite junction circulator," *IEEE Trans. Microwave Theory Tech.* **13**, 15–27 (1965).
[7.14] R. Levy, "Directional couples," in *Advances in Microwaves* (L. Young, ed.), Academic Press, New York (1966).

8
Engineering Topics: Radiation

8.1. TRANSMITTING AND RECEIVING ANTENNAS

Broadly speaking, any device able to radiate an electromagnetic field is a *transmitting antenna*. However, specific constants are usually imposed on the radiated field, such as space and time coherence, angular shaping, polarization purity, and so on. These constrains limit the class of devices that can be defined as antennas. In other words, a poorly shielded electronic cabinet responsible for unwanted radiation (*man-made electromagnetic noise*) is not considered to be a transmitting antenna according to this more restricted practical definition.

Similarly, any device able to collect an electromagnetic field is, in a broad sense, a *receiving antenna*. Again, constraints are imposed on the collecting mechanism: for instance, the received field should generate collectable voltages and currents across the terminals of a (real or equivalent) transmission line. A person exposed to an electromagnetic field is a site of induced currents and charges, but is not considered to be a receiving antenna.

A typical radiating or receiving device is the wire antenna (see Fig. 8.1). In its transmitting mode (Fig. 8.1A), a real TL connects a current (or voltage) generator to the antenna terminals, thus inducing a current distribution along the wire. These currents are a source of the electromagnetic field, which is radiated in the space surrounding the antenna. In other cases, e.g., for aperture (see Section 8.5) or reflector (see Section 8.6) antennas, no accessible terminals exist on the antenna, which is fed by devices other than a real TL, such as a waveguide. In both cases, the electromagnetic field distribution along the feeder near the transmitting antenna is not simple, as it is determined by the (usually complicated) boundary conditions specified by the antenna–feeder connection. This field can be expanded, in principle, in the modes of the feeding structure. These modes are usually all cut-off except for the fundamental one, which only exists on the feed at large distances from the antenna and is described by appropriate voltages and currents, V and I, respectively. We can therefore unambiguously model the feeding structure in terms of either a real or, most

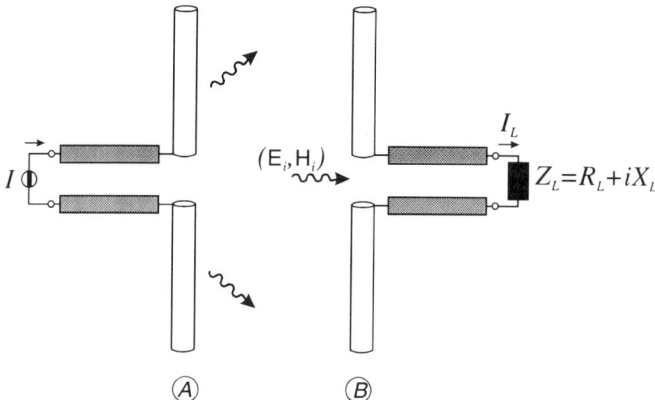

FIGURE 8.1. A wire antenna in its (A) transmitting and (B) receiving mode.

generally, an equivalent TL, excited by the transmitter and loaded by an equivalent impedance, the *input impedance* $Z_i = R_i + iX_i$ of the antenna, such that voltages and currents along the ETL are those previously defined. Voltages and currents across the input impedance are defined as the *input voltage* V_i and *input current* I_i of the antenna, respectively. However, we can freely move to any other convenient section of the ETL to set equivalent terminals of the antenna and new (equivalent) input voltages, current, and impedances by just using the simple rules provided by Eqs. (7.11b) and (7.17a). Finally, we note that the proposed model allows computation of the real power delivered to the antenna, $(\frac{1}{2})R_i|I_i|^2$ for phasor fields, and radiated in the surrounding space (except for that part dissipated on the antenna if this is lossy).

Determination of the current distribution along the wire (see Fig. 8.19) is a boundary value problem in itself (see Section 8.4). Also, in the case of other radiating structures such as apertures and reflectors, where the radiating elements are *equivalent* currents, determination of the latter requires solution of usually complicated boundary value problems. Once these real or equivalent sources have been evaluated, the radiated field can be obtained by superposition,[1] and is spatially shaped by constructive or destructive interference of contributions from the elementary constituents of the radiating currents. This angular distribution of the field and its associated density power are of paramount importance from the engineering viewpoint, because it is responsible for the quality of the radio link and of the level of the interference (*electromagnetic compatibility*), in particular (and usually) at large distances from the antenna. A number of convenient parameters have been designed to account for this description: effective length (in the transmitting mode),

---

1. This requires linearity of the surrounding medium.

radiation vector, directivity, gain (see Section 8.2). These parameters are all mutually related and fully describe the antenna radiation properties, thus providing, together with the input impedance, a complete engineering treatment of the transmitting antenna. Finally, we note that knowledge of the radiated field allows computation of the total power radiated by the antenna, thus providing an alternative way for evaluating the input resistance of a lossless antenna (see Section 8.2.3).

Operation of the antenna in the receiving mode (see Fig. 8.1B) can be simply explained as follows. Assume that a known and prescribed *incident* electromagnetic field ($\mathbf{E}_i, \mathbf{H}_i$) is present in the *absence* of the antenna. When the antenna is *present*, new boundary conditions are set up: the tangential electric field on the antenna surface (see Fig. 8.1B) should be zero if the wire is perfectly conducting. Accordingly, surface currents are induced on the antenna and generate a scattered field ($\mathbf{E}_S, \mathbf{H}_S$) subject to the radiation conditions at infinity, and such that the *total* field ($\mathbf{E}_i + \mathbf{E}_S, \mathbf{H}_i + \mathbf{H}_S$) satisfies the boundary conditions on the wire *and* on the antenna terminals connected to the collecting device. Again, the field along the latter can be expanded in the modes appropriate to the guiding structure. Should only the fundamental mode be propagating, all the others being cut-off, we can unambiguously define received voltages and currents at large distances from the physical terminals of the antenna, along a real or equivalent TL. As in the case of the transmitting antenna, a convenient section of the TL can be chosen to define equivalent receiving terminals of the antenna, where received voltages V_r and currents I_r are located. As already noted, these receiving terminals can be moved along the (real or equivalent) TL and become source terminals for the load of the antenna. We can therefore compute voltages and currents at the load terminals where the impedance $Z_L = R_L + iX_L$ is applied. In particular, the real power delivered to the receiver is given by $\frac{1}{2}R_L|I_L|^2$ for the phasor field. Finally, the currents induced on the antenna provide also a scattered field: the receiving mode is necessarily connected to a transmitting mode, in general.

As in the case of the transmitting antenna, the receiving properties of the antenna can be described by a number of simple parameters: effective length (in the receiving mode) and effective area which, together with the input impedance, fully account for the engineering approach to the problem. Transmitting and receiving properties of the antenna are determined by the induced current distribution, related to the same boundary value problem. It can be surmised that transmitting and receiving parameters are closely related. The key to the connection is presented in Section 8.1.1.

8.1.1. Reciprocity Theory and Antennas

Let us consider two antennas, 1 and 2, as depicted in Fig. 8.2, where reference is made to wire antennas only for graphical purposes. The antennas are excited by (real or equivalent) shielded TLs.

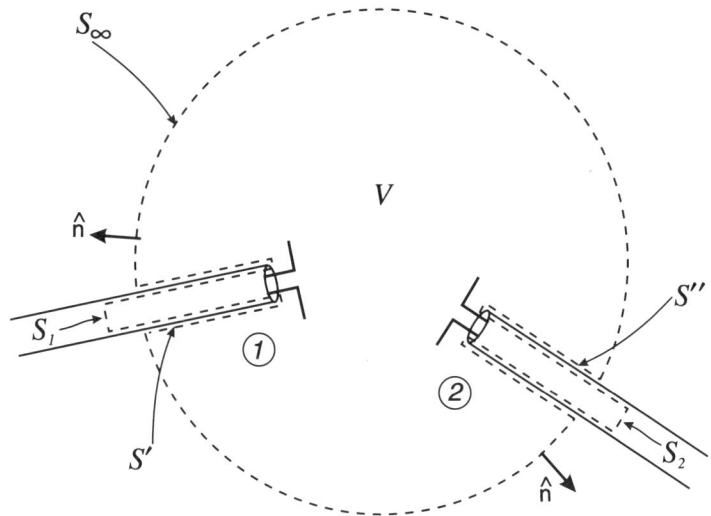

FIGURE 8.2. Application of the reciprocity theorem to antennas.

We proceed as in Section 7.6.2 and apply the reciprocity theorem (see Section 4.3.6) to the volume V limited by the surface $S_\infty + S' + S_1 + S'' + S_2$ (see Fig. 8.2).

The volume integral appearing in the reciprocity theorem [see Eq. (4.54)] is zero, because the electric field tangential to the (lossless) antennas is zero. The surface integral is zero along $S' + S''$ (perfect conductor shield) and on S_∞, if this is situated at large distances from the sources (radiation conditions). Accordingly

$$\iint_{S_1} dS\,(\mathbf{E}_1 \times \mathbf{H}_2 - \mathbf{E}_2 \times \mathbf{H}_1) \cdot \hat{\mathbf{n}} + \iint_{S_2} dS\,(\mathbf{E}_1 \times \mathbf{H}_2 - \mathbf{E}_2 \times \mathbf{H}_1) \cdot \hat{\mathbf{n}} = 0,$$

where $(\mathbf{E}_1, \mathbf{H}_1)$ and $(\mathbf{E}_2, \mathbf{H}_2)$ are the fields generated by antennas 1 and 2, respectively.

In each of the sections S_1 and S_2 a single mode is present, as discussed in Section 8.1. Let $(\mathbf{e}_1, \mathbf{h}_1)$ and $(\mathbf{e}_2, \mathbf{h}_2)$ be the corresponding vector modal functions of the two TLs. We consider the integral across the section S_1 where

$$\mathbf{E}_1 = \mathbf{e}_1 V_1, \qquad \mathbf{E}_2 = \mathbf{e}_1 V_{21}, \qquad \mathbf{H}_1 = \mathbf{h}_1 I_1, \qquad \mathbf{H}_2 = \mathbf{h}_1 I_{21},$$

and

$$\iint_{S_1} dS(\mathbf{E}_1 \times \mathbf{H}_2 - \mathbf{E}_2 \times \mathbf{H}_1) \cdot \hat{\mathbf{n}} = (V_1 I_{21} - V_{21} I_1) \iint_{S_1} dS(\mathbf{e}_1 \times \mathbf{h}_1) \cdot \hat{\mathbf{n}}.$$

By similarly operating on the section S_2 we obtain

$$\iint_{S_2} dS(\mathbf{E}_1 \times \mathbf{H}_2 - \mathbf{E}_2 \times \mathbf{H}_1) \cdot \hat{\mathbf{n}} = (V_{12} I_2 - V_2 I_{12}) \iint_{S_2} dS(\mathbf{e}_2 \times \mathbf{h}_2) \cdot \hat{\mathbf{n}}.$$

When the same normalization is used for the two TLs[2] we have

$$V_1 I_{21} - V_{21} I_1 = V_2 I_{12} - V_{12} I_2. \tag{8.1}$$

We assume now that the two antennas are fed by ideal generators, and that the feeders are lossless, so that we can always move[3] the reference plane positions to derive open circuit conditions,

$$I_{12} = I_{21} = 0,$$

and equivalent ideal current generators I_1 and I_2 there. Equation (8.1) then becomes

$$I_1 V_{21} = I_2 V_{12}, \tag{8.2}$$

which is a general relationship between two arbitrary antennas fed by ideal current generators I_1 and I_2 in the transmitting mode and collecting voltage V_{21} and V_{12} in the receiving mode. Transmitting and receiving terminals coincide for each antenna. The only limitation on the validity of Eq. (8.2) is the absence of nonreciprocal materials in the volume V (see Section 4.3.6).

Equation (8.2) is the basis on which to establish a relation between antenna parameters in the transmitting and receiving modes: for given source values I_1 and I_2, an increase in V_{21} requires a corresponding increase in V_{12}, and vice versa. An antenna which is able to satisfactorily radiate within a prescribed angular region must exhibit a similar good performance when used in the receiving mode. The opposite is also true. Quantitative statements supporting these qualitative considerations are given in Section 8.3.

2. If the vector modal functions $\mathbf{e}$ and $\mathbf{h}$ are real, then $\mathbf{e} \times \mathbf{h} = \mathbf{e} \times \mathbf{h}^*$ and the normalization of Section 7.4.4 is recovered.
3. An ideal current generator exhibits an infinite input impedance; the same (open circuit) conditions are recovered at distances $n\lambda/2$ from the generator itself.

8.2. PARAMETERS OF THE TRANSMITTING ANTENNA

In the following we present the basic parameters of the transmitting antenna and examine their mutual relationships. We always refer to far fields.

For an elementary dipole of length Δz (see Section 4.6.1), we have in the far field

$$\begin{cases} \mathbf{E} = i\zeta \dfrac{I\Delta z}{2\lambda r} \sin\theta \exp(-i\beta r)\hat{\boldsymbol{\theta}} = i\zeta \dfrac{I\mathbf{l}}{2\lambda r} \exp(-i\beta r), \\ \mathbf{H} = \dfrac{1}{\zeta}\, \hat{\mathbf{r}} \times \mathbf{E}, \end{cases}$$

as follows from Eqs. (4.93) where

$$\mathbf{l} = \Delta z \sin\theta \hat{\boldsymbol{\theta}} \tag{8.3}$$

is the *effective length* of the dipole.

For any other type of antenna where input terminals and a corresponding input current can be defined, we can always set the radiated electric far field equal *by definition* to

$$\mathbf{E}(r,\theta,\phi) = i\zeta \dfrac{I\mathbf{l}(\theta,\phi)}{2\lambda r} \exp(-i\beta r) = K\mathbf{l}(\theta,\phi), \tag{8.4}$$

with respect to a spherical system of coordinates centered at the antenna terminals. Equation (8.4) allows computation of the effective length when the far field is evaluated. This parameter is commonly used in the case of wire antennas (see Section 8.4).

It is not always convenient to use definition (8.4), thus normalizing the radiation properties of the antenna to those of the elementary dipole. An alternative definition leads to the *radiation vector*, i.e., the ratio[4] between $\exp(ikr)$ times the electric far field along the generic (θ,ϕ) direction, and the modulus at large distance of the same electric far field along a prescribed direction (θ_0,ϕ_0), usually that of the maximum radiation. Hence

$$\mathbf{e}(\theta,\phi) = \dfrac{\mathbf{E}(r,\theta,\phi)\exp(ikr)}{|\mathbf{E}_0|}, \quad \mathbf{E}_0 = \mathbf{E}(r,\theta_0,\phi_0), \tag{8.5a}$$

where the r-dependence cancels in the ratio of the two electric fields. For the

---

4. We assume k = real, i.e., the medium is lossless.

electric dipole

$$\mathbf{e}(\theta) = \frac{\mathbf{E}(r, \theta)}{|\mathbf{E}(r, 90°)|} = i \sin \theta \hat{\boldsymbol{\theta}}.$$

The radiation vector (8.5a) is usually referred to co-polar $\hat{\mathbf{p}}$ and cross-polar $\hat{\mathbf{q}}$ reference axes (see Section 4.2.2):

$$\mathbf{e}(\theta, \phi) = (\mathbf{e} \cdot \hat{\mathbf{p}}^*)\hat{\mathbf{p}} + (\mathbf{e} \cdot \hat{\mathbf{q}}^*)\hat{\mathbf{q}} = \mathbf{e}_p + \mathbf{e}_q. \tag{8.5b}$$

For the electric dipole oriented with respect to the spherical axes of Fig. 4.12, we have

$$\mathbf{E} = -i\zeta \frac{I\Delta z}{2\lambda r} \exp(-ikr)(\cos\theta \cos\phi \hat{\boldsymbol{\theta}} - \sin\phi \hat{\boldsymbol{\phi}});$$

see Section 4.7.3 and Eq. (4.97a). When we use the polarization basis reported in the first line of Table 4.2, we obtain

$$\mathbf{e}_p = -i\hat{\mathbf{p}}, \qquad \mathbf{e}_q = 0,$$

and

$$\mathbf{e}_p = -i\cos\theta\hat{\mathbf{p}}, \qquad\qquad \mathbf{e}_q = i\sin^2\theta \sin\phi \cos\phi \hat{\mathbf{q}},$$
$$\mathbf{e}_p = -i(\cos\theta \cos^2\phi + \sin^2\phi)\hat{\mathbf{p}}, \qquad \mathbf{e}_q = i(1 - \cos\theta)\sin\phi \cos\phi \hat{\mathbf{q}},$$

if reference is made to the second and third lines of Table 4.2, respectively.

For prescribed values of ϕ, the polar plots of the amplitude of the components of the radiation vector $\mathbf{e}(\theta, \phi)$ or, often, of its modulus $|\mathbf{e}(\theta, \phi)|$ are referred to as *radiation diagrams*. Popular choices are $\phi = 0, \pi$ and $\phi = \pm \pi/2$ (principal planes) and the diagonal directions in between (cross planes). Some examples of radiation diagrams of specific antennas are given in Sections 8.5 and 8.6.

The effective length and radiation vector are clearly related. The defining equations (8.4) and (8.5a) yield

$$\mathbf{E}(r, \theta, \phi) = K\mathbf{l}(\theta, \phi) \quad \text{and} \quad \mathbf{E}(r, \theta_0, \phi_0) = K\mathbf{l}(\theta_0, \phi_0),$$

$$\mathbf{e}(\theta, \phi) = \frac{K\mathbf{l}(\theta, \phi)}{|K||\mathbf{l}(\theta_0, \phi_0)|}. \tag{8.6}$$

The effective length and radiation vectors rely on field amplitudes. Additional parameters can be defined with respect to powers.

The total power radiated by the antenna is given by

$$P_r = \frac{1}{2\zeta} \oiint dS \, |\mathbf{E}(r, \theta, \phi)|^2,$$

where we are in the far field and S is a remote spherical surface centered on the radiator. The *directivity function* $D(\theta, \phi)$, or simply *directivity*, is defined as

$$D(\theta, \phi) = \frac{4\pi r^2 |\mathbf{E}(r, \theta, \phi)|^2 / 2\zeta}{P_r}. \tag{8.7}$$

For the electric dipole

$$P_r = \tfrac{1}{2}\zeta |I|^2 \left(\frac{\Delta z}{\lambda}\right)^2 \frac{2\pi}{3}$$

(see Section 4.6.1) and

$$D(\theta) = \tfrac{3}{2} \sin^2 \theta. \tag{8.8}$$

Very often, the maximum value of $D(\theta, \phi)$ is of interest. It is usually referred to as the *antenna directivity*, and the same symbol D is used when confusion with the directivity function does not arise. For the electric dipole $D = 1.5$.

A common measurement unit of the directivity is the decibel:

$$D^{\mathrm{dB}} = 10 \log D \quad [\mathrm{dB}]. \tag{8.9}$$

The *gain function* $G(\theta, \phi)$, or simply *gain*, is defined as

$$G(\theta, \phi) = \frac{4\pi r^2 |\mathbf{E}(r, \theta, \phi)|^2 / 2\zeta}{P_i}, \tag{8.10}$$

where $P_i = \tfrac{1}{2} R_i |I_i|^2$ is the input real power to the antenna. For a lossless antenna $P_i = P_r$ and $G = D$. Gain is a more general parameter than directivity, and reduces to the latter for lossless radiators. If losses are present, $P_i > P_r$ and $G < D$. Also in the case of gain, reference is often made to its maximum value, usually referred to as G again, and to its value G^{dB} in dB.

The far field is simply related to the gain:

$$\frac{1}{2\zeta}|\mathbf{E}(r,\theta,\phi)|^2 = \frac{GP_i}{4\pi r^2}, \tag{8.11}$$

as follows from Eq. (8.10). The quantity

$$EIRP = GP_i \tag{8.12}$$

is referred to as the *equivalent isotropic radiated power*, a common parameter in radio-link computations.

Amplitude and power parameters are *not* unrelated. With reference to a spherical coordinate system centered on the antenna, Eqs. (8.7) and (8.5a) yield

$$D(\theta,\phi) = \frac{4\pi|\mathbf{e}(\theta,\phi)|^2}{\int_0^{2\pi} d\phi \int_0^{\pi} d\theta \sin\theta |\mathbf{e}(\theta,\phi)|^2}, \tag{8.13a}$$

which relates directivity and radiation vector. Also, Eqs. (8.10) and (8.4) lead to

$$G = \frac{4\pi r^2(\zeta^2|I|^2|\mathbf{l}|^2/4\lambda^2 r^2)/2\zeta}{R_i|I|^2/2} \quad \text{and} \quad R_i = \pi \frac{|\mathbf{l}|^2}{\lambda^2 G}\zeta, \tag{8.13b}$$

which relates the input resistance to the radiation parameters.[5]

8.2.1. The Radiation Parameters of the Elementary Loop Antenna

The electric far field of an elementary loop antenna of radius a is determined by Eqs. (4.95) and the results of Section 4.6.2:

$$\mathbf{E}(r,\theta,\phi) = -i\omega\mu \frac{\pi a^2 I}{4\pi} \frac{ik}{r} \sin\theta \exp(-ikr)\hat{\boldsymbol{\phi}},$$

with respect to the coordinate system of Fig. 4.38. Comparison with Eq. (8.4) shows that

$$\mathbf{l} = -ik\pi a^2 \sin\theta\hat{\boldsymbol{\phi}}. \tag{8.14a}$$

---

[5] In Eq. (8.13b), the angular dependence of $|\mathbf{l}|^2$ and G is the same and cancels in their ratio. We can safely use their maximum values.

Should the loop comprise N turns, Eq. (8.14a) is replaced by

$$\mathbf{l} = -iNkA \sin\theta \hat{\boldsymbol{\phi}}, \tag{8.14b}$$

which is applicable to a loop of any shape and area A.

For the directivity, we obtain the same Eq. (8.8) pertaining to the elementary electric dipole.

8.2.2. The Radiation Parameters of the Elementary Huygens Source

The electric far field of an elementary Huygens source is readily obtained from Eq. (4.97c):

$$\mathbf{E}(r, \theta, \phi) = i\frac{E_0 \Delta S}{2\lambda r}(1 + \cos\theta)(\cos\phi\hat{\boldsymbol{\theta}} - \sin\phi\hat{\boldsymbol{\phi}})\exp(-ikr),$$

with respect to the coordinate system of Fig. 4.39. Comparison with Eq. (8.5a) shows that

$$\mathbf{e}(\theta, \phi) = i\cos^2\frac{\theta}{2}(\cos\phi\hat{\boldsymbol{\theta}} - \sin\phi\hat{\boldsymbol{\phi}}) = i\cos^2\frac{\theta}{2}\hat{\mathbf{p}},$$

when reference is made to the polarization basis of the third line of Table 4.1.

From Eq. (8.13a) we easily obtain

$$D(\theta) = 3\cos^2\frac{\theta}{2}.$$

8.2.3. Input Resistance of Elementary Antennas

For an elementary electric or magnetic antenna of effective length $|\mathbf{l}| = l\sin\theta$ in free space, the total radiated power is

$$P_r = \frac{1}{2}\zeta\frac{|I|^2 l^2}{4\lambda^2 r^2}\int_0^{2\pi} d\phi \int_0^{\pi} d\theta r^2 \sin^3\theta = \frac{1}{2}\zeta\frac{2\pi}{3}\left(\frac{l}{\lambda}\right)^2.$$

By means of the relation

$$P_r = \tfrac{1}{2}R_i|I|^2,$$

ENGINEERING TOPICS: RADIATION

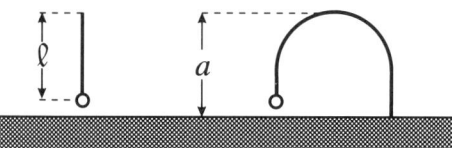

FIGURE 8.3. Elementary electric and magnetic antennas over a conducting plane.

the input resistance is derived in the form

$$R_i = \frac{2\pi}{3} \zeta \left(\frac{l}{\lambda}\right)^2 \approx 800 \left(\frac{l}{\lambda}\right)^2 \quad [\Omega], \quad (8.15)$$

where l is equal to the physical length in the case of the electric dipole and to $k\pi a^2$ in the case of the loop antenna.

Consider an electric dipole oriented vertically, and a loop antenna perpendicular, to a perfectly conducting plane (see Fig. 8.3). The lengths of these elementary antennas should be doubled when computing their associated fields (see Section 1.5.1). The radiated density power is quadrupled compared to the case of the antennas in free space, but it is integrated over only the hemisphere at infinity. Accordingly, and from Eq. (8.15),

$$R_i = \frac{\pi}{3} \zeta \left(\frac{2l}{\lambda}\right)^2 = \frac{4\pi}{3} \left(\frac{l}{\lambda}\right)^2,$$

where l is the physical length for the electric dipole and $k\pi a^2/2$ for the loop antenna.

8.2.4. Antenna Beamwidth

A popular way to define the *antenna beamwidth* makes reference to its angular rate of gain (see Fig. 8.4). For any $\phi = \bar{\phi} = \text{const}$, the *3dB antenna beamwidth* θ_{3dB} is the angular sector (across the direction of maximum radiation) within which the antenna gain does not drop below one-half of its maximum value.

As an alternative, we can define the effective beamwidth Ω_e by

$$G_0 \Omega_e = \int_{4\pi} d\Omega G(\Omega),$$

Ω being the solid angle (see Fig. 8.5) and $G_0 = G(0)$ the gain maximum. When

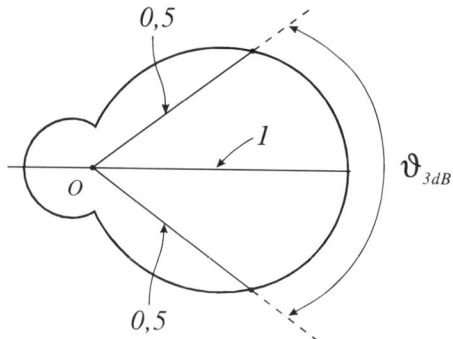

FIGURE 8.4. Definition of the antenna effective beamwidth.

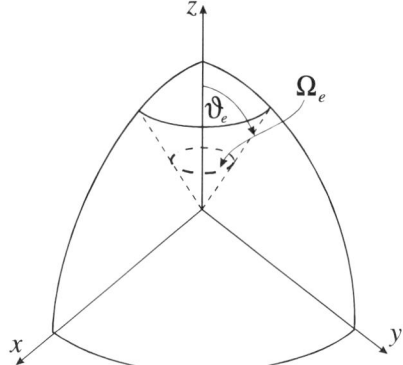

FIGURE 8.5. Relating the effective stereoangular and angular beamwidths.

the antenna is lossless, $G = D$ and

$$\int_{4\pi} d\Omega G(\Omega) = \frac{4\pi \iint dS|\mathbf{E}|^2/2\zeta}{P_r} = 4\pi, \qquad r^2 d\Omega = dS, \tag{8.16}$$

as follows from Eq. (8.7). Then

$$\Omega_e = 4\pi/G_0. \tag{8.17}$$

In addition

$$r^2 \Omega_e = \int_0^{2\pi} d\phi \int_0^{\theta_e/2} d\theta r^2 \sin\theta = 4\pi r^2 \sin^2\frac{\theta_e}{4},$$

so that

$$\sin^2 \frac{\theta_e}{4} = \frac{1}{G_0}, \qquad (8.18a)$$

when the radiation diagram is symmetric around its maximum value. The two parameters Ω_e and θ_e are referred to as the *effective stereoangular* and *angular beamwidths* of the antenna, respectively.

In a large number of applications the effective beamwidth is very small, and Eq. (8.18a) becomes

$$\theta_e^2 \approx \frac{4}{G(0)}; \qquad (8.18b)$$

if the radiation diagram is not symmetric, an effective beamwidth can be defined for any plane cut $\phi = \text{const}$. A relation which is often useful is the following:

$$\theta_m \theta_M \approx \frac{4}{G(0)}, \qquad (8.19)$$

where θ_m and θ_M are the minimum and maximum value of the effective beamwidths, respectively.

8.2.5. Mechanical Forces on Antennas

It is shown in Section 1.3.5 that a radiation pressure is associated with the electromagnetic field; its time-averaged value is given by $\mathbf{S}/c$ for phasor fields. Radiation from a source implies transmission of the total mechanical force

$$\mathbf{F} = \frac{1}{c} \iint_A dA \mathbf{S}$$

outside the closed surface A surrounding the antenna. Accordingly, the latter is subject to the radiation force $-\mathbf{F}$.

For a highly directive antenna the Poynting vector $\mathbf{S}$ is essentially confined inside the effective angular beamwidth, and parallel to the direction of maximum radiation. The resulting force exhibits the same direction and the value

$$F \approx \frac{1}{c} P_r.$$

8.3. PARAMETERS OF THE RECEIVING ANTENNA

Let us consider a wire antenna excited by an incident, locally plane field $(\mathbf{E}_i, \mathbf{H}_i)$ (see Fig. 8.6). A voltage V_r is induced at the antenna receiving terminals, which are assumed open-circuited. This (open-circuit) voltage is related to the incident field by a linear homogeneous functional: the simplest *ansatz* based on the linearity of the system is

$$V_r = -\mathbf{E}_i \cdot \mathbf{l}_r, \tag{8.20}$$

where $\mathbf{l}_r$ is the receiving effective length of the antenna.

For the elementary (open-circuited) loop of Fig. 8.7 we have

$$\oint_C dc \mathbf{E}_i \cdot \hat{\mathbf{c}} = V_r = -i\omega\mu \iint_A dA \mathbf{H}_i \cdot \hat{\mathbf{z}}$$

$$= -\frac{i\omega\mu}{\zeta} \iint_A dA (\mathbf{E}_i \times \hat{\mathbf{r}}) \cdot \hat{\mathbf{z}} = ikA \sin\theta \mathbf{E}_i \cdot \hat{\boldsymbol{\phi}} = -\mathbf{E}_i \cdot \mathbf{l}_r,$$

where we have identified the total field with the incident one. In fact, no current is induced in the *open-circuited elementary* $(A \to 0)$ loop, and the scattered field

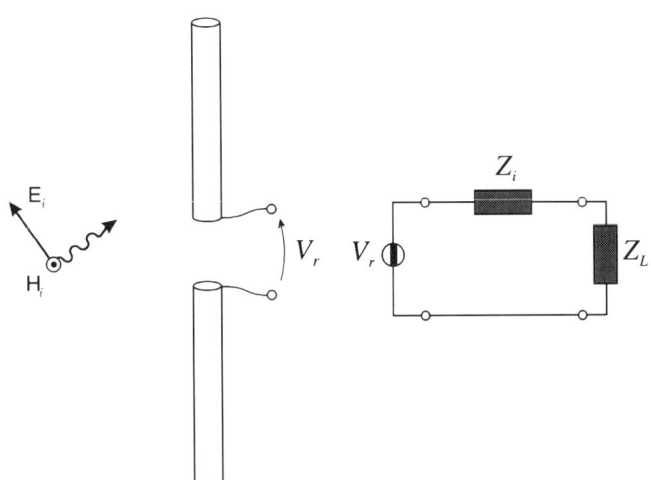

FIGURE 8.6. A receiving antenna and its associated equivalent circuit.

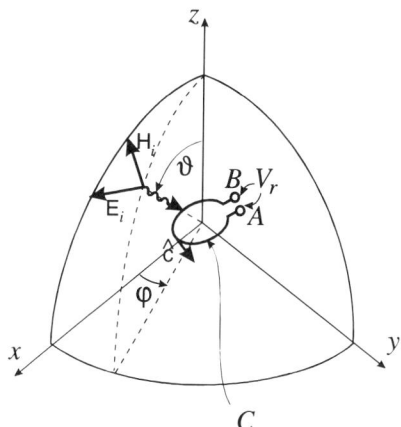

FIGURE 8.7. An elementary loop in its receiving mode.

tends to zero. Comparison with Eq. (8.20) shows that

$$\mathbf{l}_r = -ikA\sin\theta\hat{\boldsymbol{\phi}} = \mathbf{l};$$

see Eq. (8.14). The two effective lengths, in the transmitting and receiving modes, are identical.

Most generally, and for any antenna, transmitting (8.4) and receiving (8.20) effective lengths, $\mathbf{l}$ and $\mathbf{l}_r$, are related by the reciprocity relation (8.2). With reference to Fig. 8.2

$$V_{21} = -KI_2\mathbf{l}_2\cdot\mathbf{l}_{r1} \quad \text{and} \quad V_{12} = -KI_1\mathbf{l}_1\cdot\mathbf{l}_{r2},$$

where

$$K = i\frac{\zeta}{2\lambda r}\exp(-ikr);$$

r is the distance between the two antennas, each in the far field of the other. Then, Eq. (8.2) shows that

$$\mathbf{l}_2\cdot\mathbf{l}_{r1} = \mathbf{l}_1\cdot\mathbf{l}_{r2}.$$

If one of the two antennas, say number 2, is a loop antenna, $\mathbf{l}_2 = \mathbf{l}_{r2}$, and the components of $\mathbf{l}_1$ and $\mathbf{l}_{r1}$ along $\mathbf{l}_2$ are equal. This is true for all possible orientations of the loop, so that

$$\mathbf{l} = \mathbf{l}_r. \tag{8.21}$$

We conclude that the effective length of any antenna in the transmitting and receiving mode coincide. Equation (8.21) is only subject to the validity of the reciprocity theorem (see Section 4.3.6). The medium surrounding the antenna must be reciprocal.

As in the case of transmitting antennas, it is convenient to introduce an additional parameter related to the power rather than to the voltage amplitude at the antenna receiving terminals.

Let us first compute the power received when the antenna is closed on the impedance Z_L. Application of the Thevenin[6] theorem leads to the equivalent circuit of Fig. 8.6. The ideal generator coincides with the open-circuit voltage V_r of Eq. (8.20), and the generator input impedance coincides with the antenna input impedance Z_i. The (real) power delivered to the load $Z_L = R_L + iX_L$ is given by

$$P_R = \frac{1}{2} R_L \frac{|V_r|^2}{|Z_i + Z_L|^2} = \frac{1}{2} \frac{R_L}{|Z_i + Z_L|^2} |\mathbf{E}_i \cdot \mathbf{l}|^2 \qquad (8.22)$$

and can be maximized by adjusting both the load Z_L and the antenna orientation which changes the value of the scalar product $|\mathbf{E}_i \cdot \mathbf{l}|^2$.

As far as the load is concerned, the factor $R_L/|Z_i + Z_L|^2$ reaches its maximum value,

$$\frac{R_L}{|Z_i + Z_L|^2} \to \frac{1}{4R_i}, \qquad (8.23)$$

when $Z_i = Z_i^*$ (see Section 8.3.1).

As regards to the second factor, it is maximum when $\mathbf{l} \sim \alpha \mathbf{E}_i^*$, i.e., when the effective length of the receiving antenna is matched with the complex conjugate polarization of the incident field:

$$|\mathbf{E}_i \cdot \mathbf{l}|^2 \to |\mathbf{E}_i|^2 |\mathbf{l}|^2; \qquad (8.24)$$

see Section 8.3.1. Accordingly, maximum power transfer is obtained if load and input impedances are each other's complex conjugate, as well as (but for a constant) the effective lengths of the transmitting and receiving antennas.

When conditions (8.23) and (8.24) (*power and polarization matching*) are verified, we have from Eq. (8.22)

$$P_r = \frac{1}{2\zeta} |\mathbf{E}_i|^2 \frac{\zeta}{4R_i} |\mathbf{l}|^2 = S_i A,$$

$$A = \frac{\zeta}{4R_i} |\mathbf{l}|^2, \qquad (8.25a)$$

---

6. Leon Thevenin: Meaux (France), 1857–1926.

where S_i is the amplitude of the incident Poynting vector and A is, by definition, the *effective area* of the antenna.

The connection between the effective area A and its radiative counterpart G is readily obtained by comparing Eqs. (8.25a) and (8.13b):

$$G = 4\pi \frac{A}{\lambda^2}. \tag{8.26}$$

8.3.1. Power and Polarization Matching

When $Z_i = Z_L^*$, the ratio

$$\frac{R_L}{|Z_i + Z_L|^2} = \frac{R_L}{(X_i + X_L)^2 + (R_i + R_L)^2}$$

attains its maximum value $1/4R_i$, as follows by equating to zero its derivative with respect to R_L. Similarly, the product $|\mathbf{E}_i \cdot \mathbf{l}|^2$ reaches its maximum value $|\mathbf{E}_i|^2 |\mathbf{l}|^2$ when $\mathbf{l} = \alpha \mathbf{E}_i^*$ with α constant; see Eq. (4.29) and Section 4.2.1.

If

$$\frac{R_L}{|Z_i + Z_L|^2} = \frac{1}{4R_L} \xi,$$

then

$$\frac{4R_L^2}{(R_i + R_L)^2 + (X_i + X_L)^2} = \xi \leqslant 1 \tag{8.27}$$

is the *power matching factor*. Further, if

$$|\mathbf{E}_i \cdot \mathbf{l}|^2 = |\mathbf{E}_i|^2 |\mathbf{l}|^2 \chi,$$

then

$$\chi = \frac{|\mathbf{E}_i \cdot \mathbf{l}|^2}{|\mathbf{E}_i|^2 |\mathbf{l}|^2} \leqslant 1 \tag{8.28}$$

is the *polarization matching factor*. Equation (8.22) becomes

$$P_R = S_i A \xi \chi. \tag{8.25b}$$

8.3.2. The Radio Link Equation

Consider a radio link between a transmitting and a receiving antenna spaced distance r apart. If P_T and G_T are the transmitted power and gain associated with the transmitting antenna, while P_R and A_R are the received power and effective area of the receiving antenna, then

$$P_R = S_i A_R \xi \chi = \frac{P_T G_T}{4\pi r^2} A_R \xi \chi,$$

if the transmission space is lossless (*free space*). Accordingly

$$\frac{P_R}{P_T} = \frac{G_T A_R}{4\pi r^2} \xi \chi = G_T G_R \left(\frac{\lambda}{4\pi r}\right)^2 \xi \chi = \frac{A_T A_R}{\lambda^2 r^2} \xi \chi, \tag{8.29}$$

where G_R is the gain of the receiving antenna, A_T the effective area of the transmitting antenna, and use has been made of Eq. (8.26).

8.3.3. Effective Area of Elementary Antennas

The effective area

$$A = \frac{G\lambda^2}{4\pi}$$

of any elementary antenna seems to increase with no bound in the low-frequency limit $\lambda \to \infty$, if we take $G \approx D = \text{const}$. However, the latter assumption is not valid.

The electric dipole exhibits a large capacitive input reactance, and matching, which is the basis of Eq. (8.26), requires the series connection of a large inductor, whose ohmic losses reduce the gain.

The radiation resistance of the magnetic dipole,

$$R_r = \left(\frac{kA}{\lambda}\right)^2 \frac{2\pi}{3} \zeta = \frac{8\pi^3}{3} \zeta \left(\frac{A}{\lambda^2}\right)^2$$

[see Eq. (8.15)], rapidly decreases with the frequency, the contribution of the ohmic losses of the loop cannot be neglected, and the value of the gain is lowered.

8.3.4. Noise Temperature of the Antenna

Any body at absolute temperature T (Kelvin) radiates energy to and absorbs energy from the environment. The absorbed and received powers are equal when conditions of thermal equilibrium are attained. The radiating properties are described by the *brightness* B [watt/(Hz × m² × stereoradian)] of the body, which is the power radiated per unit bandwidth per unit surface of the body in a unit solid angle. The *black body* is the idealized version of the real body. At each frequency, its radiating and absorbing properties coincide. For such a body, the brightness (in the radio-frequency range) is given by

$$B = \frac{2kT}{\lambda^2}, \tag{8.30}$$

where $k = 1.38 \times 10^{-23}$ [Joule/Kelvin] is the Boltzman[7] constant. Also, the radiated power density at distance r from the elementary area dS of the black body within the bandwidth Δf is

$$\frac{2kT\,dS}{\lambda^2 r^2}\,\Delta f.$$

The related field is totally unpolarized (see Section 4.2.4), due to the random mechanism of radiation associated with the thermal motion of the elementary constituents of the body.

Consider now a receiving antenna in the presence of a black body of emitting area dS, as depicted in Fig. 8.8. The antenna receives power

$$dN = \xi \left(\frac{G\lambda^2}{4\pi}\right) \frac{1}{2} \left(\frac{2kT\Delta f}{\lambda^2}\,d\Omega\right)$$

in the bandwidth Δf, where ξ is the power matching factor given by Eq. (8.27), the factor 1/2 accounting for the polarization mismatch due to the incoherent nature of the incident field, and $dS/r^2 = d\Omega$. This power should be considered as *thermal noise*, and is collected by the antenna from the entire surrounding environment (in addition to other undesired signals, such as *man-made* noise and interferences). The total thermal noise received from the environment is given by

$$N = \xi \frac{k\Delta f}{4\pi} \int_{4\pi} d\Omega \alpha(\theta, \phi) G(\theta, \phi) T(\theta, \phi), \tag{8.31}$$

---

7. Ludwig Boltzman: Wien (Austria), 1844–Duino, Trieste, 1906.

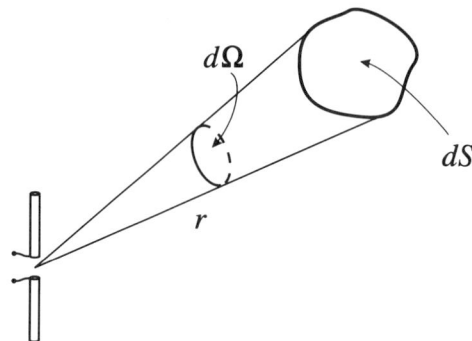

FIGURE 8.8. Receiving antenna in the presence of a black body.

where the (θ, ϕ) dependence of T accounts for the different temperatures of radiating bodies in the total solid angle, and the coefficient α accounts for the correction to Eq. (8.30), which applies to ideal, and not real, radiating bodies.

The environment temperature T changes drastically when the antenna points toward either the sky or the Earth. In the latter case we can take $T \approx 300\,\text{K}$; in the former, large differences are encountered when the antenna points to the Sun or to another empty region of the sky (*cold sky*). The sky temperature is frequency-dependent; see Fig. 8.9.

If

$$T_e = \frac{1}{4\pi} \int_{4\pi} d\Omega \alpha(\theta, \phi) G(\theta, \phi) T(\theta, \phi) \qquad (8.32)$$

is the *effective temperature* of the antenna, then Eq. (8.31) yields

$$N = \xi k T_e \Delta f.$$

For a matched antenna ($R_L = R_i$, $\xi = 1$) we have

$$N = k T_e \Delta f,$$

which is the classical expression for the *available noise power*. A quadratic average noise voltage,

$$\langle V_N^2 \rangle = 4 R_i k T_e \Delta f,$$

does appear at the antenna terminals.

ENGINEERING TOPICS: RADIATION

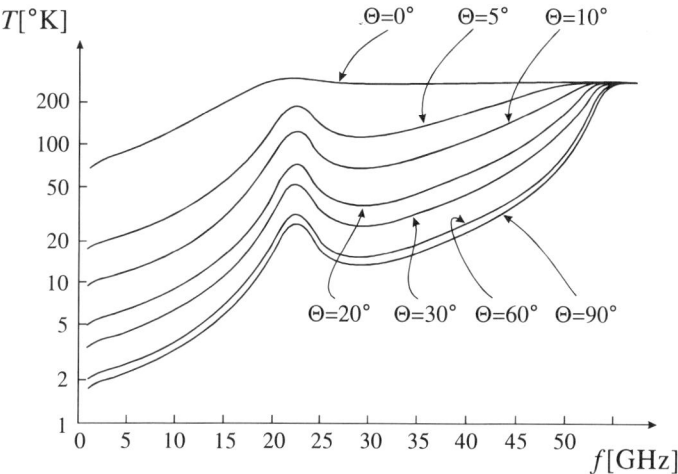

FIGURE 8.9. Sky temperature T (K) vs. frequency for several angles of elevation $\theta°$ over the horizon.

For a highly directive lossless $(G \to D)$ antenna pointing to a given direction (θ_0, ϕ_0) of the sky $(\alpha \approx 1)$, we have

$$T_e \approx T(\theta_0, \phi_0) \frac{1}{4\pi} \int_{4\pi} d\Omega D(\theta, \phi) = T(\theta_0, \phi_0);$$

see Eq. (8.13a).

A common quality parameter of a space antenna is the ratio between its gain and its noise temperature:

$$v^{dB} = \frac{G^{dB}}{T_e} \qquad [dB/K]. \tag{8.33}$$

The signal-to-noise ratio at the receiving antenna terminals is linearly proportional to v.

8.4. WIRE ANTENNAS

Consider the wire antenna depicted in Fig. 8.10 and excited by an (ideal) voltage generator V_0. We assume the antenna to be *thin*, i.e., its length $2l$ is much larger than its diameter $2a$. From a quantitative viewpoint, it is

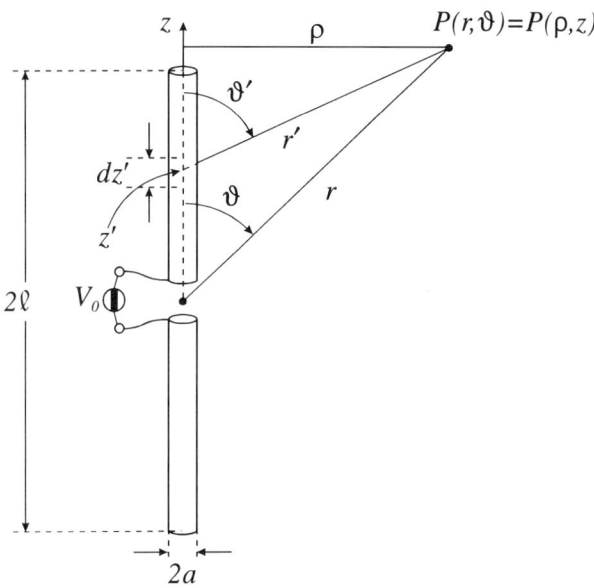

FIGURE 8.10. A wire antenna.

customary to define the *slenderness index*,

$$\Omega = 2\ln\frac{2l}{a}, \tag{8.34}$$

so the wire antenna is thin if $\Omega > 10$, i.e., $2l/a > 150$.

For a perfectly conducting cylinder, a current linear density $\mathbf{J}(z) = J(z)\hat{\mathbf{z}}$ is set along the wire, and this determines the total current $I(z')$ crossing the antenna section at the abscissa z'. We also have $I(0) = I_0$ and $I(\pm l) = 0$.[8] The far field, i.e., the field radiated at large distances from the antenna (the Fraunhofer region; see Section 4.6), can be computed by decomposing the wire in a cascade of elementary dipoles of length dz', and then invoking superposition:

$$\begin{aligned}\mathbf{E} &= i\zeta\frac{1}{2\lambda}\int_{-l}^{l} dz'\, \frac{I(z')}{r'}\sin\theta'\exp(-ikr')\hat{\boldsymbol{\theta}}' \\ &\approx i\zeta\frac{I_0\exp(-ikr)}{2\lambda r}\hat{\boldsymbol{\theta}}\sin\theta\int_{-l}^{l} dz'\,\frac{I(z')}{I_0}\exp(ikz'\cos\theta).\end{aligned} \tag{8.35}$$

---

8. This is true in the limit $a \to 0$. A less stringent condition is obtained if the antenna is realized by means of two hollow tubes: the terminal capacitances are lowered and the terminal currents approach zero.

ENGINEERING TOPICS: RADIATION

In Eq. (8.35) we set $r' \approx r - z'\cos\theta$ equal to r only in the amplitude term, but not in the phase term, because

$$kz'\cos\theta = 2\pi \frac{z'}{\lambda}\cos\theta$$

can attain large values compared to unity.

The quantity

$$\mathbf{l}(\theta) = \hat{\boldsymbol{\theta}}\sin\theta \int_{-l}^{l} dz'\, i(z')\exp(ikz'\cos\theta), \qquad \text{where } i(z') = \frac{I(z')}{I_0}, \qquad (8.36)$$

is the effective length [see Eq. (8.4)] of the wire antenna. Its knowledge, together with that of the input impedance,

$$Z_i = \frac{V_0}{I_0},$$

fully determines all transmitting and receiving parameters of the antenna. Accordingly, evaluation of the current distribution along the wire is of paramount importance.

The elementary fields which have been used to synthesize by superposition the radiation from the antenna, Eq. (8.35), already fulfil the radiation condition at infinity. Additional conditions on the wire surface should be imposed. For a perfectly conducting wire the tangential component of the electric field must be equal to zero on its surface, but in the gap region. In our model the antenna is fed at an *infinitesimal gap*, and the line integral of E_z across the gap must account for the applied voltage V_0:

$$E_z(a, z') = -V_0\delta(z'), \qquad -l \leqslant z' \leqslant l, \qquad (8.37)$$

where reference is made to a cylindrical system of coordinates centered on the wire axis.

From Eq. (4.87)

$$\mathbf{E}(\rho, z) = \frac{-i\omega\mu}{4\pi}\left[\mathbf{I} + \frac{\nabla\nabla}{k^2}\right]\cdot\mathbf{A}$$

and

$$\mathbf{A}(\rho, z) = \hat{\mathbf{z}}\int_{-l}^{l} dz'\, \frac{I(z')\exp(-ikr')}{r'}$$

$$= A(\rho, z)\hat{\mathbf{z}}, \qquad r' = \sqrt{\rho^2 + (z-z')^2},$$

in general; on the wire,

$$E_z(a, z) = -\frac{i\omega\mu}{4\pi}\left[A + \frac{1}{k^2}\frac{d^2A}{dz^2}\right] \tag{8.38a}$$

and

$$A(a, z) = \int_{-l}^{l} dz' \frac{I(z')\exp[-ik\sqrt{a^2 + (z-z')^2}]}{\sqrt{a^2 + (z-z')^2}}. \tag{8.38b}$$

Enforcement of the boundary conditions (8.37) in Eq. (8.38a) leads to the following differential equation for $A(a, z)$:

$$\frac{d^2A}{dz^2} + k^2A = -4\pi i\omega\varepsilon V_0\delta(z'), \tag{8.39}$$

whose solution, even and continuous at $z = 0$, is given by

$$A(a, z) = C\cos kz + S\sin k|z|.$$

Integration of Eq. (8.39) across the gap from $-\Delta z$ to Δz, $\Delta z \to 0$, leads to

$$\left.\frac{dA}{dz}\right|_{-\Delta z}^{\Delta z} = -4\pi i\omega\varepsilon V_0,$$

$$2Sk = -4\pi i\omega\varepsilon V_0,$$

and

$$A(a, z) = C\cos kz - 2\pi i\frac{V_0}{\zeta}\sin k|z|;$$

when this solution is substituted into Eq. (8.38b), we obtain

$$\int_{-l}^{l} dz' I(z')\frac{\exp[-ik\sqrt{a^2 + (z-z')^2}]}{\sqrt{a^2 + (z-z')^2}} = C\cos kz - 2\pi i\frac{V_0}{\zeta}\sin k|z|, \tag{8.40a}$$

which is Hallen's[9] *integral equation* for the current distribution $I(z)$.

9. Erik Hallen: Göthenburg (Sweden), 1899–Stockholm, 1975.

Equation (8.40a) is a Fredholm equation of *first kind*, unstable when a numerical solution is sought[10], and further elaboration is desirable. If

$$g(z - z') = \frac{1 - \exp[-ik\sqrt{a^2 + (z - z')^2}]}{ik\sqrt{a^2 + (z - z')^2}}$$

$$\approx \exp\left[-ik\frac{|z - z'|}{2}\right] \operatorname{sinc}\left[k\frac{z - z'}{2}\right],$$

then from Eq. (8.40a)

$$C \cos kz - 2\pi i \frac{V_0}{\zeta} \sin k|z| = \int_{-l}^{l} dz' \frac{I(z')}{\sqrt{a^2 + (z - z')^2}} - ik \int_{-l}^{l} dz' I(z') g(z - z'). \tag{8.40b}$$

When $a \to 0$ (thin antenna), the first integral in Eq. (8.40b) is mainly determined by the values of the integrand near $z' = z$. Accordingly

$$\int_{-l}^{l} dz' \frac{I(z')}{\sqrt{a^2 + (z - z')^2}} \approx I(z) \int_{-l}^{l} \frac{dz'}{\sqrt{a^2 + (z - z')^2}}$$

$$= I(z) \left[\int_{-l}^{z} \frac{dz'}{\sqrt{a^2 + (z - z')^2}} + \int_{z}^{l} \frac{dz'}{\sqrt{a^2 + (z - z')^2}} \right]$$

$$= I(z) \left[-\int_{l+z}^{0} \frac{du}{\sqrt{a^2 + u^2}} + \int_{0}^{l-z} \frac{du}{\sqrt{a^2 + u^2}} \right],$$

where the substitution $z - z' = \pm u$ has been introduced. The bracketed function can be evaluated via Eq. (C.11) and turns out to be slowly varying with z, so that its value at $z = 0$ is a reasonable approximation. This value coincides with the slenderness index Ω; see Eq. (8.34). Substitution into Eq. (8.40b) leads to

$$C \cos kz - 2\pi i \frac{V_0}{\zeta} \sin k|z| = \Omega I(z) - ik \int_{-l}^{l} dz' I(z') g(z - z'). \tag{8.41}$$

10. From a rigorous point of view, Eq. (8.40) does not admit any continuous solution for the current $I(z)$ due to the presence of the nonanalytic function at its right end side. Further elaboration is also aimed at somehow circumventing this shortcoming.

which is a Fredholm equation of the *second kind*, and stable from a numerical viewpoint. The constant C is determined by the requirement[11] that $I(l) = 0$.

A first approximate solution of Eq. (8.41) can be obtained by neglecting the integral term. In this case

$$C = 2\pi i \frac{V_0}{\zeta} \frac{\sin kl}{\cos kl},$$

and

$$I(z) \approx i \frac{2\pi V_0}{\zeta\Omega} \frac{\sin k(l - |z|)}{\cos kl} = I_0 \frac{\sin k(l - |z|)}{\sin kl}, \tag{8.42}$$

which shows that the current distribution is *sinusoidal*. This approximation models the thin antenna in terms of an open-circuited transmission line of characteristic impedance $\zeta\Omega/2\pi$.[12]

Although Eq. (8.42) can conveniently be used for computing the radiated field (and the radiation resistance, see Section 8.2.3), it is not adequate for computing the input impedance of the antenna because we obtain a pure reactance. Equation (8.42) should be regarded as the limiting value that the current distribution attains when $\Omega \to \infty$, i.e., when the antenna is infinitesimally thin. Improved solutions of Eq. (8.41) can be obtained numerically, or by means of an iterative procedure. The lower-order determination of $I(z)$ is used for computing the integral which appears in Eq. (8.41), thus generating the subsequent value of $I(z)$. The correction terms are very small in the case of thin antennas, thus validating the sinusoidal result as a very good approximation for the antenna current, except for the input impedance computation (see Section 8.4.2).

8.4.1. Short Antennas

An antenna is *short* if

$$kl = \pi \frac{2l}{\lambda} \ll 1.$$

Its current distribution in this case is triangular,

$$I(z) \approx I_0 \frac{(l - |z|)}{l}$$

---

11. The other condition $I(-l) = 0$ is automatically enforced by the symmetry of the solution $I(-z) = I(z)$.
12. In the sinusoidal approximation for the current, $I_0 = 0$ when $kl = n\pi$ and $V_0 = 0$ when $kl = (2n + 1)\pi/2$, respectively. Accordingly, the second or first expression for $I(z)$ should be used.

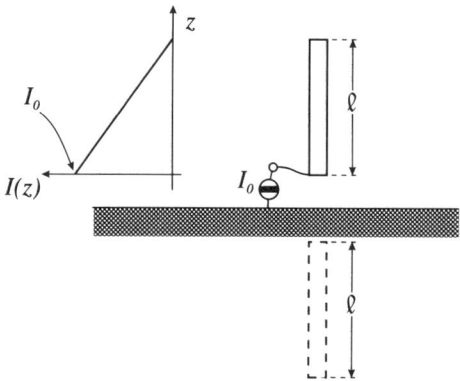

FIGURE 8.11. A short antenna over a conducting half-plane.

[see Eq. (8.42)], and from Eq. (8.36)

$$\mathbf{l}(\theta) \approx \hat{\mathbf{\theta}} \sin\theta \int_{-l}^{l} dz \, \frac{l - |z|}{l} = \hat{\mathbf{\theta}} l \sin\theta,$$

showing that its effective length is reduced by a factor of one-half compared to that of an elementary dipole [see Eq. (8.3)]. The directivity of the short antenna coincides with that of the elementary dipole.

The short antenna is usually implemented by means of a vertical wire over a conducting plane (see Fig. 8.11), in many instances the Earth surface (the Marconi[13] antenna), which is a good approximation of a conductor in the low-frequency regime. The input impedance of a short antenna is the series connection of a large reactive component and a resistor accounting for radiation and ohmic losses of the wire. The radiation resistance is given by

$$R_r = 1600 \left(\frac{l}{\lambda}\right)^2 \quad [\Omega];$$

see Section 8.2.3.

8.4.2. The Half-Wave Dipole Antenna

The half-wave dipole antenna is a wire antenna of length $2l = \lambda/2$, i.e., $kl = \pi/2$. Its effective length is given by

$$\mathbf{l}(\theta) = \hat{\mathbf{\theta}} \sin\theta \int_{-l}^{l} dz \, \sin[k(l - |z|)] \exp(ikz \cos\theta)$$
$$= \frac{2}{\pi} 2l \, \frac{\cos[(\pi/2)\cos\theta]}{\sin\theta} \, \hat{\mathbf{\theta}}. \tag{8.43}$$

---

13. Guglielmo Marconi: Bologna (Italy), 1874–Roma (Italy), 1937.

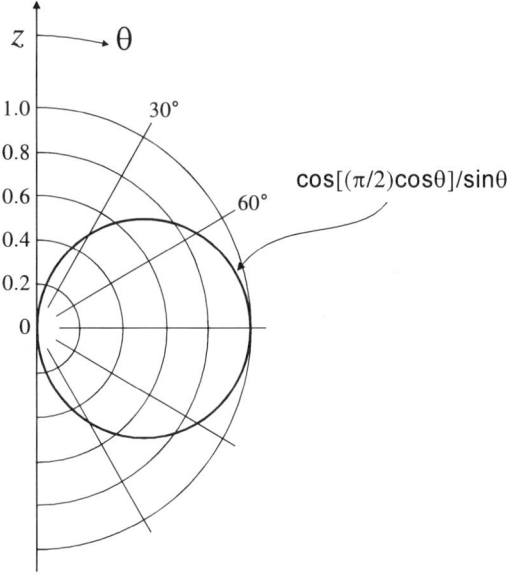

FIGURE 8.12. Polar radiation diagram of a half-wave dipole antenna.

The polar diagram of the modulus of the radiation vector [see Eq. (8.6)] is given in Fig. 8.12.

For computing the directivity, we need a more manageable expression for the radiation pattern in the interval $(0, \pi)$. By means of the Fourier series expansion in the period $(0, \pi)$,

$$\frac{\cos[(\pi/2)\cos\theta]}{\sin\theta} = \sum_{0}^{n} a_n \sin[(2n+1)\theta],$$

we obtain

$$a_0 = \frac{2}{\pi} \int_0^\pi d\theta \cos\left(\frac{\pi}{2}\cos\theta\right) = 2J_0\left(\frac{\pi}{2}\right) \approx 0.945,$$

where use has been made of Eq. (C.12). We approximate the far-field pattern with its $n = 0$ space harmonic and derive from Eq. (8.13a)

$$D_M = \frac{4\pi}{(0.945)^2 \int_0^{2\pi} d\phi \int_0^\pi d\theta \sin^3\theta} \approx 1.63.$$

ENGINEERING TOPICS: RADIATION

For the radiation resistance, we obtain from Eq. (8.13b) and $G = D$ (lossless antenna)

$$R_i = \pi \frac{\lambda^2}{\pi^2 \lambda^2 D_M} \zeta = 73.6 \quad [\Omega].$$

This value should be compared to that obtainable from the zeroth-order approximation for the antenna current, Eq. (8.42), which provides $Z_i = V_0/I(0) = 0$.[14] As already noted in Section 8.4, and improved expression for the current is needed, obtainable from Eq. (8.41) by first neglecting the integral when evaluating the first-order current $I(z)$,

$$I(z) = \frac{1}{\Omega}\left(C \cos kz - 2\pi i \frac{V_0}{\zeta} \sin k|z|\right),$$

and then using this value to compute the integral in Eq. (8.41):

$$C\left[1 + \frac{ik}{\Omega}\int_{-l}^{l} dz' \cos kz' g(z-z')\right] \cos kz$$
$$- 2\pi i \frac{V_0}{\zeta}\left[1 + \frac{ik}{\Omega}\int_{-l}^{l} dz' \sin k|z'| g(z-z')\right] \sin k|z| = \Omega I(z).$$

The integrals in this expression can be evaluated via Eqs. (C.13) and (C.14) if

$$g(u) = \exp\left(-ik\frac{|u|}{2}\right)\operatorname{sinc}\left(k\frac{u}{2}\right) \approx \exp\left(-ik\frac{|u|}{2}\right),$$

which is allowed because $|ku/2| \leqslant \pi/2$. The value of C is determined by setting $I(l) = 0$. For $kl \approx \pi/2$, we have $\cos kl \ll 1$, $\sin kl \approx 1$, and $1/\Omega \ll 1$. Hence

$$I(0) \approx \frac{2\pi i V_0}{\zeta \Omega} \frac{\sin kl}{\cos kl + (4i/3\Omega)(1+i)}$$

and

$$Z(0) = Z_i = \frac{V_0}{I(0)} \approx \frac{2\zeta}{3\pi}\left[1 + i\left(1 - \frac{3\Omega}{4}\cos kl\right)\right] = R_i + iX_i.$$

14. When $kl = \pi/4$, Eq. (8.42) shows that the current remains finite only if $V_0 \to 0$.

At *resonance*

$$\cos k_r l = \frac{4}{3\Omega}, \quad \text{so } Z(0) = R_i = \frac{2\zeta}{3\pi} \approx 80\,\Omega,$$

and this improved solution of Hallen's equation (8.41) correctly predicts a resonant length slightly shorter than $\lambda/2$,

$$k_r l \approx \frac{\pi}{2} - \frac{4}{3\Omega},$$

and an input resistance value not too far from $73\,\Omega$. If ω_r is the resonant angular frequency, where $\omega_r\sqrt{\varepsilon\mu} = k_r$, we also have near ω_r

$$\cos kl \approx \cos k_r l - (k - k_r)l \sin k_r l \approx \frac{4}{3\Omega} - (kl - k_r l)$$

and

$$Z_i \approx \frac{2\zeta}{3\pi} + i\,\frac{2\zeta}{3\pi}\frac{3\Omega}{4}(kl - k_r l) = R_i + i\,\frac{\zeta\Omega}{2\pi}\frac{\omega - \omega_r}{\omega_r}k_r l,$$

which models the antenna as the series resonant circuit of Fig. 8.13:

$$Z_i = R_i + iX_i, \quad R_i = \frac{2\zeta}{3\pi},$$

$$X_i = \omega L - \frac{1}{\omega C} = (\omega - \omega_r)L + \omega_r L - \frac{1}{(\omega - \omega_r)C + \omega_r C}$$

$$\approx (\omega - \omega_r)L + \omega_r L - \frac{1}{\omega_r C} + \frac{(\omega - \omega_r)}{\omega_r^2 C} = 2(\omega - \omega_r)L,$$

$$L = \frac{\zeta\Omega}{4\pi}\frac{k_r l}{\omega_r} = \frac{\Omega\mu l}{4\pi}, \quad C = \frac{1}{\omega_r^2 L} \approx \frac{16\varepsilon l}{\pi\Omega}\left(1 + \frac{16}{3\pi\Omega}\right).$$

A convenient way to define the *antenna bandwidth* makes reference to the power matching factor, Eq. (8.27). Within the bandwidth $\Delta\omega$ this factor should not be lower than 0.5, so that the received power loss is smaller than 3 dB. For a resistive load matched at resonance we have $R_L = R_i$. As we move away and

ENGINEERING TOPICS: RADIATION

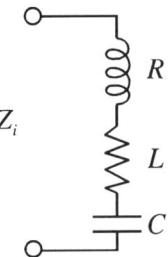

FIGURE 8.13. Equivalent circuit for the input impedance of a half-wave antenna near resonance: $R = R_i = 2\zeta/3\pi$, $L = (\mu\Omega/4\pi)l$, $C = (16\varepsilon/\pi\Omega)l[1 + (16/3\pi\Omega)]$.

around the resonance we require that

$$\frac{4R_i^2}{4R_i^2 + X_i^2} \geqslant \frac{1}{2}, \quad \text{i.e., } |X_i| \leqslant 2R_i,$$

which defines the bandwidth $\pm \Delta\omega/2$ across ω_r:

$$\frac{\zeta\Omega}{2\pi}\frac{\Delta\omega}{2\omega_r}\frac{\pi}{2} = \frac{4\zeta}{3\pi}, \quad \text{so} \quad \frac{\Delta\omega}{\omega_r} = \frac{32}{3\pi}\frac{1}{\Omega}.$$

We conclude that slender antennas ($\Omega \gg 1$) are narrowband: if Ω is decreased, the bandwidth increases accordingly.

8.4.3. The Traveling-Wave Antenna

Consider a wire antenna with progressive current distribution

$$I(z) = I_0 \exp(-ikz),$$

as depicted in Fig. 8.14. Its effective length is given by

$$\mathbf{l}(\theta) = \hat{\boldsymbol{\theta}} \sin\theta \int_{-l}^{l} dz \exp(-ikz) \exp(ikz\cos\theta)$$

$$= \hat{\boldsymbol{\theta}} 2l \sin\theta \operatorname{sinc}\left[2kl \sin^2\frac{\theta}{2}\right]. \tag{8.44}$$

The polar diagram of the modulus of the radiation vector is given in Fig. 8.14.

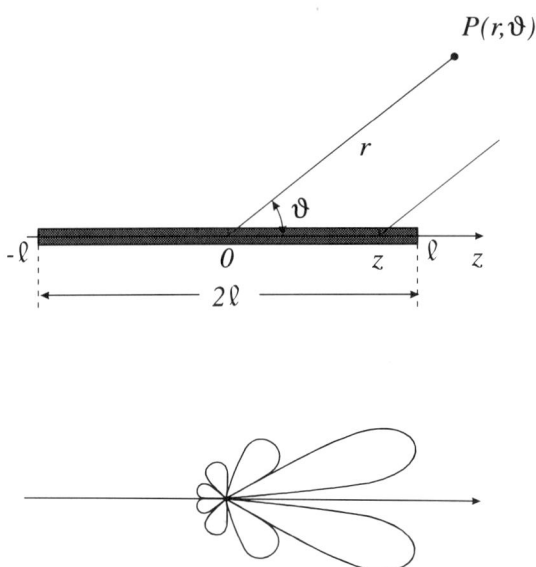

FIGURE 8.14. The traveling wave antenna and the polar plot of the modulus of its associated radiation vector.

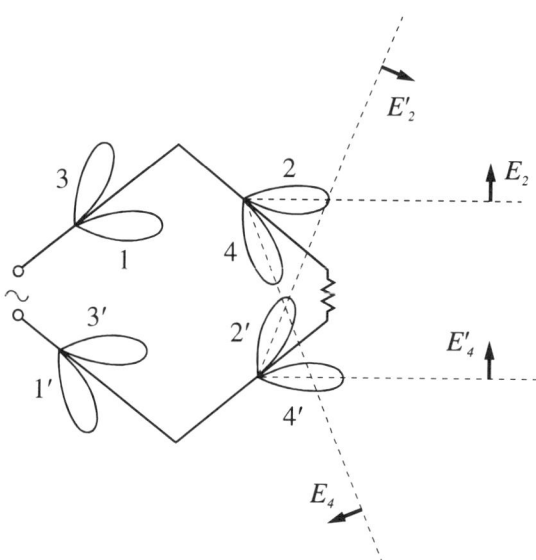

FIGURE 8.15. The rhombic antenna.

When kl is large, the maximum radiation occurs at an angle θ_M close to $\theta = 0$. A series expansion of Eq. (8.44) near $\theta = 0$ yields

$$\mathbf{l} \approx 4l \sin \frac{\theta}{2} \left[1 - \frac{(2kl)^2 \sin^4(\theta/2)}{6} \right] \hat{\boldsymbol{\theta}},$$

and we obtain

$$\sin^4 \frac{\theta_M}{2} = \frac{6}{5} \frac{1}{(2kl)^2}, \quad \text{so } \mathbf{l}(\theta_M) = 4l \frac{4}{5} \sqrt[4]{\frac{6}{5}} \sqrt{\frac{1}{2kl}} \hat{\boldsymbol{\theta}} = 1.67 \times 2l \sqrt{\frac{1}{2kl}} \hat{\boldsymbol{\theta}}$$

by setting equal to zero the derivative of l with respect to $\sin(\theta/2)$.

A practical implementation of this type of radiator is the rhombic antenna shown in Fig. 8.15. The antenna is matched at its end, so that only a progressive wave is present. The angle between and the length of the four arms of the antenna are chosen in such a way as to enhance radiation in the forward direction.

8.4.4. Mutual Impedance

We consider two wire antennas, 1 and 2, driven by currents I_1 and I_2, respectively. Input voltages, V_1 and V_2, are related to the currents by the relations

$$\begin{cases} V_1 = Z_{11}I_1 + Z_{12}I_2 \\ V_2 = Z_{21}I_1 + Z_{22}I_2 \end{cases}$$

due to linearity of the system; see Section 7.6.

The quantity Z_{11} is the *self-impedance* of antenna 1, and is equal to the ratio V_1/I_1 when the second antenna is open-circuited ($I_2 = 0$). In most, if not all, practical situations this impedance is unaffected by the presence of the other (open-circuited) antenna, and can be computed ignoring the latter. Identical reasonings apply for the self-impedance Z_{22}.

The quantity Z_{21}, which equals Z_{12} due to reciprocity (see Section 7.6.2), is the *mutual impedance* between the two antennas and equals the ratio V_2/I_1 when the second antenna is open-circuited ($I_2 = 0$). A formal expression that allows its computation is derived by again resorting to the reciprocity theorem applied to the configuration of Fig. 8.16. Antenna 2 is open-circuited and antenna 1 has been physically removed, leaving only the current sheet originally present on the wire surface, so that its field $(\mathbf{E}_1, \mathbf{H}_1)$ remains

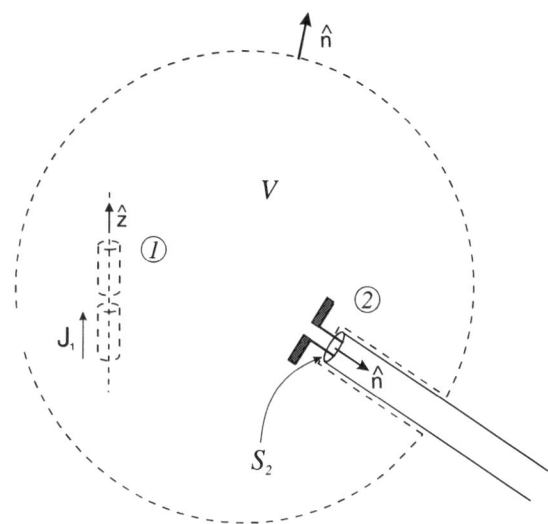

FIGURE 8.16. Computation of the mutual impedance between two wire antennas.

unchanged. Proceeding as in Section 8.1.1 we have

$$-\iiint_V dV \mathbf{E}_2 \cdot \mathbf{J}_1 = \iint_{S_2} dS(\mathbf{E}_1 \times \mathbf{H}_2 - \mathbf{E}_2 \times \mathbf{H}_1) \cdot \hat{\mathbf{n}}, \quad (8.45a)$$

because $\mathbf{E}_1 \cdot \mathbf{J}_2 = 0$ on the perfectly conducting wire antenna 2. We note further that $\mathbf{E}_2 \times \mathbf{H}_1 \cdot \hat{\mathbf{n}} = \mathbf{E}_2 \cdot (\mathbf{H}_1 \times \hat{\mathbf{n}}) = 0$ on the section S_2, because it corresponds to an open circuit, and that the volume integral can be reduced to a line integral if antenna 1 is very thin ($a \to 0$). Accordingly, Eq. (8.45a) reduces to

$$-\int dz I_1(z) \mathbf{E}_{21}(z) \cdot \hat{\mathbf{z}} = V_{12} I_2(0) \iint_{S_2} dS \mathbf{e}_2 \times \mathbf{h}_2 \cdot \hat{\mathbf{n}} = V_{12} I_2(0), \quad (8.45b)$$

where V_{12} is the voltage induced by antenna 1 on (open-circuited) antenna 2, and $I_2(0)$ is the input current of antenna 2 (when transmitting) that generates the electric field $\mathbf{E}_2(z) \to \mathbf{E}_{21}(z)$ over the axis of antenna 1 (being removed).[15] We finally obtain

$$Z_{12} = \frac{V_{12}}{I_1(0)} = -\frac{1}{I_1(0) I_2(0)} \int dz I_1(z) \mathbf{E}_{21}(z) \cdot \hat{\mathbf{z}}. \quad (8.46)$$

15. The usual normalization for the modal vector functions $\mathbf{e}_2$ and $\mathbf{h}_2$ has been implemented.

ENGINEERING TOPICS: RADIATION

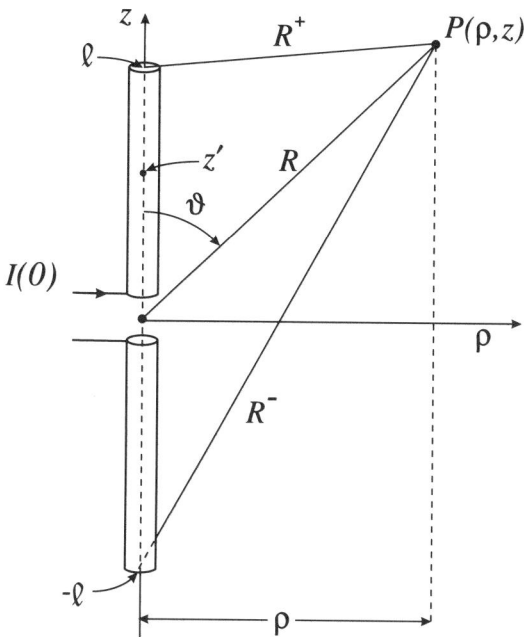

FIGURE 8.17. Computation of the field associated with a wire antenna in cylindrical coordinates.

To proceed further in the case of parallel wire antennas, we must compute the z-component of the associated electric field. Reference is made to Fig. 8.17 and Eq. (4.87), which yields

$$E_z(\rho, z) = -\frac{i\omega\mu}{4\pi} \int_{-l}^{l} dz'\, I(z') \left[I + \frac{1}{k^2} \frac{d^2}{dz^2} \right] g(z - z'),$$

with

$$g(z - z') = \frac{\exp[-ik\sqrt{\rho^2 + (z - z')^2}]}{\sqrt{\rho^2 + (z - z')^2}}.$$

Since

$$\frac{d^2}{dz^2} \to \frac{d^2}{dz'^2},$$

then integration by parts with the current distribution of Eq. (8.42) leads to

$$E_z(\rho, z) = -\frac{i\zeta I(0)}{4\pi \sin kl}\left[\frac{\exp(-ikR^+)}{R^+} + \frac{\exp(-ikR^-)}{R^-}\right]$$
$$- \frac{i\omega\mu}{4\pi}\int_{-l}^{l} dz'\left[I(z') + \frac{1}{k^2}\frac{d^2I(z')}{dz'}\right]g(z-z'). \quad (8.47)$$

But Eq. (8.42) shows that

$$\frac{d^2 I}{dz^2} + k^2 I = -2kI(0)\frac{\cos kl}{\sin kl}\delta(z),$$

as follows by integration across a small segment centered at $z = 0$. Hence Eq. (8.47) simplifies to

$$E_z(\rho, z) = -\frac{i\zeta I(0)}{4\pi \sin kl}\left[\frac{\exp(-ikR^+)}{R^+} + \frac{\exp(-ikR^-)}{R^-} - 2\cos kl\frac{\exp(-ikR)}{R}\right]. \quad (8.48)$$

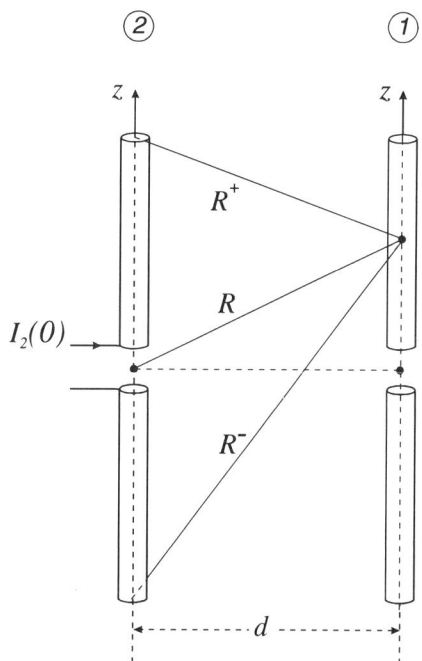

FIGURE 8.18. Computation of mutual impedance between two parallel half-wave dipoles.

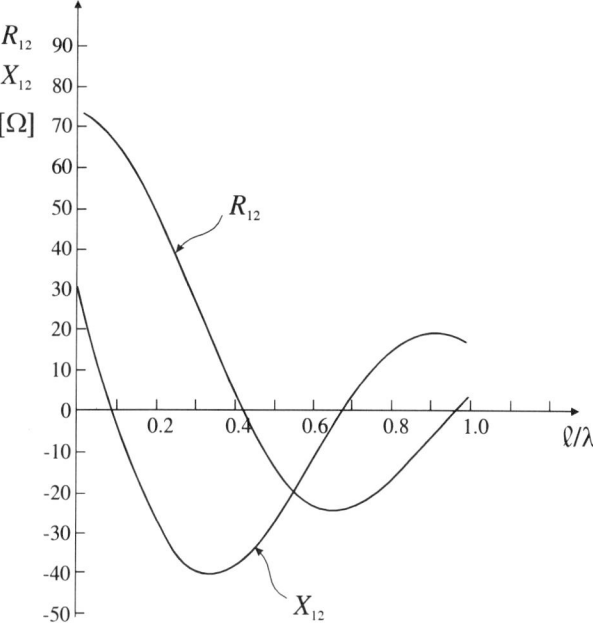

FIGURE 8.19. Mutual impedance between parallel wire antennas.

In the case of two parallel half-wavelength dipoles ($kl = \pi/2$), we obtain (see Fig. 8.18) with the current distribution of Eq. (8.42) and from Eq. (8.46)

$$Z_{21} = i \frac{\zeta}{4\pi} \int_{-l}^{l} dz \sin[k(l - |z|)] \left\{ \frac{\exp(-ikR^+)}{R^+} + \frac{\exp(-ikR^-)}{R^-} \right\},$$

$$R^\pm = \sqrt{d^2 + (l \mp z)^2}.$$

Graphs of mutual impedance between parallel wire antennas are given in Fig. 8.19.

8.5. APERTURE ANTENNAS

Consider a volume V enclosed by a surface S which is all metallized but for an aperture S'; see Fig. 8.20A. Electromagnetic sources inside the volume provide a field $(\mathbf{E}_0, \mathbf{H}_0)$ over the aperture. The field outside the volume V can be computed by applying the equivalence theorem of Section 4.3.7. The new

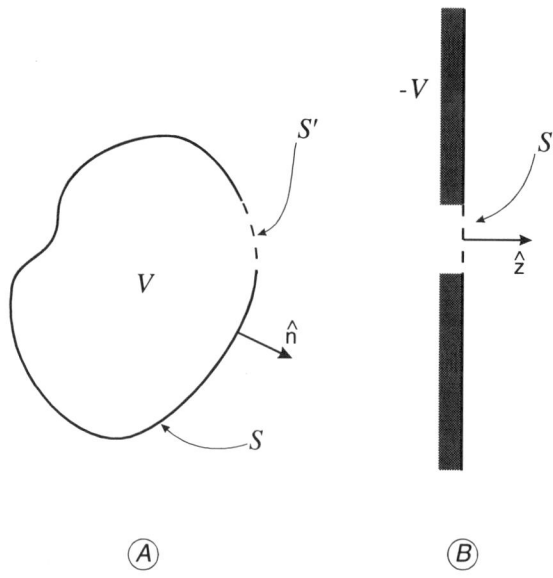

FIGURE 8.20. Computation of radiation from apertures.

sources are the equivalent surface currents

$$\mathbf{K}_S = -\hat{\mathbf{n}} \times \mathbf{E}_0, \qquad \mathbf{J}_S = \hat{\mathbf{n}} \times \mathbf{H}_0$$

over the aperture S', in addition to the induced surface currents over the remaining metal shield.

Although the above-mentioned technique is exact, its rigorous application renders this procedure useless. In fact, exact knowledge of the field over S is required, which calls for a field solution everywhere in space, *both inside and outside V*, so equivalent sources are no longer necessary.

However, the situation can be placed on a different footing if an approximation can be tolerated. If a reasonable guess of the field over S, hence of the new sources, is available, then computation of the field outside V is straightforward. This is the case if the aperture dimension is *large* compared to the wavelength and the field therein is properly phased to generate a focused beam. Then, illumination of the outside screen becomes negligible and the contribution of the surface currents induced there can be neglected. In addition, the field over the aperture may be often approximated by the *incident* one, i.e., the field that would be present in the absence of the screen (the *unperturbed field*), thus neglecting aperture edge diffraction.

Such an approximation would lead to an excellent prediction of the aperture far field within the angular regions where the radiated field is intense

ENGINEERING TOPICS: RADIATION

(main lobe), i.e., it has been focalized. Degradation is expected in low-field regions (low sidelobes, nulls) where radiation from the aperture tends to cancel and become comparable (or even negligible) with respect to the field scattered by the screen.

A particularly interesting case of an aperture antenna is that in Fig. 8.20B, where the screen is planar. We can rely on a modified form of the equivalence theorem (see Section 4.3.7) by metallizing the aperture and doubling the equivalent magnetic currents, i.e., the electric field $\mathbf{E}_0(x, y)$ there. The far field can be easily computed via Eq. (3.85) in terms of the spectrum, i.e., the double FT of the aperature field $\mathbf{E}_0(x, y)$ evaluated along the radiation directions.

8.5.1. The Rectangular Aperture

Consider a rectangular aperture in a ground plane as depicted in Fig. 8.21 with electric field distribution

$$\mathbf{E}_a = E_0 \cos\left(\frac{\pi}{a} x\right) \text{rect}\left(\frac{x}{2a}\right) \text{rect}\left(\frac{y}{2b}\right) \hat{\mathbf{y}}. \tag{8.49a}$$

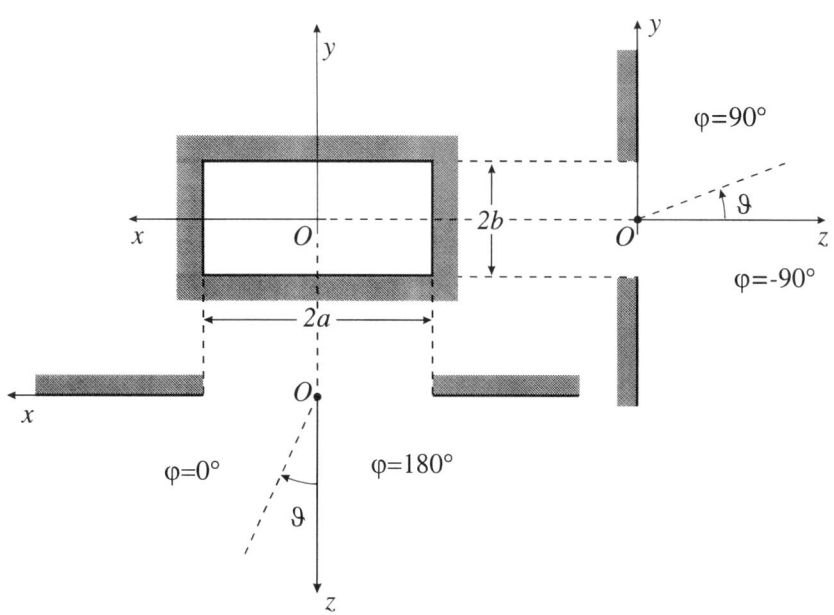

FIGURE 8.21. Rectangular aperture.

The corresponding plane-wave spectrum has been computed in Section 3.5.3[16]:

$$\hat{\mathbf{E}}(u, v) = \frac{2}{\pi} 4abE_0 \operatorname{sinc}(vb) \frac{\cos ua}{1 - (2ua/\pi)^2} \left(\hat{\mathbf{y}} + \frac{v}{w} \hat{\mathbf{z}} \right).$$

The radiated far field is given by Eq. (3.85):

$$\mathbf{E}(\mathbf{r}) = ik \cos\theta \, \frac{\hat{\mathbf{E}}(-k\sin\theta\sin\phi, -k\sin\theta\cos\phi)}{2\pi r} \exp(-ikr). \quad (8.49b)$$

With these values of (v, w) we obtain

$$\cos\theta \left(\hat{\mathbf{y}} + \frac{v}{w} \hat{\mathbf{z}} \right) = \cos\theta \hat{\mathbf{y}} - \sin\theta\sin\phi \hat{\mathbf{z}} = \sin\phi \hat{\boldsymbol{\theta}} + \cos\theta\cos\phi \hat{\boldsymbol{\phi}}$$

with the aid of Eqs. (A.45), so the radiation vector is given by

$$\mathbf{e}(\theta, \phi) = i \operatorname{sinc}(ka\sin\theta\sin\phi) \frac{\cos(kb\sin\theta\cos\phi)}{1 - (2kb\sin\theta\cos\phi/\pi)^2} \frac{(\sin\phi \hat{\boldsymbol{\theta}} + \cos\theta\cos\phi \hat{\boldsymbol{\phi}})}{\sqrt{\sin^2\phi + \cos^2\theta\cos^2\phi}}, \quad (8.50)$$

with respect to a spherical system of coordinates (r, θ, ϕ). We note that the polarization state of the radiated field coincides with that of a magnetic dipole oriented along the x-axis; see Section 4.2.2 and the first row of Table 4.2.

The two principal planes of the radiation diagram are $\phi = \pm 90°$ (E-plane) and $\phi = 0°, 180°$ (H-plane). In the former the radiaton vector is oriented along $\pm \hat{\boldsymbol{\theta}}$ and the radiation diagram is given by

$$e_E(\theta) = i \operatorname{sinc}(kb\sin\theta), \quad \text{in the E-plane.} \quad (8.51a)$$

In the latter plane the radiation vector is oriented along $\pm \hat{\boldsymbol{\phi}}$, i.e., along $\hat{\mathbf{y}}$, and the radiation diagram is given by

$$e_H(\theta) = i \frac{\cos(ka\sin\theta)}{1 - (2ka\sin\theta/\pi)^2} \cos\theta, \quad \text{in the H-plane.} \quad (8.51b)$$

The 3 dB beamwidths are equal to $\sin(\theta_E/2) = 0.44\lambda/2b$ and $\sin(\theta_H/2) = 0.60\lambda/2a$, and the first sidelobes to -13.46 dB in the E-plane and -23.52 dB in the H-plane, respectively. Note the enlargement of the radiation diagram

---

16. Unlike in Section 3.5.3, the electric field in the aperture is polarized along the y-axis.

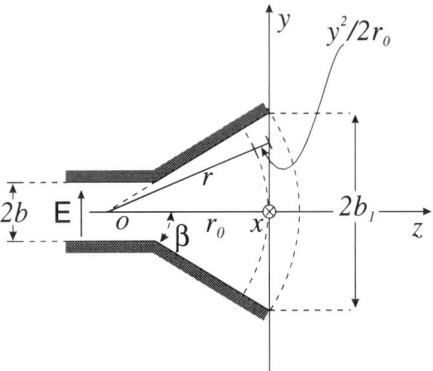

FIGURE 8.22. The pyramidal horn.

and the decrease of sidelobes in the H-plane, where the aperture distribution is tapered.

The directivity is obtained simply via Eq. (8.7) by assuming $H_0 \approx E_0/\zeta$ and then computing the flux of the Poynting vector across the aperture area:

$$D = \frac{4\pi r^2 |E(\theta = 0)|^2}{\int_{-a}^{a} dx \int_{-b}^{b} dy |E_0(x)|^2} = \frac{4\pi}{\lambda^2} A \frac{8}{\pi^2},$$

where $A = 4ab$ is the aperture area. Comparison with Eq. (8.26) shows that the effective area of this aperture is $8/\pi^2 \approx 0.8$ times its geometrical area.

The aperture field distribution (8.49a) is the same as the TE_{10} transverse field of a rectangular waveguide. In many applications the rectangular waveguide is continued in a *pyramidal horn*, as shown in Fig. 8.22. In this way the aperture dimension can be enlarged, but the phase distribution over the aperture is no longer uniform.

For the E-plane horn (see Fig. 8.22),

$$r = \sqrt{r_0^2 + y^2} \approx r_0 + (y^2/2r_0), \qquad b_1/r_0 < 1,$$

and Eq. (8.49a) should be multiplied by the additional factor[17]

$$\exp(-iky^2/2r_0), \quad \text{with } r_0 \tan\beta = b_1, \quad \beta < 45°. \tag{8.52}$$

---

17. Note that the additional geometric attenuation factor r_0/r is compensated by the projection factor r/r_0.

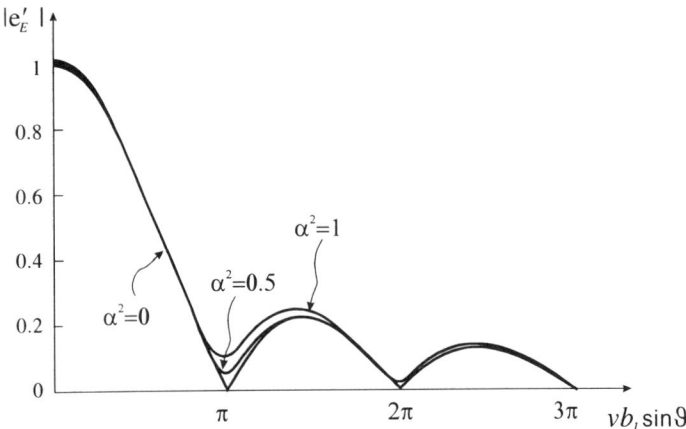

FIGURE 8.23. Radiation diagram of a rectangular aperture with quadratic phase tapering $\alpha^2 = kb_1^2/r_0$.

The E-plane radiation diagram is proportional to the FT of the product of rect$(y/2b_1)$ times Eq. (8.52), i.e., it is the convolution of the corresponding FTs:

$$e'_E(vb_1) = 2b_1 \sqrt{\frac{r_0}{2\pi k b_1^2}} \exp\left(-i\frac{\pi}{4}\right) \cdot \int_{-\infty}^{+\infty} d\xi \, \text{sinc}(vb_1 - \xi) \exp\left(i\frac{r_0}{2kb_1^2} \xi^2\right), \tag{8.53a}$$

where use has been made of Eq. (C.17) for computing the FT of Eq. (8.52).

Equation (8.53a) represents the new (nonnormalized) radiation diagram and depends on the *phase tapering* parameter

$$\alpha^2 = \frac{kb_1^2}{r_0} = kb_1 \cotan \beta, \tag{8.54}$$

which quantifies departure from a constant phase over the aperture. When $\alpha \to 0$,

$$\sqrt{\frac{1}{2\pi}} \frac{1}{\alpha} \exp\left(-i\frac{\pi}{4}\right) \exp\left(i\frac{\xi^2}{2\alpha^2}\right) \to \delta(\xi)$$

[see Eq. (3.7a) with $\varepsilon \to 2i\alpha^2$] and the result of Eq. (8.51a) is recovered. This suggests a series expansion of $\text{sinc}(vb_1 - \xi)$ in Eq. (8.53a). When computing the resulting series of integrals, note that

$$\int_{-\infty}^{+\infty} d\xi \exp\left(i\frac{\xi^2}{2\alpha^2}\right) = \sqrt{2\pi}\,\alpha \exp\left(i\frac{\pi}{4}\right) = f(\alpha)$$

[see Eq. (C.5)] and

$$\int_{-\infty}^{+\infty} d\xi \exp\left(i\frac{\xi^2}{2\alpha^2}\right)\xi = 0,$$

$$\int_{-\infty}^{+\infty} d\xi \exp\left(i\frac{\xi^2}{2\alpha^2}\right)\xi^2 = \frac{2}{i}\frac{d}{d(1/\alpha^2)}\,f(\alpha) = i\sqrt{2\pi}\,\alpha^3 \exp\left(i\frac{\pi}{4}\right).$$

Accordingly, Eq. (8.53a) becomes

$$e'_E = 2b_1\left[\operatorname{sinc}(vb_1) + i\frac{\alpha^2}{2}\frac{d^2\operatorname{sinc}(u)}{du^2}\bigg|_{vb_1}\right], \quad \text{with } v = k\sin\theta, \quad (8.53\text{b})$$

whose normalized modulus is presented in Fig. 8.23. The first-order effect of the quadratic phase tapering is to fill in the radiation nulls.

Finally, we note that a rectangular aperture exhibits a radiation diagram which is *not* equal in the two principal planes. A more symmetric radiation diagram would be obtained with a similar electric field distribution in *both* the aperture planes, which amounts to a change in the boundary conditions in the H-plane from those pertinent to a perfect electric conductor to those pertinent to a perfect magnetic conductor. One way to accomplish this requirement is

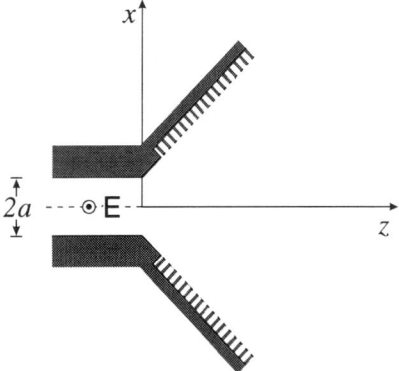

FIGURE 8.24. The corrugated horn.

shown in Fig. 8.24. The walls of the horn are corrugated, the length of each slot chosen in such a way as to generate an open circuit at its input. If there are many slots per horn wavelength, the boundary conditions resemble those of a magnetic wall (at least within a limited frequency band) and the tangential electric field is not forced to be zero there. This type of aperture antenna is usually referred to as a *corrugated horn* or, sometimes, scalar feed.

8.5.2. The Circular Aperture

Let us consider the circular aperture in Fig. 8.25 with electric field excitation

$$\mathbf{E}_a(\rho, \alpha) = E_0 \left[\frac{a}{q_{11}\rho} J_1\left(q_{11} \frac{\rho}{a}\right) \sin\alpha \hat{\boldsymbol{\rho}} + J_1'\left(q_{11} \frac{\rho}{a}\right) \cos\alpha \hat{\boldsymbol{\alpha}} \right], \quad (8.54a)$$

where q_{11} is the first zero of the derivative of the Bessel function $J_1(x)$; see Table 4.4. This aperture distribution coincides (but for the normalization) with the transverse electric field of the TE_{11} mode in a circular waveguide with

$$\psi_{11}(\rho, \alpha) = J_1\left(q_{11} \frac{\rho}{a}\right) \cos\alpha;$$

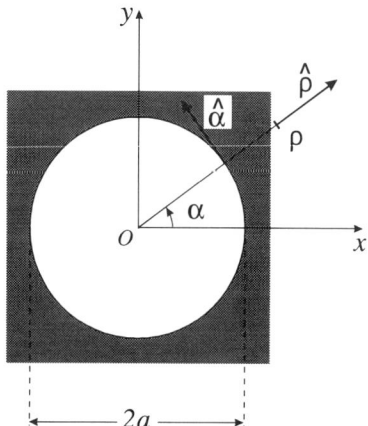

FIGURE 8.25. The circular aperture.

ENGINEERING TOPICS: RADIATION

see Section 7.4.2. In rectangular coordinates

$$\mathbf{E}_a(\rho, \alpha) = E_0 \left[\frac{a}{q_{11}\rho} J_1\left(q_{11} \frac{\rho}{a}\right) - J_1'\left(q_{11} \frac{\rho}{a}\right) \right] \sin \alpha \cos \alpha \hat{\mathbf{x}}$$

$$+ E_0 \left[\frac{a}{q_{11}\rho} J_1\left(q_{11} \frac{\rho}{a}\right) \sin^2\alpha + J_1'\left(q_{11} \frac{\rho}{a}\right) \cos^2\alpha \right] \hat{\mathbf{y}}.$$

With the aid of the known relationships

$$\sin \alpha \cos \alpha = \frac{\sin 2\alpha}{2}, \qquad \sin^2\alpha = \frac{1 - \cos 2\alpha}{2}, \qquad \cos^2\alpha = \frac{1 + \cos 2\alpha}{2},$$

$$\frac{J_1(t)}{t} - J_1'(t) = J_2(t), \qquad \frac{J_1(t)}{t} + J_1'(t) = J_0(t)$$

[see Eq. (D.7)], we have

$$\mathbf{E}_a(\rho, \alpha) = \frac{E_0}{2} \left[J_0\left(q_{11} \frac{\rho}{a}\right) \hat{\mathbf{y}} - J_2\left(q_{11} \frac{\rho}{a}\right) \cos 2\alpha \hat{\mathbf{y}} + J_2\left(q_{11} \frac{\rho}{a}\right) \sin 2\alpha \hat{\mathbf{x}} \right].$$

(8.54b)

If

$$u = k_t \cos \phi, \qquad v = k_t \sin \phi, \qquad k_t = k \sin \theta,$$
$$x = \rho \cos \alpha, \qquad y = \rho \sin \alpha, \qquad \rho = r \sin \theta,$$

with the usual spherical coordinate system (r, θ, ϕ) centered in the aperture and with the z-axis normal to it, the transverse part of the plane-wave spectrum is readily obtained (see Section 3.5):

$$\mathbf{E}_t(u, v) = \int_0^a d\rho \int_0^{2\pi} d\alpha \exp[-ik_t \rho \cos(\alpha - \phi)] \rho \mathbf{E}_a(\rho, \alpha).$$

To proceed further we use Eq. (C.16) to expand the exponential and integrate with respect to α:

$$\mathbf{E}_t(u, v) = \pi \left(\frac{a}{q_{11}}\right)^2 E_0 \hat{\mathbf{y}} \int_0^{q_{11}} dt J_0(t) t J_0\left(\frac{k_t a}{q_{11}} t\right)$$

$$- \pi \left(\frac{a}{q_{11}}\right)^2 E_0(\hat{\mathbf{x}} \sin 2\phi - \hat{\mathbf{y}} \cos 2\phi) \int_0^{q_{11}} dt J_2(t) t J_2\left(\frac{k_t a}{q_{11}} t\right),$$

where the change of variable $q_{11}\rho/a \to t$ has also been introduced. The radiated field is now discussed.

The first integral dominates when $k_t a$ is not large, i.e., in the angular region of the main radiation [see Eq. (8.55) below], and can be computed via Eq. (C.19):

$$\mathbf{E}_t(u, v) = \pi a^2 E_0 \hat{\mathbf{y}} \frac{q_{11} J_1(q_{11}) J_0(k_t a) - k_t a J_1(k_t a) J_0(q_{11})}{q_{11}^2 - (k_t a)^2}$$

$$= \pi a^2 E_0 \hat{\mathbf{y}} J_0(q_{11}) \left[J_0(k_t a) + (k_t a)^2 \frac{J_1'(k_t a)}{q_{11}^2 - (k_t a)^2} \right],$$

where Eq. (D.7) has been employed. Note that the transverse spectrum remains finite for $k_t a \to q_{11}$ because $J_1'(q_{11}) = 0$, and that it is dominated by the first factor in the square brackets inside the angular region of the main radiation. With this approximation we compute the complete plane-wave spectrum (see Section 3.5) via Eq. (3.82) and the radiated far field via Eq. (3.85):

$$\mathbf{E}(r, \theta, \phi) = ik\pi a^2 J_0(q_{11}) \frac{E_0}{2\pi r} J_0(ka \sin\theta) \cdot (\sin\phi \hat{\boldsymbol{\theta}} + \cos\theta \cos\phi \hat{\boldsymbol{\phi}}) \exp(-ikr).$$

(8.55)

The polarization of the radiated field equals that generated by the rectangular aperture [see Eq. (8.50)].

The total radiated power is

$$\frac{1}{2\zeta} \int_0^a d\rho \int_0^{2\pi} d\alpha |\mathbf{E}_a(\rho, \alpha)|^2 \rho$$

$$= \frac{|E_0|^2}{2\zeta} \frac{\pi}{2} \int_0^a d\rho \left[J_0^2\left(q_{11}\frac{\rho}{a}\right) + J_2^2\left(q_{11}\frac{\rho}{a}\right) \right] \rho$$

$$= \frac{|E_0|^2}{2\zeta} \frac{\pi a^2}{4} [J_0^2(q_{11}) + J_1^2(q_{11}) + J_2^2(q_{11}) - J_1(q_{11}) J_3(q_{11})],$$

where use has been made of Eq. (C.20). The directivity is computed as in Section 8.5.1 to yield

$$D = \frac{4\pi}{\lambda^2} A \frac{4 J_0^2(q_{11})}{J_0^2(q_{11}) + J_1^2(q_{11}) + J_2^2(q_{11}) - J_1(q_{11}) J_3(q_{11})} = \frac{4\pi}{\lambda^2} A_e,$$

where $A = \pi a^2$ is the aperture area and

$$A_e \approx 0.71 A$$

is its effective area.

8.5.3. The Patch Antenna

The *patch antenna* consists of a thin metallic film bonded to a grounded dielectric substrate (see Fig. 8.26) whose dielectric constant is usually relatively low ($\varepsilon_r \leq 4$). The patch can be of any shape, although it is usually a square, a rectangle, or a circle. The patch can be fed by a strip line or by a coaxial cable (see Fig. 8.26). Its ability to be integrated makes the patch a very attractive antenna for conformal arrays.

A very simple, yet approximate, model of the patch antenna is now presented. Consider the rectangular patch in Fig. 8.27 whose physical dimension along z has been doubled to account for its image with respect to the

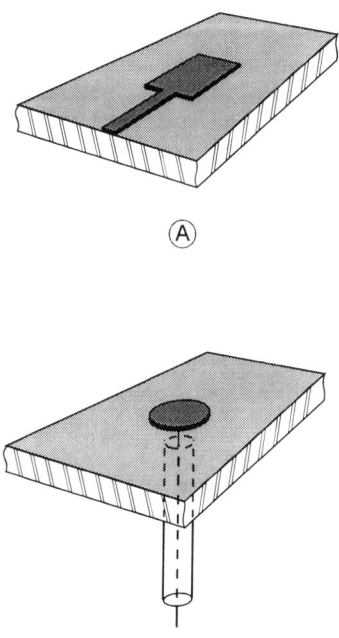

FIGURE 8.26. Patch antennas: (A) rectangular fed by a strip line and (B) circular fed by a coaxial line.

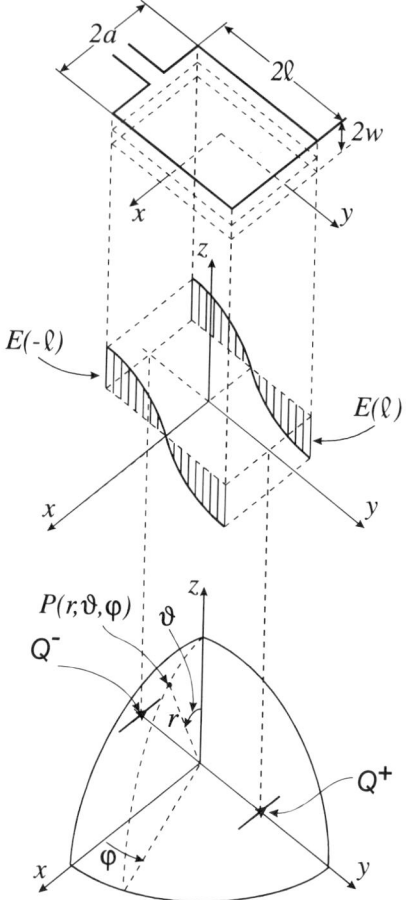

FIGURE 8.27. Computation of radiation from a rectangular patch.

ground plane $z = 0$. The electromagnetic field inside the patch cavity is determined by the excitation and by the boundary conditions on the top and bottom plates and on the four sides of the patch box. We approximate the latter by magnetic walls, so that the only tangential field there is the electric one. We assume the field to be uniform along x, and to possess a cosinusoidal pattern along y with a phase change of π from one to the other wall of the patch. We now apply the equivalence theorem (see Section 4.3.7) by setting equivalent magnetic surface currents,

$$\mathbf{K}_S = -\hat{\mathbf{n}} \times \mathbf{E} = -\hat{\mathbf{n}} \times \hat{\mathbf{z}} E,$$

ENGINEERING TOPICS: RADIATION

on the patch lateral walls, $\hat{\mathbf{n}}$ being the unit normal to them. These currents have the same sense on the walls $y = \pm l$ and opposite sense on the walls $x = \pm a$. If the dimension a is small compared to the (free-space) wavelength, radiation from the latter can be neglected and only the former are effective. These, in turn, can be modeled as elementary magnetic dipoles whose current is given by

$$I_m(\pm l) = \left[\mp \int_{-w}^{w} dz \hat{\mathbf{y}} \times \hat{\mathbf{z}} E(\pm l, z) \right] \cdot \hat{\mathbf{x}}$$
$$= \mp \int_{-w}^{w} dz E(\pm l, z) = 2V, \qquad (8.56)$$

where $V = V(l) = -V(-l)$ is the voltage at $y = l$ across the patch to the ground plane.

The far field of a magnetic dipole at the origin of the (spherical) coordinate system and oreinted along the x-axis is given by

$$\mathbf{E} = -\frac{\omega U}{2\lambda r} \exp(-ikr)(\sin\phi \hat{\boldsymbol{\theta}} + \cos\theta \cos\phi \hat{\boldsymbol{\phi}}),$$

with

$$-I_m 2a + i\omega U = 0;$$

see the first row of Table 4.2 and Fig. 4.13. For the two magnetic dipoles displaced $2l$ along the y-axis (see Fig. 8.27) we assume the polarization of the radiated field to remain unchanged, and we only account for the phase difference

$$|PQ^{\pm}| = \sqrt{(r\sin\theta\cos\phi)^2 + (r\sin\theta\sin\phi \mp l)^2 + (r\cos\theta)^2}$$
$$\approx r \mp l\sin\theta\sin\phi.$$

Accordingly, the far field radiated by the patch antenna is given by

$$\mathbf{E}(r, \theta, \phi) = i\frac{2Va}{\lambda r} \cos(kl\sin\theta\sin\phi) \exp(-ikr)(\sin\phi \hat{\boldsymbol{\theta}} + \cos\theta\cos\phi \hat{\boldsymbol{\phi}}),$$

where the magnetic currents (8.56) have been doubled to account for the presence of their images with respect to the magnetic wall.

8.6. REFLECTOR ANTENNAS

Any large metal surface that scatters an electromagnetic field radiated by *primary sources* can be regarded as a *reflector* antenna (see Fig. 8.28). In practical applications, however, the metal scatterer is shaped so as to conveniently modify the radiation diagram of the primary sources, usually by focusing the incoming radiation within narrow angular regions of space.

There are several techniques with which to evaluate the radiation parameters of a reflector antenna, based on the assumption that the reflector is large in terms of the incident wavelength. These techniques can be grouped into two classes, by the use of a *wave model* or a *ray model* for the incident as well as for the scattered field. A combination of these two extreme points of view may often be employed.

The wave model proceeds along three steps: determination or postulation of a primary wave; interaction of the field with the reflector and evaluation of the induced surface currents; computation of the field scattered by the reflector as a superposition of those radiated by the elementary surface currents. The last two steps may be implemented by substituting the reflector with a current sheet,

$$\mathbf{J}_S = \hat{\mathbf{n}} \times (\mathbf{H}_2 - \mathbf{H}_1) \tag{8.57}$$

(see Fig. 8.29), that radiates in free space. Very often, if not always, these currents are evaluated by assuming that the tangential magnetic field, $\hat{\mathbf{n}} \times \mathbf{H}$,

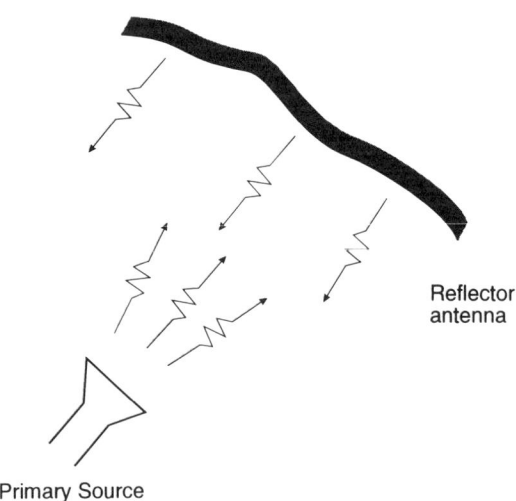

FIGURE 8.28. A reflector antenna.

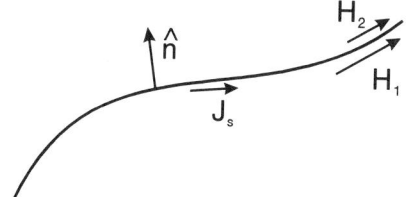

FIGURE 8.29. Equivalent currents for computing the radiation from a reflector antenna.

on the metal reflector is zero on its dark side, and is twice as large as the tangential component of the *incident* field on its illuminated side:

$$\mathbf{J}_S = -2\hat{\mathbf{n}} \times \mathbf{H}_i \quad \text{on the illuminated side,}$$
$$\mathbf{J}_S = 0 \quad \text{on the dark side.} \quad (8.58)$$

These equations are exact for a planar metal surface of infinite extent and are usually referred to as the physical optics (PO) approximation (see also Section 5.4). They provide excellent results when the scattered field is computed in regions of intense radiation.

The ray model proceeds in three steps as well: determination or postulation of a primary ray congruence; interaction of the rays with the reflector surface and evaluation of reflection and diffraction points; computation of the diffracted ray congruence by applying to each ray the appropriate reflection or diffraction matrix (see Sections 5.3.3 and 5.4.2).

There is a fundamental distinction between the two procedures. In the wave approach, the field scattered to a prescribed point is computed as an integral over *all* the reflecting surface, unlike the ray approach, where it is obtained as a superposition of rays from a (usually very) limited number of points over the reflector (but in caustic regions). Clearly, the PO integral is dominated by those integration regions over the reflector from which the (reflected and diffracted) rays are coming. From this viewpoint, the wave approach is somewhat inefficient. But the ray approach has its shortcomings, too: determination of the (reflection and diffraction) points may be not an easy task, requiring extensive *ray tracing*. This implies either checking whether *all* incident rays generate a corresponding (reflected or diffracted) ray to the considered point; or whether *all* rays from this point can reach the source via the reflector.

A mixed ray–wave procedure leads to the *equivalent aperture approach* (see Fig. 8.30). First, the ray approach is used to compute the field scattered by the reflector over a convenient, usually planar, surface. Then, this surface is

FIGURE 8.30. Equivalent aperture approach to compute the field scattered by a reflector antenna.

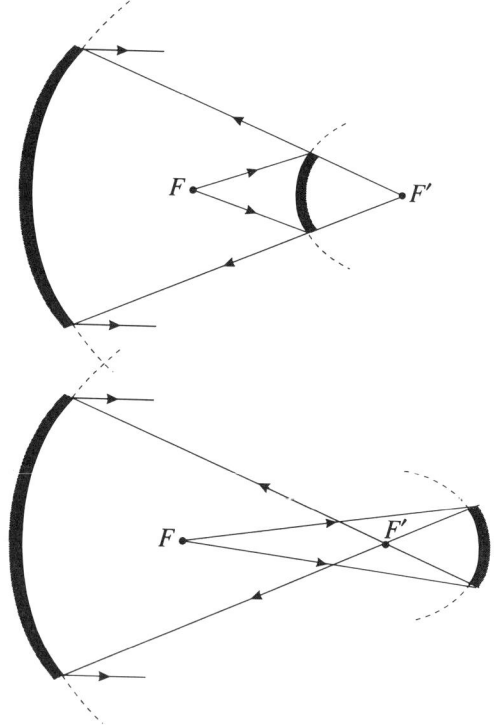

FIGURE 8.31. Multiple reflector antennas: Cassegranian (A) and Gregorian (B) antennas.

treated as an aperture for computing the scattered field. This procedure is particularly useful when several reflectors are used in a cascade arrangement. Two important cases are depicted in Fig. 8.31: the *cassegranian antenna*, where the feed is located at one focus of a hyperboloid subreflector; and the *gregorian antenna*, where the feed is located at one focus of an elliptic subreflector. In both cases, the other focus of the subreflectors operates as the virtual focus for the main parabolic reflector: ray tracing from the primary source to the subreflector and then to the reflector is used to compute the aperture field over the main reflector mouth.

8.6.1. The Parabolic Dish

Let us consider the parabolic dish depicted in Fig. 8.32 with focal illumination. A proper design of the feed radiation pattern enables a convenient amplitude distribution to be obtained over the dish aperture (plane $z = 0$), the phase being constant. We further assume the aperture field to be azimuthally independent and linearly polarized along the y-axis.

The simplest assumption is to take the aperture field $f(\rho)$ constant:

$$f(\rho) = \text{rect}\left(\frac{\rho}{2a}\right). \tag{8.59a}$$

Alternatively, an amplitude taper, convenient for subsequent analysis, is the following:

$$f(\rho) = \left[1 - \left(\frac{\rho}{a}\right)^2\right]^n \text{rect}\left(\frac{\rho}{2a}\right). \tag{8.60a}$$

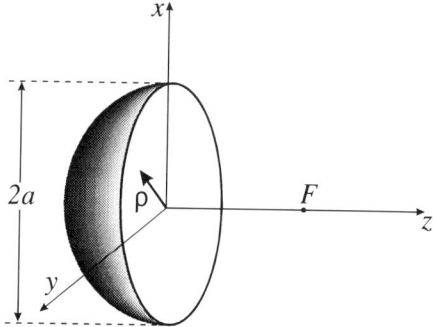

FIGURE 8.32. A parabolic dish.

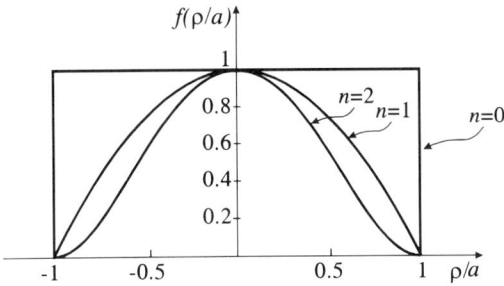

FIGURE 8.33. Several aperture distributions for the illumination of a parabolic dish.

This equation reduces to Eq. (8.59a) for $n = 0$ and is depicted graphically in Fig. 8.33 for $n = 0, 1, 2$.

To compute the transverse spectrum we proceed as in Section 8.5.2:

$$\mathbf{E}_t(k_t) = E_0 \hat{\mathbf{y}} \int_0^{2\pi} d\phi \int_0^a d\rho \exp[-ik_t\rho\cos(\phi' - \phi)]\rho f(\rho)$$

$$= 2\pi E_0 \hat{\mathbf{y}} \int_0^a d\rho J_0(k_t\rho)\rho f(\rho) = E(k_t)\hat{\mathbf{y}},$$

where $k_t = k\sin\theta$. The usual spherical coordinate system centered at the dish center is used and integration with respect to ϕ performed via Eq. (C.18).

For the constant aperture distribution, Eq. (8.59a), we obtain

$$E_t(k_t) = \pi a^2 E_0 \frac{2J_1(k_t a)}{k_t a} \tag{8.59b}$$

[see Eq. (C.21)], and for the tapered one, Eq. (8.60a),

$$E_t(k_t) = \pi a^2 E_0 \frac{2^{n+1} n! J_{n+1}(k_t a)}{(k_t a)^{n+1}} \tag{8.60b}$$

[see Eq. (C.22)], which reduces to (8.59b) for $n = 0$.

The far field is computed by first evaluating all the spectrum,

$$\mathbf{E}(\mathbf{k}_t) = \left(\hat{\mathbf{y}} + \frac{v}{w}\hat{\mathbf{z}}\right) E_t(k_t),$$

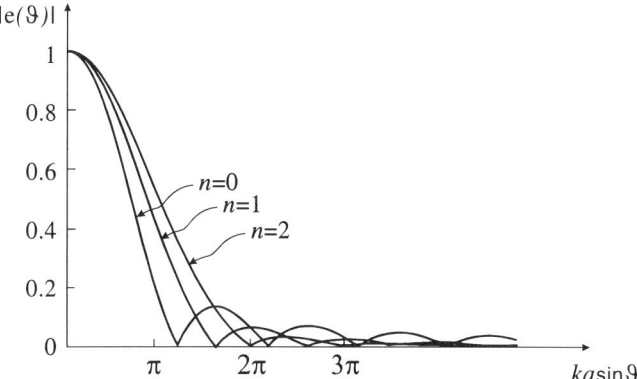

FIGURE 8.34. The normalized radiation diagram of a parabolic dish without ($n = 0$) and with ($n = 1, 2$) aperture amplitude taper.

and then employing Eq. (3.85) to obtain

$$\mathbf{E}(r, \theta, \phi) = ik\, \frac{\pi a^2}{n+1}\, E_0\, \frac{1}{2\pi r}\, \exp(-ikr)\mathbf{e}(\theta, \phi),$$

where

$$\mathbf{e}(\theta, \phi) = \frac{2^{n+1}(n+1)!\, J_{n+1}(ka\sin\theta)}{(ka\sin\theta)^{n+1}}\, (\sin\phi\hat{\boldsymbol{\theta}} + \cos\theta\cos\phi\hat{\boldsymbol{\phi}}),$$

which includes the case of a constant aperture field ($n = 0$).

Normalized radiation diagrams in the E-plane $\phi = \pm 90°$ are given in Fig. 8.34.

The radiated power is given by

$$P_r = \frac{|E_0|^2}{2\zeta}\, 2\pi \int_0^a d\rho [1 - (\rho/a)^2]^n \rho = \frac{\pi a^2}{n+1}\, |E_0|^2$$

and the directivity by

$$D = \frac{4\pi}{\lambda^2}\, A\, \frac{1}{n+1} = \frac{4\pi}{\lambda^2}\, A_e,$$

where $A = \pi a^2$ is the aperture area of the dish and $A_e = A/(n+1)$ its effective area.

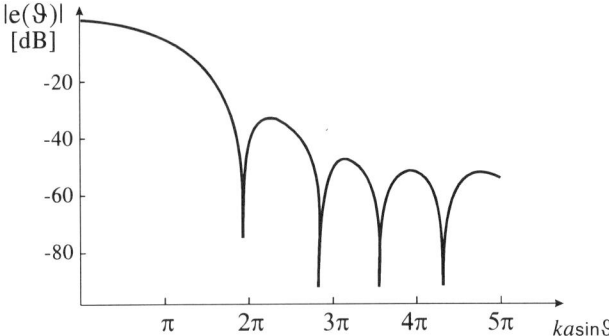

FIGURE 8.35. The normalized radiation diagram of a parabolic dish with a pedestal illumination $\alpha = 0.18$, $\beta = 0.82$.

The uniform and tapered amplitude distributions can be combined to generate a *pedestal*-type aperture illumination:

$$f(\rho) = \left\{\alpha + \beta\left[1 - \left(\frac{\rho}{a}\right)^2\right]^n\right\}\mathrm{rect}\left(\frac{\rho}{2a}\right), \quad \text{with } \alpha + \beta = 1.$$

Examination of Fig. 8.34 shows that in the higher lobe region the radiation diagrams for $n = 0$ and $n = 2$ tend to be out of phase with each other; a proper combination of α and β would reduce them. An example is given in Fig. 8.35. Frequent use of a pedestal is known practice in circuit theory to design filters with a smooth response.

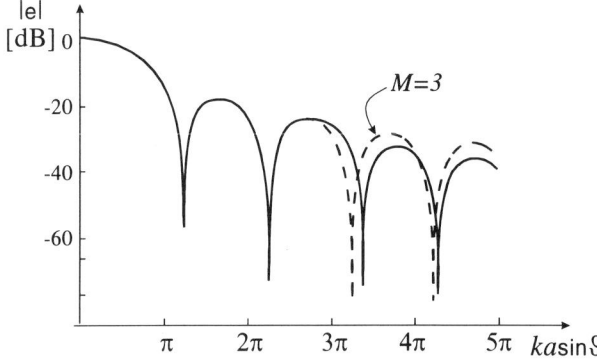

FIGURE 8.36. Sampling reconstruction (dashed line) of the radiation diagram (solid line) of a parabolic dish with uniform illumination. The sampling series is truncated to $M = 3$, with oversampling factor $\chi = 1.2$ and E-plane pattern.

ENGINEERING TOPICS: RADIATION

Finally, we recall that the radiation diagram can be calculated by applying the sampling expansion; see Section 4.6.4, Eq. (4.98a). For instance, in the E-plane

$$e(\theta) = \sum_{m} e(v_m) \operatorname{sinc}(\chi v a - m\pi), \tag{8.61a}$$

$$va = ka \sin \theta, \qquad v_m a = ka \sin \theta_m = m\pi, \tag{8.61b}$$

and χ is the oversampling factor [Eq. (4.18)]

An example of reconstruction of the radiation diagram starting from its *samples* $e(v_m)$ is given in Fig. 8.36. Note the excellent performance of the sampling expansion up to $v \approx v_M$, where M is the truncated index of the series. The reconstruction is highly improved by using values of χ even slightly larger than unity.

8.6.2. Computation of the Reflector Radiation Diagram via the Current Integration Method

In this section, as well as in the next two, we consider cylindrical reflectors instead of real dishes, to simplify the pertinent discussion. Extension to the three-dimensional case only complicates the mathematics.

Consider the parabolic cyclindrical reflector depicted in Fig. 8.37, whose plane cut by $y = 0$ is described by the equation

$$x^2 = 4f(f + z), \qquad |x| \leqslant a. \tag{8.62}$$

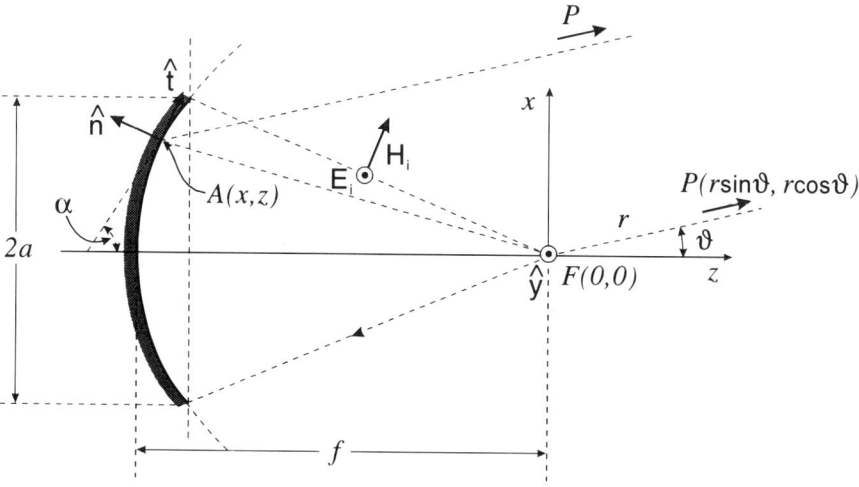

FIGURE 8.37. Computation of the field radiated by a parabolic cylinder via the current integration method.

The primary source, located at the focal point F, is linearly polarized with the electric field along the y-axis.

In order to compute the field scattered by the reflector under the PO approximation of Eqs. (8.58), we must compute the surface current density over the illuminated side of the cylinder. We have, from Eq. (8.62) and for $x \geqslant 0$,

$$\frac{dx}{dz} = \frac{\sqrt{f}}{\sqrt{f+z}} = \frac{1}{x/2f} = \tan \alpha,$$

and

$$\hat{\mathbf{t}} = \sin \alpha \hat{\mathbf{x}} + \cos \alpha \hat{\mathbf{z}} = \frac{1}{\sqrt{1+(x/2f)^2}} \hat{\mathbf{x}} + \frac{x/2f}{\sqrt{1+(x/2f)^2}} \hat{\mathbf{z}}, \quad (8.63a)$$

$$\hat{\mathbf{n}} = \hat{\mathbf{y}} \times \hat{\mathbf{t}} = \frac{x/2f}{\sqrt{1+(x/2f)^2}} \hat{\mathbf{x}} - \frac{1}{\sqrt{1+(x/2f)^2}} \hat{\mathbf{z}}, \quad (8.63b)$$

so that

$$\mathbf{H}_i = \frac{E_i}{\zeta} \left[-\frac{z}{\sqrt{x^2+z^2}} \hat{\mathbf{x}} + \frac{x}{\sqrt{x^2+z^2}} \hat{\mathbf{z}} \right] = \frac{E_i}{\zeta} \left[\frac{1-(x/2f)^2}{1+(x/2f)^2} \hat{\mathbf{x}} + \frac{x/f}{1+(x/2f)^2} \hat{\mathbf{z}} \right],$$

where E_i is the electric field incident over the point $P(x, z)$ of the reflector surface. For the induced currents

$$\mathbf{J}_S = -\hat{\mathbf{n}} \times \mathbf{H}_i = \frac{E_i}{\zeta} \frac{1}{\sqrt{1+(x/2f)^2}} \hat{\mathbf{y}}.$$

In order to compute the scattered field, we need a formal expression for the distance from the focal point to the reflector, $|FA|$, and from the reflector to the field point, $|AP|$ (see Fig. 8.37):

$$|FA| = \sqrt{x^2 + z^2} = 2f + z, \quad (8.64a)$$

$$|FA| + |AP| = \sqrt{x^2 + z^2} + \sqrt{(r \sin \theta - x)^2 + (r \cos \theta - z)^2}$$
$$\approx 2f + z + r - x \sin \theta - z \cos \theta$$
$$= r + f(1 + \cos \theta) - x \sin \theta + \frac{x^2}{4f}(1 - \cos \theta). \quad (8.64b)$$

ENGINEERING TOPICS: RADIATION

The field $\mathbf{E}(\theta) = E(\theta)\hat{\mathbf{y}}$ scattered by the parabolic cylinder is obtained by superposition:

$$E(\theta) = \frac{E_0}{\sqrt{2kr}} \exp(-ikr) \exp[-ikf(1 + \cos\theta)] F(ka \sin\theta), \qquad (8.65a)$$

$$F(v) = \int_{-a}^{a} dx \, \frac{ds}{dx} e(x) \exp\left[ikx \sin\theta - \frac{ikx^2}{4f}(1 - \cos\theta)\right]$$

$$= \int_{-1}^{1} dt \, e(t) \exp\left[ivt - i\frac{ka^2}{4f}(1 - \cos\theta)t^2\right]. \qquad (8.65b)$$

In Eq. (8.65b), v and t are the normalized variables given by

$$v = ka \sin\theta \quad \text{and} \quad t = \frac{x}{a},$$

s is the curvilinear coordinate along the parabola, and $E_i = E_0 e(x)$ where E_0 is the value of the incident electric field at $x = 0$ and $e(x) \to e(t)$ describes the illumination of the reflector surface by the primary source. Should the latter be isotropic, then

$$e(x) = \frac{1}{\sqrt{1 + (x/2f)^2}}.$$

It is convenient to expand $e(t)$ as a Fourier series between $(-1, 1)$:

$$e(t) = \tfrac{1}{2} \sum_n f_n \exp(-in\pi t), \qquad (8.66a)$$

$$f_n = \int_{-1}^{1} dt \exp(in\pi t) e(t). \qquad (8.66b)$$

The expansion coefficients f_n in Eq. (8.66a) can be very conveniently computed, because they coincide with 2π times the values of the FT of $e(t)$,

$$f(v) = \frac{1}{2\pi} \int_{-1}^{1} dt \exp(ivt) e(t),$$

evaluated at $v_n = n\pi$:

$$f_n = f(v_n) = f(n\pi).$$

Accordingly, efficient FFT codes can be used.

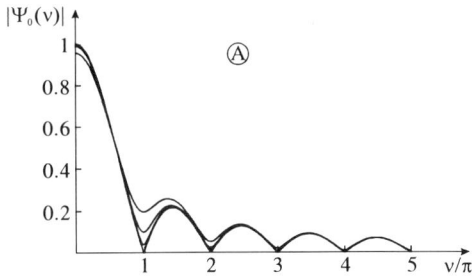

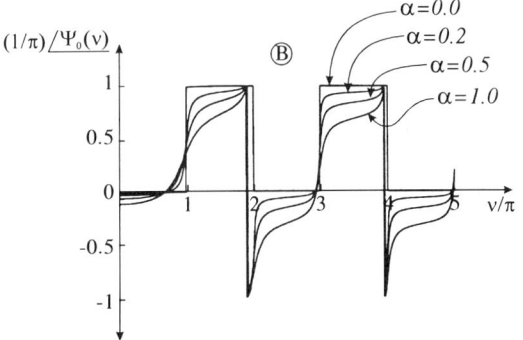

FIGURE 8.38. Modulus (A) and phase (B) of the sampling-like function $\psi_0(v, \alpha)$ for $\alpha = (ka^2/4f)(1 - \cos\theta)$ equal to 0, 0.2, 0.5, and 1.

Substitution of Eq. (8.66a) into Eq. (8.65b) yields

$$F(v) = \sum_n f_n \psi_n(v), \tag{8.67a}$$

$$\psi_n(v) = \frac{1}{2} \int_{-1}^{1} dt \exp[i(v - n\pi)t] \exp\left[-i\frac{ka^2}{4f}(1 - \cos\theta)t^2\right] \tag{8.67b}$$

which is a *sampling-like* representation for the scattered field. As a matter of fact, expansion (8.67) resembles Eq. (8.61) and reduces to it for $f \to \infty$, i.e., for a flat aperture. The sampling-like base functions (8.67b) can be computed in terms of Fresnel integrals. The function $\psi_0(v)$ is presented in Fig. 8.38 for a few values of the parameter

$$\alpha = \frac{ka^2}{4f}(1 - \cos\theta).$$

The other functions $\psi_n(v)$, $n > 0$, are obtained simply by translation of $n\pi$ over the v-axis. For $\alpha = 0$, $\psi_0(v)$ coincides with sinc(v), the parabolic cylinder degenerates into a flat strip, and sampling reconstruction of the radiation diagram is recovered (see Section 4.6.4). As α is increased, the sampling-like function $\psi_0(v)$ departs from the sinc(v) function and the alternative representation (8.67a) corresponding to the parabolic cylinder is needed.

8.6.3. Computation of the Reflector Radiation Diagram via Optical Techniques

Consider the same parabolic cylinder reflector excited by a focal illumination, as depicted in Fig. 8.37. From the optical viewpoint, the scattered field is represented in terms of the rays reflected by the cylinder surface and of the rays diffracted by the reflector edges.

The reflected rays are all parallel to the z-axis in the case of focal illumination (see Section 5.2.1). However, they cannot be used directly to predict the field intensity, because this direction is a caustic. Alternative techniques must be used (see Section 8.6.2).

The diffracted rays depart from the cylinder edges. We locally approximate the cylinder with its tangent planes at the edges, in order to use the half-plane diffraction coefficients, Eq. (5.51a) of Section 5.4.2. For the primary electric field polarized along the y-axis (see Fig. 8.39),

$$D(\phi^\pm, \phi_0) = \exp\left(-i\frac{\pi}{4}\right) \frac{\sin(\phi^\pm/2)\sin(\phi_0/2)}{\cos\phi^\pm + \cos\phi_0}. \tag{8.68}$$

For evaluating the radiation diagram it is necessary to relate the angles $\phi^\pm$ and ϕ_0 to the observation angle θ (see Fig. 8.39). Hence

$$\phi_0 = \pi - (\alpha_0 + \beta_0), \qquad \phi^\pm = \pi - \alpha_0 \pm \theta, \tag{8.69a}$$
$$r^\pm = r + |FA|\cos(\beta_0 \pm \theta). \tag{8.69b}$$

The parameters $|FA|$, α_0, and β_0 are now computed. We have

$$|FA| = f[1 + (a/2f)^2] \tag{8.70}$$

by using Eq. (8.62) specified at $x = a$. On the other hand,

$$\sin\alpha_0 = \frac{1}{\sqrt{1 + (a/2f)^2}} \quad \text{and} \quad \cos\alpha_0 = \frac{a/2f}{\sqrt{1 + (a/2f)^2}}, \tag{8.71a}$$

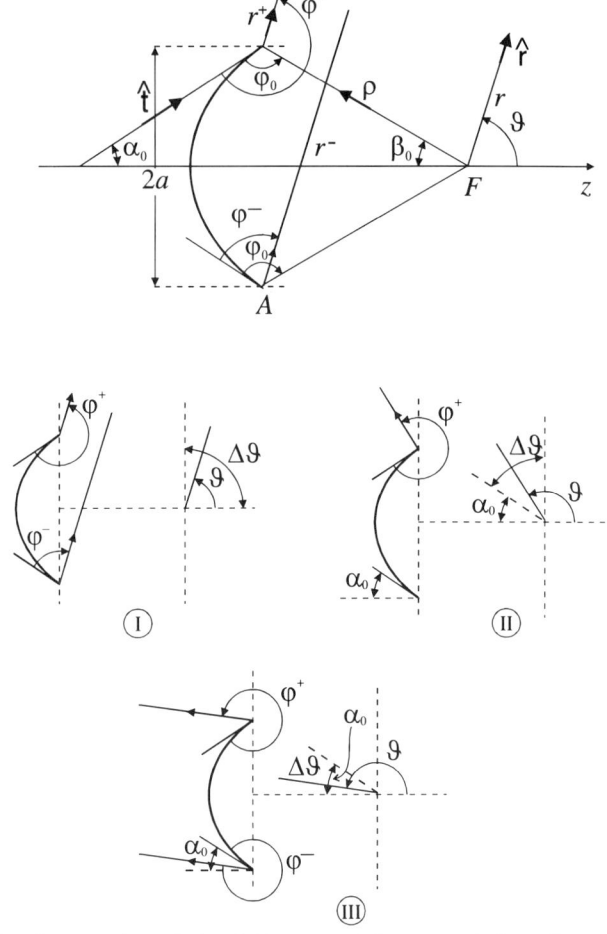

FIGURE 8.39. Computation of the field scattered by a parabolic cylinder via optical techniques.

from Eq. (8.63a) specified at $x = a$, and

$$\sin \beta_0 = \frac{a}{|FA|} = \frac{a/f}{1 + (a/2f)^2} \quad \text{and} \quad \cos \beta_0 = \frac{1 - (a/2f)^2}{1 + (a/2f)^2}. \quad (8.71b)$$

Furthermore

$$\sin \frac{\beta_0}{2} = \sqrt{\frac{1 - \cos \beta_0}{2}} = \frac{a/2f}{\sqrt{1 + (a/2f)^2}} = \cos \alpha_0,$$

as follows from Eq. (8.71). Accordingly

$$\alpha_0 = \frac{\pi}{2} - \frac{\beta_0}{2}, \quad \text{or } 2\alpha_0 + \beta_0 = \pi. \tag{8.72}$$

It is convenient to divide the angular region $0 \leqslant \theta \leqslant 180°$ into three subspaces, as shown in Fig. 8.39. In the first region (I), $0 \leqslant \theta \leqslant \pi/2$, both edges contribute to the scattered field. In the second region (II), $\pi/2 = \theta < \pi - \alpha_0$, only one edge contributes to the scattered field, the other being shielded. In the third region (III), $\pi - \alpha_0 \leqslant \theta \leqslant \pi$, again both edges contribute to the scattered field.

If E_0 is the amplitude of the incident field on the edge, then the scattered field $E_d(\theta)$ [see Eq. (5.50a)] is given by

$$E_d(\theta) = \sqrt{\frac{2}{\pi k r}} \exp(-ikr) E_0 \{ D^+(\theta) \cdot \exp[ikaq^+(\theta)] + D^-(\theta) \exp[ikaq^-(\theta)] \} \tag{8.73a}$$

in regions I and III, and by

$$E_d(\theta) = \sqrt{\frac{1}{\pi k r}} \exp(-ikr) E_0 D^+(\theta) \exp[ikaq^+(\theta)] \tag{8.73b}$$

in region II, for $0 \leqslant \theta \leqslant 180°$, and

$$D^\pm(\theta) = \exp\left(-i\frac{\pi}{4}\right) \frac{\sin[\alpha_0/2] \cos[(\alpha_0 \mp \theta)/2]}{\cos \alpha_0 - \cos(\alpha_0 \mp \theta)}, \tag{8.74a}$$

$$q^\pm(\theta) = \frac{f}{a}\left[1 + \left(\frac{a}{2f}\right)^2\right] \cos(2\alpha_0 \mp \theta) = -\frac{\cos(2\alpha_0 \mp \theta)}{\sin 2\alpha_0}, \tag{8.74b}$$

with

$$\tan \alpha_0 = \frac{2f}{a}. \tag{8.75}$$

Equations (8.74) are obtained substituting Eqs. (8.69a), (8.70), and (8.72) into Eqs. (8.68); Eq. (8.75) is obtained from Eq. (8.63a) specified at $x = a$.

We note that the diffraction coefficient, Eq. (8.74a), blows up at $\theta = 0$ (reflection boundary) and at $\theta = 2\alpha_0$ (shadow boundary); see also Section 5.4.1. The asymptotic evaluation of the diffracted field integral is invalid near these directions, and transition functions are needed for the correct evaluation of the field (see Section 5.4.3). Additional comments are in order.

The use of the transition function in Eq. (5.54) requires a choice for the distance from the field point to the antenna. This is inconvenient, because we desire distance-independent (far-field) antenna parameters. Alternative regularization procedures can be explored.

A first possibility is based on the observation that the primary field is incident from the parabola focus, i.e., a finite distance $R = a/\sin \alpha_0$ from the cylinder edge. This implies that the additional phase term $-(k \sin^2 \phi_0 / 2R) x^2$ appears in the integral of Eq. (5.46a). If we set $r \to \infty$, we obtain the same transition function, Eq. (5.54), with

$$\eta \to \sqrt{kR} \, \frac{\cos \phi + \cos \phi_0}{\sin \phi_0}.$$

The result is a cylindrical wave also in the transition region.

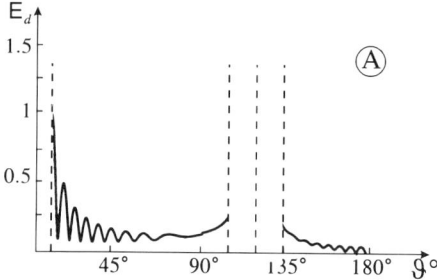

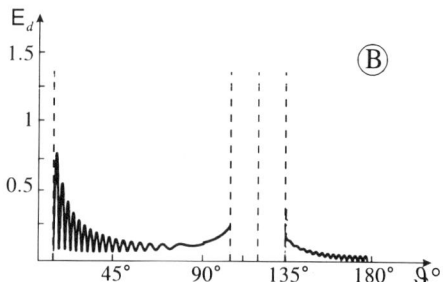

FIGURE 8.40. A field (in arbitrary units) scattered by the edges of a parabolic cylinder computed using optical techniques: (A) $a/\lambda = 5$ and (B) $a/\lambda = 10$. In both cases $a/2f = \sqrt{3}$. Blank regions denote angular ranges where the simple geometrical model is no longer appropriate.

ENGINEERING TOPICS: RADIATION

As an alternative, we see that the illumination is not really localized in the parabola focus, but rather it extends over a segment of length $2b$ along the x-axis, thus generating a space-varying illumination over the edge. From Section 5.4.4 we have $\delta e_i/\delta x \sim b/R$ and Eq. (5.58a) becomes

$$D(\phi, \phi_0)E_0 + \frac{\partial D(\phi, \phi_0)}{\partial \cos\phi_0} \frac{1}{ik} \frac{\partial e_i}{\partial x} \approx D(\phi, \phi'_0)E_0,$$

$$\phi'_0 = \phi_0 + i\gamma \frac{b}{R},$$

with γ a constant. The angle ϕ_0 becomes complex and the difffraction coefficient remains finite in the transition regions.

It must be recognized that all the *regularization* procedures are not only probably questionable, but also not in favor of the transparent physical picture which is the strength of the optical model. It may be more convenient to make use of a simple model where it is valid, and resort to other techniques in the *forbidden* regions. With this in mind, transition functions may be used for a quantitative assessment of size in such regions, where the pure optical model cannot be used.

An example of a field scattered by the edges of a parabolic cylinder is given in Fig. 8.40.

8.7. ARRAYS

A cluster of antennas can be used to synthesize a desired radiation diagram with appropriate choice of antenna excitation and spacing (see Fig. 8.41). The far field of the array is given by

$$\mathbf{E}(P) = K \sum_n I_n \mathbf{l}_n \exp(ik|\mathbf{r} - \mathbf{r}_n|), \qquad (8.76a)$$

where I_n and $\mathbf{l}_n$ are the input current and effective length of the generic antenna, respectively, and $K\ [\Omega/m^2]$ is a constant (see Section 8.2).[18]

If the antennas are all identical and equally oriented, then $\mathbf{l}_n = \mathbf{l}$ and Eq. (8.76a) is factorized as follows:

$$\mathbf{E}(p) = K\mathbf{l} \sum_n a_n \exp(ik|\mathbf{r} - \mathbf{r}_n|)$$
$$= K\mathbf{l}(\theta, \phi) F(\theta, \phi), \qquad (8.76b)$$

18. It may be convenient to use the radiation vector instead of the effective length. In Eq. (8.76a) the factor $I_n \mathbf{l}_n$ is replaced by $A_n \mathbf{e}_n$, A_n being an appropriate excitation coefficient.

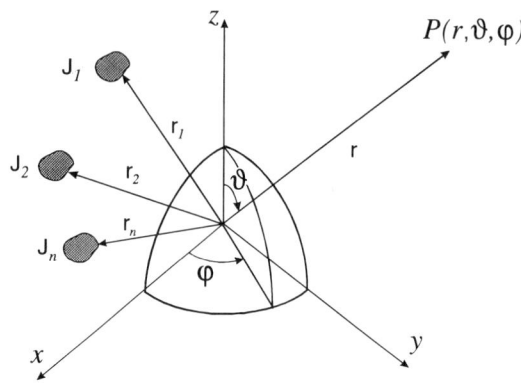

FIGURE 8.41. The array antenna.

where I is a reference current, a_n the (normalized) excitation coefficients of the generic antenna, and $F(\theta, \phi)$ the *array factor*. This type of array is referred to as *homogeneous* and its radiating properties are described by the product of the *element factor* **l** and the array factor F.

A *planar array* has all its elements lying on a plane, and a *linear array* along a line (see Fig. 8.42). If the element spacings are all equal, the array is

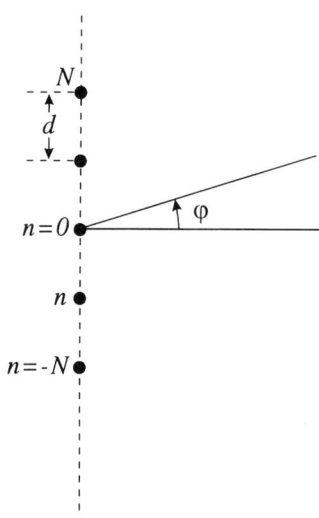

FIGURE 8.42. A linear array.

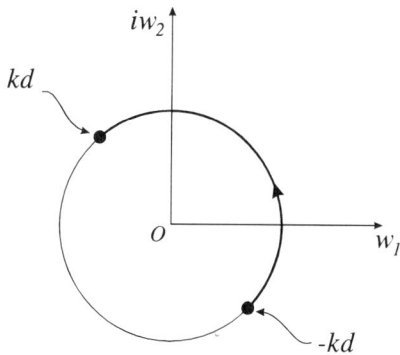

FIGURE 8.43. The Schelkunoff unit circle.

referred to as *uniform* and the array factor in the plane $\theta = 90°$ is given by

$$F(\phi) = \sum_{-N}^{N} {}_a a_n \exp(in\gamma), \quad \text{where } \gamma = kd \sin \phi, \quad (8.77a)$$

for an odd number $2N + 1$ of elements, and

$$F(\phi) = \sum_{1}^{N} {}_n a_n \exp[i(n + 1/2)\gamma] + \sum_{1}^{N} {}_n a_n \exp[-i(n + 1/2)\gamma] \quad (8.77b)$$

for an even number $2N$.

In the case of a uniform linear array, we set $w = \exp(in\gamma)$ and the array factor becomes equal to

$$F(\gamma) = \sum a_n w^n, \quad (8.78)$$

i.e., a polynomial of degree equal to the number of array elements minus one. As $\sin \phi$ changes from -1 to $+1$, and γ from $-kd$ to $+kd$, w moves along a circle (see Fig. 8.43), the *Schelkunoff*[19] *circle*, with a total excursion of $2kd$ radians. In addition

$$F(\gamma) = \sum_a a_n w^n = \prod_n (w - w_n), \quad (8.79)$$

by the fundamental theorem of algebra, which shows that the array factor has a number of zeros w_n equal to the number of array elements minus one. Should these zeros lie on the Schelkunoff circle (see Fig. 8.44), they appear as points

---

19. Sergei A. Schelkunoff: Samara (Russia), 1897–Highstown (New Jersey), 1992.

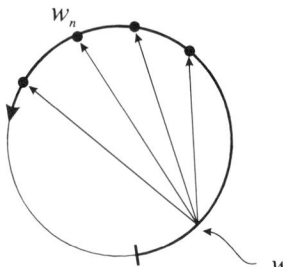

FIGURE 8.44. Graphical computation of the array factor, when the array nulls lie on the Schelkunoff circle.

of null radiation on the array factor and interlace with its (relative) maxima. At any angle ϕ the array factor is represented by the product of all the phasors starting from $w = kd \sin \phi$ and ending on w_n. It is evident that reduction of the amplitude of the radiation lobes in a prescribed angular region can be accomplished by increasing the density of null points therein. For a fixed number of array elements this is possible at the expense of wider null spacing in the remaining (visible) part of the Schelkunoff circle, while the main radiation lobe is enlarged. This is consistent with the general result of Section 3.5.3.

8.7.1. Array with Uniform Excitation

Consider an array with uniform excitation $a_n = 1$ and $2N + 1$ antennas spaced d apart. The coordinate center is located at the site of the element $n = 0$ and Eq. (8.77a) is valid for computing the array factor:

$$F(\phi) = \sum_0^N \exp(in\gamma) + \sum_0^N \exp(-in\gamma) - 1$$

$$= \frac{1 - \exp[i(N+1)\gamma]}{1 - \exp[i\gamma]} + \frac{1 + \exp[-i(N+1)\gamma]}{1 - \exp[i\gamma]} - 1$$

$$= \frac{2\cos(N\gamma/2)\sin[(N+1)\gamma/2]}{\sin(\gamma/2)} - 1 = \frac{\sin[(2N+1)\gamma/2]}{\sin(\gamma/2)},$$

which is the same result as in Section 4.6.5 for the broadside configuration. The array exhibits nulls when

$$\frac{2N+1}{2} kd \sin \phi = n\pi, \qquad n = 1, 2, \ldots,$$

i.e., at angles

$$\sin \phi_n = \frac{n\lambda}{(2N+1)d} = n\frac{\lambda}{L},$$

where L is the array dimension. These nulls are equispaced over the Schelkunoff circle. The 3-dB beamwidth in the $\theta = 90°$ plane is approximately coincident with that of the E-plane of a rectangular aperture [see Eq. (8.51a)] and we have $\Delta\phi \approx 0.88\lambda/L$; similarly, the first sidelobe is 13.46 dB down the principal one.

When computing the directivity of the array it is necessary to evaluate the field radiated throughout the space, not only on the $\theta = 90°$ plane. If the array is aligned along the y-axis, then the distance from each array element to the far-field point $P(r, \theta, \phi)$ is given by $r \sim nd \sin\theta \sin\phi$, in which case

$$\mathbf{E}(r, \theta, \phi) = i\zeta \frac{Il}{2\lambda r} \exp(-ikr) \frac{\sin[(kL\sin\theta\sin\phi)/2]}{\sin[(kd\sin\theta\sin\phi)/2]}. \tag{8.80}$$

This equation exhibits two symmetric radiation maxima for $\phi = 0, \pi$ (broadside configuration). The radiation diagram can be made asymmetric by placing the array at a distance $\lambda/4$ from a conducting plane (see Fig. 8.45), as follows readily from image theory. In a practical configuration the array is backed by a conducting screen of finite extent, and we have

$$\mathbf{E} \approx -2\zeta \frac{Il}{2\lambda r} \exp(-ikr) \sin\left(\frac{\pi}{2}\sin\theta\cos\phi\right) \frac{\sin[(kL\sin\theta\sin\phi)/2]}{\sin[(kd\sin\theta\sin\phi)/2]}.$$

The total radiated power is obtained by computing the flux of the Poynting vector over the hemisphere $|\phi| \leq \pi/2$ at infinity. When $kL \gg 1$, we can take $\sin\phi \approx \phi$ and $\cos\phi \approx 1$ and extend the ϕ integration limits to infinity. For

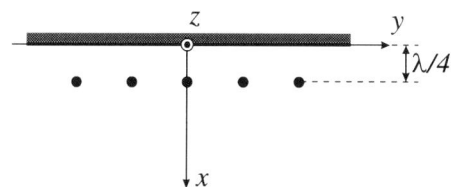

FIGURE 8.45. Broadside array in front of a reflecting mirror.

azimuthally independent effective lengths $\mathbf{l} = \mathbf{l}(\theta)$,

$$P \approx \frac{1}{2} \zeta \frac{|I|^2}{\lambda^2} \int_0^\pi d\theta \sin\theta \sin^2[(\pi/2)\sin\theta]|\mathbf{l}(\theta)|^2 \int_{-\infty}^{+\infty} d\phi \frac{\sin^2[(kL\phi\sin\theta)/2]}{[(kd\phi\sin\theta)/2]^2}$$

$$= \frac{1}{2} \zeta \frac{|I|^2}{\lambda^2} \pi \frac{2L}{kd^2} \int_0^\pi d\theta \sin^2[(\pi/2)\sin\theta]\,|\mathbf{l}(\theta)|^2, \qquad (8.81a)$$

where use has been made of Eq. (C.6).

A popular element in a linear array is the half-wave dipole antenna (see Section 8.4.2). We have already approximated its effective length by means of the first Fourier harmonic,

$$\mathbf{l}(\theta) \approx \frac{\lambda}{\pi} 2J_0\left(\frac{\pi}{2}\right)\sin\theta\hat{\boldsymbol{\theta}} = 0.945\,\frac{\lambda}{\pi}\sin\theta\hat{\boldsymbol{\theta}}. \qquad (8.82a)$$

A similar procedure and approximation for the factor $\sin[(\pi/2)\sin\theta]$ is employed, leading to

$$\sin[(\pi/2)\sin\theta] \approx 2J_1(\pi/2)\sin\theta = 1.13\sin\theta, \qquad (8.82b)$$

where use has been made of Eq. (C.25).

A final substitution into Eq. (8.81a) yields

$$P = \frac{1}{2}\zeta|I|^2\,\frac{3L}{4kd^2}\,1.16, \qquad (8.81b)$$

where use has been made of Eq. (C.26).

The directivity can be immediately computed[20] via Eq. (8.7):

$$D = \frac{16kL}{3\pi\cdot 1.16} = 9.2\,\frac{L}{\lambda}.$$

If the element spacing d equals $\lambda/2$, as is often the case, then the directivity is approximately equal to the number of elements times 4.

8.7.2. Array with Tapered Excitation

Let us consider an array of $2N+1$ elements and total length $L = (2N+1)d$, with excitation coefficients

$$a_n = \cos\left(\frac{\pi}{L}nd\right),$$

---

20. Although radiation is limited to only one half-space, the general definition of directivity is used.

so that the input current on each element is (cosinusoidally) tapered from the center to the two edges of the array. Equation (8.77a) then gives

$$F(\phi) = \frac{1}{2}\sum_{-N}^{N}\exp\left[in\left(\gamma + \frac{\pi d}{L}\right)\right] + \frac{1}{2}\sum_{-N}^{N}\exp\left[in\left(\gamma - \frac{\pi d}{L}\right)\right]$$
$$= \frac{1}{2}\frac{\sin[(2N+1)(\gamma + \pi d/L)/2]}{\sin[(\gamma + \pi d/L)/2]} + \frac{1}{2}\frac{\sin[(2N+1)(\gamma - \pi d/L)/2]}{\sin[(\gamma - \pi d/L)/2]}$$
$$= \frac{1}{2}\cos\left[\frac{2N+1}{2}\gamma\right]\frac{\sin[(\gamma - \pi d/L)/2] - \sin[(\gamma + \pi d/L)/2]}{\sin[(\gamma + \pi d/L)/2]\sin[(\gamma - \pi d/L)/2]},$$

because $(2N+1)\pi d/L = \pi$. When $\pi d/L \ll 1$, a series expansion is allowed and we have

$$F(\phi) \approx -\cos\left[\frac{2N+1}{2}\gamma\right]\frac{(\pi d/2L)\cos(\gamma/2)}{\sin^2(\gamma/2) - (\pi d/2L)^2\cos^2(\gamma/2)}$$
$$\approx \frac{2}{\pi}(2N+1)\frac{\cos[(2N+1)\gamma/2]\cos(\gamma/2)}{1 - (2L/\pi d)^2\sin^2(\gamma/2)}, \qquad (8.82)$$

which is equivalent to the array of the radiation diagram of the aperture with the same tapering, Eq. (8.51b).

The array exhibits nulls[21] when

$$\frac{2N+1}{2}\gamma_m = (2m+1)\frac{\pi}{2}, \qquad m = 1, 2, \ldots,$$

which are again equispaced over the Schelkunoff circle. The 3-dB beamwidth in the $\theta = 90°$ plane is given approximately by $\Delta\phi \approx 1.2\lambda/L$, and the first sidelobe is -23.52 dB down the principal one [see Eq. (8.51b)]. Note again that a tapered excitation reduces the amplitude of the sidelobes at the expense of beamwidth enlargement.

8.7.3. The Binomial Array

Consider the array factor

$$F(\phi) = (1+w)^N, \qquad w = kd\sin\phi. \qquad (8.83)$$

---

21. The zero in the numerator of Eq. (8.82) when $m = 0$ is compensated by the zero in the denominator when $\sin(\gamma/2) \approx \gamma/2$.

By means of a binomial expansion we have

$$F(\phi) = \binom{N}{0} + \binom{N}{1} w + \binom{N}{2} w^2 + \cdots + \binom{N}{N} w^N,$$

which shows that the array factor (8.83) can be realized by means of $(N + 1)$ equispaced antennas with excitation coefficients

$$a_n = \binom{N}{n}.$$

For instance, when $N = 6$

$$a_0 = a_6 = 1, \qquad a_1 = a_5 = 6, \qquad a_2 = a_4 = 15, \qquad a_3 = 20.$$

Suppose

$$F(\phi) = \exp\left(i \frac{Nkd}{2} \sin \phi\right) 2^N \cos^N\left(\frac{kd}{2} \sin \phi\right).$$

If $kd \leq \pi$, i.e., $d \leq \lambda/2$, the array factor is a monotonically decreasing function from $\phi = 0$ to $\phi = \pm\pi/2$ with no zeros in the considered interval.

Since

$$\cos^{2N}\left(\frac{kd}{2} \sin \phi\right) \approx \left[1 - \frac{1}{2}\left(\frac{kd}{2}\right)^2 \sin^2 \phi\right]^{2N} \approx 1 - N\left(\frac{kd}{2}\right)^2 \sin^2 \phi$$

around $\phi = 0$, the 3-dB beamwidth is approximately given by $\Delta\phi \approx \lambda/(\sqrt{2N}\pi d)$. The price paid for the absence of sidelobes is a beamwidth which is inversely proportional to the square root (instead of the first power) of the number of array elements, and the large dynamics of the excitation coefficients.

8.7.4. Sum and Difference Patterns

Consider an array of $2N$ elements with uniform excitation. The two signals received by the two halves of the array can be either summed or subtracted from each other, generating the two array factors

$$F^{\pm}(\phi) = \frac{\sin(N\gamma/2)}{\sin(\gamma/2)} \left[\exp\left(i\frac{N}{2}\gamma\right) \pm \exp\left(-i\frac{N}{2}\gamma\right)\right],$$

$$F^+(\phi) = 2 \frac{\sin(N\gamma)}{\sin \gamma} \cos(\gamma/2), \qquad F^-(\phi) = 2i \frac{\sin^2(N\gamma/2)}{\sin(\gamma/2)};$$

ENGINEERING TOPICS: RADIATION

see Eq. (8.77b). These two patterns exhibits a maximum and a null at $\phi = 0$, respectively, and are important in radar applications: the sum pattern is used to detect the target, with an accuracy limited by its beamwidth $\Delta\phi \sim \lambda/2Nd$; the difference pattern is aimed at improved tracking, whose accuracy is related to the slope of the pattern at the origin,

$$dF^-/d\phi = 2\pi N^2 d/\lambda = \pi N/\Delta\phi.$$

8.8. SUMMARY AND SELECTED REFERENCES

In this chapter, parameters that characterize the radiation properties of antennas are presented and discussed. They are categorized as those relative to the transmitting mode: effective length, radiation vector, directivity, and gain (Section 8.2); and those relative to the receiving mode: input impedance, effective length, beamwidth, and effective area (Section 8.3). The two categories are related by the classical application of the reciprocity theorem (Section 8.1.1). All this material is available in many textbooks [8.1–8.8]. Section 8.3.4 is devoted to the noise and equivalent temperature of the antenna, and Section 8.3.1 to matching and polarization issues. All the concepts introduced are illustrated with reference to elementary antennas, the electric and magnetic dipole, and the Huygens source.

Wire antennas are examined in Section 8.4; Hallen's integral equation [8.9] is derived, providing its second-order solutions and the equivalent input circuit for the half-wave dipole (Section 8.4.2). The traveling-wave antenna [8.10] is discussed in Section 8.4.3, citing its application to rhombic antenna design [8.11]. A full discussion about the mutual impedance [8.12] between wire antennas is given in Section 8.4.4.

Section 8.5 is devoted to aperture antennas. Illustrative examples include rectangular and circular apertures (Sections 8.5.1 and 8.5.2), as well as the patch antenna [8.13] (Section 8.5.3).

Reflectors are treated in Section 8.6. The parabolic dish, and its simpler two-dimensional version, the parabolic cylinder, are used to illustrate available analysis techniques: the equivalent aperture (Section 8.6.1), current integration (Section 8.6.2), and quasi-optical methods (Section 8.6.3). Sampling and sampling-like techniques [8.14 and 8.15] for efficient computation of the far field are also provided.

Arrays [8.16–8.17] are studied in Section 8.7: uniform [8.18] (Section 8.7.1) as well as tapered excitation (Section 8.7.2) are considered. As other examples, the binomial array (Section 8.7.3) and sum and difference patterns (Section 8.7.4) are illustrated. More material with extension to the synthesis problem may be found in [8.18].

References

[8.1] C. A. Balanis, *Antenna Theory: Analysis and Design*, Harper and Row, New York (1982).
[8.2] R. E. Collin and F. J. Zucker, *Antenna Theory*, McGraw-Hill, New York (1969).
[8.3] R. S. Elliott, *Antenna Theory and Design*, Prentice-Hall, New York (1981).
[8.4] S. K. Shelkunoff and H. T. Friis, *Antenna Theory and Practice*, Wiley, New York (1952).
[8.5] S. Silver, *Microwave Antenna Theory and Design*, Dover Publications, New York (1965).
[8.6] F. K. Lee, *Principles of Antenna Theory*, McGraw-Hill, New York (1984).
[8.7] E. A. Wolf, *Antenna Analysis*, Wiley, New York (1966).
[8.8] W. L. Stutzman and G. A. Thiele, *Antenna Theory and Design*, Wiley, New York (1981).
[8.9] E. Hallen, "Theoretical investigation into the transmitting and receiving qualities of antennae," *Nova Acta Reg. Soc. Scient. Upsalensis* **11**, 1–44 (1938).
[8.10] C. H. Walter, *Travelling Wave Antennae*, McGraw-Hill, New York (1965).
[8.11] A. E. Harper, *Rhombic Antenna Design*, Van Nostrand, New York (1941).
[8.12] R. W. P. King, *Theory of Linear Antennas*, Harvard University Press, Cambridge, Mass (1956).
[8.13] Y. T. Lo, D. Solomon, and W. F. Richards, "Theory and experiments on microstrip antennas," *IEEE Trans. Antennas Propagat.* **AP-27**, 137–145 (1979).
[8.14] G. Franceschetti, "Sampling and sampling-like techniques in field representation," *Ann. Telecom.* **39**, 52–59 (1984).
[8.15] O. M. Bucci and G. Franceschetti, "On the spatial bandwidth of scattered fields," *IEEE Trans. Antennas Propagat.* **AP-35**, 1445–1455 (1987).
[8.16] N. Amitay, V. Galindo-Israel, and C. P. Wu, *Theory and Analysis of Phased Array Antennas*, Wiley Interscience, New York (1972).
[8.17] R. S. Elliott, "The theory of antenna arrays," in *Microwave Scanning Antennas* (R. C. Hansen, ed.), Academic Press, New York (1966).
[8.18] H. Bach and J. E. Hansen, "Uniformly spaced arrays," in *Antenna Theory* (R. E. Collin and F. J. Zucker, eds.), Part 1, McGraw-Hill, New York (1969).

9
Engineering Topics: Scattering

9.1. INTERIOR RESONANCE

We consider a volume V bounded by a perfectly conducting surface S, as depicted in Fig. 9.1. Prescribed sources $\mathbf{J}$ excite the electromagnetic field $(\mathbf{E}, \mathbf{H})$ inside the volume V, which is usually referred to as an *electromagnetic cavity*. The field in the cavity is the solution of Eqs. (4.45) with boundary condition $\hat{\mathbf{n}} \times \mathbf{E} = 0$ over the surface S. The sources radiate inside the cavity, and the cavity field can be modeled as a superposition of the direct field and the multiply diffracted field due to the cavity wall. For this reason the problem can be addressed as *interior scattering*.

For an isotropic homogeneous cavity we proceed as in Section 3.2.9 by taking the curl of Eq. (4.46a) and substituting the curl of the magnetic field from Eq. (4.46b):

$$\nabla \times \nabla \times \mathbf{E} - k^2 \mathbf{E} = -i\omega\mu\mathbf{J}, \qquad k^2 = \omega^2 \varepsilon \mu. \tag{9.1}$$

Let us examine the source-free case, thus computing the *resonant fields*, the solution of the *boundary value problem*

$$\nabla \times \nabla \times \mathbf{E}_n - k_n^2 \mathbf{E}_n = 0, \qquad \hat{\mathbf{n}} \times \mathbf{E}_n = 0 \quad \text{on } S, \tag{9.2}$$

whose solution differs from zero only for a (discrete) set[1] of *eigenvalues* k_n^2. Dot multiplication of Eq. (9.2) by $\mathbf{E}_n^*$, integration over the cavity volume, and use of Eq. (A.27) in Appendix A shows that

$$\iiint_V dV (\nabla \times \mathbf{E}_n) \cdot (\nabla \times \mathbf{E}_n)^* - \oiint_S dS\,\hat{\mathbf{n}} \cdot (\mathbf{E}_n^* \times \nabla \times \mathbf{E}_n) - k_n^2 \iiint_V dV\, \mathbf{E}_n \cdot \mathbf{E}_n^* = 0, \tag{9.3a}$$

---

1. We use a shorthand notation: for three-dimensional cavities each eigenvalue is characterized by three indexes.

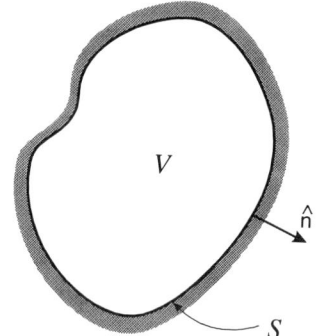

FIGURE 9.1. An electromagnetic cavity.

so that

$$k_n^2 = \frac{\iiint\limits_V dV |\nabla \times \mathbf{E}_n|^2}{\iiint\limits_V dV |\mathbf{E}_n|^2} \geq 0, \qquad (9.3b)$$

because

$$\oiint\limits_S dS\hat{\mathbf{n}} \cdot (\mathbf{E}_n^* \times \nabla \times \mathbf{E}_n) = \oiint\limits_S dS(\hat{\mathbf{n}} \times \mathbf{E}_n)^* \cdot (\nabla \times \mathbf{E}_n) = 0$$

due to the boundary conditions on S. Accordingly, the eigenvalues are real positive for this completely lossless cavity.

We now consider two resonant field distributions, or *eigenfields*, $\mathbf{E}_n$ and $\mathbf{E}_m$, corresponding to distinct eigenvalues k_n and k_m:

$$\nabla \times \nabla \times \mathbf{E}_n - k_n^2 \mathbf{E}_n = 0 \quad \text{and} \quad \nabla \times \nabla \times \mathbf{E}_m^* - k_m^2 \mathbf{E}_m^* = 0.$$

Dot multiplication of the first equation by $\mathbf{E}_m^*$, of the second by $\mathbf{E}_n$, subtraction of one equation from the other, and integration over the cavity volume lead to

$$\iiint\limits_V dV [\mathbf{E}_m^* \cdot \nabla \times \nabla \times \mathbf{E}_n - \mathbf{E}_n \cdot \nabla \times \nabla \times \mathbf{E}_m^*] - (k_n^2 - k_m^2) \iiint\limits_V dV \mathbf{E}_m^* \cdot \mathbf{E}_n = 0,$$

i.e.,

$$(k_n^2 - k_m^2) \iiint_V dV \mathbf{E}_m^* \cdot \mathbf{E}_n = 0,$$

upon applying Eq. (A.28) and the boundary conditions. We conclude that the resonant fields form an *orthogonal set*[2]; see Section 7.4.4. These eigenfields can be conveniently normalized,

$$\iiint dV \mathbf{E}_n^* \cdot \mathbf{E}_m = \delta_{nm}, \qquad (9.4)$$

and the set is referred to as *orthonormal*.

It can be shown that the set $\mathbf{E}_n$ is *complete*, this implying that a generic electromagnetic field satisfying the boundary conditions inside the cavity can be represented as an appropriate combination of resonant modes.[3] This implies also that the field in a lossless cavity can oscillate only at the *resonant (angular) frequencies*:

$$\omega \to \omega_n = k_n/\sqrt{\varepsilon\mu} = k_n c. \qquad (9.5)$$

At resonance, Eq. (9.3b) yields

$$k_n^2 = \frac{\omega^2 \mu^2 \iiint_V dV |\mathbf{H}_n|^2}{\iiint_V dV |\mathbf{E}_n|^2} = k_n^2 \frac{\mu \iiint_V dV |\mathbf{H}_n|^2}{\varepsilon \iiint_V dV |\mathbf{E}_n|^2},$$

and time-averaged *magnetic and electric energies are equal*:

$$W_n^e = \frac{1}{4}\varepsilon \iiint_V dV |\mathbf{E}_n|^2 = W_n^m = \frac{1}{4}\mu \iiint_V dV |\mathbf{H}_n|^2; \qquad (9.6)$$

see Eq. (4.49b).

The above results are based on the assumption that the cavity is completely lossless, which is an ideal situation. Losses may be present in the

---

2. If $k_n = k_m$ with $\mathbf{E}_n \neq \mathbf{E}_m$, the two eigenfunctions are degenerate, but their appropriate linear combination can be made orthogonal; see Section 7.4.4.
3. Completeness requires the definition of a norm: with a proper choice of the coefficients of the modal expansion, the norm of the difference between the field and its resonant mode representation is rendered equal to zero. A convenient norm is the quadratic one and the resonant fields form the bases of a linear Hilbert space.

volume of the cavity because the filling material may not be a perfect dielectric, and are always present in the metal walls of the cavity because this material is never a perfect conductor.

In the case of volume losses, k_n remains real; see Eq. (9.3b). For dielectric losses, $\varepsilon \to \varepsilon_1 - i\varepsilon_2$,

$$k_n^2 = \omega_n^2 \mu (\varepsilon_1 - i\varepsilon_2),$$

and the resonant frequency is forced to become complex:

$$(\omega_n + i\omega_n')^2 \mu (\varepsilon_1 - i\varepsilon_2) = k_n^2, \qquad (9.7a)$$

i.e., in the case of small losses $\varepsilon_2 \ll \varepsilon_1$:

$$\begin{cases} \omega_n' = \omega_n \dfrac{\sqrt{\varepsilon_1^2 + \varepsilon_2^2} - \varepsilon_1}{\varepsilon_2} \approx \omega_n \dfrac{\varepsilon_2}{2\varepsilon_1}, \\ 2\omega_n \omega_n' \mu \dfrac{\varepsilon_1^2 + \varepsilon_2^2}{\varepsilon_2} \approx \omega_n^2 \mu \varepsilon_1 = k_n^2, \end{cases} \qquad (9.7b)$$

by equating the imaginary and real parts of Eq. (9.7a).

Equations (9.7b) show that small volume losses do not appreciably change the (real part of the) resonant frequency, and generate a small imaginary part proportional to the ratio of the (time-averaged) dissipated power P_d and the total (time-averaged) stored energy $W_n = W_n^e + W_n^m$:

$$\omega_n' = \dfrac{\tfrac{1}{2}\omega_n \varepsilon_2 \iiint\limits_V dV |\mathbf{E}_n|^2}{2 \cdot 2 \left[\tfrac{1}{4}\varepsilon_1 \iiint\limits_V dV |\mathbf{E}_n|^2 \right]} = \dfrac{P_{dn}}{2W_n}.$$

If

$$Q_n = \omega_n \dfrac{W_n}{P_{dn}}. \qquad (9.8)$$

is the *quality*, or *Q-factor*, of the cavity, then also

$$\omega_n' = \dfrac{\omega_n}{2Q_n}. \qquad (9.9)$$

Let us examine wall losses. Their simplest model is obtained by changing the

ENGINEERING TOPICS: SCATTERING   505

boundary conditions to those of impedance-type, the Leontovič boundary conditions,

$$\hat{n} \times E_n \approx Z_M \hat{n} \times H_n \times \hat{n}, \quad (9.10)$$

where Z_M is the metal intrinsic impedance [see Eqs. (7.68) and (7.69)] and $\hat{n}$ the unit normal toward the wall. In this case the eigenvalue k_n^2 becomes complex, because in Eq. (9.3a) the surface integral is no longer equal to zero:

$$\oint_S dS(\hat{n} \times E_n)^* \cdot \nabla \times E_n = -i\omega_n \mu Z_M^* \oint_S dS|\hat{n} \times H_n|^2 = -2i\omega_n \mu P_n, \quad (9.11)$$

where use has been made of Eq. (9.10). In Eq. (9.11)

$$P_n = P_{rn} + iP_m = \frac{1}{2} Z_M^* \oint_S dS|\hat{n} \times H_n|^2$$

is the complex power injected in the cavity walls.

On substituting Eq. (9.11) into Eq. (9.3a) and solving for k_n^2 we obtain

$$k_n^2 = (\omega_n + i\omega_n')^2 \varepsilon \mu = \frac{|\omega_n|^2 \mu^2 \iiint_V dV |H_n|^2 + 2i\omega_n \mu P_n}{\iiint_V dV |E_n|^2}, \quad (9.12)$$

and equate real and imaginary parts of Eq. (9.12). When the conductivity of the cavity walls is very large, the (real part of the) resonant frequency is not appreciably changed while its imaginary part differs from zero and is given by

$$2\omega_n \omega_n' \varepsilon \mu = \frac{2\omega_n \mu P_{dn}}{\iiint_V dV |E_n|^2} = \frac{2\omega_n \varepsilon \mu P_{dn}}{4W_n^e}, \quad \text{with } \omega_n' = \frac{P_{dn}}{2W}, \quad (9.13)$$

because we still have $W_n^e \approx W_n^m$. Equation (9.13) is formally in agreement with Eq. (9.8): we must only use the power dissipated in the walls of the cavity instead of that lost in the cavity-filling material.

When both volume and wall losses are present, the total dissipated power is the sum of the two individual powers, and the inverse of the resulting Q_n factor is the sum of the inverses of the two individual factors.

We conclude that in the absence of sources, the resonant fields inside the cavity decay as $\exp(-\omega_n t/2Q_n)$ and the energy as $\exp(-\omega_n t/Q_n)$. The quantity

$$\tau = \frac{Q_n}{\omega_n} \tag{9.14}$$

is the energy *relaxation time*.

A final comment is in order. For a lossless cavity, only resonant fields may be excited, composed of the line spectral distribution given by Eq. (9.5). In a more realistic model the dispersion equation (9.5) is replaced by Eqs. (9.7) and (9.12) for volume and wall losses, respectively. This allows for continuous variation of the resonant frequencies around the original line values, with a wider excursion as the losses are increased. Such an approach generates resonance bands (see Section 9.1.6), which may coalesce in an almost continuous spectrum in the case of oversized high-loss cavities (see Section 9.1.3).

9.1.1. The Parallelepiped Cavity

Consider the empty parallelepiped cavity depicted in Fig. 9.2 with $a > b$. Its ETL consists of a line section of length l short-circuited at its two ends. In resonant conditions no source is necessary to obtain a field distribution in the cavity, no equivalent generator is applied to the ETL, voltages and currents are continuous, and the transverse resonance condition does apply (see Section 7.3.2). We choose the section AA in Fig. 9.2 to enforce this condition and Eq.

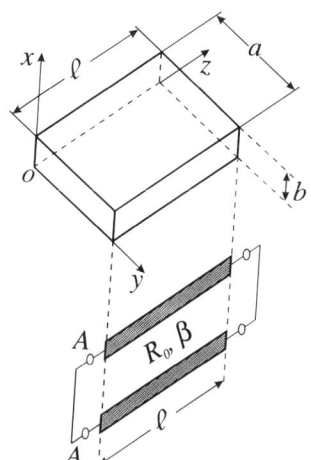

FIGURE 9.2. The parallelepiped cavity.

ENGINEERING TOPICS: SCATTERING

(7.37) becomes

$$iR_0 \tan \beta_s l = 0, \qquad \beta_s l = s\pi, \qquad s = 1, 2, \ldots,$$

i.e.,

$$k_{nms}^2 = \omega_{nms}^2 \varepsilon\mu = \left(\frac{n\pi}{a}\right)^2 + \left(\frac{m\pi}{b}\right)^2 + \left(\frac{s\pi}{l}\right)^2. \tag{9.15}$$

Equation (9.15) defines the *resonant (angular) frequencies* of the cavity. The lowest one is relative to the TE_{101} resonant mode:

$$\omega_{101}^2/c^2 = \left(\frac{\pi}{a}\right)^2 + \left(\frac{\pi}{l}\right)^2, \qquad l > b,$$

and the field is TE with respect to the z-axis.[4] For this mode the electric field is given by

$$\mathbf{E} = \mathbf{e}(x)V(z),$$

$$\mathbf{e}(x) = -\sqrt{\frac{2}{ab}} \sin\left(\frac{\pi}{a}x\right)\hat{\mathbf{y}}, \qquad V(z) = -iR_0 I_0 \sin\left(\frac{\pi}{l}z\right)$$

(see Section 7.4.1), and the stored energy by

$$W = 2W_e = \frac{\varepsilon}{2}\int_0^a dx \int_0^b dy |\mathbf{e}(x)|^2 \int_0^l dz |V(z)|^2$$

$$= \frac{\varepsilon}{2} R_0^2 |I_0|^2 \int_0^l \sin^2\left(\frac{\pi}{l}\right) dz = \frac{R_0^2 |I_0|^2 \varepsilon l}{4}. \tag{9.16a}$$

As stated earlier, the power P_d dissipated on the cavity walls is obtained by using the Leontovič boundary conditions [see Eq. (7.68)]:

$$P_d = \frac{2}{2\sigma\delta}\left[|I_0|^2 \int_0^a dx \int_0^b dy |\mathbf{h}(x)|^2 + \frac{1}{(\omega\mu)^2}\int_0^b dy |\nabla_t \cdot \mathbf{h}(0)|^2 \int_0^l dz |V(z)|^2 \right.$$
$$\left. + \int_0^a dx |\mathbf{h}(x)|^2 \int_0^l dz |I(z)|^2 + \frac{1}{(\omega\mu)^2}\int_0^a dx |\nabla_t \cdot \mathbf{h}(x)|^2 \int_0^l dz |V(z)|^2 \right],$$

$$\mathbf{h} = \hat{\mathbf{z}} \times \mathbf{e}, \qquad I(z) = I_0 \cos\left(\frac{\pi}{l}z\right),$$

where the integrals account for the cavity wall losses: the first on the wall at

---

4. Should l be smaller than b, the lowest resonant mode would be TE_{110}.

$z = 0$, the second on the wall at $x = 0$, and the last two on the wall at $y = 0$; the wall losses at $z = l$, $x = a$, $y = b$ are equal to those at $z = 0$, $x = 0$, $y = 0$, respectively. Hence

$$P_d = \frac{|I_0|^2}{\sigma \delta} \left[1 + \frac{R_0^2}{(\omega \mu a)^2} \frac{\pi^2 l}{a} + \frac{l}{2b} + \frac{R_0^2}{(\omega \mu a)^2} \frac{\pi^2 l}{2b} \right]. \tag{9.16b}$$

Equations (9.16) are now used to compute the Q of the cavity. It is convenient to express the characteristic resistance in terms of the cavity geometry, $R_0 = \omega \mu / \beta = \omega \mu l / \pi$, so

$$Q = \frac{\zeta \sigma \delta}{4\pi^2} \frac{(k_0 l)^3}{1 + (l/2b) + (l/a)^3 [1 + (a/2b)]}, \tag{9.17a}$$

$$k_0 = \omega_{101}/c.$$

However

$$k_0^2 = \left(\frac{\pi}{a}\right)^2 + \left(\frac{\pi}{l}\right)^2,$$

therefore Eq. (9.17a) can also be written as

$$Q = \frac{\pi \zeta \sigma \delta}{4} \frac{[1 + (l/a)^2]^{3/2}}{1 + (l/2b) + (l/a)^3 [1 + (a/2b)]}. \tag{9.17b}$$

9.1.2. The Loaded Coaxial Cavity

Let us examine the loaded coaxial cable shorted at its ends and depicted in Fig. 9.3. By applying the transverse resonance condition, Eq. (7.37), at $z = 0$ we have for TEM resonant modes

$$iR_0 \tan kl_1 + i \frac{R_0}{\sqrt{\varepsilon_r}} \tan(k\sqrt{\varepsilon_r} l_2) = 0, \tag{9.18}$$

where R_0 and k are the characteristic resistance and the propagation constant of the empty section of the coaxial line.

A first case is considered by assuming $l_1 = l_2 = l$ and $\varepsilon_r = 4$. Equation (9.18) becomes

$$\tan kl + \tfrac{1}{2} \tan 2kl = 0, \quad k^2 = \omega^2 \varepsilon \mu,$$

i.e.,

$$\tan kl + \frac{\tan kl}{1 - \tan^2 kl} = 0,$$

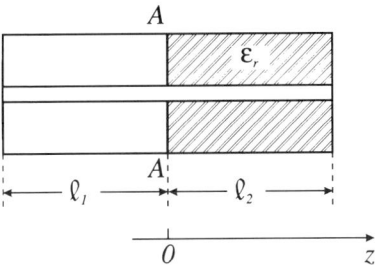

FIGURE 9.3. The loaded coaxial cavity.

which provides the resonant solutions

$$\tan kl = 0, \quad k_n l = n\pi, \quad n = 1, 2, \ldots$$

$$\tan^2 kl = 2, \quad \begin{cases} k_m l = 0.304\pi + m\pi, & m = 0, 1, \\ k_{m'} l = -0.304\pi + m'\pi, & m' = 1, 2, \ldots. \end{cases}$$

The first resonant (angular) frequency is given by

$$\omega_0/c = \frac{0.304\pi}{l}$$

and corresponds to the index $m = 0$. Successive resonances are obtained for $m' = 1$, $n = 1$, $m = 1$. The voltage distribution along the cavity is continuous at $z = 0$:

$$V_1(z) = V_0 \sin k(l + z), \quad z < 0,$$

$$V_2(z) = V_0 \sin kl \frac{\sin 2k(l - z)}{\sin 2kl}, \quad z > 0,$$

and is presented in Fig. 9.4 for the first four modes.[5]

---

5. When $\sin kl = 0$, we enforce the condition that the voltage derivative is continuous at $z = 0$, as the permeability is the same in the two media. We obtain

$$V_1(z) = V_0 \sin k(l + z), \quad z > 0,$$

$$V_2(z) = -V_0 \frac{\cos kl}{2 \cos kl} \sin 2k(l - z), \quad z > 0.$$

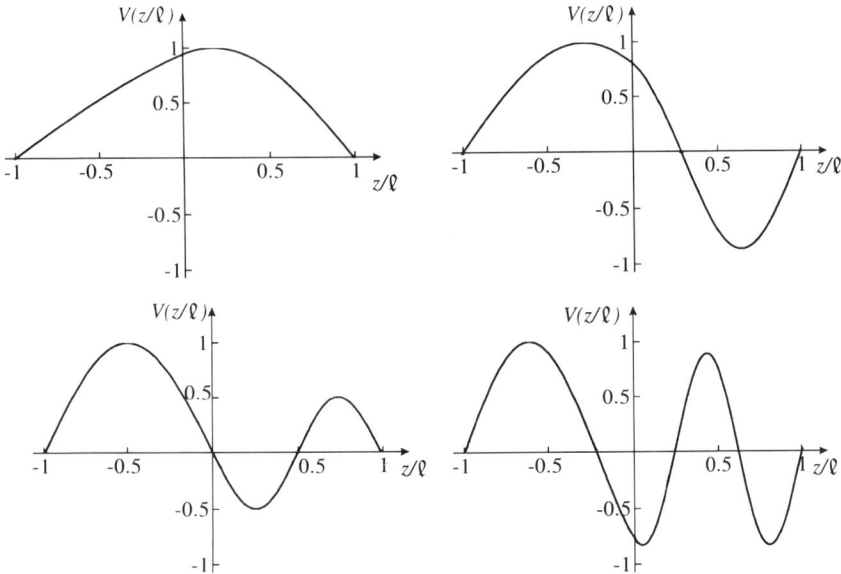

FIGURE 9.4. Voltage distribution for the first four resonant modes of the loaded coaxial cavity of Fig. 9.3 with $l_1 = l_2 = l$ and $\varepsilon_y = 4$. The voltages are normalized to their maximum value.

The stored energy is given by

$$W = \frac{2}{4}\int_{-l}^{l} dz\, C(z)|V(z)|^2,$$

where C is the coaxial cable capacity per unit length (see Section 7.4.3), different in the two sections of the cavity.

In some cases the volume losses in the dielectric may dominate the wall losses. If $\tan \Lambda$ is the loss tangent of the dielectric, then

$$Q = \omega \frac{\frac{1}{4}\int_{-l}^{0} dz|V_1(z)|^2 + \int_{0}^{l} dz|V_2(z)|^2}{\omega \tan \Lambda \int_{0}^{l} dz|V_2(z)|^2}$$

For the fundamental mode

$$Q_0 \tan \Lambda = 1 + \cos^2 kl\, \frac{1 - \mathrm{sinc}(2kl)}{1 - \mathrm{sinc}(4kl)} = 1.145,$$

$$kl = 0.304\pi,$$

ENGINEERING TOPICS: SCATTERING

and the resonant frequency is only slightly larger than the value corresponding to the fully loaded cavity.

A second interesting case is obtained by setting $l_1 = l$ and $l_2 = l\sqrt{\varepsilon_r}$. Equation (9.18) becomes

$$iR_0 \tan kl + i\frac{R_0}{\sqrt{\varepsilon_r}} \tan kl = 0, \qquad (9.19)$$

whose solutions are

$$k_n l = n\pi, \qquad n = 1, 2, \ldots,$$

and also

$$k_m l = (2m + 1)\frac{\pi}{2}, \qquad m = 0, 1, \ldots,$$

because the two terms appearing in Eq. (9.19) may compensate each other at the poles of the tangent.[6]

Note that the resonant frequencies of this cavity coincide with those of the empty section either open or short-circuited at its right end. The additional loaded section may be used to control the Q of the cavity.

9.1.3. Multimode Cavities

Consider a cavity whose dimensions are large compared to the wavelength, implying that a large number of modes can be excited. This is the case if either some lossy material is present inside the cavity or its boundaries slightly oscillate. In the first case resonances are complex (see Section 9.1), each resonance exhibits a bandwidth (see Section 9.1.6), and the spectrum tends to become continuous. In the second case vibration of the boundaries or insertion of a rotating arm inside the cavity forces new modes to be continuously generated to instantaneously match the time-varying boundary conditions. The relaxation time of these modes (see Section 9.1.5) should be large compared to vibration or rotating arm period; the cavity is said to be in the *mixed mode* condition.

---

6. To confirm this result, it suffices just to apply the transverse resonance condition in terms of the input admittances,

$$\overleftarrow{Y} + \overrightarrow{Y} = 0,$$

instead of impedances.

Implementation of the first case leads to microwave ovens, while the second case leads to *reverberating rooms*, which are aimed at generating a homogeneous, isotropic field distribution inside the cavity. In all practical realizations the walls of the reverberating room are lossy and it is important to compute the cavity Q, and also to ascertain that the corresponding relaxation time (see Section 9.1.5) is at least comparable with the inverse of the vibration frequency of the mixing mechanism. Computation of this Q is now treated.

We refer to a cubic cavity of side a. Due to the multitude of modes and their random generation, we can reasonably assume them to be fully incoherent. To proceed further a simple model is considered: six wave streams, moving back and forth along the three axes of the cubic cavity. Let S be the Poynting vector of each stream. The corresponding energy density is S/c (see also Section 1.3.5) and the power density loss in the cavity wall is

$$S_M = \text{Re}\left(\frac{4\zeta\zeta_M}{|\zeta + \zeta_M|^2}\right) S \approx \frac{4S}{\zeta}\sqrt{\frac{\omega\mu}{2\sigma}},$$

where ζ_M is the metal intrinsic impedance [see Eq. (7.69)] and the power transmission coefficient for normal incidence has been used (see Section 4.5.1).

The energy of each stream is given by

$$\frac{S}{c} a^3 = S\sqrt{\varepsilon\mu}\, a^3$$

and the dissipated power by

$$\frac{4S}{\zeta}\sqrt{\frac{\omega\mu}{2\sigma}}\, a^2 = 2S\omega\sqrt{\varepsilon\mu}\,\delta a^2,$$

where δ is the skin depth [see Eq. (4.79)] of the cavity walls. Due to the postulated incoherent nature of the field, we sum over the six streams and compute the cavity quality factor:

$$Q = \frac{3a}{\delta}. \qquad (9.20)$$

This expression can be generalized to cavities of most general shape by substituting $3a \to a_1 + a_2 + a_3$, a_i being the effective widths of the cavity along the three axes.

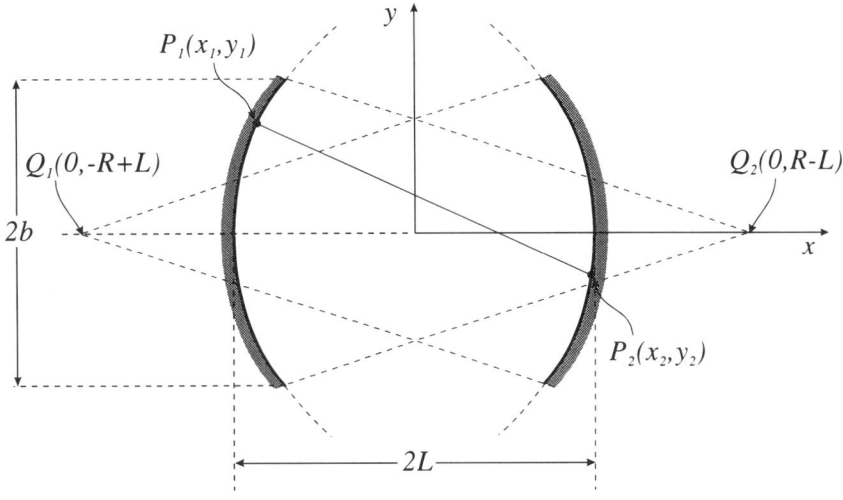

FIGURE 9.5. Geometry of an open cavity.

9.1.4. Open Cavities

In high-frequency bands, monomode cavities become too small and the quality factor Q too low. For this reason *open cavities* are used which derive conceptually from the Fabry–Perot[7] resonator widely used in the optical range.

We refer in the following to a simple geometry, the open cyclindrical[8] cavity of Fig. 9.5: two incomplete metal cylinders of equal radius R and spacing $2L$.

The equations of the two cylindrical surfaces are

$$(x_1 - R + L)^2 + y_1^2 = R^2, \qquad |y_1| \leqslant b,$$

and

$$(x_2 + R - L)^2 + y_2^2 = R^2, \qquad |y_2| \leqslant b.$$

When $b \ll R$

$$x_1 - R + L = -\sqrt{R^2 - y_1^2} \approx -R + \frac{y_1^2}{2R}, \qquad |y_1| \leqslant b \qquad (9.21a)$$

---

7. Charles Fabry: Marseilles (France), 1867–Paris, 1945.
   Alfreed Perot: Metz (France), 1863–Paris, 1925.
8. Cylindrical cavities are used instead of planar ones, because in the latter case it is difficult to assure exact parallelism of the walls.

and

$$x_2 + R - L = \sqrt{R^2 - y_2^2} \approx +R - \frac{y_2^2}{2R}, \qquad |y_2| \leq b, \qquad (9.21b)$$

where the choice of the square-root sign allows the correct limit result when $R \to \infty$.

Let $J_1(y_1)\hat{z}$ be the surface current density over the left wall, or *mirror*, of the cavity. We use Eq. (5.41), PO approximation (8.58), and superposition to compute the surface current density $J_2(x_2)\hat{z}$ over the right mirror. To this end we need a manageable expression for the distance $r = |P_1 P_2|$ (see Fig. 9.5):

$$\begin{aligned}
r &= \sqrt{(x_2 - x_1)^2 + (y_2 - y_1)^2} \\
&\approx 2L + \left(\frac{1}{2L} - \frac{1}{R}\right)\frac{y_1^2}{2} + \left(\frac{1}{2L} - \frac{1}{R}\right)\frac{y_2^2}{2} - \frac{y_1 y_2}{2L} \\
&= 2L + \frac{y_1^2}{2a} + \frac{y_2^2}{2a} - \frac{y_1 y_2}{2L}, \qquad \frac{1}{a} = \frac{1}{2L} - \frac{1}{R},
\end{aligned}$$

where use has been made of Eqs. (9.21). Hence

$$\mathbf{J}_2(y_2) = 2\hat{n} \times \mathbf{H}(y_2) \approx -\frac{2}{\zeta}\mathbf{E}(y_2), \qquad \hat{n} \approx \hat{x}$$

and

$$\begin{aligned}
J_2(y_2) = &-\frac{1}{2}\sqrt{\frac{k}{\pi L}}\exp\left(i\frac{\pi}{4}\right)\exp(-2ikL) \\
&\cdot \int_{-b}^{b} dy_1 \exp\left(-ik\frac{y_1^2 + y_2^2}{2a}\right) \cdot \exp\left(ik\frac{y_1 y_2}{2L}\right) J_1(y),
\end{aligned}$$

where we have retained the y-dependence of the distance $r(y)$ only in the phase term.

For a resonant mode the current density $J_2(y_2)$ must be a replica (but for the sign) of the current density $J_1(y_2)$, for any $|y_2| \leq b$:

$$J_2(y) = \pm J_1(y), \qquad (9.22)$$

which is the wave counterpart of the ray consistency condition (see Section 5.2.3). When lengths are normalized to (half) the cavity transverse dimension

ENGINEERING TOPICS: SCATTERING

b, we obtain (omitting the subscript)

$$J(\eta) = \pm \frac{1}{2}\sqrt{\frac{kb^2}{\pi L}} \exp\left(i\frac{\pi}{4}\right) \exp(-2ikL)$$

$$\cdot \int_{-1}^{1} d\xi \exp\left(-ikb^2 \frac{\xi^2 + \eta^2}{2a} + ikb^2 \frac{\xi\eta}{2L}\right) J(\xi), \qquad (9.23)$$

which is an *integral equation* in the unknown current density $J(y) \to J(\eta)$ on the mirrors.

The solution of Eq. (9.23) provides the resonant modes of the cavity and usually requires a numerical effort. In some cases an analytical solution can be gained, e.g., when $kb^2/2a \approx 0$. Note that this condition can be met exactly by letting $2L = R$ (*confocal cavity*). We neglect the quadratic terms in Eq. (9.23), in which case

$$J(\eta) = \pm \frac{1}{2}\sqrt{\frac{kb^2}{\pi L}} \exp\left(i\frac{\pi}{4}\right) \exp(-2ikL) \int_{-1}^{1} d\xi \exp\left(i\frac{kb^2}{2L}\eta\xi\right) J(\xi),$$

which shows that the resonant modes $J(\eta)$ should be a scaled replica of their truncated FT. The solution to this equation is provided in terms of the (angular) *prolate spheroidal functions*, $S(t, \chi)$,

$$(1 - t^2)\frac{d^2 S}{dt^2} - 2t\frac{dS}{dt} + (\chi - c^2 t^2)S = 0, \qquad c^2 = \frac{kb^2}{2L};$$

this equation exhibits continuous solutions in the closed interval $(-1, 1)$ only for a discrete set of positive values $0 < \chi_0(c) < \chi_1(c) < \cdots < \chi_n(c)$ of the parameter $\chi(c)$. These, in turn, determine the eigenvalues of the integral equation for the surface current density and the resonant frequencies of the cavity.

9.1.5. Energy Decay in the Cavity

Consider a cavity with stored energy W and losses P_d. As time elapses (and in the absence of sources) the energy decreases, so we have

$$-dW = P_d dt.$$

From Eq. (9.8)

$$-dW = \frac{\omega W}{Q} dt,$$

or

$$W(t) = W(0) \exp\left[-\frac{\omega t}{Q}\right],$$

which is consistent with Eq. (9.14).

9.1.6. Equivalent Circuit of the Cavity

We examine the simple TL[9] depicted in Fig. 9.6, short-circuited at its ends and driven by the current generator I. The voltage distribution, which is continuous at $z = 0$ and satisfies boundary conditions at $z = -l_1$ and $z = l_2$, is given by

$$\overleftarrow{V}(z) = V_0 \frac{\sin k(l_1 + z)}{\sin kl_1} \quad \text{and} \quad \overrightarrow{V}(z) = V_0 \frac{\sin k(l_2 - z)}{\sin kl_2}, \quad (9.24a)$$

$V(0) = V_0$, for negative and positive values of z, respectively.
The current distribution is readily obtained from Eq. (7.10):

$$\overleftarrow{I}(z) = \frac{iV_0}{R_0} \frac{\cos k(l_1 + z)}{\sin kl_1} \quad \text{and} \quad \overrightarrow{I}(z) = -\frac{iV_0}{R_0} \frac{\cos k(l_2 - z)}{\sin kl_2}. \quad (9.24b)$$

These latter equations yield

$$I = -\overleftarrow{I}(0) + \overrightarrow{I}(0) = -\frac{iV_0}{R_0} \frac{\sin kl}{\sin kl_1 \sin kl_2}. \quad (9.25)$$

The transverse resonance condition (7.37) applied at the section $z = 0$ reads

$$iR_0 \tan kl_1 + iR_0 \tan kl_2 = iR_0 \frac{\sin kl}{\cos kl_1 \cos kl_2} = 0,$$

---

9. In the more elaborate case of a cavity we should consider the *equivalent* transmission line and the *equivalent* generator, but final results are essentially the same.

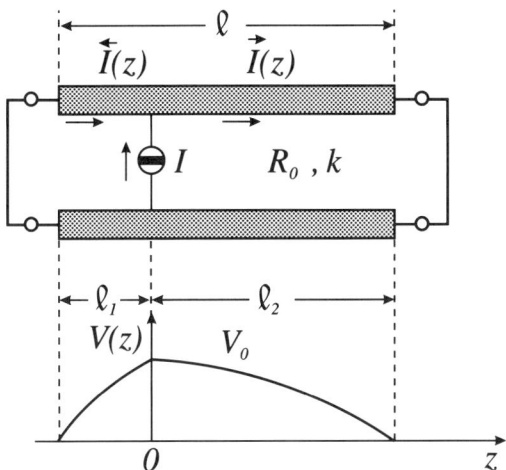

FIGURE 9.6. A resonant transmission line.

i.e.,
$$k_n l = n\pi,$$

as expected. At resonance $I = 0$, $\overleftarrow{I}(0) = \overrightarrow{I}(0)$ [see Eq. (9.25)], and not only the voltage but also the current (as well as their derivatives) are continuous at $z = 0$.

Always at resonance the stored energy is given by

$$W = 2W_e = \frac{1}{2}C\left[\int_{-l_1}^{0} dz|\vec{V}(z)|^2 + \int_{0}^{l_2} dz|\vec{V}(z)|^2\right]$$
$$= \frac{Cl}{4}\frac{|V_0|^2}{\sin^2 k_n l_2}. \tag{9.26}$$

This equation is obtained by noting that

$$\sin^2 k_n l_1 = \sin^2 k_n (l - l_2) = \cos^2 k_n l_2 \sin^2 k_n l_2 = \sin^2 k_n l_2,$$

and employing the substitution $l_1 = (l - l_2)$ in the result of the integration.

We now depart slightly from the resonant (angular) frequency ω_n and obtain

$$kl = k_n l + (k - k_n)l = n\pi + (\omega - \omega_n)\sqrt{LC}\, l,$$
$$\sin kl \approx \cos k_n l \cdot (\omega - \omega_n)\sqrt{LC}\, l, \tag{9.27}$$

and the current distribution is no longer continuous at $z = 0$. Accordingly, a current source generator is needed; see Eq. (9.25). Note that the voltage remains continuous (with a discontinuous derivatives) at $z = 0$.

The stored electric and magnetic energies are no longer equal, because we are (slightly) off resonance. The source generator provides the reactive power P_r to compensate for the magnetic and electric stored energy imbalance:

$$P_r = \frac{1}{2} V_0 I^* = \frac{i}{2} \frac{|V_0|^2}{R_0} \frac{\sin kl}{\sin kl_1 \sin kl_2}$$

$$= \frac{i}{2} \frac{|V_0|^2}{\sin^2 k_n l_2} (\omega - \omega_0) Cl, \quad (9.28a)$$

where use has been made of Eqs. (9.25), (9.27), and of the relation

$$\sin kl_1 \approx \sin k_n l_1 = \sin k_n(l - l_2) = -\cos k_n l \sin k_n l_2.$$

At this stage the cavity is lossless. Inclusion of the losses can be accounted for, as usual, by perturbation procedure (see, for instance, Section 9.1.1). We use the field of the lossless cavity to compute the losses in its lossy elements. In our case the field distribution is taken as that at the resonant frequency. Again, the small frequency deviation has not appreciably changed the field as far as loss computation is concerned, and we can reasonably utilize Eqs. (9.8) and (9.26) to yield

$$P_d = \frac{\omega_n W}{Q_n} = \frac{\omega_n Cl}{4 Q_n} \frac{|V_0|^2}{\sin^2 k_n l_2}. \quad (9.28b)$$

We now compute the complex power $P = P_1 + iP_2$ associated with the resonant (parallel) circuit of Fig. 9.7 near the resonance,

$$P_1 = \frac{1}{2} \frac{|V_0|^2 / N^2}{R_e},$$

$$P_2 = \frac{1}{2} \left(\omega C_e - \frac{1}{\omega L_e} \right) \frac{|V_0|^2}{N^2}$$

$$= \frac{1}{2} \frac{[\omega_n + (\omega - \omega_n)]^2 L_e C_e - 1}{\omega_n L_e + (\omega - \omega_n) L_e} \frac{|V_0|^2}{N^2}$$

$$\approx (\omega - \omega_n) C_e \frac{|V_0|^2}{N^2}, \quad \omega_n^2 L_e C_e = 1,$$

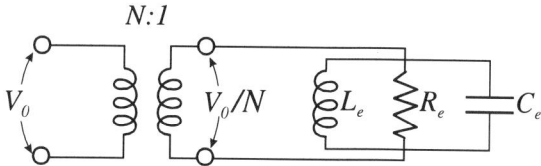

FIGURE 9.7. Equivalent (parallel) circuit, L_e, R_e, and C_e, for a resonant cavity. The circuit includes an ideal transformer with voltage ratio $N:1$.

and identify it with that given by Eqs. (9.28). Hence

$$N = \sin k_n l_2, \qquad R_e = \frac{2Q_n}{\omega_n Cl} = \frac{2Q_n R_0}{n\pi},$$

$$C_e = C\frac{l}{2}, \qquad L_e = \frac{1}{\omega_n^2 C_e} = \frac{2Ll}{(n\pi)^2}. \qquad (9.29)$$

These expressions provide the parameters with which to model the cavity around the resonance as a parallel resonant circuit. Note the possibility of changing the input admittance of the cavity,

$$Y = G + iB \approx \frac{1}{N^2}\left[\frac{1}{R_e} + 2i(\omega - \omega_n)C_e\right],$$

by moving the coupling point, i.e., the source position, along the TL.

The modulus of the current, $|I| = |YV_0|$, attains its minimum value at resonance, and is increased by the factor $\sqrt{2}$ when

$$2|\omega - \omega_n|C_e = 2\frac{\Delta\omega}{2}C_e = \frac{1}{R_e} \qquad \text{and} \qquad \frac{\Delta\omega}{\omega_n} = \frac{1}{Q_n}. \qquad (9.30)$$

Equation (9.30) defines the *relative bandwith* of the cavity and is inversely proportional to the Q-factor.

9.2. EXTERIOR RESONANCE

The exterior resonance problem is the *dual* of the interior one, and is depicted in Fig. 9.8. The volume under consideration is that outside the surface S, which is assumed perfectly conducting. The field is excited by sources present in the volume V, and is the solution of Maxwell's equations (4.45) with

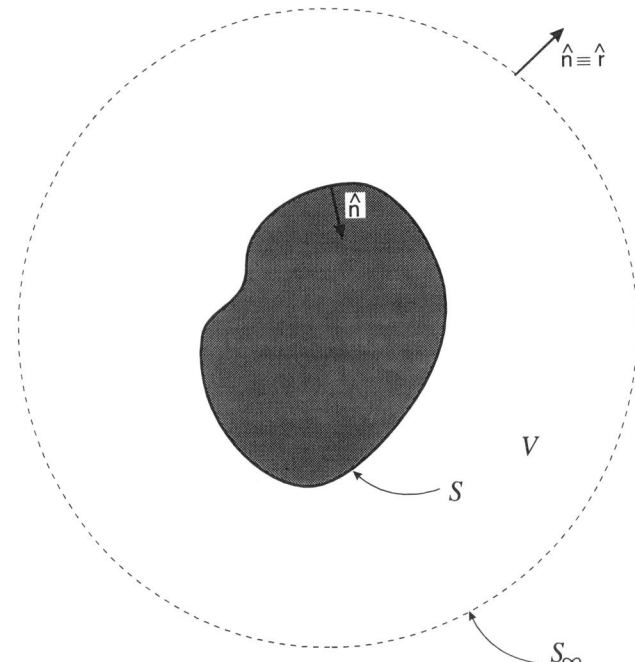

FIGURE 9.8. The exterior resonance problem.

boundary conditions $\hat{\mathbf{n}} \times \mathbf{E} = 0$ on S and radiation condition, Eq. (1.62), on the sphere S_∞ at infinity. Again, a convenient model for studying the problem is to regard the total field as the superposition of that radiated by the sources and that scattered by the metal body. For this reason the problem can be addressed as *exterior scattering*.

In some cases the incident field is replaced by an appropriate excitation over the scatterer, which amounts to prescribing the (incident) tangential electric or magnetic field over S. This field is determined by slots or gaps distributed over the body surface and excited by the field inside the surface S. The tangential field can be modeled by equivalent currents over the body surface, and is used to synthesize a prescribed scattered field distribution in the volume V. The body behaves as an antenna, and the engineering characterization of the scattered field may rely on the antenna parameters introduced in Chapter 8. In fact, this is the same as the problem treated there, i.e., a radiation problem.

In other cases (and for other applications) the incident field is prescribed and the scattered field is used to detect the presence, shape, and possibly cinematic attributes of the body (if this is moving). This is a problem different from those considered in Chapter 8, and use of new, appropriate parameters is

ENGINEERING TOPICS: SCATTERING

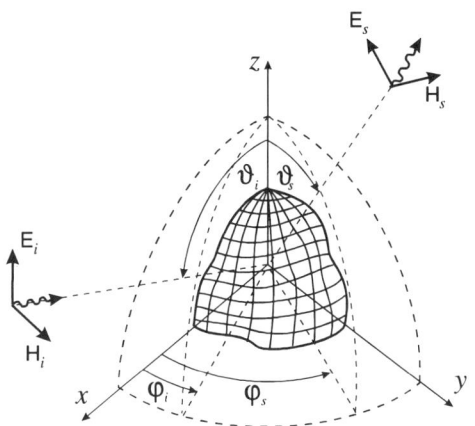

FIGURE 9.9. Definition of the scattering cross section.

desirable. A convenient parameter is the *scattering cross section* σ_0 of the object, sometimes referred to as the *radar cross section* (RCS), whose definition is now examined.

Consider the spherical system of coordinates centered on the scattered body (see Fig. 9.9). Let $\mathbf{E}_i(0, \theta_i, \phi_i)$ be the incident field, i.e., the field excited at $r = 0$ by the external sources with the body removed; and let $\mathbf{E}_s(r, \theta_s, \phi_s)$ be the scattered field at distances r and in the direction (θ_s, ϕ_s). We further assume the incident field to be (locally) plane and employ the relation

$$\frac{1}{2\zeta}|\mathbf{E}_i(0, \theta_i, \phi_i)|^2 \sigma_0(\theta_i, \phi_i; \theta_s, \phi_s) = \frac{1}{2\zeta} 4\pi r^2 |\mathbf{E}_s(r \to \infty, \theta_s, \phi_s)|^2,$$

namely

$$\sigma_0(\theta_i, \phi_i; \theta_s, \phi_s) = 4\pi r^2 \frac{|\mathbf{E}_s(r \to \infty, \theta_s, \phi_s)|^2}{|\mathbf{E}_i(0, \theta_i, \phi_i)|^2}, \qquad (9.31)$$

to define the *bistatic* cross section of the body.

Equation (9.31) provides information about the visibility of the scatterer (*target detection*) in the direction (θ_s, ϕ_s) when the latter is illuminated from the direction (θ_i, ϕ_i). When $\theta_s = \theta_i$ and $\phi_s = \phi_i$,

$$\sigma_0(\theta_i, \phi_i; \theta_i, \phi_i) \to \sigma_B(\theta_i, \phi_i), \qquad (9.32a)$$

and is referred to as the *monostatic* or *backscattering* cross section. Similarly, when $\theta_s = \pi - \theta_i$ and $\phi_s = \pi + \phi_i$,

$$\sigma_0(\theta_i, \phi_i; \pi - \theta_i, \pi + \phi_i) \to \sigma_F(\theta_i, \phi_i), \qquad (9.32b)$$

and is referred to as the *forward scattering* cross section.

In order to characterize the field scattered in the volume V we proceed as in Section 9.1 and first study the source-free field, i.e., the resonant solutions $\mathbf{E}_n$. We are lead to the same equation (9.3a), where the surface S is that of the body augmented by S_∞ (see Fig. 9.8). On the sphere at infinity we make use of the radiation condition,

$$\hat{\mathbf{r}} \times \mathbf{E}_n = \zeta \hat{\mathbf{r}} \times \mathbf{H}_n \times \hat{\mathbf{r}},$$

and obtain

$$\hat{\mathbf{r}} \cdot \mathbf{E}_n^* \times \nabla \times \mathbf{E}_n = \hat{\mathbf{r}} \times \mathbf{E}_n^* \cdot (-i\omega\mu\mathbf{H}_n) = i\omega\mu\zeta |\hat{\mathbf{r}} \times \mathbf{H}_n|^2.$$

The surface integral in Eq. (9.3a) is no longer zero and the eigenvalues are no longer positive real. Accordingly, the resonant solutions do not form the bases of an orthogonal set, as for the interior scattering, and field orthogonality can be recovered in a more restricted sense; see the subsequent discussion and Sections 9.2.1 and 9.2.2.

The possibility of an *analytical* determination of the resonant modes is related to the availability of *separable solutions* to Eq. (9.2), or to the equivalent one

$$\nabla^2 \mathbf{E} + k^2 \mathbf{E} = 0, \qquad (9.33)$$

which are known to exist only for six coordinate systems[10]: rectangular, circular, elliptic and parabolic cylindrical, spherical, and conical.

The simplest case of exterior scattering is obtained when the surface S is the infinite plane $z = 0$, and the volume V is defined by $z \geqslant 0$. This problem was considered in Section 3.5 and is usually referred to as the plane-wave expansion. The eigenfunctions, the solution of Eq. (9.33) with the radiation condition for $z \to \infty$, are

$$\mathbf{E}(\mathbf{k}) \exp(-i\mathbf{k} \cdot \mathbf{r}), \qquad \mathbf{k} = u\hat{\mathbf{x}} + v\hat{\mathbf{y}} + w\hat{\mathbf{z}}, \qquad (9.34)$$

$$w = +\sqrt{k^2 - (u^2 + v^2)}, \qquad u^2 + v^2 \leqslant k^2,$$

$$w = -i\sqrt{(u^2 + v^2) - k^2}, \qquad u^2 + v^2 > k^2.$$

---

10. The scalar wave equation is separable in an additional five systems: oblate and prolate spheroidal, parabolic, paraboloidal, and ellipsoidal. All these eleven systems can be derived from the most general ellipsoidal coordinates.

Orthogonality of the eigenfunctions (9.34) is *not* available throughout the space $z \geqslant 0$, but only on the *transverse coordinate surfaces* $z = z_0 = \text{const}$, where

$$\frac{1}{(2\pi)^2} \int_{-\infty}^{+\infty} dx \int_{-\infty}^{+\infty} dy \exp(i\mathbf{k}\cdot\mathbf{r})(-i\mathbf{k}'\cdot\mathbf{r})$$

$$= \frac{1}{(2\pi)^2} \exp[i(w-w')z_0] \int_{-\infty}^{+\infty} dx \int_{-\infty}^{+\infty} dy \exp[i(u-u')x + i(v-v')y]$$

$$= \exp[i(w'-w)z_0]\delta(u-u')\delta(v-v') = \delta(u-u')\delta(v-v'). \tag{9.35}$$

This equation can be regarded as the generalization for the continuous spectrum of the usual orthogonality relationship pertinent to a discrete set of eigenfunctions.

Other coordinate systems are treated in Sections 9.2.1 and 9.2.2.

9.2.1. Cylindrical Coordinates

Let us consider z-independent resonant fields outside a metal cylinder of radius a. We refer to a cylindrical coordinate system (r, ϕ, z) centered on the cylinder axis.

A complete (orthogonal) expansion set on any coordinate surface $r = \text{const}$ is provided by the eigenfunctions $\exp(in\phi)$. We postulate this ϕ-dependence for all field components[11] and so Maxwell's equations (4.46a–b) in cylindrical coordinates read as follows:

$$\frac{in}{r} E_z = -i\omega\mu H_r, \tag{9.36a}$$

$$-\frac{\partial E_z}{\partial r} = -i\omega\mu H_\phi, \tag{9.36b}$$

$$-\frac{in}{r} E_r + \frac{1}{r}\frac{\partial}{\partial r}(rE_\phi) = -i\omega\mu H_z, \tag{9.36c}$$

$$\frac{in}{r} H_z = i\omega\varepsilon E_r, \tag{9.36d}$$

$$-\frac{\partial H_z}{\partial r} = i\omega\varepsilon E_\phi, \tag{9.36e}$$

$$-\frac{in}{r} H_r + \frac{1}{r}\frac{\partial}{\partial r}(rH_\phi) = i\omega\varepsilon E_z, \tag{9.36f}$$

11. The transverse coordinate dependence of each field component should be generally expanded by means of a linear functional of the transverse eigenfunctions (see Section 9.2.2). This is not necessary for the case at hand.

where use has been made of Eqs. (A.41) with $\partial/\partial\phi \to in$ (and $\partial/\partial z \to 0$ for the assumed z-independent fields).

Inspection of Eqs. (9.36) immediately shows that the assumed $\exp(in\phi)$ dependence for *all* field components is acceptable:

$$\mathbf{E}(r, \phi) = \exp(in\phi)\mathbf{e}(r), \qquad \mathbf{H}(r, \phi) = \exp(in\phi)\mathbf{h}(r), \qquad (9.37)$$

and provides (as anticipated in Section 9.2) an expansion set which is orthogonal on any coordinate surface $r = \text{const}$.

We first set $h_z = 0$ and obtain from Eqs. (9.36d) and (9.36e) $e_r = e_\phi = 0$. Then multiply Eq. (9.36b) by r, take the derivative with respect to r, and substitute for h_ϕ from Eq. (9.36f) to obtain

$$\frac{d}{dr} r \frac{de_z}{dr} = i\omega\mu(i\omega\varepsilon e_r + inh_r).$$

We finally substitute for h_r from Eq. (9.36a) and have

$$\frac{1}{r}\frac{d}{dr} r \frac{de_z}{dr} + \left(k^2 - \frac{n^2}{r^2}\right)e_z = 0, \qquad (9.38)$$

which is the Bessel equation whose solutions satisfying the radiation condition at infinity are the Hankel function of second kind. Accordingly

$$\begin{cases} E_z = \sum_n a_n \exp(in\phi) H_n^{(2)}(kr), \\ \zeta H_r = -\frac{1}{kr}\sum_n a_n n \exp(in\phi) H_n^{(2)}(kr), \\ \zeta H_\phi = -i\sum_n a_n \exp(in\phi)[H_n^{(2)}(kr)]'. \end{cases} \qquad (9.39a)$$

where a prime implies the derivative with respect to kr and Eqs. (9.36a) and (9.36b) have been employed. Equations (9.39a) are referred to as the *parallel polarized* resonant modes. The dual *perpendicular polarized* resonant modes are readily obtained by setting $e_z = 0$:

$$\begin{cases} \zeta H_z = \sum_n b_n \exp(in\phi) H_n^{(2)}(kr), \\ E_r = \frac{1}{kr}\sum_n b_n n \exp(in\phi) H_n^{(2)}(kr), \\ E_\phi = i\sum_n b_n \exp(in\phi)[H_n^{(2)}(kr)]'. \end{cases} \qquad (9.39b)$$

Consider now a plane wave normally incident on the cylinder. For parallel polarization (incident electric field parallel to the cylinder axis) we have

$$\mathbf{E}_i(r, \phi) = E_0 \hat{\mathbf{z}} \exp(ikr \cos \phi),$$

whose ϕ dependence can be expanded in the set:

$$\exp(ikr \cos \phi) = \sum_n c_n \exp(in\phi),$$

where

$$c_n = \frac{1}{2\pi} \int_0^{2\pi} d\phi \, \exp(ikr \cos \phi) \exp(-in\phi) = i^n J_n(kr),$$

and use has been made of Eq. (C.18) with $n \to -n$. Accordingly

$$E_i = E_0 \sum_n i^n J_n(kr) \exp(in\phi).$$

We use the resonant set (9.39a) for matching each harmonic of the incident field to the corresponding harmonic of the resonant field in such a way as to have a zero tangential electric field over the cylinder surface:

$$a_n H_n^{(2)}(ka) + E_0 i^n J_n(ka) = 0,$$

with the scattered field given by

$$E_z(r, \phi) = -E_0 \sum_n i^n \exp(in\phi) \frac{J_n(ka)}{H_n^{(2)}(ka)} H_n^{(2)}(kr). \tag{9.40a}$$

When $ka \ll 1$, i.e., the cylinder radius is small compared to the incident wavelength, the series expansions (D.17) and (D.18) for $J_n(ka)$ and $H_n^{(2)}(ka)$ are allowed. The dominant term is the $n = 0$ term and the scattering cross section[12] [m] per unit cylinder length is given by

$$\sigma_0(\phi) = \frac{2\pi r |E_0|^2 |J_0(ka)/H_0^{(2)}(ka)|^2 |H_0^{(2)}(kr \to \infty)|^2}{|E_0|^2}$$

$$\approx \frac{\pi^2 a}{ka[\ln(0,89ka)]^2},$$

which may be significant even for very small values of ka.

---

12. Equation (9.31) is suitable for this two-dimensional case.

As ka is increased,

$$\left|\frac{J_0(ka)}{H_0^{(2)}(ka)}\right|^2 = \frac{1}{1 + [Y_0(ka)/J_0(ka)]^2}$$

attains its maximum value when $Y_0(ka) = 0$, i.e., $ka \approx 0.9$. When $ka = 0.9$, the subsequent series terms may still be neglected and the scattering cross section is equal to

$$\sigma_0 \approx \frac{4}{k} = \frac{4a}{0.9} = 4.44a, \qquad 2\pi a = 0.9\lambda,$$

which is 2.2 times larger than the cylinder backscattering cross section in the high-frequency regime (see Section 5.3.2).

In the case of perpendicular polarization (incident magnetic field parallel to the cylinder axis) we have

$$\mathbf{H}_i(r, \phi) = H_0 \hat{\mathbf{z}} \exp(ikr \cos \phi),$$

and the boundary conditions on the cylinder surface require that

$$b_n [H_n^{(2)}(ka)]' + i^n [J_n(ka)]' = 0,$$

$$H_z(r, \theta) = -H_0 \sum_n i^n \exp(in\phi) \frac{[J_n(ka)]'}{[H_n^{(2)}(ka)]'} H_n^{(2)}(ka). \qquad (9.40b)$$

As in the case of parallel polarization, Eq. (9.40b) has been obtained by computing the modal expansion of the ϕ-component of the incident and scattered field, and by matching each mode to obtain a zero tangential electric field over the cylinder surface. Again, the dominant contribution for $ka \ll 1$ is provided by the $n = 0$ term to yield

$$\sigma_0 = \frac{\pi^2}{4}(ka)^3 a.$$

with the aid of Eqs. (D.7) with (D.17b) and (D.18b). The cylinder becomes rapidly *invisible* to the incident radiation as ka is reduced.

When ka is increased, the $n = 0$ coefficient of the series (9.40b) attains its first maximum when $Y_1(ka) = 0$, namely $ka \approx 2.2$, a condition that does not allow one to neglect the subsequent terms of series (9.40b) in computing the scattering cross section. Note that this maximum occurs at a frequency larger than the corresponding one of the parallel polarization case.

ENGINEERING TOPICS: SCATTERING

TABLE 9.1. The First Legendre Polynomials

n	$P_n(x)$	$P_n(\cos\theta)$
0	1	1
1	x	$\cos\theta$
2	$\frac{1}{2}(3x^2 - 1)$	$\frac{1}{4}(3\cos 2\theta + 1)$
3	$\frac{1}{2}(5x^3 - 3x)$	$\frac{1}{8}(5\cos 3\theta + 3\cos\theta)$
4	$\frac{1}{8}(35x^4 - 30x^2 + 3)$	$\frac{1}{64}(35\cos 4\theta + 20\cos 2\theta + 9)$

9.2.2. Spherical Coordinates

Let us consider ϕ-independent resonant fields outside a metal sphere of radius a. We refer to a spherical coordinate system (r, θ, ϕ) centered on the sphere. A complete (orthogonal) expansion set on any coordinate surface $r = $ const is provided by the Legendre[13] polynomials (see Table 9.1), which are the solution of the differential equation

$$\frac{1}{\sin\theta}\frac{d}{d\theta}\sin\theta\frac{dP_n}{d\theta} + n(n+1)P_n = 0 \tag{9.41}$$

[see Eq. (D.23b)]. We specify Maxwell equations (4.46a–b) in spherical coordinates with $\partial/\partial\phi \to 0$:

$$\frac{1}{r\sin\theta}\frac{\partial}{\partial\theta}(\sin\theta E_\phi) = -i\omega\mu H_r, \tag{9.42a}$$

$$-\frac{1}{r}\frac{\partial}{\partial r}rE_\phi = -i\omega\mu H_\theta, \tag{9.42b}$$

$$-\frac{1}{r}\frac{\partial E_r}{\partial\theta} + \frac{1}{r}\frac{\partial}{\partial r}(rE_\theta) = -i\omega\mu H_\phi, \tag{9.42c}$$

$$\frac{1}{r\sin\theta}\frac{\partial}{\partial\theta}(\sin\theta H_\phi) = i\omega\varepsilon E_r, \tag{9.42d}$$

$$-\frac{1}{r}\frac{\partial}{\partial r}rH_\phi = i\omega\varepsilon E_\theta, \tag{9.42e}$$

$$-\frac{1}{r}\frac{\partial H_r}{\partial\theta} + \frac{1}{r}\frac{\partial}{\partial r}(rH_\theta) = i\omega\varepsilon E_\phi. \tag{9.42f}$$

---

13. Adrien-Marie Legendre, Paris (France), 1752–Autenil, 1833.

The θ-dependence of the field components is factorized if

$$E_r = P_n(\theta)e_r(r), \qquad E_\theta = \frac{dP_n}{d\theta} e_\theta(r), \qquad E_\phi = \frac{dP_n}{d\theta} e_\phi(r), \qquad (9.43a)$$

$$H_r = P_n(\theta)h_r, \qquad H_\theta = \frac{dP_n}{d\theta} h_\theta(r), \qquad H_\phi = \frac{dP_n}{d\theta} h_\phi(r), \qquad (9.43b)$$

and Eqs. (9.42) transform for each resonant mode as follows:

$$\begin{cases} -\dfrac{n(n+1)}{r} e_\phi = -i\omega\mu h_r, & (9.44a) \\[4pt] -\dfrac{1}{r}\dfrac{d}{dr}(re_\phi) = -i\omega\mu h_\theta^-, & (9.44b) \\[4pt] -\dfrac{1}{r} e_r + \dfrac{1}{r}\dfrac{d}{dr}(re_\theta) = -i\omega\mu h_\phi, & (9.44c) \\[4pt] -\dfrac{n(n+1)}{r} h_\phi = i\omega\varepsilon e_r, & (9.44d) \\[4pt] -\dfrac{1}{r}\dfrac{d}{dr}(rh_\phi) = i\omega\varepsilon e_\theta, & (9.44e) \\[4pt] -\dfrac{1}{r} h_r + \dfrac{1}{r}\dfrac{d}{dr}(rh_\theta) = i\omega\varepsilon e_\phi, & (9.44f) \end{cases}$$

where Eq. (9.41) has been employed to derive Eqs. (9.44a) and (9.44d). Again, we obtain a field expansion which is orthogonal on any coordinate surface $r = \text{const}$.

The complete solution to Eqs. (9.44) is obtained as a superposition of the $e_r = 0$ and $h_r = 0$ modes, i.e., modes TE and TM with respect to the radial direction r. In the latter case Eqs. (9.44a–b) yield

$$e_\phi = h_\theta = 0,$$

and

$$i\omega\varepsilon \frac{d}{dr} re_\theta = k^2 rh_\phi - \frac{n(n+1)}{r} h_\phi \qquad (9.45)$$

ENGINEERING TOPICS: SCATTERING

TABLE 9.2. Spherical Hankel Functions of the Second Kind

n	$h_n^{(2)}(x)$
0	$\dfrac{i}{x}\exp(-ix)$
1	$\left(\dfrac{i}{x^2} - \dfrac{1}{x}\right)\exp(-ix)$
2	$\left(\dfrac{6i}{x^3} - \dfrac{3}{x^2} - \dfrac{i}{x}\right)\exp(-ix)$
3	$\left(\dfrac{90i}{x^4} - \dfrac{30}{x^3} - \dfrac{6i}{x^2} + \dfrac{i}{x}\right)\exp(-ix)$

from Eqs. (9.44c–d). If we substitute Eq. (9.44e) into Eq. (9.45), we finally derive

$$\frac{1}{r}\frac{d^2}{dr^2}(rh_\phi) + \left[k^2 - \frac{n(n+1)}{r^2}\right]h_\phi = 0, \qquad (9.46)$$

which is the *spherical Bessel equation* whose solutions satisfying the radiation condition at infinity are spherical Hankel functions of the second kind, the first of which are reported in Table 9.2.

As an example, consider the spherical antenna excited by the voltage V_0 across an equatorial gap as depicted in Fig. 9.10. The scattered field is ϕ-

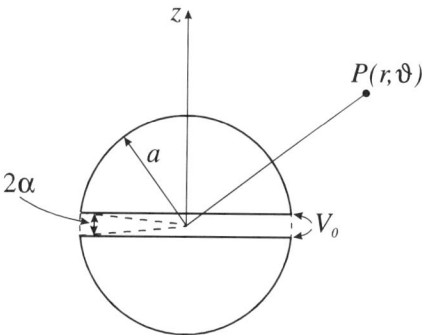

FIGURE 9.10. The spherical antenna.

independent and is represented by superposition of the resonant modes:

$$\begin{cases} E_r = i\dfrac{\zeta}{kr}\sum_{0}^{n} b_n P_n(\theta) n(n+1) h_n^{(2)}(kr), & (9.47a) \\[2mm] E_\theta = i\dfrac{\zeta}{kr}\sum_{0}^{n} b_n \dfrac{dP_n}{d\theta}[krh_n^{(2)}(kr)]', & (9.47b) \\[2mm] H_\phi = \sum_{0}^{n} b_n \dfrac{dP_n}{d\theta} h_n^{(2)}(kr). & (9.47c) \end{cases}$$

The electric field applied across the gap,

$$\mathbf{E}_0 = E_0(\theta)\hat{\boldsymbol{\theta}}, \quad \text{with } -a\int_0^\pi d\theta E_0(\theta) = V_0,$$

is expanded in terms of $dP_n/d\theta$:

$$E_0(\theta) = \sum_{1}^{n} c_n \frac{dP_n}{d\theta}, \qquad (9.48)$$

where

$$c_n = \frac{2n+1}{2n(n+1)}\int_0^\pi d\theta E_0(\theta)\frac{dP_n}{d\theta}\sin\theta = \frac{2n+1}{2n(n+1)}\frac{V_0}{a}v_n$$

with

$$v_n = -\frac{\int_0^\pi d\theta E_0(\theta)(dP_n/d\theta)\sin\theta}{\int_0^\pi d\theta E_0(\theta)},$$

and use has been made of Eq. (C.23). We now set $E_\theta(a,\theta) = E_0(\theta)$, equating Eqs. (9.47b) and (9.48) term by term. Hence

$$b_n = -ika\frac{2n+1}{2n(n+1)}\frac{V_0/\zeta}{a}\frac{v_n}{[kah_n^{(2)}(ka)]'}. \qquad (9.49)$$

We readily obtain a modal expansion for the antenna input admittance,

$$Y = \frac{2\pi a H_\phi(a,\pi/2)}{V_0}, \qquad (9.50)$$

ENGINEERING TOPICS: SCATTERING

when Eq. (9.49) is substituted into Eqs. (9.47c), and for its effective length $\mathbf{l}(\theta)$ by means of the expression

$$E_\theta(r \to \infty, \theta)\hat{\boldsymbol{\theta}} = i\zeta \frac{YV_0 k^2 \mathbf{l}(\theta)}{4\pi kr} \exp(-ikr) \qquad (9.51)$$

[see Eq. (8.4)].

Examination of Eq. (9.48) shows that only odd values of n provide modal voltages $v_n \neq 0$, because $dP_{2n}/d\theta$ is odd with respect to $\theta = \pi/2$ (see Table 9.1). Accordingly, the series summation in Eqs. (9.47) can be extended to odd values of n only.

When $n\alpha \ll 1$, we can take $dP_n/d\theta$ constant within the gap region (see Table 9.1 again) and Eq. (9.48) becomes

$$v_n \approx \left.\frac{dP_n}{d(\cos\theta)}\right|_0. \qquad (9.52)$$

This approximation is no longer valid when $n\alpha \geq 1$, i.e., for large values of n.

Computation of the input admittance of the antenna via its expansion (9.50) is not easy, because the series does not converge rapidly from one side, and also because we cannot use the simple expression (9.52) for v_n over the whole range of n values. We note that in the static limit $ka \to 0$ the series expansion of the spherical Bessel functions can be employed [see Eqs. (D.4c) and (D.18b)] to yield

$$\frac{h_n^{(2)}(ka)}{[kah_n^{(2)}(ka)]'} = \frac{H_n^{(2)}(ka)}{\frac{1}{2}H_n^{(2)}(ka) + (ka)H_n^{(2)'}(ka)} \to -\frac{1}{n},$$

a result which remains valid for any $n \gg ka$. Substitution into Eq. (9.50) leads to

$$Y_s = i\omega\varepsilon a 2\pi \sum_1^\infty \frac{4n-1}{4n(2n-1)^2} v_{2n-1} \left.\frac{dP_{2n-1}}{d\theta}\right|_{\pi/2}$$

$$= i\omega C_s, \qquad (9.53a)$$

which provides the static capacitance C_s of the gap of the spherical antenna.

We now subtract the static term, Eq. (9.53a), from Eq. (9.50), which amounts to subtracting from the series its asymptotic expression for large values of the summation index. This procedure is usually referred to as *renormalization*. It improves the series convergence and provides in our case

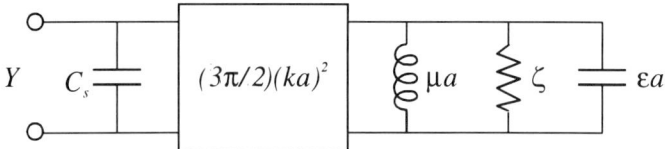

FIGURE 9.11. Equivalent circuit for the input admittance of a spherical antenna near the first resonance $(ka)^2 = 1$.

the *dynamic* part of the antenna input admittance:

$$Y_d = Y - Y_s$$

$$= \frac{ika}{\zeta} 2\pi \sum_1^n \frac{4n-1}{4n(2n-1)} \left[\frac{dP_{2n-1}}{d\theta}\right] v_{2n-1} \cdot \left[\frac{h_{2n-1}^{(2)}(ka)}{[kah_{2n-1}^{(2)}(ka)]'} + \frac{1}{(2n-1)}\right]. \quad (9.54)$$

The series (9.54) is now rapidly convergent and can be truncated to its first terms, depending on the values of ka. For this reason the approximation specified by Eq. (9.52) can be tolerated and we obtain

$$Y_d = -\frac{ika}{\zeta} 2\pi \sum_1^n \frac{4n-1}{4n(2n-1)^2} \left[\frac{dP_{2n-1}}{d\theta}\right]^2 \cdot \left[\frac{h_{2n-1}^{(2)}(ka)}{[kah_{2n-1}^{(2)}(ka)]'} + \frac{1}{(2n-1)}\right]. \quad (9.53b)$$

The first term $n = 1$ leads to the value

$$Y_d^{(1)} = \frac{(3\pi/2)(ka)^2}{\zeta + i\omega\mu a + (1/i\omega\varepsilon a)}$$

for the *dynamic* part of the antenna input admittance which resonates for $(ka)^2 = 1$. At the neighbors of this frequency the input admittance of the spherical antenna is properly described by the equivalent circuit of Fig. 9.11: an admittance inverter loaded by a resonant parallel circuit, in shunt connection with the static capacitance of the gap. At resonance, the effective length of the antenna is given by

$$\mathbf{l}_1(\theta) = 2a \exp(ika)\hat{\boldsymbol{\theta}}.$$

9.3. EXTERIOR SCATTERING VIA ASYMPTOTIC TECHNIQUES

High-frequency scattering is defined as reradiation by bodies whose typical dimension is large compared to the incident wavelength. In this situation, the modal expansion techniques considered in Sections 9.1 and 9.2 are of limited interest, if any, because the number of series terms to be retained for an accurate description of the scattered field is exceedingly large. For instance, consider the far field reradiated by a circular cylinder excited by a z-polarized plane wave. From Eq. (9.40a) we have

$$E_z(r, \phi) = -E_0 \sqrt{\frac{2}{\pi k r}} \exp\left(-ikr + i\frac{\pi}{4}\right) \sum_0^n (-)^n \exp(in\phi) \frac{J_n(ka)}{H_n^{(2)}(ka)},$$

where use has been made of Eq. (D.21). The amplitude of the series terms begin to steadily decrease only when $n > ka$: if $a = 20\lambda$, $n > 125$. Alternative computational techniques are necessary.

The most natural way to study high-frequency scattering is to employ the results of Chapter 5 by following two possible procedures. One possibility is to model the incident field in terms of a ray congruence and then to *locally* approximate the scattering body by simple geometrical shapes: wedges, cylinders, spheres, etc. The scattered field is modeled again in terms of a ray congruence, related linearly to the incident field by reflection and diffraction matrixes. Examples are given in Sections 5.2.1 and 8.6.3.

A second possibility is to use PO approximation (see Section 8.6) to estimate the surface currents induced on the scattering body,[14] then to compute the field scattered by each elementary surface current and obtain the total scattered field as a superposition integral over the body surface. Asymptotic techniques as described in Sections 3.5.2 and 5.4 can be applied for an efficient evaluation of the resulting integral.

In some cases the surface properties of the scattering body can only be described in a statistical way and average estimates are necessary to characterize the scattered field. An example of such a problem is given below in Sections 9.3.1 and 9.3.2.

9.3.1. Rough Surfaces

Consider a surface $z = z(x, y)$ as shown in Fig. 9.12. In many applications the height profile $z(x, y)$ cannot be described in a deterministic way. This is so in the case of *rough surfaces*, whose departure from a polished shape is provided by surface undulations whose local amplitude and frequency spatial changes cannot be predicted by simple mathematical expressions.[15] In these situations the height profile should be regarded as an element of a statistical ensemble and described by its statistical properties. In the subsequent analysis

---

14. A perfectly conducting body is assumed. Should this not be the case, a more refined procedure is necessary.
15. In some cases a convenient mathematical description of the surface can be accomplished with fractal functions.

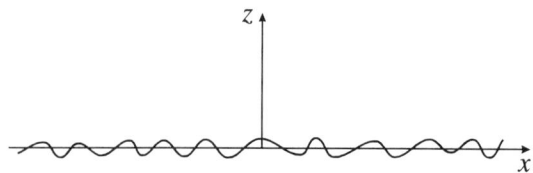

FIGURE 9.12. Geometry of a rough surface.

these are specified by the *self* and *joint probability density functions* (pdf), $p(z)$ and $p(z_1, z_2)$, respectively, which characterize the statistical process (to second order).

The self pdf, $p(z)$, provides an estimate of the amplitude of the surface roughness, where $p(z)\Delta z$ represents the probability that the surface height lies between z and $z + \Delta z$. For a planar surface $p(z) \to \delta(z)$; on the contrary a wide $p(z)$ describes rough surfaces with large amplitude excursions.

The joint pdf, $p(z_1, z_2)$, provides an estimate of the rate of spatial oscillation of the surface roughness, where $p(z_1, z_2)\, \Delta z_1 \Delta z_2$ represents the joint probability that the occurrence of surface height between z_1 and $z_1 + \Delta z_1$ requires also the existence of sites at which the height lies between z_2 and $z_2 + \Delta z_2$. An important parameter, the *correlation coefficient*, is present in the joint pdf, and depends on the coordinates (x_1, y_1) and (x_2, y_2) of z_1 and z_2, often only on the distance between these two points. A large correlation implies bounds on the rate of the change in height, while a small correlation allows more intense (spatial) frequency oscillations.

The reference coordinate system can usually be chosen such that

$$\langle z \rangle = 0, \tag{9.55}$$

and the *random process* z is characterized by zero mean.

A popular self pdf of a zero-mean random process is the gaussian function

$$p(z) = \frac{1}{\sqrt{2\pi}\,\sigma} \exp\left(-\frac{z^2}{2\sigma^2}\right). \tag{9.56}$$

The height σ is the *mean-square deviation*; its squared value σ^2 is called the *variance*.[16] The probability that z lies within $(-\sigma, \sigma)$ is given by

$$\int_{-\sigma}^{\sigma} dz\, p(z) \approx 0.68;$$

the probability increases to ≈ 0.95 for $-2\sigma \leqslant z \leqslant 2\sigma$ and is obviously equal to unity when z spans all possible values between $-\infty$ and $+\infty$. It is clear that

---

16. The variance equals the average of z^2, i.e., $\langle z^2 \rangle = \sigma^2$.

the larger σ, the more pronounced is the change in height of the surface. Also, σ is a sort of boundary between values of $|z|$ with large ($|z| < \sigma$) and small ($|z| > \sigma$) probability of occurrence. For this reason, σ can be identified as the surface *effective amplitude oscillation*.

The gaussian joint pdf is given by

$$p(z_1, z_2) = \frac{1}{2\pi\sigma^2\sqrt{1-C^2}} \exp\left[-\frac{z_1^2 - 2Cz_1z_2 + z_2^2}{2\sigma^2(1-C^2)}\right], \tag{9.57}$$

where $0 \leq C \leq 1$ is the *correlation coefficient*.[17] If $C \to 0$, then

$$p(z_1, z_2) \to \frac{1}{\sqrt{2\pi}\,\sigma} \exp\left(-\frac{z_1^2}{2\sigma^2}\right) \frac{1}{\sqrt{2\pi}\,\sigma} \exp\left(-\frac{z_2^2}{2\sigma^2}\right)$$

$$= p(z_1)p(z_2),$$

and no constraint is imposed on the joint values of z_1 and z_2, which can freely vary from one to another site of the surface, and surface oscillations are most pronounced. On the contrary, when $C \to 1$ it is convenient to express Eq. (9.57) as

$$p(z_1, z_2) \to \frac{1}{\sqrt{2\pi}\,\sigma\sqrt{1-C^2}} \exp\left(-\frac{(z_1-z_2)^2}{2\sigma^2(1-C^2)}\right) \cdot \frac{1}{\sqrt{2\pi}\,\sigma} \exp\left(-\frac{2z_1z_2}{(1+C)2\sigma^2}\right), \tag{9.58}$$

which shows that the probability of height variations larger than $\sigma\sqrt{1-C^2}$ is small, and surface oscillations are limited.

The correlation coefficient C is often a function of the *distance* between any two points on the surface. If $z_1 = z(x_1, y_1)$ and $z_2 = z(x_2, y_2)$, a popular expression for the correlation coefficient is

$$C = \exp\left[-\frac{(x_1 - x_2)^2 + (y_1 - y_2)^2}{2\rho^2}\right], \tag{9.59}$$

for surfaces with isotropic statistical behavior. The parameter ρ is referred to as the *correlation length*, and is an estimate of the space interval within which changes in height are limited. A more precise statement is obtained by

---

17. For a zero-mean random process, the correlation coefficient is the ratio between the correlation function $W = \langle z_1, z_2 \rangle$ and the variance $\langle z_1^2 \rangle$. For the gaussian case $W = \sigma^2 C$.

expanding Eq. (9.59) for $r^2 = (x_1 - x_2)^2 + (y_1 - y_2)^2 < \rho^2$:

$$1 - C^2 \approx r^2/\rho^2, \quad r \leqslant \rho, \tag{9.60}$$

and substituting Eq. (9.60) into Eq. (9.58). This shows that the mean-square deviation of the *rate of change* of the surface undulation, $|z_1 - z_2|/r$, is equal to σ/ρ. Accordingly, an estimate of the surface height change over the distance r is given by

$$|z_1 - z_2| \leqslant \sigma \frac{r}{\rho},$$

which shows that this change is limited when we move over a region whose dimension is smaller than the correlation length.

The *roughness parameter* σ/ρ provides a statistical estimate of the change in slope of the surface and is a quantitative overall index of the statistical behavior of the surface. It is *smooth* or *rough* according to whether the values of σ/ρ are smaller or larger than unity.

9.3.2. Scattering by a Planar Rough Surface

Consider metal plane whose surface is rough; the plane is excited by a current line I at point T, as depicted in Fig. 9.13. We consider a two-dimensional problem: the excitation is y-independent, as well as the statistical parameters of the plane. The latter are described by gaussian pdfs and characterized by an effective amplitude oscillation σ and a correlation length ρ [see Eqs. (9.56) and (9.59)].

Proceeding as in Section 5.4.1, the field scattered at point P is computed as a superposition of the fields radiated by the surface current density elements induced over the surface. The latter are computed in the PO approximation [see Eqs. (8.58)] and are twice as large as the tangent component of the *incident* magnetic field. However, the presence of undulations over the surface requires one to modify the usual computational procedure which is applied to flat surfaces.

First, the two distances, $|TP|$ from the source to the generic scattering point and $|PQ|$ from the latter to the receiving point, must account for the height profile:

$$|TP| = \sqrt{(x_p - x)^2 + (D - z)^2} \approx R - \frac{D}{R}z, \tag{9.61a}$$

$$|PQ| = \sqrt{(x_q - x)^2 + (d - z)^2} \approx r - \frac{d}{r}z, \tag{9.61b}$$

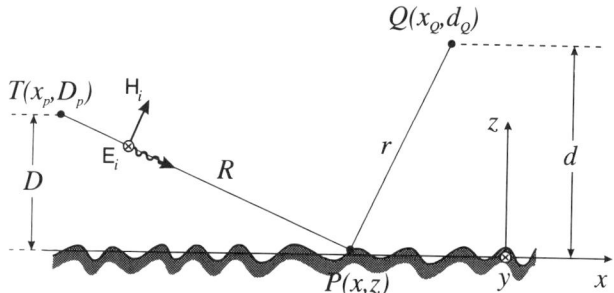

FIGURE 9.13. Scattering from a rough planar surface.

where

$$R(x) = \sqrt{(x_p - x)^2 + D^2} \quad \text{and} \quad r(x) = \sqrt{(x_q - x)^2 + d^2} \quad (9.61c)$$

are the distances from the average (flat) surface. Both the distances D and d are supposed to be large compared to σ.

As a second modification, the component of the incident magnetic field tangent to the surface is given by

$$\frac{E_0}{\zeta}\left(\frac{D}{R}\hat{\mathbf{x}} + \frac{|x_p - x|}{R}\hat{\mathbf{z}}\right) \cdot \frac{\hat{\mathbf{x}} + (dz/dx)\hat{\mathbf{z}}}{\sqrt{1 + (dz/dx)^2}} = \frac{E_0}{\zeta}\left[(D/R) + \frac{(|x_p - x|/R)(dz/dx)}{\sqrt{1 + (dz/dx)^2}}\right],$$

where E_0 is the amplitude of the incident electric field at P and is given by

$$E_0 = -\frac{\zeta}{4}I\sqrt{2ik/\pi R}\exp(-ikR);$$

see Eq. (5.41). The first vector describes the polarization of the incident magnetic field and the second is tangent to the surface $z = z(x)$.

We use superposition and the expression for the curvilinear element of the scattering surface,

$$ds = dx\sqrt{1 + (dz/dx)^2},$$

to obtain an integral expression for the field scattered to the point Q:

$$E = \zeta\frac{I}{2}\frac{ik}{2\pi}\int dx \exp\frac{[-ik(R+r)]}{\sqrt{Rr}}\exp[ikqz]\cdot\left[\frac{D}{R} + \frac{|x_p - x|}{R}\frac{dz}{dx}\right] \quad (9.62)$$

with

$$q(x) = \frac{D}{R(x)} + \frac{d}{r(x)}, \tag{9.63}$$

where the integration is extended to all the scattering surface.

Before proceeding further, a slight modification of the integral (9.62) is in order. We have

$$\exp[ikqz]\frac{dz}{dx} \approx \frac{1}{ikq}\frac{d}{dx}\exp[ikqz],$$

and integration by parts generates two edge contributions that are of no concern here,[18] plus an integral containing the derivative of the phase term,

$$\frac{d(r+R)}{dx},$$

which is zero in the subsequent asymptotic evaluation of the integral. Accordingly

$$E = \zeta \frac{I}{2}\frac{ik}{2\pi}\int dx \frac{D}{R}\frac{\exp[-ik(R+r)]}{\sqrt{Rr}}\exp[ikqz]. \tag{9.64}$$

Note that in the application of the PO approximation we ignored possible shadowed areas of the surface.

The statistical average of the field is obtained by computing the (ensemble) average of the z-dependent factor in Eq. (9.64):

$$\langle\exp(ikqz)\rangle = \int dz \exp(ikqz)p(z) = \exp\left(-\frac{k^2\sigma^2q^2}{2}\right), \tag{9.65}$$

by using Eqs. (9.56) and (C.1).

Two cases are of interest: $k\sigma < 1$ and $k\sigma > 1$. In the former case $(k\sigma)^2 \ll 1$, the mean-square deviation of the surface undulations is small compared to the incident wavelength, and the factor of Eq. (9.65) may be neglected. For a surface of finite extent, assume that the positions of T and Q are such that a specular point exists inside the reflecting surface (see Fig. 9.14). Then, a series

---

18. These edge contributions disappear if the plane is of infinite extent.

ENGINEERING TOPICS: SCATTERING

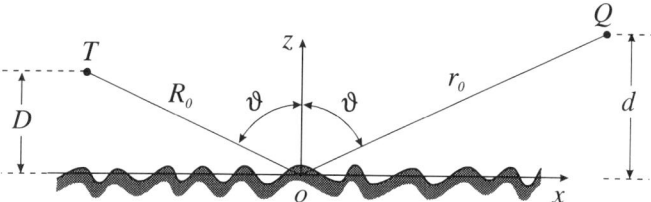

FIGURE 9.14. Scattering from a rough surface (continued).

expansion of $r + R$ around $x = 0$ leads to

$$R + r \approx R_0 + r_0 + \frac{1}{2}\cos^2\theta \left(\frac{1}{R_0} + \frac{1}{r_0}\right)x^2, \tag{9.66}$$

$$R_0 = \sqrt{x_p^2 + D^2}, \quad r_0 = \sqrt{x_q^2 + d^2},$$

as shown in Section 5.4, and so Eq. (9.64) reduces to

$$E \sim \frac{\zeta I}{2}\frac{ik}{2\pi}\frac{\cos\theta}{\sqrt{R_0 r_0}}\exp[-ik(R_0 + r_0)] \cdot \int dx \exp\left[-ik\left(\frac{1}{R_0} + \frac{1}{r_0}\right)\cos^2\theta \frac{x^2}{2}\right], \tag{9.67}$$

where the integration is limited to the extent of the plane.

When the plane is infinite, the integration limits run from $-\infty$ to $+\infty$ in which case

$$E \sim \frac{\zeta I}{2}\sqrt{\frac{ik}{2\pi(R_0 + r)}}\exp[-ik(R_0 + r_0)], \tag{9.68}$$

thus recovering the expression for the field reflected by a smooth plane. Note that Eq. (9.68) is consistent with image theory.

Consider, on the contrary, a plane of extent $2a$ across the reflection point, and values a_n satisfying

$$\frac{2\pi}{\lambda}\frac{R_0 + r_0}{R_0 r_0}\cos^2\theta \frac{a_n^2}{2} = n\pi.$$

For $a \leqslant a_1$ the modulus of Eq. (9.67) is a steadily increasing function of a, as follows from a geometric representation of the intergrand in the phase plane (see Fig. 9.15). The modulus then decreases for $a_1 \leqslant a \leqslant a_2$, increases again for

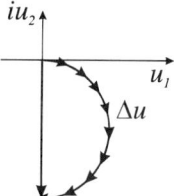

FIGURE 9.15. Geometrical representation of the integral $u = \int_0^\beta dx \exp[-i\alpha x^2]$, $\Delta u = \exp[-i\alpha x^2]\Delta x$. In this case $\alpha\beta^2 = \pi$.

$a_2 \leq a \leq a_3$, and so on, to approach the limit value given by Eq. (9.68). The values of a_n,

$$a_n = \sqrt{n \frac{R_0 r_0 \lambda}{(R_0 + r_0)\cos^2\theta}}, \qquad (9.69)$$

define successive *Fresnel zones*: when

$$a > a_4, \qquad (9.70)$$

then $E \approx -E_0$ with an error of about 10% in the modulus and less than 0.1 rad in the phase.

Let us now turn to the case $(k\sigma)^2 \gg 1$, i.e., the effective amplitude oscillations of the surface are large in terms of the wavelength. The value of Eq. (9.65) then approaches a very small number, and the field statistical average is close to zero. This means that the elementary fields scattered by different regions of the surface tend to cancel each other. The physical reason is the rapid change in the phase of the field scattered by even close points over the rough surface; the coherent addition of these fields is impaired. The statistical counterpart is the significant variation in the field associated with different elements of the ensemble of the scattering surfaces, when we exchange space by ensemble average (the *ergodic assumption*).

A more stable, meaningful quantity is the statistical average of the field density power, proportional to EE^*, which requires computation of the ensemble average

$$\langle \exp(ikq_1 z_1 - ikq_2 z_2) \rangle = \int dz_1 \exp(ikq_1 z_1) \int dz_2 \exp(-ikq_2 z_2) p(z_1, z_2)$$

$$= \exp\left[-\frac{k^2\sigma^2}{2}(q_1^2 + q_2^2 - 2Cq_1 q_2)\right], \qquad (9.71)$$

ENGINEERING TOPICS: SCATTERING

with

$$q_1 = q(x_1) = \frac{D}{R(x_1)} + \frac{d}{r(x_1)} \quad \text{and} \quad q_2 = q(x_2) = \frac{D}{R(x_2)} + \frac{d}{r(x_2)},$$

where we have used Eq. (9.57) and Eq. (C.1) twice. In this assumption $k^2\sigma^2 \gg 1$, Eq. (9.71) differs appreciably from zero only for $|x_1 - x_2| < \rho$, i.e., C is close to unity, and $q_1 \approx q_2$, i.e., we move in the neighborhood of $x = 0$. Accordingly, we set $q_1 = q_2 = q$ and $q = q(0)$, and expand Eq. (9.59) to yield

$$q_1^2 + q_2^2 - 2Cq_1q_2 \approx q^2 \frac{(x_1 - x_2)^2}{\rho^2},$$

so

$$\exp\left[-\frac{k^2\sigma^2}{2}(q_1^2 + q_2^2 - 2Cq_1q_2)\right] \approx \exp\left[-\frac{k^2\sigma^2}{2\rho^2} q^2(x_1 - x_2)^2\right]$$

$$= \exp\left[-\frac{2k^2\sigma^2}{\rho^2} \cos^2\theta (x_1 - x_2)^2\right].$$

Again, the case of the specular reflection point inside the scattering area is of interest. Proceeding as in the smooth-plane case we have

$$\langle EE^* \rangle \sim \frac{|V|^2 k^2 \cos^2\theta}{(2\pi)^2 R_0 r_0} \int dx_1 \int dx_2 \exp\left[-ik \frac{x_1^2 - x_2^2}{2}\left(\frac{1}{R_0} + \frac{1}{r_0}\right)\cos^2\theta\right]$$

$$\cdot \exp\left[-\frac{2k^2\sigma^2}{\rho^2} \cos^2\theta (x_1 - x_2)^2\right], \tag{9.72}$$

where $V = \zeta I/2$. If the rotation of coordinates,

$$\begin{cases} x_1 - x_2 = \sqrt{2}\,\xi, \\ x_1 + x_2 = \sqrt{2}\,\eta, \end{cases}$$

is introduced, then the integral (9.72) transforms as follows:

$$\langle EE^* \rangle = |V|^2 \frac{k^2 \cos^2\theta}{(2\pi)^2 R_0 r_0} \int d\xi \exp\left(-\frac{4k^2\sigma^2}{\rho^2} \cos^2\theta\,\xi^2\right) \cdot \int d\eta \exp\left(-ik\xi \frac{R_0 + r_0}{R_0 r_0} \eta\right).$$

$$\tag{9.73}$$

If the scattering surface is of infinite extent, then

$$\int_{-\infty}^{+\infty} d\eta \exp\left(-ik\xi \cos^2\theta \frac{R_0 + r_0}{R_0 r_0} \eta\right) = 2\pi\delta\left(k\xi \cos^2\theta \frac{R_0 + r_0}{R_0 r_0}\right),$$

in which case

$$\langle EE^* \rangle = \frac{k|V|^2}{2\pi(R_0 + r_0)}, \tag{9.74}$$

which shows that the incident power is uniformly scattered in the half-space $z \geqslant 0$. Note that Eq. (9.74) is consistent with the laws of geometrical optics, which would predict the same density power in the direction of specular reflection.

In many cases the rough surface may extend over a limited width $2a$ across the specular point. In this case the η-integration limits should run from $-(\sqrt{2}a - |\xi|)$ to $(\sqrt{2}a - |\xi|)$ to yield

$$\langle EE^* \rangle = \frac{2|V|^2}{(2\pi)^2} \frac{k^2 \cos^2\theta}{R_0 r_0} \int_{-\sqrt{2}a}^{\sqrt{2}a} d\xi (\sqrt{2}a - |\xi|) \operatorname{sinc}\left[k \cos^2\theta \frac{R_0 + r_0}{R_0 r_0} \xi(\sqrt{2}a - |\xi|)\right]$$
$$\cdot \exp\left(-\frac{4k^2\sigma^2}{\rho^2} \cos^2\theta \xi^2\right).$$

In the case of a rough surface

$$\frac{4\sigma^2 \cos^2\theta}{\rho^2} 2k^2 a^2 \gg 1,$$

and the integral is dominated by values of ξ close to the origin, i.e., $x_1 \approx x_2$. Hence $\sqrt{2}a - |\xi| \approx \sqrt{2}a$, the integration limits can be extended to infinity, and the coordinate change

$$k \cos^2\theta \frac{R_0 + r_0}{R_0 r_0} \sqrt{2}a\xi \to \xi$$

leads to

$$\langle EE^* \rangle = \frac{2|V|^2 k}{(2\pi)^2 (R_0 + r_0)} \int_{-\infty}^{+\infty} d\xi \operatorname{sinc}(\xi) \exp\left[-\frac{1}{2}\left(\frac{2\sigma}{\rho \cos\theta} \frac{R_0 r_0}{R_0 + r_0} \frac{1}{a}\right)^2 \xi^2\right]. \tag{9.75}$$

If

$$a > \frac{R_0 r_0}{R_0 + r_0} \frac{2\sigma}{\rho \cos \theta}, \qquad (9.76)$$

the result provided by Eq. (9.74) is recovered, as follows by approximating the exponential function with unity and evaluating the remaining integral via Eq. (C.24).

Equation (9.76) for the rough plane is analogous to Eq. (9.70) for the smooth plane, and defines the minimal extent of the scattering surface necessary to obtain a scattered power density in the specular direction consistent with the laws of geometrical optics.

In the opposite case,

$$a < \frac{R_0 r_0}{R_0 + r_0} \frac{2\sigma}{\rho \cos \theta},$$

we can approximate to unity the sinc(ξ) function, and derive from Eq. (9.75)

$$\langle EE^* \rangle = \frac{|V|^2 k}{2\pi R_0} \frac{\rho a \cos \theta}{\sqrt{2\pi}\,\sigma r_0} = |E_i|^2 \frac{\rho a \cos \theta}{\sqrt{2\pi}\,\sigma r_0},$$

where E_i is the incident field at $x = 0$.

Equation (9.31), suitably adjusted to this two-dimensional scattering case, yields the forward scattering cross section (per unit length) of the rough surface:

$$\sigma_0 = \sqrt{2\pi} \frac{\rho}{\sigma} a \cos \theta. \qquad (9.77)$$

This is valid for the case of rough surfaces ($\sigma/\rho \gg 1$) and shows that scattering in the forward direction is reduced compared to the smooth-surface case.

It is enlightening to derive the condition provided by Eq. (9.76) from an alternative viewpoint, based on a field rather than a power approach. We resort again to Eq. (9.64) and employ Eqs. (9.66) and (9.63) to obtain the expression for the phase term at the neighbors of $x = 0$:

$$\psi \approx -k\left(R_0 + r_0 + \frac{1}{2}\cos^2\theta \frac{R_0 + r_0}{r_0 R_0} x^2 - 2z \cos \theta\right). \qquad (9.78)$$

The value of the integral (9.64) is determined by the contribution of stationary points,

$$\frac{d\psi}{dx} = -k\left(\cos^2\theta \frac{R_0 + r_0}{R_0 r_0} x - 2\cos\theta \frac{dz}{dx}\right) = 0. \qquad (9.79)$$

These contributions sum incoherently in the rough-surface case, as already noted, so that the field (ensemble) average tends to zero and computation of the power density (ensemble) average is most appropriate. However, Eq. (9.79) is interesting inasmuch as it states an implicit relation between the surface slope and the phase stationary points:

$$x_s = \frac{R_0 r_0}{R_0 + r_0} \frac{2}{\cos\theta} \frac{dz}{dx}\bigg|_{x_s}. \qquad (9.80)$$

The derivative of a gaussian process of variance σ^2 is still gaussian with variance[19] σ^2/ρ^2. If the statistical estimate of the surface slope is taken to have the value

$$\left|\frac{dz}{dx}\right| \approx \frac{\sigma}{\rho},$$

then Eq. (9.80) provides an estimate of the extent around the specular point which includes the relevant stationary points, and we recover condition (9.76). Accordingly, this is the area which is responsible (and necessary) for generating the same scattered field as in the case of a surface of infinite extent. As this region includes all the relevant stationary points, it can be called *Fermat's region*.

9.4. SUMMARY AND SELECTED REFERENCES

This chapter presents both internal and external scattering problems: the former (Section 9.1) consist of interior resonance, the latter (Section 9.2) of scattering in open regions.

Interior resonance, i.e., cavities, are studied systematically by the application of Green's vector identities. Their properties are derived in Section 9.1, introducing such engineering parameters as quality factor, equivalent circuit, and bandwidth. Application is made to parallelepiped and multisection coaxial cavities (Sections 9.1.1 and 9.1.2), as well as to multimode (Section 9.1.3) and open [9.1] cavities.

---

19. The variance of the derivative of the gaussian process is given by minus the second derivative of the correlation function $W(x_1 - x_2) = \sigma^2 C(x_1 - x_2)$ at the origin.

Exterior resonance (Section 9.2) leads to the eigenfield expansion [9.2]: plane (already considered in Section 3.5), cylindrical (Section 9.2.1), and spherical [9.3] (Section 9.2.2) waves are presented. Application is made to the computation of the scattering cross section of metal cylinders and of the equivalent circuit of a spherical antenna [9.4].

Scattering can be conveniently studied by employing asymptotic techniques (see Chapter 5) when the considered objects are large compared to the wavelength of the incident radiation. Systematic application is made to rough surfaces (Section 9.3.1) [9.5–9.7] whose statistical parameters are discussed in detail [9.8]. Application is also made to scattering from a planar rough surface (Section 9.3.2), presenting an appealing physical model of the phenomenon [9.9] and computing the forward scattering cross section [9.10] of the rough surface. This is important in remote-sensing applications [9.11].

References

[9.1] G. D. Boyd and J. P. Gordon, "Confocal multimode resonators for millimeter through optical wavelength maser," *Bell System Tech. J.* **40**, 489–508 (1961).

[9.2] J. J. Bowman, T. B. A. Senior, and P. L. E. Uslenghi, *Electromagnetic and Acoustic Scattering by Simple Shapes*, North-Holland, Amsterdam (1969).

[9.3] N. Marcuvitz, "Field representation in spherically stratified regions," *Comm. Pure Appl. Math.* **4**, 263–315 (1951).

[9.4] G. Franceschetti, "A canonical problem in transient radiation. The spherical antenna," *IEEE Trans. Antennas Propagat.*, **AP-26**, 551–555 (1978).

[9.5] P. Beckman and A. Spizzichino, *The Scattering of Electromagnetic Waves by Rough Surfaces*, Artech House, Norwood, Mass. (1987).

[9.6] A. Ishimaru, *Wave Propagation and Scattering in Random Media*, Academic Press, New York (1978).

[9.7] D. E. Barrik and W. H. Peake, "A review of scattering from surfaces with different roughness scale," *Radio Sci.* **3**, 365–368 (1968).

[9.8] W. B. Davenport and W. L. Root, *An Introduction to the Theory of Random Signals and Noise*, McGraw-Hill, New York (1958).

[9.9] R. B. Dutt, G. Franceschetti, N. Naraghi, and J. Tatoian, "Image theory for a rough surface," *J. Electromagnetic Waves and Appl.* **8**, 961–972 (1994).

[9.10] G. T. Ruck, D. E. Barrik, W. D. Stuart, and C. K. Krickbaum, *Radar Cross-Section Handbook*, Vols. 1 and 2, McGraw-Hill, New York (1970).

[9.11] C. Elachi, *Introduction to the Physics and Techniques of Remote Sensing*, Wiley, New York (1987).

Appendixes

APPENDIX A. VECTOR ANALYSIS

In the following $\mathbf{A}$ and $\mathbf{B}$ are vector functions, while Φ and Ψ are scalar functions.

A.1. Vector Multiplication

$$\mathbf{A} \cdot (\mathbf{B} \times \mathbf{C}) = \mathbf{B} \cdot (\mathbf{C} \times \mathbf{A}) = \mathbf{C} \cdot (\mathbf{A} \times \mathbf{B}), \tag{A.1}$$

$$\mathbf{A} \times (\mathbf{B} \times \mathbf{C}) = \mathbf{B}(\mathbf{A} \cdot \mathbf{C}) - \mathbf{C}(\mathbf{A} \cdot \mathbf{B}), \tag{A.2}$$

$$\mathbf{A} \times (\mathbf{B} \times \mathbf{C}) - \mathbf{C} \times (\mathbf{B} \times \mathbf{A}) = \mathbf{B} \times (\mathbf{A} \times \mathbf{C}), \tag{A.3}$$

$$(\mathbf{A} \times \mathbf{B}) \cdot (\mathbf{C} \times \mathbf{D}) = (\mathbf{A} \cdot \mathbf{C})(\mathbf{B} \cdot \mathbf{D}) - (\mathbf{A} \cdot \mathbf{D})(\mathbf{B} \cdot \mathbf{C}). \tag{A.4}$$

A.2. Differential Relationships

$$\nabla(\Phi + \Psi) = \nabla\Phi + \nabla\Psi, \tag{A.5}$$

$$\nabla \cdot (\mathbf{A} + \mathbf{B}) = \nabla \cdot \mathbf{A} + \nabla \cdot \mathbf{B}, \tag{A.6}$$

$$\nabla \times (\mathbf{A} + \mathbf{B}) = \nabla \times \mathbf{A} + \nabla \times \mathbf{B}, \tag{A.7}$$

$$\nabla(\Phi\Psi) = \Phi\nabla\Psi + \Psi\nabla\Phi, \tag{A.8}$$

$$\nabla(\mathbf{A} \cdot \mathbf{B}) = \mathbf{A} \times \nabla \times \mathbf{B} + \mathbf{B} \times \nabla \times \mathbf{A} + (\mathbf{B} \cdot \nabla)\mathbf{A} + (\mathbf{A} \cdot \nabla)\mathbf{B}, \tag{A.9}$$

$$\nabla \cdot (\Phi \mathbf{A}) = \Phi \nabla \cdot \mathbf{A} + \nabla\Phi \cdot \mathbf{A}, \tag{A.10}$$

$$\nabla \cdot (\mathbf{A} \times \mathbf{B}) = \mathbf{B} \cdot \nabla \times \mathbf{A} - \mathbf{A} \cdot \nabla \times \mathbf{B}, \tag{A.11}$$

$$\nabla \times (\Phi \mathbf{A}) = \nabla\Phi \times \mathbf{A} + \Phi \nabla \times \mathbf{A}, \tag{A.12}$$

$$\nabla \times (\mathbf{A} \times \mathbf{B}) = \mathbf{A}\nabla \cdot \mathbf{B} - \mathbf{B}\nabla \cdot \mathbf{A} + (\mathbf{B} \cdot \nabla)\mathbf{A} - (\mathbf{A} \cdot \nabla)\mathbf{B}, \tag{A.13}$$

$$\nabla \cdot \nabla\Phi = \nabla^2\Phi, \tag{A.14}$$

$$\nabla \times \nabla \Phi = 0, \tag{A.15}$$

$$\nabla \cdot \nabla \times \mathbf{A} = 0, \tag{A.16}$$

$$\nabla \times \nabla \times \mathbf{A} = \nabla \nabla \cdot \mathbf{A} - \nabla^2 \mathbf{A}, \tag{A.17}$$

$$\nabla f(\Phi) = f'(\Phi) \nabla \Phi, \quad f'(\Phi) = df/d\Phi, \tag{A.18}$$

$$\nabla^2 (\Phi \Psi) = \Phi \nabla^2 \Psi + 2 \nabla \Phi \cdot \nabla \Psi + \Phi \nabla^2 \Psi, \tag{A.19}$$

$$\nabla^2 (\Phi \mathbf{A}) = \Phi \nabla^2 \mathbf{A} + \mathbf{A} \nabla^2 \Phi + 2(\nabla \Phi \cdot \nabla) \mathbf{A}, \tag{A.20}$$

$$\nabla \nabla \cdot (\Phi \mathbf{A}) = (\nabla \Phi) \nabla \cdot \mathbf{A} + \Phi \nabla \nabla \cdot \mathbf{A} + \nabla \Phi \times \nabla \times \mathbf{A}$$
$$+ (\mathbf{A} \cdot \nabla) \nabla \Phi + (\nabla \Phi \cdot \nabla) \mathbf{A}, \tag{A.21}$$

$$\nabla \times \nabla \times (\Phi \mathbf{A}) = \nabla \Phi \times \nabla \times \mathbf{A} - A \nabla^2 \Phi + (\mathbf{A} \cdot \nabla) \nabla \Phi + \Phi \nabla \times \nabla \times \mathbf{A}$$
$$+ \nabla \Phi \nabla \cdot \mathbf{A} - (\nabla \Phi \cdot \nabla) \mathbf{A}. \tag{A.22}$$

A.3. Integral Relationships

Volumes V, surfaces S, lines C, and unit vectors $\hat{\mathbf{n}}$ and $\hat{\mathbf{c}}$ are defined in Fig. A.1. All surfaces and curves are sufficiently regular that tangent and normal vectors can be defined unambiguously.

$$\iint_S dS \hat{\mathbf{n}} \cdot \nabla \times \mathbf{A} = \int_C dc A \cdot \hat{\mathbf{c}}, \quad \text{(Stokes theorem)} \tag{A.23}$$

$$\iiint_V dV \nabla \cdot \mathbf{A} = \oiint_S dS \mathbf{A} \cdot \hat{\mathbf{n}}, \quad \text{(Gauss theorem)} \tag{A.24}$$

$$\oiint_S dS \Phi \frac{\partial \Psi}{\partial n} = \iiint_V dV (\Phi \nabla^2 \Psi + \nabla \Phi \cdot \nabla \Psi),$$

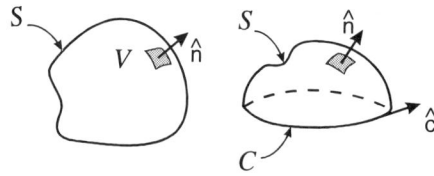

FIGURE A.1. Volumes, surfaces, and lines of integration.

$$\oiint_S dS f\Phi \frac{\partial \Psi}{\partial n} = \iiint_V dV[\Phi\nabla\cdot(f\nabla\psi) + f\nabla\Phi\cdot\nabla\Psi],$$

(1st scalar Green's identity and its generalization) (A.25)

$$\oiint_S dS\left(\Psi\frac{\partial\Phi}{\partial n} - \Phi\frac{\partial\Psi}{\partial n}\right) = \iiint_V dV(\Phi\nabla^2\Psi - \Psi\nabla^2\Phi),$$

$$\oiint_S dSf\left(\Psi\frac{\partial\Phi}{\partial n} - \Phi\frac{\cup\partial\Psi}{\partial \pi}\right) = \iiint_V dV[\Phi\nabla\cdot(f\nabla\Psi) - \Psi\nabla\cdot(f\nabla\phi)],$$

(2nd scalar Green's identity and its generalization) (A.26)

$$\oiint_S dS\hat{n}\cdot(\mathbf{A}\times\nabla\times\mathbf{B}) = \iiint_V dV(\nabla\times\mathbf{A}\cdot\nabla\times\mathbf{B} - \mathbf{A}\cdot\nabla\times\nabla\times\mathbf{B}),$$

$$\oiint_S dS\hat{n}\cdot(\mathbf{A}\times\nabla\times\mathbf{B})f = \iiint_V dV[f\nabla\times\mathbf{A}\cdot\nabla\times\mathbf{B}$$
$$- \mathbf{A}\cdot\nabla\times(f\nabla\times\mathbf{B})],$$

(1st vector Green's identity and its generalization) (A.27)

$$\oiint_S dS\hat{n}\cdot(\mathbf{A}\times\nabla\times\mathbf{B} - \mathbf{B}\times\nabla\times\mathbf{A}) = \iiint_V dV(\mathbf{B}\cdot\nabla\times\nabla\times\mathbf{A} - \mathbf{A}\cdot\nabla\times\nabla\times\mathbf{B}),$$

$$\oiint_S dS\hat{n}\cdot(\mathbf{A}\times\nabla\times\mathbf{B} - \mathbf{B}\times\nabla\times\mathbf{A})f = \iiint_V dV[\mathbf{B}\cdot\nabla\times(f\nabla\times\mathbf{A})$$
$$- \mathbf{A}\cdot\nabla\times(f\nabla\times\mathbf{B})],$$

(2nd vector Green's identity and its generalization (A.28)

$$\iiint_V dV\nabla\Phi = \oiint_S dS\Phi\hat{n}, \qquad (A.29)$$

$$\iiint_V dV \nabla \times \mathbf{A} = \oiint_S dS \hat{\mathbf{n}} \times \mathbf{A}, \qquad (A.30)$$

$$\iint_S dS \hat{\mathbf{n}} \times \nabla \Phi = \oint_C dc \Phi \hat{\mathbf{c}}. \qquad (A.31)$$

A.4. Cartesian Coordinates

Cartesian coordinates are depicted in Fig. A.2.

$$\nabla \Phi = \frac{\partial \Phi}{\partial x} \hat{\mathbf{x}} + \frac{\partial \Phi}{\partial y} \hat{\mathbf{y}} + \frac{\partial \Phi}{\partial z} \hat{\mathbf{z}}, \qquad (A.32)$$

$$\nabla \cdot \mathbf{A} = \frac{\partial A_x}{\partial x} + \frac{\partial A_y}{\partial z} + \frac{\partial A_z}{\partial z}, \qquad (A.33)$$

$$\nabla \times \mathbf{A} = \left(\frac{\partial A_z}{\partial y} - \frac{\partial A_y}{\partial z} \right) \hat{\mathbf{x}} + \left(\frac{\partial A_x}{\partial z} - \frac{\partial A_z}{\partial x} \right) \hat{\mathbf{y}} + \left(\frac{\partial A_y}{\partial x} - \frac{\partial A_x}{\partial y} \right) \hat{\mathbf{z}}, \qquad (A.34)$$

$$\nabla^2 \Phi = \frac{\partial^2 \Phi}{\partial x^2} + \frac{\partial^2 \Phi}{\partial y^2} + \frac{\partial^2 \Phi}{\partial z^2}, \qquad (A.35)$$

$$\nabla^2 \mathbf{A} = (\nabla^2 A_x) \hat{\mathbf{x}} + (\nabla^2 A_y) \hat{\mathbf{y}} + (\nabla^2 A_z) \hat{\mathbf{z}}. \qquad (A.36)$$

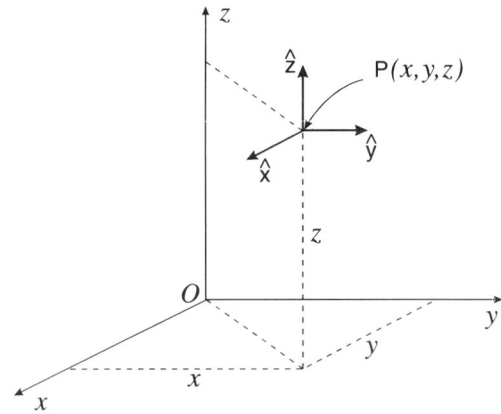

FIGURE A.2. Cartesian coordinates.

A.5. Cylindrical Coordinates

Cylindrical coordinates are depicted in Fig. A.3.

$$\begin{cases} \hat{\mathbf{r}} = \hat{\mathbf{x}}\cos\phi + \hat{\mathbf{y}}\sin\phi, \\ \hat{\mathbf{y}} = \hat{\mathbf{x}}\sin\phi + \hat{\mathbf{y}}\cos\phi, \\ \hat{\mathbf{z}} = \hat{\mathbf{z}}; \end{cases} \quad (A.37)$$

$$\begin{cases} \hat{\mathbf{x}} = \hat{\mathbf{r}}\cos\phi - \hat{\boldsymbol{\phi}}\sin\phi, \\ \hat{\mathbf{y}} = \hat{\mathbf{r}}\sin\phi + \hat{\boldsymbol{\phi}}\cos\phi, \\ \hat{\mathbf{z}} = \hat{\mathbf{z}}; \end{cases} \quad (A.38)$$

$$\nabla\Phi = \frac{\partial\Phi}{\partial r}\hat{\mathbf{r}} + \frac{1}{r}\frac{\partial\Phi}{\partial\phi}\hat{\boldsymbol{\phi}} + \frac{\partial\Phi}{\partial z}\hat{\mathbf{z}}, \quad (A.39)$$

$$\nabla\cdot\mathbf{A} = \frac{1}{r}\frac{\partial}{\partial r}(rA_r) + \frac{1}{r}\frac{\partial A_\phi}{\partial\phi} + \frac{\partial A_z}{\partial z}, \quad (A.40)$$

$$\nabla\times\mathbf{A} = \left(\frac{1}{r}\frac{\partial A_z}{\partial\phi} - \frac{\partial A_\phi}{\partial z}\right)\hat{\mathbf{r}} + \left(\frac{\partial A_r}{\partial z} - \frac{\partial A_z}{\partial r}\right)\hat{\boldsymbol{\phi}} + \frac{1}{r}\left[\frac{\partial}{\partial r}(rA_\phi) - \frac{\partial A_r}{\partial\phi}\right]\hat{\mathbf{z}}, \quad (A.41)$$

$$\nabla^2\Phi = \frac{1}{r}\frac{\partial}{\partial r}\left(r\frac{\partial\Phi}{\partial r}\right) + \frac{1}{r^2}\frac{\partial^2\Phi}{\partial\phi^2} + \frac{\partial^2\Phi}{\partial z^2}, \quad (A.42)$$

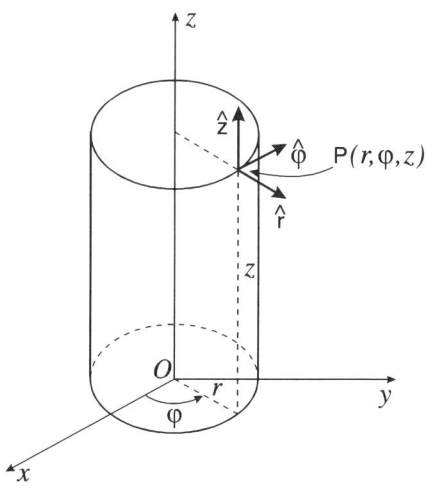

FIGURE A.3. Cylindrical coordinates.

$$\nabla^2 \mathbf{A} = \left(\nabla^2 A_r - \frac{A_r}{r^2} - \frac{2}{r^2} \frac{\partial A_\phi}{\partial \phi} \right) \hat{\mathbf{r}} + \left(\nabla^2 A_\phi - \frac{A_\phi}{r^2} + \frac{2}{r^2} \frac{\partial A_r}{\partial \phi} \right) \hat{\boldsymbol{\phi}} + (\nabla^2 A_z) \cdot \hat{\mathbf{z}}.$$
(A.43)

A.6. Spherical Coordinates

Spherical coordinates are depicted in Fig. A.4.

$$\begin{cases} \hat{\mathbf{r}} = \hat{\mathbf{x}} \sin\theta \cos\phi + \hat{\mathbf{y}} \sin\theta \sin\phi + \hat{\mathbf{z}} \cos\theta, \\ \hat{\boldsymbol{\theta}} = \hat{\mathbf{x}} \cos\theta \cos\phi + \hat{\mathbf{y}} \cos\theta \sin\phi - \hat{\mathbf{z}} \sin\theta, \\ \hat{\boldsymbol{\phi}} = \hat{\mathbf{x}} \sin\phi + \hat{\mathbf{y}} \cos\phi; \end{cases}$$
(A.44)

$$\begin{cases} \hat{\mathbf{x}} = \hat{\mathbf{r}} \sin\theta \cos\phi + \hat{\boldsymbol{\theta}} \cos\theta \cos\phi - \hat{\boldsymbol{\phi}} \sin\phi, \\ \hat{\mathbf{y}} = \hat{\mathbf{r}} \sin\theta \sin\phi + \hat{\boldsymbol{\theta}} \cos\theta \sin\phi + \hat{\boldsymbol{\phi}} \cos\phi, \\ \hat{\mathbf{z}} = \hat{\mathbf{r}} \cos\theta - \hat{\boldsymbol{\theta}} \sin\theta; \end{cases}$$
(A.45)

$$\nabla \Phi = \frac{\partial \Phi}{\partial r} \hat{\mathbf{r}} + \frac{1}{r} \frac{\partial \Phi}{\partial \theta} \hat{\boldsymbol{\theta}} + \frac{1}{\sin\theta} \frac{\partial \Phi}{\partial \phi} \hat{\boldsymbol{\phi}},$$
(A.46)

$$\nabla \cdot \mathbf{A} = \frac{1}{r^2} \frac{\partial}{\partial r} (r^2 A_r) + \frac{1}{r \sin\theta} \frac{\partial}{\partial \theta} (\sin\theta A_\theta) + \frac{1}{r \sin\theta} \frac{\partial A_\phi}{\partial \phi},$$
(A.47)

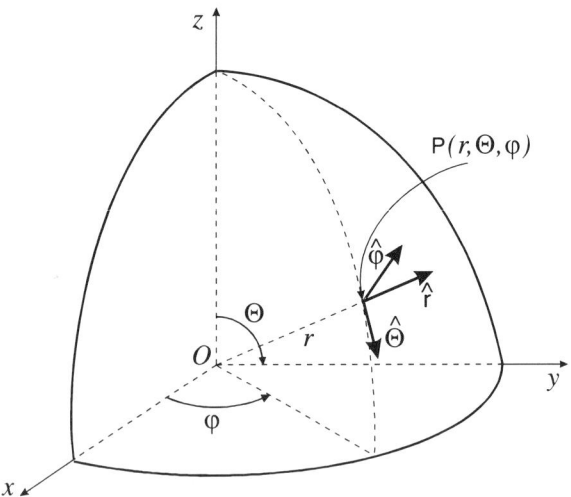

FIGURE A.4. Spherical coordinates.

$$\nabla \times \mathbf{A} = \frac{1}{r \sin \theta} \left[\frac{\partial}{\partial \theta} (\sin \theta A_\phi) - \frac{\partial A_\theta}{\partial \phi} \right] \hat{\mathbf{r}}$$

$$+ \frac{1}{r} \left[\frac{1}{\sin \theta} \frac{\partial A_r}{\partial \phi} - \frac{\partial}{\partial r} (rA_\phi) \right] \hat{\boldsymbol{\theta}} + \frac{1}{r} \left[\frac{\partial}{\partial r} (rA_\theta) - \frac{\partial A_r}{\partial \theta} \right] \hat{\boldsymbol{\phi}}, \quad (A.48)$$

$$\nabla^2 \Phi = \frac{1}{r^2} \frac{\partial}{\partial r} \left(r^2 \frac{\partial \Phi}{\partial r} \right) + \frac{1}{r^2 \sin \theta} \frac{\partial}{\partial \theta} \left(\sin \theta \frac{\partial \Phi}{\partial \theta} \right) + \frac{1}{r^2 \sin^2 \theta} \frac{\partial^2 \Phi}{\partial \phi^2}, \quad (A.49)$$

$$\nabla^2 \mathbf{A} = \left(\nabla^2 A_r - \frac{2 A_r}{r^2} - \frac{2 \operatorname{ctg} \theta}{r^2} A_\theta - \frac{2}{r^2} \frac{\partial A_\theta}{\partial \theta} - \frac{2}{r^2 \sin \theta} \frac{\partial A_\phi}{\partial \phi} \right) \hat{\mathbf{r}}$$

$$+ \left(\nabla^2 A_\theta + \frac{2}{r^2} \frac{\partial A_r}{\partial \theta} - \frac{A_\theta}{r^2 \sin^2 \theta} - \frac{2 \cos \theta}{r^2 \sin^2 \theta} \frac{\partial A_\phi}{\partial \phi} \right) \hat{\boldsymbol{\theta}}$$

$$+ \left(\nabla^2 A_\phi + \frac{2}{r^2 \sin \theta} \frac{\partial A_r}{\partial \phi} - \frac{1}{r^2 \sin^2 \theta} A_\phi + \frac{2 \cos \theta}{r^2 \sin^2 \theta} \frac{\partial A_\theta}{\partial \phi} \right) \hat{\boldsymbol{\phi}}. \quad (A.50)$$

APPENDIX B. DYADIC ANALYSIS

In the following C and D are dyadic functions, while $\mathbf{A}$ and $\mathbf{B}$ are vector functions and Φ a scalar function, as in Appendix A.

B.1. Definition of a Dyad

$$C = \mathbf{AB} = A_x B_x \hat{\mathbf{x}}\hat{\mathbf{x}} + A_x B_y \hat{\mathbf{x}}\hat{\mathbf{y}} + A_x B_z \hat{\mathbf{x}}\hat{\mathbf{z}} + A_y B_x \hat{\mathbf{y}}\hat{\mathbf{x}} + A_y B_y \hat{\mathbf{y}}\hat{\mathbf{y}} + A_y B_z \hat{\mathbf{y}}\hat{\mathbf{z}}$$
$$+ A_z B_x \hat{\mathbf{z}}\hat{\mathbf{x}} + A_z B_y \hat{\mathbf{z}}\hat{\mathbf{y}} + A_z B_z \hat{\mathbf{z}}\hat{\mathbf{z}}. \quad (B.1)$$

We have also

$$C = \hat{\mathbf{x}} \mathbf{C}_x^r + \hat{\mathbf{y}} \mathbf{C}_y^r + \hat{\mathbf{z}} \mathbf{C}_z^r, \quad (B.2)$$

where

$$\mathbf{C}_x^r = A_x B_x \hat{\mathbf{x}} + A_x B_y \hat{\mathbf{y}} + A_x B_z \hat{\mathbf{z}}, \quad (B.3)$$
$$\cdots \quad \cdots \quad \cdots$$

are the row vectors of the dyad. Similarly

$$C = \mathbf{C}_x^c \hat{\mathbf{x}} + \mathbf{C}_y^c \hat{\mathbf{y}} + \mathbf{C}_z^c \hat{\mathbf{z}}, \quad (B.4)$$

where

$$\mathbf{C}_x^c = A_x B_x \hat{\mathbf{x}} + A_y B_x \hat{\mathbf{y}} + A_z B_z \hat{\mathbf{z}}, \qquad (B.5)$$

$$\cdots \qquad \cdots \qquad \cdots$$

are the column vectors of the dyad.

B.2. Dyadic Multiplication

$$\mathbf{C} \cdot \mathbf{A} = \mathbf{C}_x^c A_x + \mathbf{C}_y^c A_y + \mathbf{C}_z^c A_z, \qquad (B.6)$$

$$\mathbf{A} \cdot \mathbf{C} = A_x \mathbf{C}_x^r + A_y \mathbf{C}_y^r + A_z \mathbf{C}_z^r, \qquad (B.7)$$

$$\widetilde{\mathbf{A} \cdot \mathbf{C}} = \tilde{\mathbf{C}} \cdot \mathbf{A}, \qquad (B.8)$$

$$(\mathbf{A} \cdot \mathbf{C}) \cdot \mathbf{B} = \mathbf{A} \cdot (\mathbf{C} \cdot \mathbf{B}) = \mathbf{A} \cdot \mathbf{C} \cdot \mathbf{B}, \qquad (B.9)$$

$$(\mathbf{A} \times \mathbf{B}) \cdot \mathbf{C} = \mathbf{A} \cdot (\mathbf{B} \times \mathbf{C}) = -\mathbf{B} \cdot (\mathbf{A} \times \mathbf{C}), \qquad (B.10)$$

$$(\mathbf{C} \times \mathbf{A}) \cdot \mathbf{B} = \mathbf{C} \cdot (\mathbf{A} \times \mathbf{B}) = -(\mathbf{C} \times \mathbf{B}) \cdot \mathbf{A} \neq (\mathbf{C} \cdot \mathbf{A}) \times \mathbf{B}, \qquad (B.11)$$

$$(\mathbf{A} \times \mathbf{C}) \cdot \mathbf{B} = \mathbf{A} \times (\mathbf{C} \cdot \mathbf{B}), \qquad (B.12)$$

$$(\mathbf{A} \cdot \mathbf{C}) \times \mathbf{B} = \mathbf{A} \cdot (\mathbf{C} \times \mathbf{B}), \qquad (B.13)$$

$$(\mathbf{A} \times \mathbf{C}) \times \mathbf{B} = \mathbf{A}(\mathbf{C} \times \mathbf{B}) = \mathbf{A} \times \mathbf{C} \times \mathbf{B}, \qquad (B.14)$$

$$\mathbf{A} \times (\mathbf{B} \times \mathbf{C}) = \mathbf{B}(\mathbf{A} \cdot \mathbf{C}) - \mathbf{C}(\mathbf{A} \cdot \mathbf{B}), \qquad (B.15)$$

$$\widetilde{\mathbf{C} \cdot \mathbf{D}} = \tilde{\mathbf{D}} \cdot \tilde{\mathbf{C}}, \qquad (B.16)$$

$$(\mathbf{A} \cdot \mathbf{C}) \cdot \mathbf{D} = \mathbf{A} \cdot (\mathbf{C} \cdot \mathbf{D}) = \mathbf{A} \cdot \mathbf{C} \cdot \mathbf{D}, \qquad (B.17)$$

$$(\mathbf{C} \cdot \mathbf{D}) \cdot \mathbf{A} = \mathbf{C} \cdot (\mathbf{D} \cdot \mathbf{A}) = \mathbf{C} \cdot \mathbf{D} \cdot \mathbf{A}, \qquad (B.18)$$

$$(\mathbf{A} \times \mathbf{C}) \cdot \mathbf{D} = \mathbf{A} \times (\mathbf{C} \cdot \mathbf{D}) = \mathbf{A} \times \mathbf{C} \cdot \mathbf{D}, \qquad (B.19)$$

$$(\mathbf{C} \cdot \mathbf{D}) \times \mathbf{A} = \mathbf{C} \cdot (\mathbf{D} \times \mathbf{A}) = \mathbf{C} \cdot \mathbf{D} \times \mathbf{A}, \qquad (B.20)$$

$$(\mathbf{C} \times \mathbf{A}) \cdot \mathbf{D} = \mathbf{C} \cdot (\mathbf{A} \times \mathbf{D}), \qquad (B.21)$$

$$\mathbf{A} \cdot \mathbf{C} \cdot \mathbf{B} = \mathbf{B} \cdot \tilde{\mathbf{C}} \cdot \mathbf{A}, \qquad (B.22)$$

$$(\mathbf{I} \times \mathbf{A}) \cdot \mathbf{B} = \mathbf{A} \cdot (\mathbf{I} \times \mathbf{B}) = \mathbf{A} \times \mathbf{B}, \qquad (B.23)$$

$$(\mathbf{I} \times \mathbf{A}) \cdot \mathbf{C} = \mathbf{A} \times \mathbf{C} = (\mathbf{A} \times \mathbf{I}) \cdot \mathbf{C}, \qquad (B.24)$$

$$\mathbf{I} \times (\mathbf{A} \times \mathbf{B}) = \mathbf{B}\mathbf{A} - \mathbf{A}\mathbf{B}, \qquad (B.25)$$

$$\mathbf{I} \times \mathbf{A} = \mathbf{A} \times \mathbf{I}. \qquad (B.26)$$

B.3. Differential Relationships

$$\nabla A = \hat{x}\frac{\partial A}{\partial x} + \hat{y}\frac{\partial A}{\partial y} + \hat{z}\frac{\partial A}{\partial z}, \tag{B.27}$$

$$\nabla \cdot C = (\nabla \cdot C_x^c)\hat{x} + (\nabla \cdot C_y^c)\hat{y} + (\nabla \cdot C_z^c)\hat{z} = \frac{\partial C_x^r}{\partial x} + \frac{\partial C_y^r}{\partial y} + \frac{\partial C_z^r}{\partial z}, \tag{B.28}$$

$$\nabla \times C = (\nabla \times C_x^c)\hat{x} + (\nabla \times C_y^c)\hat{y} + (\nabla \times C_z^c)\hat{z}, \tag{B.29}$$

$$\nabla^2 C = \nabla\nabla \cdot C - \nabla \times \nabla \times C = \frac{\partial^2 C}{\partial x^2} + \frac{\partial^2 C}{\partial y^2} + \frac{\partial^2 C}{\partial z^2}, \tag{B.30}$$

$$\nabla \cdot \nabla \times C = 0, \tag{B.31}$$

$$\nabla \times \nabla A = 0, \tag{B.32}$$

$$\nabla(A \times B) = (\nabla A) \times B - (\nabla B) \times A, \tag{B.33}$$

$$\nabla(\Phi B) = \nabla\Phi B + \Phi\nabla B, \tag{B.34}$$

$$\nabla \cdot (AB) = (\nabla \cdot A)B + (A \cdot \nabla)B, \tag{B.35}$$

$$\nabla \cdot (\Phi C) = \nabla\Phi \cdot C + \Phi\nabla \cdot C, \tag{B.36}$$

$$\nabla \cdot (A \times C) = (\nabla \times A) \cdot C - A \cdot \nabla \times C, \tag{B.37}$$

$$\nabla \times (\Phi C) = \nabla\Phi \times C + \Phi\nabla \times C. \tag{B.38}$$

B.4. Miscellaneous

$$A^{-1} \cdot A = I, \tag{B.39}$$

$$\widetilde{(A^{-1})} = (\tilde{A})^{-1}, \tag{B.40}$$

$$(A \cdot B)^{-1} = B^{-1} \cdot A^{-1}, \tag{B.41}$$

$$\widetilde{(A \cdot B)}^{-1} = (\tilde{A})^{-1} \cdot (\tilde{B})^{-1}. \tag{B.42}$$

APPENDIX C. USEFUL INTEGRALS AND SERIES

$$\int_{-\infty}^{+\infty} dx \exp(-a^2 x^2 - 2bx) = \frac{\sqrt{\pi}}{a} \exp\left(\frac{b^2}{a^2}\right), \tag{C.1}$$

$$\frac{1}{2\pi i}\int_{-i\infty}^{+i\infty} dp \exp(pt)[1 - \exp(bp - b\sqrt{p^2 + a^2})] = ab\frac{J_1(a\sqrt{t^2 + 2bt})}{\sqrt{t^2 + 2bt}} U(t), \tag{C.2}$$

$$\int_0^{+\infty} dx \exp(-ax^2) \cos bx = \frac{1}{2}\sqrt{\frac{\pi}{a}} \exp\left(-\frac{b^2}{4a}\right), \tag{C.3}$$

$$\frac{1}{2\pi i}\int_{-i\infty}^{+i\infty} dp \exp(pt)[\sqrt{p^2+a^2}-p] = a\frac{J_1(at)}{t} U(t), \tag{C.4}$$

$$\int_{-\infty}^{+\infty} \exp(ia^2t^2)dt = \frac{\sqrt{\pi}\exp(i\pi/4)}{a}, \tag{C.5}$$

$$\int_{-\infty}^{+\infty} dx\, \operatorname{sinc}^2(ax) = \frac{\pi}{a}, \tag{C.6}$$

$$\int_{-\infty}^{+\infty} dx\, \operatorname{sinc}[ax-n\pi]\cdot\operatorname{sinc}[ax-m\pi] = 0, \qquad n\neq m, \tag{C.7}$$

$$\int_{-\infty}^{+\infty} dx\, \operatorname{sinc}[ax-\alpha]\operatorname{sinc}[bx-\beta] = \frac{\pi}{a}\operatorname{sinc}[b\alpha-a\beta)/a], \qquad a>b, \tag{C.8}$$

$$\int_0^1 dx\, xJ_n^2(q_{nm}x) = \frac{1}{2}\frac{q_{nm}^2-n^2}{q_{nm}^2} J_n^2(q_{nm}), \tag{C.9}$$

$$\int_0^1 dx\, xJ_n^2(p_{nm}x) = \frac{1}{2} J_{n+1}(p_{nm}), \tag{C.10}$$

$$\int \frac{dx}{\sqrt{a^2+x^2}} = \frac{1}{a}\ln[x+\sqrt{a^2+x^2}], \tag{C.11}$$

$$\frac{1}{\pi}\int_0^\pi dx\, \cos(z\cos x)\cos nx = J_n(z), \tag{C.12}$$

$$\int dx\, \exp(i\alpha x)\sin\beta x = \frac{\exp(i\alpha x)}{\beta^2-\alpha^2}[i\alpha\cos\beta x+\beta\sin\beta x], \tag{C.13}$$

$$\int dx\, \exp(i\alpha x)\cos\beta x = \frac{\exp(i\alpha x)}{\beta^2-\alpha^2}[i\alpha\sin\beta x-\beta\cos\beta x], \tag{C.14}$$

$$\int_0^1 dx\, \frac{x\sin(ax)}{\sqrt{1-x^2}} = \frac{\pi}{2} J_1(a), \tag{C.15}$$

$$\exp(ix\cos\theta) = \sum_{n=-\infty}^{+\infty} i^n J_n(x)\exp(in\theta), \tag{C.16}$$

$$\int_{-\infty}^{+\infty} dx\, \exp(iax^2+ibx) = \frac{\sqrt{\pi}}{\sqrt{a}}\exp\left(-i\frac{b^2}{4a}\right)\exp\left(i\frac{\pi}{4}\right), \tag{C.17}$$

$$\int_0^{2\pi} d\theta\, \exp[iz\cos\theta]\exp[in\theta] = (-i)^n 2\pi J_n(z), \tag{C.18}$$

$$\int dx\, J_n(ax) x J_n(bx) = \frac{bx J_n(ax) J_{n-1}(bx) - ax J_{n-1}(ax) J_n(bx)}{a^2 - b^2}, \qquad (C.19)$$

$$\int dx\, J_n^2(x) x = \frac{x^2}{2}\left[J_n^2(x) - J_{n-1}(x) J_{n+1}(x)\right], \qquad (C.20)$$

$$\int_0^1 dx\, J_0(ax) x = \frac{J_1(a)}{a}, \qquad (C.21)$$

$$\int_0^1 dx\, (1 - x^2)^n x J_0(ax) = 2^n n!\, a^{-(n+1)} J_{n+1}(a), \qquad (C.22)$$

$$\int_0^\pi d\theta\, \frac{dP_n}{d\theta} \frac{dP_m}{d\theta} \sin\theta = \frac{2n(n+1)}{2n+1} \delta_{nm}, \qquad (C.23)$$

$$\int_{-\infty}^{+\infty} dx\, \mathrm{sinc}(ax) = \frac{\pi}{a}, \qquad (C.24)$$

$$\frac{1}{\pi} \int_0^\pi dx\, \sin(z \sin x) \sin x = J_1(z), \qquad (C.25)$$

$$\int_0^\pi d\theta\, \sin^4\theta = \frac{3\pi}{8}, \qquad (C.26)$$

$$\frac{i}{\pi} \int_{-\infty}^{+\infty} dx\, \frac{\exp(-i\sqrt{z^2 + x^2})}{\sqrt{z^2 + x^2}} = H_0^{(2)}(z), \qquad z > 0. \qquad (C.27)$$

APPENDIX D. SPECIAL FUNCTIONS AND ASYMPTOTIC EVALUATIONS

D.1. Bessel and Related Functions

Bessel functions of integer order n are solutions of the differential equation

$$\frac{1}{x}\frac{d}{dx} x \frac{dB_n}{dx} + \left(1 - \frac{n^2}{x^2}\right) B_n = 0, \qquad (D.1)$$

which exhibits two independent solutions: the *Bessel* functions of the *first kind* $J_n(x)$ and of the *second kind* $Y_n(x)$. Note that x is a nondimensional independent variable.

A linear combination of these two solutions leads to the Bessel function of the third kind, or *Hankel functions* of the *first kind*,

$$H_n^{(1)}(x) = J_n(x) + i Y_n(x), \qquad (D.2a)$$

and of the *second kind*,

$$H_n^{(2)}(x) = J_n(x) - iY_n(x). \tag{D.2b}$$

If the argument is made complex, $x \to ix$, we obtain the *modified Bessel functions* of the *first kind*,

$$I_n(x) = \exp\left(-in\frac{\pi}{2}\right)J_n(ix), \tag{D.3a}$$

and of the *second kind*,

$$K_n(x) = \frac{\pi i}{2}\exp\left(in\frac{\pi}{2}\right)H_n^{(1)}(ix), \tag{D.3b}$$

also referred to as the *Kelvin function*.

When the index is equal to an integer plus one-half, $n \to n + 1/2$, we are lead to the *spherical Bessel function* of the *first kind*,

$$j_n(x) = \sqrt{\frac{\pi}{2x}}\, J_{n+1/2}(x) = x^n\left[-\frac{1}{x}\frac{d}{dx}\right]^n \frac{\sin x}{x}, \tag{D.4a}$$

of the *second kind*,

$$y_n(x) = \sqrt{\frac{\pi}{2x}}\, Y_{n+1/2}(x) = -x^n\left[-\frac{1}{x}\frac{d}{dx}\right]^n \frac{\cos x}{x}, \tag{D.4b}$$

and of the *third kind*,

$$h_n^{(1,2)}(x) = \sqrt{\frac{\pi}{2x}}\, H_{n+1/2}^{(1,2)}(x). \tag{D.4c}$$

D.1.1. Recursion Relationships

For Bessel functions of the first, second, and third kind

$$B_{n-1}(x) + B_{n+1}(x) = \frac{2n}{x} B_n(x) \tag{D.5}$$

and

$$B_{n-1}(x) - B_{n+1}(x) = 2dB_n(x)/dx. \tag{D.6}$$

Also

$$\frac{dB_n(x)}{dx} = B_{n-1}(x) - n\frac{B_n(x)}{x} = -B_{n+1}(x) + n\frac{B_n(x)}{x}. \quad (D.7)$$

For the modified Bessel function

$$I_{n-1}(x) - I_{n+1}(x) = \frac{2n}{x} I_n(x), \quad (D.8)$$

$$I_{n-1}(x) + I_{n+1}(x) = 2dI_n(x)/dx, \quad (D.9)$$

$$K_{n-1}(x) - K_{n+1}(x) = -\frac{2n}{x} K_n(x), \quad (D.10)$$

$$K_{n-1}(x) + K_{n+1}(x) = -2dK_n(x)/dx. \quad (D.11)$$

D.1.2. Functional Relationships

$$J_n[x\exp(i\pi)] = \exp[in\pi]J_n(x), \quad (D.12)$$

$$H_n^{(1)}[x\exp(i\pi)] = -\exp[-in\pi]H_n^{(2)}(x), \quad (D.13a)$$

$$H_n^{(2)}[x\exp(-i\pi)] = -\exp[in\pi]H_n^{(1)}(x), \quad (D.13b)$$

$$J_{-n}(x) = (-)^n J_n(x), \quad (D.14a)$$

$$Y_{-n}(x) = (-)^n Y_n(x), \quad (D.14b)$$

$$H_{-n}^{(1)}(x) = \exp(in\pi)H_n^{(1)}(x), \quad (D.15a)$$

$$H_{-n}^{(2)}(x) = \exp(-in\pi)H_n^{(2)}(x), \quad (D.15b)$$

$$K_{-n}(x) = K_n(x). \quad (D.16)$$

D.1.3. Expansion for Small Values of the Argument

$$J_0(x) \approx 1, \quad Y_0(x) \approx \frac{2}{\pi}\left(\ln\frac{x}{2} + C\right), \quad (D.17a)$$

$$H_0^{(1,2)}(x) \approx \pm\frac{2i}{\pi}\left(\ln\frac{x}{2} + C\right), \quad (D.18a)$$

$$K_0(x) \approx \ln\frac{x}{2} + C, \quad (D.19a)$$

where $C = 0.5772$ is Euler's[1] constant.

---

1. Leonhard Euler (Eulero): Basel (Switzerland), 1707–St. Pietersburg, 1783.

For $n > 0$

$$J_n(x) \approx \frac{(x/2)^n}{n!}, \qquad Y_n(x) \approx -\frac{(n-1)!}{\pi}\left(\frac{2}{x}\right)^n, \tag{D.17b}$$

$$H_n^{(1,2)} \approx \mp i \frac{(n-1)!}{\pi}\left(\frac{2}{x}\right)^n, \tag{D.18b}$$

$$K_n(x) \approx -\frac{(n-1)!}{2}\left(\frac{2}{x}\right)^n. \tag{D.19b}$$

D.1.4. Asymptotic Expansion

For large values of the argument

$$J_n(x) \sim \sqrt{\frac{2}{\pi x}} \cos\left(x - \frac{n\pi}{2} - \frac{\pi}{4}\right), \tag{D.20a}$$

$$Y_n(x) \sim \sqrt{\frac{2}{\pi x}} \sin\left(x - \frac{n\pi}{2} - \frac{\pi}{4}\right), \tag{D.20b}$$

$$H_n^{(1,2)}(x) \sim \sqrt{\frac{2}{\pi x}} \exp\left[\pm i\left(x - \frac{n\pi}{2} - \frac{\pi}{4}\right)\right], \tag{D.21}$$

$$K_n(x) \sim \sqrt{\frac{\pi}{2x}} \exp(-x). \tag{D.22}$$

D.2. Legendre Polynomials

Legendre polynomials are a solution of the differential equation

$$(1 - x^2)\frac{d^2 L_n(x)}{dx^2} - 2x\frac{dL_n(x)}{dx} + n(n+1)L_n(x) = 0, \tag{D.23a}$$

or of its equivalent form ($x = \cos\theta$)

$$\frac{1}{\sin\theta}\frac{d}{d\theta}\sin\theta\frac{dL_n(\theta)}{d\theta} + n(n+1)L_n(\theta) = 0, \tag{D.23b}$$

which exhibits two independent solutions, the *Legendre functions* of the *first kind*, $P_n(\theta)$, and of the *second kind* $Q_n(\theta)$.

APPENDIXES

The Legendre functions of the first kind are polynomials,

$$P_n(x) = \frac{1}{2^n n!} \frac{d^n}{dx^n} (x^2 - 1)^n, \tag{D.24}$$

a particular ($m = 0$) form of the *associated Legendre polynomials*

$$P_n^m(x) = (1 - x^2)^{m/2} \frac{d^m}{dx^m} P_n(x), \qquad m \geq 0. \tag{D.25}$$

D.2.1. Recursion Relationships

$$(n + 1)P_{n+1}(x) + nP_{n-1}(x) = (2n + 1)xP_n(x), \tag{D.26}$$

$$\begin{aligned}(x^2 - 1) \frac{dP_n(x)}{dx} &= -n[P_{n-1}(x) - xP_n(x)] \\ &= (n + 1)[P_{n+1}(x) - xP_n(x)] \\ &= \frac{n(n + 1)}{2n + 1} [P_{n+1}(x) - P_{n-1}(x)].\end{aligned} \tag{D.27}$$

D.2.2. Asymptotic Expansion for Large Values of the Index

For large values of the index

$$P_n(\cos\theta) \sim \sqrt{\frac{2}{\pi n \sin\theta}} \cos\left[\left(n + \frac{1}{2}\right)\theta - \frac{\pi}{4}\right], \tag{D.28a}$$

$$Q_n(\cos\theta) \sim \sqrt{\frac{2}{\pi n \sin\theta}} \cos\left[\left(n + \frac{1}{2}\right)\theta + \frac{\pi}{4}\right]. \tag{D.28b}$$

D.2.3. Special Values

$$P_n(1) = 1, \tag{D.29}$$

$$P_{2n-1}(0) = 0, \tag{D.30}$$

$$P_{2n}(0) = (-)^n \frac{(2n - 1)!!}{2^n n!}. \tag{D.31}$$

D.3. Parseval's Theorem

$$\int_{-\infty}^{+\infty} dt\, a^2(t) = \frac{1}{2\pi} \int_{-\infty}^{+\infty} d\omega\, |A(\omega)|^2. \tag{D.32}$$

Most generally

$$\int_{-\infty}^{+\infty} dt\, a(t) b^*(t) = \frac{1}{2\pi} \int_{-\infty}^{+\infty} d\omega\, A(\omega) B^*(\omega). \tag{D.33}$$

D.4. Schwartz's Inequality

$$\left| \int_{-\infty}^{+\infty} dt\, a(t) b^*(t) \right|^2 \leq \int_{-\infty}^{+\infty} dt\, |a(t)|^2 \cdot \int_{-\infty}^{+\infty} dt\, |b(t)|^2, \tag{D.34}$$

the equality being attained for $b(t) = \alpha a(t)$, α being a constant.

D.5. Asymptotic Evaluation of Integrals

Consider the integral

$$I(\Omega) = \int_C dz\, f(z) \exp[\Omega q(z)], \tag{D.35}$$

where $z = x + iy$, C is the integration contour in the complex z-plane, and Ω is a parameter. We wish to evaluate the integral (D.35) *asymptotically* as $\Omega \to \infty$.

Should the contour C coincide with the real x-axis and $q(z) \to iq(z)$, then

$$I(\Omega) \to \int_{-\infty}^{+\infty} dx\, f(x) \exp[i\Omega q(x)], \tag{D.36}$$

and the *stationary phase evaluation* of the integral is recovered. For $\Omega \to \infty$, only the integration region close to the stationary phase point $x = x_s$,

$$\frac{dq}{dx} = q'(x_s) = 0, \tag{D.37}$$

provides a significant contribution to the integral, in which case

$$I(\Omega) \sim \exp[i\Omega q(x_s)] \int_{-\infty}^{+\infty} dx f(x) \exp\left[i\Omega \frac{q''(x_s)}{2}(x-x_s)^2\right]$$

$$\sim \exp[i\Omega q(x_s)] f(x_s) \sqrt{\frac{2\pi i}{\Omega q''(x_s)}}, \quad (D.38)$$

where use has been made of Eq. (C.5). Note that the assumption $f(x) \approx f(x_s)$ is possible only if $f(x)$ is regular near $x = x_s$.

The *steepest dependent path* (SDP) technique for the asymptotic evaluation of integral (D.35) is the generalization of the stationary phase integration method when $q(z)$ is complex and the integration contour C extends in the complex z-plane. But its rationale is essentially the same. Again, the only significant contribution to the integral is provided by a small region near the *saddle point* $z = z_s$,

$$\frac{dq}{dz} = q'(z_s) = 0, \quad (D.39)$$

which is a stationary point for the function $q(z)$. In the following we take $z_s = 0$, which implies setting the origin of coordinates at the stationary point.

Let

$$q(x, y) = u(x, y) + iv(x, y). \quad (D.40)$$

At the stationary point $x = y = 0$, we have

$$\frac{\partial u}{\partial x} = \frac{\partial u}{\partial y} = \frac{\partial v}{\partial x} = \frac{\partial v}{\partial y} = 0, \quad (D.41)$$

and

$$\frac{\partial^2 u}{\partial x^2} = -\frac{\partial^2 u}{\partial y^2}, \quad \frac{\partial^2 v}{\partial x^2} = -\frac{\partial^2 v}{\partial y^2}, \quad (D.42)$$

by using the Cauchy–Riemann conditions.

Suppose the x-axis is chosen in such a way that $v(x, 0) = v(0, 0)$ near the stationary point: $y = 0$ is a *locally* constant phase path and $\partial^2 v/\partial x^2$ is zero there. Then, Eq. (D.42) shows that the phase term $v(x, y)$ is locally constant also along the y-axis, and that $u(x, y)$ is increasing along one path, say $x = 0$, i.e., $\partial^2 u/\partial y^2 > 0$ there, and decreasing along the other, say $y = 0$, where

$\partial^2 u/\partial x^2 < 0$. Accordingly, along the latter path

$$q''(0,0) < 0, \qquad q(x,0) \approx q(0,0) - \frac{|u''(0,0)|}{2} x^2,$$

and the original integration contour of the integral (D.36) can be deformed along this SDP $y = 0$, thus obtaining

$$\begin{aligned} I(\Omega) &\sim \exp[\Omega q(0,0)] f(0,0) \int_{-\infty}^{+\infty} dx \exp\left[-\frac{|u''(0,0)|}{2} x^2\right] \\ &\sim \frac{\exp[\Omega q(0,0)] f(0,0)}{\sqrt{2\pi |u''(0,0)|}}, \end{aligned} \qquad (D.43)$$

in addition to possible contributions arising from the singularities of $f(x, y)$ which have been crossed in the deformation of C to the SDP. As in the case of stationary phase evaluation, these singularities should not approach the stationary point z_s, otherwise the function $f(x, y)$ cannot be taken outside the integral.

If $u''(0,0) = 0$, above derivation is invalid and alternative asymptotic procedures must be used.

Index

Absorbing boundary conditions, 321–324
Ampère equivalence: see Dipole
Ampère–Faraday law, 3
Antenna beamwidth, 437–439
Apertures
   circular, 470–473
   radiation from, 133–135, 463–473
   rectangular aperture, 465–470
Arrays
   binomial, 497–498
   linear, 222–225
   sum and difference pattern, 498–499
   with tapered excitation, 496–497
   with uniform excitation, 494–496

Bandlimited signals, 151–155
Bayard–Bode relations, 91–92
Bloch waves, 419
Boundary conditions, 30–32
Bounded waves, 62–65
Brewster angle, 200
Brillouin diagram, 182

Cavity
   coaxial loaded, 508–511
   energy decay, 515–516
   equivalent circuit, 516–519
   multimode, 511–512
   open, 513–515
   parallelepiped, 506–508
   Q-factor, 505–506
Cerenkov cone: see Radiation
Cerenkov radiation: see Radiation
Circuits: see Equivalent circuits
Circulator, 409–410
Coaxial-cable, 376–378
Coherence, 168–170
Complex vectors, 155–162
Compressible plasma, 116–118
Consistency conditions, 249

Current density equation
   in frequency domain, 87
   in time domain, 4
Cut-off frequency, 203
Coulomb law, 3

Debye circle, 94–94
Dielectric
   plasma, 19–20, 93–94, 116–118
   polar, 17–19, 93
Dielectric slab: see Propagation
Dipole
   electric, 8
   equivalence between electric and magnetic dipoles, 72–74, 216–217
   magnetic, 72–74
   radiation from, 68–74, 212–217
Dirac functions, 84–86
Directional coupler, 413–419
Directivity, 434
   of broadside array, 496
   of circular aperture, 472
   of elementary antennas, 434–436
   of half-wave dipole, 454
   of parabolic dish, 481
   of rectangular aperture, 467
Dispersive media, 17–21, 92–96, 116–118
Duality transformation, 39–40

Edge condition, 35–37
Effective area, 442
   of circular aperture, 473
   of elementary antennas, 444
   of parabolic dish, 481
   of rectangular aperture, 467
Effective length, 432
   of elementary antennas, 435–436
   of half-wave dipole, 453
   of receiving antennas, 440–441
   of travelling-wave antenna, 457
   of wire antennas, 449

Eikonal equation, 228
EIRP, 435
Elementary sources: *see* Dipole
Huygens source, 217–220
Energy theorem, 173–174
Equivalence theorem, 178–179
Equivalent circuits
  of cavity, 516–519
  of circulator, 409–410
  of directional coupler, 410–413
  of half-wave dipole, 455–457
  matrix representations, 400–405
  periodic structures, 418–421
  properties, 405–409
  of spherical antenna, 529–532
Expansion
  cylindrical wave, 523–527
  plane-wave, 126–133
  sampling, 153, 483, 496–497
  sampling like, 485–487
  spherical wave, 527–532

Fermat
  principle, 241
  zone, 544
Fibre, 388–397; *see also* Propagation
Finite Difference Method, 297, 324–329
Finite Elements Method, 296–297, 310–320
Fredholm's equations, 303, 451
Frénet equations, 238
Fresnel
  coefficients, 198
  region, 212
  zones, 540

Gain, 435
Gaussian beams, 135–138
Geometrical optics, 246
Geometrical theory of diffraction, 273
Green's function
  dyadic, 108–109
  in free-space, 74–76, 109–110
  in wavenumber-domain, 112–114
Group velocity, 183

Half-wave dipole antenna, 453–457
Heaviside condition, 56, 200, 339
Hertz potential: *see* Potential
Huygens source, 217–220

Image theory, 41
Impedance
  characteristic, 338

Impedance (*cont.*)
  half-wave antenna, 456
  intrinsic, 46
  mutual, 459–463
Initial conditions, 30–32
Integral equations
  electric field, 305–308
  magnetic field, 305–308
Iris: *see* Waveguides

Jacobi method, 299
Junctions: *see* Equivalent circuits

Keller's cone, 274
Kramers–Kronig relations, 90

Lagrange invariant, 240–241
Leapfrog scheme, 325
Lens antennas, 246–248
Lentz–Neuman law, 3, 4–6
Leontovic boundary conditions, 384, 505
Limit angle, 200
Lorentz force equation, 1

Magnetization, 6–8
Matrix
  condition, 301–303
  eigenvalues, 299–301
  inversion, 298–299
Maxwell equations
  in frequency-domain, 87
  in full spectral-domain, 115
  for sinusoidal fields, 171
  in time-domain, 1
  in wavenumber-domain, 111
Maxwell stress tensor, 24–26
Mechanical forces on antennas, 439–440
Media
  dispersive, 12; *see also* Plasma, Polar dielectric
  local, 11
Method of moments, 295–296, 303–305
Microstrip, 398
  effective dielectric constant, 399–400
Moving charge: *see* Radiation

Narrowband signals, 141–144
Noise temperature, 445–447
Numerical dispersion, 326–328
Numerical stability, 326–328

Parabolic dish, 479–483
  cassegranian, 479
  gregorian, 479

INDEX 567

Parallel-plate guide: *see* Propagation
Patch antenna, 473–475
Periodic structures, 418–421
Phase velocity, 181
Phasors, 144–151; *see also* Complex vectors
Physical Optics approximation, 264, 477
    asymptotic, 272
Planar guides: *see* Microstrip
Plane-wave expansion: *see* Expansion
Plane waves
    in conductive medium, 102–104
    in frequency-domain, 96–100
    in plasma medium, 101–102, 188–192
    for sinusoidal fields, 179–184
    in time-domain 45–54
Plasma, *see also* Dielectric
    cyclotron frequency, 95
    plasma frequency, 95
Poincaré sphere: *see* Polarization
Polarization, 6–8
    bases, 163
    matching, 443
    Pointcaré sphere, 163–167
    states, 162–163
    Stokes parameters, 163–167
Potential
    Hertz, 67
    scalar, 76–78
    vector, 76–78
Power matching, 443
Poynting theorem
    for anisotropic media, 26–27
    for dispersive media, 27–28
    for sinusoidal fields, 171–173
    in time-domain, 22–24
Propagation
    along circular pipe, 206–208, 373–376
    along dielectric slab, 205–206, 250
    along fibre, 209–210, 250–251, 388–397
    in layered medium, 235–238, 359–360
    in magnetized plasma, 188–192
    inside parallel-plate guide, 205–206, 249
    transient, in dispersive media, 96–100
    of wavepacket, 184–188
Pulsed dipole, 70–71

Q factor: *see* Cavity

Radiation
    condition, 33–35
    in dispersive media, 107–109
    from moving charge, 78–81

Radiation (*cont.*)
    from prescribed sources
        from planar sources, 220–222
        for sinusoidal fields, 210–212
        in time domain, 65–68
        in wavenumber domain, 111–115
    pressure, 28–30
Radiation vector, 432
    of elementary antennas, 435–437
    of parabolic dish, 481
    of rectangular aperture, 466
Radio link equation, 444
Ray
    coordinate system, 251–255
    creeping, 284–292
    edge, 272–275
    in homogeneous medium, 233–235
    lateral, 279–284
    polarization change, 228–229
    properties, 239–245
Reciprocity theorem, 176–179
    use of, 405–408, 429–431, 459–461
Reflection antenna
    parabolic cylinder, 245–246, 483–491
    parabolic dish, 479–483
Reflector coefficient
    in frequency domain, 105__
    Fresnel, 198
    for the half plane, 274–275
    matrix, 257
    in time-domain, 55
Reflector transformation, 39; *see also* Image theory
    at space discontinuity
        plane waves in frequency domain, 104–107
        plane waves in time-domain, 54–57, 61–62
        plane waves with sinusoidal excitation, 195–201
    at time discontinuity, 58–61
Resonance
    exterior, 519–523
    interior, 501–506
Riccati equation, 353

Saddle point, 263, 563
Sampling expansion: *see* Expansion
Scattering
    by conducting half-plane, 266–272
    by metal cylinder, 257–258, 523–527
    by metal strip, 308–310
    by rough surface, 536–544
    by spherical antenna, 527–532
Scattering cross section, 521–522
Shelkunoff circle, 493

Skin depth, 200
Slope diffraction coefficient, 277–279
Smith's chart, 345
Snell's law, 197
  local, 243
Sommerfeld integration contour, 262
Standing wave ratio, 342
Steepest descent path, 262, 563–564
Stokes parameters: *see* Polarization
Stripline, 398
Surface wave, 199

Time reversal, 37–38
Transition functions, 275–277
Transmission coefficient
  in frequency domain, 105
  Fresnel, 198
  in time-domain, 55
Transmission lines, 332–336
  equivalent, 356–359
  generators, 355–356
  impedance, 340–345
  matching, 345–347
  multiconductor, 353–355
  multisections, 347–350
  nonuniform, 350–353
  reflection coefficient, 340–345
  telegraphists' equation, 326–340

Transport equation, 232–233
Transverse resonance, 361–362
Travelling-wave antenna, 457–459
  mutual impedance, 462–463

Uniqueness theorem
  for sinusoidal fields, 175–176
  in time-domain, 30–32

Volterra equation, 303

Yee scheme, 328–329

Watson transform, 292
Waveguide
  circular, 373–377
  excitation, 381–383
  inhomogeneous, 386–388
  losses, 383–386
  mode orthogonality, 378–381
  obstacles in, 421–425
  rectangular, 370–373
Waves, *see also* Plane waves; Reflection at space discontinuity; Reflection at time discontinuity
  Bounded, 62–65
WKB solution, 351